AF335172

PROCESS PLANT LAYOUT

PROCESS PLANT LAYOUT

Edited by

J.C. Mecklenburgh

Department of Chemical Engineering
University of Nottingham

A HALSTED PRESS BOOK

JOHN WILEY & SONS
New York

Published in the UK by
George Godwin
an imprint of:
Longman Group Limited
Longman House, Burnt Mill, Harlow
Essex CM20 2JE, England
in association with the Institution of Chemical Engineers
Associated companies throughout the world

Published in the U.S.A. by Halsted Press,
a Division of John Wiley & Sons, Inc., New York

First published 1985

ISBN 0-470-20238-6

Printed in Great Britain
at **The Bath Press, Avon**

CONTENTS

Part II. DETAILED SITE AND PLOT LAYOUT

Part III. DETAILED LAYOUT OF EQUIPMENT AND PIPEWORK

Part IV APPENDICES

LIST OF ABBREVIATIONS

American Conference of Government Industrial Hygienists	ACGIH
Acceptance quota level	AQL
Automated layout design program	ALDEP
Boiling liquid expanding vapour explosion	BLEVE
Computer aided design	CAD
Computerized relationship layout planning	CORELAP
Computerized relative allocation of facilities technique	CRAFT
Cathode ray tube	CRT
Critical path analysis	CPA
Design code allowable	DCA
Emergency exposure limit	EEL
Fatal accident rate	FAR
Gas phase chromatography	GPC
Hazard and operability studies	HAZOP
Health and Safety Executive	HSE
Immediately dangerous to life and health	IDLH
Intermediate bulk container	IBC
Long-term exposure limit	LTEL
Lower flammable limit	LFL
Materials take off	MTO
National Electrical Manufacturers Association	NEMA
Net-positive suction head	NPSH
Nominal diameter in mm	DN
Plant design and management system	PDMS
Programme evaluation and review technique	PERT
Polytetrafluoroeth(yl)ene	PTFE
Short-term exposure limit	STEL
Systems Reliability Service	SRS
Time weighted average	TWA
Unconfined vapour cloud explosion	UVCE
Upper flammable limit	UFL
Visual display unit	VDU

ACKNOWLEDGEMENTS

This work was prepared 1978/83 for the Engineering Practice Committee of the Institution of Chemical Engineers by the following Working Party:

Chairman

J. C. Mecklenburgh	University of Nottingham

Members

D. Armour	Babcock Woodall Duckham
K. Banks	Norsk Hydro Fertilizers
P. J. Comer	Technica
S. D. Green	BP Chemicals
D. J. Gunn	University College, Swansea
W. G. High	ICI, Petrochemicals and Plastics Division
J. Madden	Isopipe
M. J. Marks	·Humphreys and Glasgow

The Working Party are grateful to the following lecturers on the Plant Layout Continuing Education Courses, for use of their lectures in the book: C. L. Bell, Courtaulds; W. B. Bennett, Boots; W. Elliott, G. E. Guidoboni, John Brown Engineering and Construction; V. C. Marshall, Bradford University, J. Rollinson, Isopipe; D. H. Slater, Technica.

The Working Party also thanks the many other individuals and organizations that provided material photographs, diagrams and comments. Also thanks to Susan Close of Nottingham University for so patiently typing and retyping the numerous drafts and to Steven Dowling of Humphreys and Glasgow and Caroline Brayley of Nottingham University for preparing many of the diagrams.

PART I

GENERAL PRINCIPLES

1

INTRODUCTION

Layout is concerned with the spatial arrangement of process plant and its interconnections, such as piping. Good layout practice achieves a balance between the requirements for safety, economics, the protection of the public and the environment, construction, maintenance, operation, space for future expansion and process needs.

It is necessary to distinguish between the layout of the various plants in a *site*, the arrangement of process vessels, piping, etc. in a plant on a *plot* and finally the detailed arrangements of both *equipment* and *piping*. Thus, in this book, the term *plant layout* has been given a generic meaning covering all aspects of layout. In the earlier Institution of Chemical Engineers publication[1] the term *plant* was also used synonymously for *plot* reflecting the common occurrence where a plant occupies one plot. However, it has been found helpful to define the terminology more precisely.

Since the first book the principal addition to the subject has been hazard assessment. Also there has been an increase (though not as much as expected) in the use of computer-aided design in layout. With the growth of project size it has become recognized that layout execution must be formally organized along with other design activities. The amount of detailed layout information available has also grown (see, for example, Kern[2]). Consequently, the size of this book is much larger than the first book and it was thought desirable to provide introductory Chapters 1–4 giving general principles before going into detail from Chapter 5 onwards.

Chapter 2 is concerned with the general discipline of layout and details on the various approaches are presented in Chapters 5–8, which encompass planning, layout conception, aids to layout and hazard assessment of layouts.

Chapter 3 provides the principles of site layout whilst Chapter 9 discusses the transportation requirements of a site, and Chapters 10, 11 and 12 look at storage and warehousing. The layout features of effluent facilities, utilities and central services are examined in Chapters 13, 14 and 15.

The basic principles of plot layout are outlined in Chapter 4 and further details are given in Chapter 17. The special features of plant layout within enclosed buildings are listed in Chapter 18. Chapter 16 discusses construction, including the recent development of modular construction.

The layout of individual items of equipment is covered in Chapters 19–30 with piping layout occupying Chapter 31.

Besides distinguishing between site, plot and equipment layout it is necessary to differentiate between preliminary layout before contract or approval and detailed layout afterwards.

Design has become a three-stage process:

Stage One: pre-design sanction
Stage Two: between design and project sanctions
Stage Three: after project sanction.

'Preliminary' covers Stages One and Two and 'detailed' Stage Three. In the past, sanction and planning permission have been sought and given on the basis of Stage One. Stage Two has been combined with Stage Three.[3] These three stages are now used because of the escalating penalties of not having accurate cost and hazard assessments when commitment to the project is decided.

Preliminary layout involves conception, evaluation and modification with the last two being repeated until a satisfactory solution is achieved. Detailed layout involves developing the minutiae of the preliminary layout. Chapter 6 will indicate that computers have been, and will become, increasingly successful for evaluation and detailing but that process and project experience remains best for layout conception and modification. The engineer assigned to detailed layout is also involved with project planning especially since the introduction of computers for planning control. This problem will be discussed in Chapter 5.

The training, skills and experience of the chemical engineer are applied to hazard assessment which has become an essential part of preliminary layout. In the first book it was implied that layout was the province of the design office with the chemical engineer in the background. However, hazard assessment and layout are now very much a partnership between layout engineer and chemical engineer. In the first book a formal critical examination method was included though it was largely ignored and designers preferred the 'devil's advocate' committee method of examination. Since then the technique of hazard and operability studies (HAZOP) have been devised for critically examining flowsheets and, no doubt, a similar type of method will be developed for layout. The legal requirement for providing environmental impact assessment and hazard surveys of potentially dangerous processes will promote such development.

When the first book was prepared, separation distances as outlined in codes of practice, such as the Institute of Petroleum Codes, were sacrosanct. Now they are regarded as guidelines only for preliminary design and are being superseded in detailed layout by the development of methods based on mathematical models of processes such as leakage, evaporation, cloud drift and dispersion, vapour cloud explosions, thermal radiation, etc. Chapter 8 outlines the various types of calculation involved.

In this connection, the Working Party had some difficulty because this branch of chemical engineering is now developing rapidly. Appendix B gives a summary of the position in 1983 but the reader will need to familiarize himself with the latest developments. This issue is further complicated because knowledge of the behaviour after loss of containment is in itself insufficient. It is also necessary to assess what the probabilities are of a leak occurring and in order to distance the plants, to judge what risk of damage, injuries, etc. society will tolerate. The quality of data on plant reliability and on public acceptability is continually improving and will assist engineers in achieving better design solutions to layout problems.

Thus this book is intended to be a guide to good practice and although the contents give spacings and arrangements, it must be remembered that these are only typical and not mandatory. They may have to be altered to suit local conditions, plant owners' requirements and established safe practices. In particular the guide has to be largely phrased in terms of a new or 'greenfield' site, whereas most projects are involved with modifications and extensions where existing site constraints inevitably make observance of good practice more difficult.

REFERENCES

1. Mecklenburgh, J. C. (ed.), *Plant layout*. Leonard Hill/I.Chem. E., 1973.
2. Kern, R., 'Plant layout', *Chem Engng*, 12 parts, 1977–78.
    I 'How to manage plant design to obtain minimum cost', 23 May, 130, 1977.
    II 'Specifications are the key to successful plant design', 4 July, 123, 1977.
    III 'Layout arrangements for distillation columns', 15 Aug., 153, 1977.
    IV 'How to find optimum layout for heat exchangers', 12 Sept., 169, 1977.
    V 'Arrangements of process and storage vessels', 7 Nov., 93, 1977.
    VI 'How to get the best process plant layouts for pumps and compressors', 5 Dec., 131, 1977.
    VII 'Piperack design for process plants', 30 Jan, 105, 1978.
    VIII 'Space requirements and layout for process furnaces', 27 Feb., 117, 1978.
    IX 'Instrument arrangements for ease of maintenance and convenient operation', 10 Apr., 127, 1978.
    X 'How to arrange the plot plan for process plants', 8 May, 191, 1978.
    XI 'Arranging the housed chemical process plant', 17 July, 123, 1978.
    XII 'Controlling the cost factors in plant design, 14 Aug., 141, 1978.
3. Mackenzie, G., 'The time and resource aspects of project management in the construction of chemical plants', *Chem. Engr, Lond.* **209,** CE 118, 1967.

THE DISCIPLINE OF LAYOUT

2.1 THE NATURE OF LAYOUT PRACTICE

Plant layout is the spatial arrangement of items of process vessels and equipment and their connection by pipes, ducts, or conveyors or vehicular transportation.

Layout engineers have to satisfy several criteria in their designs:

(a) Efficient, reliable and safe plant operations.
(b) Safe and convenient maintenance of items, or components of process equipment either by removal or *in situ* repair.
(c) Minimum acceptable hazard and nuisance to the public.
(d) Safe and efficient construction.
(e) Effective and economical use of space.

The supply of services to the plant and access to the periphery of the plant for maintenance, construction and emergency services are affected by layout of the site. In a new or 'greenfield' factory, the site layout will reflect the known needs of the process plant to be constructed. More typically, a plant has to be put on a particular plot or a number of plants in an existing site where the requirements of a new plant may not have been foreseen at the time of the original site layout. Although changes to the site may be feasible, at least some requirements of access that would normally be readily arranged on a new site have to be accommodated by the layout engineers in the plot layout.

Three broad divisions may thus be recognized.

(a) Plots in relation to each other within the site and to activities outside the site, called 'site layout'.
(b) Process units in relation to each other within a plot, called 'plot layout'.
(c) Accessories around a process unit, called 'equipment layout'.

An ideal site is split up into plots by its principal road system with additional access roads for the larger plots (Fig. 2.1). Often a complete set of process units (known as a plant) fits onto a plot, although bigger plants may need two or more plots.

Fig. 2.1   A site split into plots by its road system (Courtesy: BP Chemicals)

2.2   CONSIDERATIONS IN LAYOUT

The layout is based upon the plant flowsheet (flow diagram) with process units usually arranged initially in the order of processing. Adjacent vessels and equipment are separated by distances that are sufficient to permit access for satisfactory operation and maintenance without wasting space. The layout of some plants may follow the process flow sequence closely through to the final stages, but in practice there are several frequent features that require the layout sequence to differ from this process sequence. These include:

(a) Process requirements – one vessel may be required to be placed above another to provide gravity flow.

(b) Economics – have two items sharing the same supports and so minimize structural work or place heavy plant on good load-bearing ground, thus reducing the need for piles.

(c) Ease of operation – valves and instruments should be easily accessible to the operator.

(d) Ease of maintenance – a process unit should be capable of being dismantled and if necessary removed for repair.

(e) Ease of construction – locations should be accessible for items of process equipment delivered late without having to remove plant already erected.

(f) Ease of commissioning – extra facilities especially installed for commissioning should be accessible.

(g) Ease of future expansion and extension – any foreseeable expansion should be possible with minimum interruption of production.
(h) Ease of escape and firefighting – in an emergency operators must be able to leave quickly and fire-tenders must be able to approach close to the plant by more than one route.
(i) Operator safety – the operator must be protected from injury on protrusions, moving machinery, hot surfaces, dripping acid, etc.
(j) Hazard containment – an explosion, fire or toxic release occurring in one plant should be prevented, where reasonably practicable, from spreading to other plants, to offices, etc. or outside the site.
(k) Environmental impact – obtrusive or noisy plants should not be placed near communities at site boundaries but should be screened, if possible, by pleasant offices with trees in a landscaped setting.

Any change to the initial layout because of these or other considerations may result in extra pipework or transportation costs and additional site or building areas. The changes must therefore be economically justifiable. Also, checks should continually be made to see that layout changes due to one reason do not adversely affect another requirement.

A detailed knowledge of the physics and chemistry of process materials is needed to ascertain the requirements of hazard containment, which must therefore be carried out by suitably experienced process engineers. They do not directly undertake the other activities which need less academic knowledge and training but require the ability:

(a) To identify and use appropriate statutory and in-house regulations, design standards and codes of practice.
(b) To appreciate the needs of operation, maintenance and construction.
(c) To apply engineering experience and common sense.

Instead, this work is carried out by designers supervised by engineers experienced in plant layout.

To achieve an effective layout thus involves close co-operation between the process and layout engineers. It also requires good co-operation between all the technical, scientific and engineering disciplines, in order that all relevant factors are correctly incorporated in the layout design.

2.3   RELATION OF LAYOUT TO OTHER ACTIVITIES

The relation of layout design to other activities is shown in Fig. 2.2, a simplified representation of the life of a project.

A project starts with an idea for a process or recognition of a market need or opportunity. The process may require development in the laboratory and pilot plant.

The process engineer will then produce a flowsheet and process data on the basis of the laboratory and/or pilot investigations and/or other published information. This information specifies:

(a) Equipment that is needed.
(b) Volumetric capacity, essential process dimensions and materials of construction.
(c) Conditions of operation, e.g. temperature, pressure, and composition.
(d) Material and energy flows and balances.
(e) Flow areas and materials of construction of the pipes and conveyors connecting the units.
(f) Basic process instrumentation.

In general, the relative positions of the process units are not fixed at this stage, although a specific process requirement, such as gravity feed, may be indicated.

Plant layout is the next stage of design, and the civil, mechanical, electrical and instrument engineering design follow closely.

The process layout and engineering design work are divided into three parts as shown in Fig. 2.2.

Stage One design establishes the feasibility of the project so that design sanction can be obtained from the sponsor to finance Stage Two design work. In the case of new sites, outline planning permission is also sought from the

Fig. 2.2 Life span of a typical plant

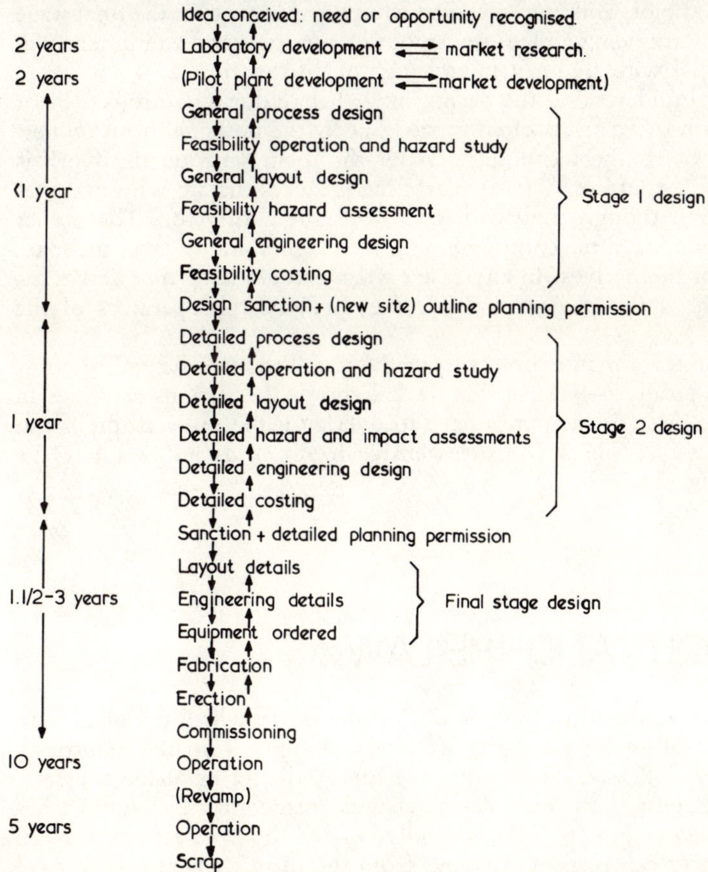

regulatory authorities at this time and if received, the site is purchased. It is in Stage One that the layout is first conceived, then adjusted in a semi-quantitative way to accommodate the various requirements listed in the section 2.2.

The purpose of Stage Two design is to provide sufficient information so that the correct amount of capital can be sanctioned by the sponsor. It also enables detailed planning permission to be obtained from the regulatory authorities at approximately the same time as sanction. In Stage Two the proposed layout is subjected to detailed analyses and optimization, e.g. hazard assessment, in order to fix the positions of all principal units (plant items, buildings, site services, etc.) and their major connections (pipes, cables, sewers, roads, etc.). The detailed layout and position of every item together with all ancillaries, connections, supports and access are determined and specified. No changes in the concept of the layout should be made in Stage Three as they can be costly in both delays and wasted effort.

The order of working can differ in the three stages. The starting point in Stage One is a rough equipment schedule, which has the size and scope of equipment, including a judgement allowance for access and ancillaries. A rough plot layout can then be determined from the equipment followed by a rough site plan although an existing site will constrain the plot layout. After design sanction the Stage Two commences with the detailed site layout followed by detailed plot, and then, detailed equipment layout. In the final stage when fine details are being added, the layout of site, plot and equipment will proceed in parallel with frequent interchange of information.

The personnel undertaking the design may change during a project. Prior to design sanction the work is often carried out by the potential plant owner, especially if he owns the technology. After the main sanction the work is usually done by a contractor. Particular circumstances dictate who executes Stage Two design though the trend is towards the contractor. The owner resumes responsibility after commissioning although some owners insist on carrying this out themselves. In any case owners should keep themselves informed throughout the project, because they will be the operators of the plant.

The lead time for a major project from the initial idea to the end of commissioning is typically 5–9 years. Figure 2.2 shows the various activities in sequence. In practice, the activities have to overlap if the project time is not to become excessive. There is also extensive feedback as indicated by the double arrows in Fig. 2.2.

2.4 LAYOUT AND THE LAW

Any layout must conform to the law of the country in which the plant is to be located. The following principles are likely to apply to many countries.

Firstly, under civil law, the layout must follow the terms of the contract. Most areas of dispute occur over design changes made after the contract has been signed (section 5.8, p. 51). Secondly, under the civil law of tort an injured third party can recover damages from the plant owners if they have been negligent. This would arise if safety and hazard aspects had not been considered properly in the layout and an accident had resulted. A litigant would have a good case, either if the owners had been found guilty on the appropriate criminal charges or if they had ignored, without good reason, pertinent codes of practice or other standard reference texts. Thirdly, the legal rights of neighbouring landowners and any historical legal constraints on the site itself must be respected.

The relevant aspects of criminal law deal with:

(a) Health, safety and employment.
(b) Environmental protection.
(c) Planning and control.

As well as the legislation itself, there are regulations, carrying legal force, which have been issued by regulatory authorities appointed under the legislation. Violations of the laws and regulations are in themselves offences, even if no incident has occurred.

The health, safety and employment laws can impinge on layout in a number of ways, both directly and indirectly.

(a) *Specific equipment and materials*: e.g. extra space for the guarding of machinery and pumps, storage and use of toxic materials, flameproof zones, ventilation.
(b) *Planning for emergencies*: e.g. means of escape, firefighting and fire protection.
(c) *Responsibilities to the public*: e.g. emergencies, pollution.
(d) *Worker amenities*: e.g. protective equipment and clothing, washrooms, canteens, medical facilities.
(e) *Management*: e.g. safety policy, management responsibilities for written instructions and manuals, training, inspection and enforcement procedures, worker responsibilities and rights in participating in safety procedures.
(f) *Design*: e.g. evidence of a safety audit (hazard and operability studies) of the design.

As well as there being a requirement for the layout engineer to introduce all reasonable safety precautions into the layout he also has the obligation not to compromise the safety features incorporated in the process design. The layout engineer must, in addition, ensure that he provides sufficient advice to other engineers, fabricators, constructors, etc. so that the layout can be safely engineered.

It is likely that the design, including the layout, of a proposed plant will undergo some form of formal inspection by the regulating authorities. It is also possible there will be legal rights for worker representatives to inspect the design. Similar inspections could also occur during construction, startup and operation. If an incident results in litigation the design could be examined by the court. Good documentation is therefore essential.

Environmental protection laws relating to effluent control probably have more impact on the process design and site selection than on layout, which may itself be affected by appearance and noise requirements. Obtaining planning permission to build a plant will probably mean the proposed layout being studied by the planning authorities for its general impact on their district or region. Other planning and control laws to observe concern the importation of equipment and materials, re-export of construction equipment, use of expatriate labour, etc. The penalty of violating planning laws is not so much the legal one but the cost of resultant delays and alterations.

In all legal matters it is advisable to employ local lawyers who are familiar with not only the laws and regulations themselves but how the courts interpret them and the regulatory and planning authorities administer them.

2.5   THE IMPORTANCE OF LAYOUT

Good layout practice plays a vital part in the commercial success of a project. This it does by providing a plant that is safe and efficient to construct, operate and maintain, whilst making effective use of the land available. A well thought-out layout also contributes to successful planning of both engineering design and construction stages.

A good layout will not compensate either for bad process design or bad engineering design. However, a bad layout will probably lead to an unsuccessful and unsafe venture.

REFERENCE

1. Rose, J. C., Wells, G. L. and Yeats, B. M. *A Guide to Project Procedure.* George Godwin/I.Chem.E., 1978.

SITE LAYOUT PRINCIPLES

3.1 SITE LAYOUT OBJECTIVES

The purpose of a good site layout is to provide safe and economical flow of materials and people and a socially acceptable environment for people working in the plant and living in the adjoining community. Possible disasters must be foreseen, and plans for containment at source must ensure that fire and accidental harmful releases do not spread throughout and beyond the site, and can be quickly controlled.

Selection and layout of a new site is one of the less common design operations because the majority of projects are carried out as additions or extensions to an existing site. However, when the need arises to plan a new development on new land (often called a 'greenfield' site) great care must be taken:

(a) To balance the ideal layout with the characteristics of available sites.
(b) To plan carefully the layout of the chosen site so that overall site operation and future developments can be carried out effectively.

For an existing site, the layout characteristics are, for better or worse, already established. Nevertheless, care must be taken to see that the layout of an extension does not violate the separation standards of the original layout. This is a common failing of many site-infill extensions.

A first layout is usually based on a materials flowsheet for the site (Fig. 3.1) which allows the processes to be positioned relative to one another. The layout is then altered to accommodate the various constraints outlined in the rest of this chapter and discussed in detail in Chapters 9–15. A typical site layout is shown in Fig. 3.2 as well as in Figs 2.1 (p. 7), 7.4 (p. 72) and 16.5 (p. 228).

3.2 SEGREGATION

In many industrial activities different parts of a manufacturing plant are purposely placed adjacent to each other to minimize interstage transport. Most

Fig. 3.1 Diagram of gas flow through a terminal (Courtesy: British Gas)

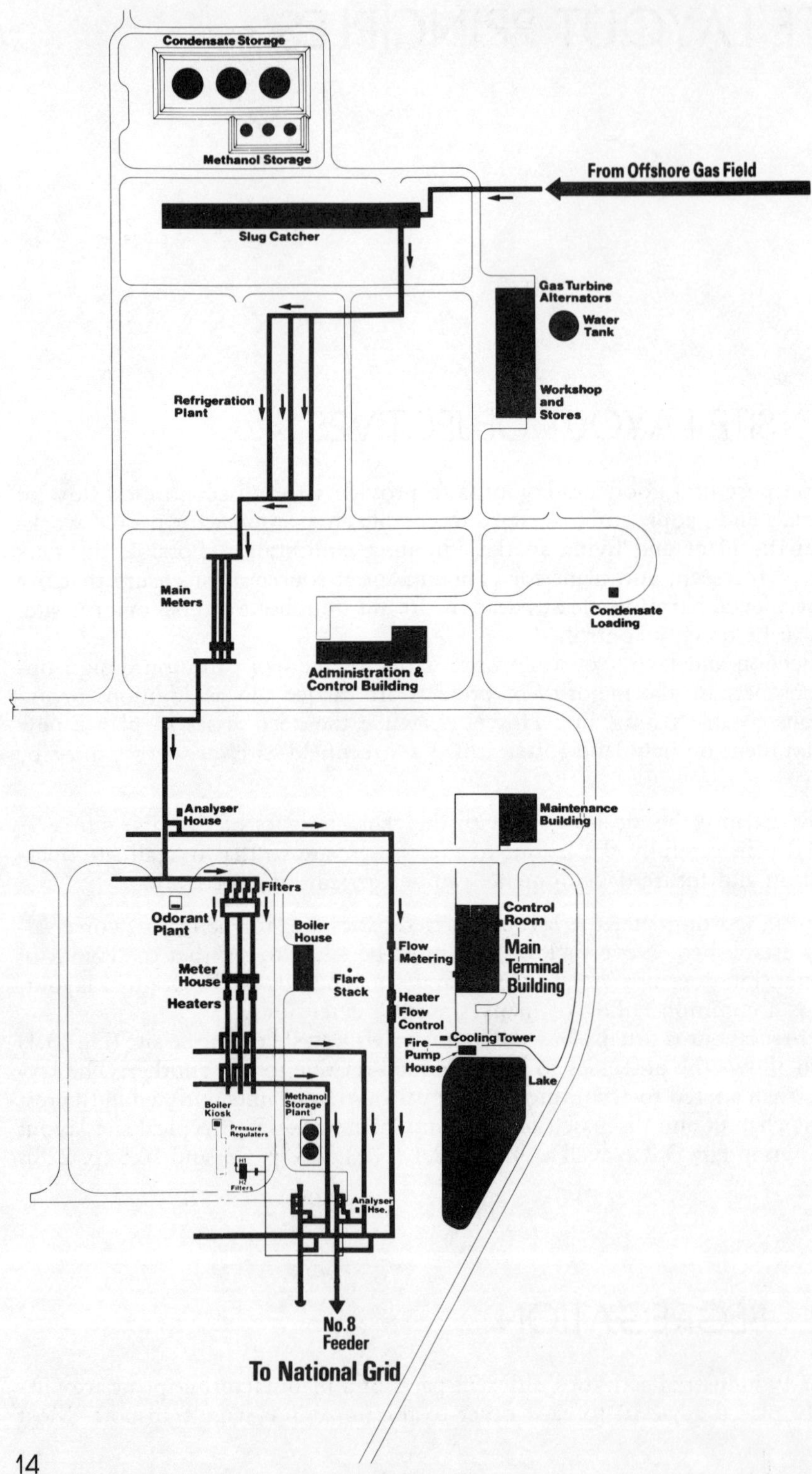

Fig. 3.2 Typical site layout (Courtesy: British Gas)

process industry sites, though, are laid out with plant units deliberately segregated, albeit at some extra transport cost. The main considerations for this policy of segregation are:

(a) Safety and loss prevention.
(b) Housekeeping (cross-contamination).
(c) Access for construction and maintenance.

It is necessary to test for these considerations positively because otherwise they can easily be overridden by purely economic considerations.

Segregation for safety and loss prevention aims to prevent a major accident or fire in one plant from spreading to another plant. Good housekeeping may dictate segregation of plants which release fumes or dust if these emissions (although within statutory limits) would contaminate the products in adjacent plants. Access spaces around plants, normally intended to allow crane access, lifting clearance or space for temporary placing of equipment during erection or maintenance, catalyst containers, etc. are generally less than those dictated by safety and housekeeping requirements. In order to control entry to restricted access or hazardous areas, it may be necessary to erect a security fence around a plot, and institute a 'dematching' or 'official access only' system.

It is essential that hazardous processes should be segregated from the public, adjoining industrial plant or vegetation that can catch fire. Also, hazardous plants should be segregated from neighbours' hazardous plants and in particular they should fall clear of each other in the event of collapse. In any case, hazardous processes should, as far as possible, be sited away from the site boundary because although it may be safe now to have a hazardous process at the boundary, future development on the other side may alter the situation. At the least this could prevent further development of the process. Thus, in positioning plants and buildings, early and ongoing consultation with the local authority is necessary.

A hazardous operation must be kept away from the public. Onlookers to a fire can interfere with firefighting and even be injured themselves.

Segregation of plants costs money in extra transportation, longer pipework connections and higher manning levels and must, therefore, be justified and based on the best data available. Where this justification is not valid, various preventive measures may be possible to reduce distances, e.g. reduced inventories or operating pressures and temperatures, subdivision of plants with remotely controlled isolating valves, improved design and fabrication standards, better instrumentation, and stronger buildings. Chapter 8 discusses the methods of hazard assessment.

3.3   EMERGENCIES

Emergencies and potential emergencies on the site can arise from a wide range of causes and it is necessary for the plant owner's company management to develop plans to deal with all emergencies up to and including a major incident. Hardware to implement the company's emergency plans will be built into the site layout, e.g. access, control and water supply.

3.3.1   ACCESS

The road system should be adequate for firefighting and damage-control vehicles and allow them to approach ideally from all directions within a reasonable distance of any risk and without hazard to factory traffic. Water supplies should be available at these places. This means in many cases that the site should have a peripheral road with access to the public road system at a minimum of two points.

The internal road system should be checked so that no area is likely to be cut-off by debris, fumes or leakage. Section 9.3 (p. 143) discusses further emergency access requirements.

3.3.2   CONTROL

A major incident control point should be planned. This, together with the fire station, medical centre, telephone exchange, emergency stores and main entrance must be located away from hazardous sources.

It is desirable to have a number of additional subcontrol points about the site. Also safe areas with shelter from weather, easily reached on foot, should be set aside as emergency assembly points for staff. Accounting for personnel after an incident is then both simpler and quicker.

In planning the layout, for emergencies the effect of an incident spreading to or from a neighbouring site must be considered and it is advisable to consult with the local police.

Section 15.5 (p. 220) gives further details on emergency services and control.

3.3.3   WATER

Firefighting water is usually taken from public mains, supplemented from natural sources or large-scale static storage reservoirs. Access to these sources must always be maintained.

Firefighting mains are preferably buried and should survive both natural and incident-caused shockwaves. The layout of mains and hydrants and details of connectors should be reviewed with the local fire authority officers to obtain their advice and to ensure their staff are suitably equipped and trained for tackling the likely fires. The larger and more complex sites may elect to have their own firefighting services, manned by both full-time and part-time personnel. These will fight the fires until the fire service arrives and then assist.

Special provisions may be necessary for disposal of water used in firefighting operations to prevent the spread of fire by carrying floating flammables and to stop the mixing of fire effluents and normal effluents which may hamper the emergency services by producing toxic fumes, etc.

Water supplies are considered further in section 14.3 (p. 210) and liquid disposal in section 13.2 (p. 195).

3.4   CENTRAL FACILITIES

The locations of plants and buildings housing the central services are determined initially by reference to the service flows into and around the site. It is economical to minimize the lengths of supply cables and piping to the largest users. However, care is needed to ensure that a central service generation or distribution plant is not exposed to serious interference by fire, explosion, or natural occurrences such as flood or wind. The serious consequences of total loss of any services are obvious. Substations, transformers, switch houses, pumping stations, etc. should also be put in areas which permit standard electrical equipment to be used unless they form an integral part of a plant, in which case flameproof equipment may be needed. In locating the boiler house, the effects of a troublesome or prevailing wind on the stack emission or of dust from the fuel pile should be taken into account. Where possible, direct access, avoiding the process areas, should be provided for the fuel supplies and ash removal. Boiler houses as sources of ignition should be well away from plants containing flammables.

Cooling towers should be sited so that water droplets will not restrict visibility or cause exterior corrosion or ice formation on other parts of the plant, roads, rail or public amenities. Siting should prevent the entrainment into the cooling towers of vapours and dusts from adjacent plants, chimneys, flare stacks, etc. The ground must be suitable for the provision of basins and in the case of natural draught designs, substantial foundations are required.

Forced-draught coolers can be noisy, and transformers can emit an annoying hum, so both should be placed away from residential areas, offices and laboratories, unless specially-designed, less noisy, models are used.

It is desirable that site expansion plans should be known when positioning central services plants. They are often centrally placed (but not necessarily together) so that site expansion can proceed in all directions, providing the services will not later become part of a hazardous area. Service distribution should run parallel to roadways and not pass through plant areas. Most piped services (steam, water, gas condensate) with some medium- and low-voltage cables and some instrumentation lines are run on piperacks. It should be remembered, though, that pipebridges across roads can seriously affect access for future construction work (Fig. 16.1 p. 222).

Water mains may be buried deeply enough to prevent freezing, though within a plant complex there may be sufficient throughput to avoid this. Most electric power and telephone cables, etc. are run in sand-filled trenches. Open pipe trenches may be used in places where there is no risk of flammable vapours collecting in them or of the material in the pipe freezing, e.g. steam mains. It is cheaper to have pipes at ground level but this method should be confined to areas where the resulting hindrance to access is unimportant.

Details on the layout of utilities are given in Chapter 14 and of piping in Chapter 31.

Administration buildings should be located on the public or safe side of the security point and close to the main entrance if possible. All buildings should be away from any plant capable of venting fumes to the atmosphere. Adequate car parking facilities should be provided. Canteens, shops for use by employees and medical centres should be located in a safe area as should workshops and general stores (i.e. not process materials). The last two should preferably be within easy access of the process units. Direct access should be provided for supplies traffic, especially if heavy, which should not pass through process areas. Off-loading of stores and supplies should not interfere with other traffic. Laboratories and workshops should be central to the plants they serve. This is not always possible except for new sites, but for both old and new sites, they must be in safe areas.

More details on administration buildings, etc. will be given in Chapter 15.

3.5 EFFLUENT

Effluent can be one of the most visible and least desirable of all the effects of a site on the surrounding community if not handled carefully and responsibly. The appropriate regulatory authorities should be consulted at the design stage to ascertain all existing and reasonably foreseeable effluent disposal standards so that design can meet these criteria. Individual plants for effluent treatment and disposal are sited in the same way as any other plant, namely by reference to the sources, volumes and transportation of effluents to the treatment plants, the storage and handling of treatment chemicals and the disposal of treated effluents from plant to sewer or gate and access for sampling and monitoring.

Incinerators should, where practicable, be sited next to the process supplying the effluent. Solids for dumping should be loaded directly from process to transport. If intermediate storage on site cannot be avoided it should be situated to avoid nuisance to the public and other plants from dust, smell, fire risk, seepage, etc.

Gaseous effluent should be burnt or discharged at such a height and location so that offensive fumes are not a public nuisance. Flare stacks may need to be on a distant site if the nuisance of thermal radiation and noise from the flame cannot be eliminated by suitable design.

Storm water and harmless aqueous plant effluent can be run in open trenches or sewers but obnoxious aqueous effluents must be in an enclosed sewer. Care should be taken in layout to avoid the flooding of sensitive areas such as pump pits and bunded areas. Liquid effluent must not be allowed to run off plants onto adjacent property, or vice versa. Extra precautions are necessary if the site slopes or contains natural watercourses.

Consideration should be given to the problems associated with those flammable effluents which are immiscible with water. There is always a possibility of burning liquid entering the effluent system (particularly open trenches) and spreading fire over long distances, unless traps and gullies are installed.

Effluent routes should run parallel to the road system and should be alongside the road so that any disruption in the sewer is less likely to close the road. The various parts of the site should be graded so that storm water goes to a specific drain. Storm drainage from undeveloped areas and from buildings, etc. having no spillable liquids (or solids!), can go direct to the community sewer providing there is the capacity.

Rain from plant areas should be treated along with the aqueous process effluent. When different effluents meet, care must be taken that no undesired reactions can take place, e.g. release of toxic fumes. To achieve these objectives, advantage should be taken of the natural grade in siting the draining systems and the effluent treatment plants. For good community relations, an effluent plant should not be sited adjacent to residential or similar property.

It should be noted that toxic or flammable fumes can easily build up in a drain from relatively small amounts of material and so proper venting should be installed.

Further details on effluent disposal will be given in Chapter 13.

3.6 TRANSPORTATION

A good site layout limits the distances materials have to flow either to or from store or during processing. It separates the raw material unloading facilities from the product loading areas. These two objectives can be met by laying out the site in the same order as the site material and flow diagram. However, this arrangement may be changed, particularly on sites with high traffic density when it may be desirable to keep the external raw material and product (including waste) traffic away from each other and from all other traffic needed for engineering, canteen, construction, personnel, etc. With lesser traffic densities the layout should be such that site roads are open for all traffic but that plot roads should only have traffic with business on that plot. An important factor that alters traffic flow is the segregation of hazardous processes, already discussed in section 3.2.

Ideally, unloading and loading areas should be situated on the edge of the site near the road entrances, rail spur or dock. However, if the materials are unpleasant or hazardous these areas cannot be near neighbours. It is usual for the storage areas to be alongside the loading and unloading areas in order to control the positioning of materials in store. Ideally, the plant should then be next to the store on the other side from the shipment area, though this is not desirable for hazardous materials.

There should be adequate parking (or rail siding) space for vehicles waiting to load or unload, to use the weighbridge or to receive clearance to enter or leave the site. Lorries should not create excessive noise, especially at night, in passing through residential areas to reach the works and should not have to queue on the public road to gain entrance to the site.

Internal transportation of materials can be by pipeline, conveyors or vehicles. Pipelines should be run parallel to the road system in the same way as the utilities are distributed. The routes for vehicles for on-site transfer of materials should be planned. In particular, road and rail traffic should not go through plot areas other than at their destinations and, even then, hazardous-area classifications must not be violated.

Adequate access must be provided to plants where equipment or materials must be brought in for maintenance purposes and for firefighting. There should be sufficient turning space for tankers and vehicles, though it is safer if these manoeuvres can be avoided.

Dual-direction roads should allow the safe passage of two vehicles and blind spots on roads must be avoided. Room must be allowed for adjacent drainage channels at the sides of roads.

Pedestrian pathways adjacent to roads should be provided when needed. Car and bus parks should be in safe areas and outside security points. Factory gates should be sited so that the effect of personnel coming off duty on the outside traffic is small and safe.

Rail tracks within works and rail links with the national system should be laid out in consultation with the local railway authorities and with the regulatory authorities (for unloading/loading safety). Account should be taken of hazardous-area classification and any special requirements for rail tanker loading and unloading. A rail track can be an obstruction to operation and maintenance and only those plants needing rail transport should be placed near the railway.

For details of layout for transportation see Chapter 9, for storage and loading areas see Chapters 10–12 and of conveyors see Chapter 29. Typical road and rail dimensions are given in Appendix C.

3.7 SECURITY

Site security facilities should maintain a secure and stable environment for the planned site activities. In particular, three conditions must be met:

(a) Company and employee property must be protected against damage, loss or theft.
(b) Unauthorized access to the site must be prevented, to fulfil the company's legal obligations to protect property and the public and to prevent trespassing.
(c) All persons on site, whether staff or visitors, must be accounted for in event of a site emergency.

A security fence around the plants is needed though this can exclude some parts of the property which may be unusable or be left for future developments. An over-long fence is difficult to supervise, particularly at night and in remote areas. Major entry points to the site are normally at road accesses and have gatehouses, weighbridges and waiting spaces for trucks and will occupy substantial areas. Minor entrances are normally not staffed and need very little space. Access to them can be controlled by magnetic cards, etc. In some cases, roads parallel to the fence, lighting of the fence, closed-circuit television and armouring (e.g. barbed wire) of the fences are employed. Emergency planning may require extra emergency gates.

Historically, security against intrusion has been concerned principally with individuals; more recently, mass-picketing or intrusion in furtherance of a labour dispute or political aim has become a problem. In the case of chemical plant this could be potentially disastrous. Layout of main gates and gatehouses should avoid creating a large or sheltered area where crowds can congregate. Key buildings such as telephone exchanges, control centres and electrical substations or those containing hazardous materials should be located and constructed so that they are difficult to damage or occupy. Earthworks may be needed to screen storage tanks of dangerous materials from the site fence.

Security needs to start when construction starts and the construction site must be treated for security purposes as if it were like a fully operational plant (see section 16.1, p. 222). Thus, it is necessary to protect the temporary offices, workshops and fabricating areas, stores buildings, storage areas and vehicle parking. Commissioning will also require office space, laboratories and, perhaps, materials storage and these will need to be made secure. The temporary facilities for construction and commissioning take a surprising amount of space which must be allowed for in the layout.

3.8 ENVIRONMENTAL ASPECTS

Designers and layout engineers have a social responsibility to take into account the impact the site will have on the local environment when applying for planning permission which will have to be sought from the appropriate authorities. In some cases it may be desirable to carry out a formal environmental impact assessment. For major new sites this may become mandatory.

Positive factors, such as the creation of jobs and inflow of cash will be strong selling points and will be readily accepted by the local population. Public attention is likely to focus on the effects the site will have on the natural environment – people, amenities, wildlife and ecology.

The most immediate impacts of the site will be on the people living around it who may be, or may believe they will be, affected by effluent, smells, noise, traffic congestion and the appearance of the site. The need to contain effluent discharges within legal limits is obvious. Additionally, effluent of all kinds must be checked for smell, smoke, dust or spray drift, and for visual eye-sores, such as solids stockpiles, drum stores, unwanted or scrap equipment, froth or stains at liquid outfalls and dark smoke. The location of site main gates and their attendant heavy vehicle and foot traffic should relate to traffic flows on the public roads. Site traffic should not cause congestion or hazards on public roads. Personnel access by foot, cycle, or vehicle should be distant from housing zones in order to minimize large traffic volumes disturbing residents at shift-change times. Siting of rail spurs, heavy goods vehicle loading points, site roads and rail lines which can cause noise particularly at night, should be kept away from housing zones. Similarly, areas of very high illumination such as compounds, marshalling yards or elevated plant-operating platforms should not be near neighbouring houses.

Although noise suppression begins with the appropriate design of plant and process equipment the layout can also reduce the amount of noise transmitted across the site boundary by locating noisy items (especially those running at night) inside enclosures well within the site. If noisy items have to be near the boundary, screening by walls, earthworks or noise deflectors may be needed. Buildings such as administration blocks, non-shift workshops and stores, etc. can be near housing and can also act as noise screens. Section 13.4 (p. 202) discusses noise problems further.

Wildlife and ecological systems have very high tolerances to noise and traffic but are critically affected by plant effluent. The obvious principal hazards are pollution of watercourses by liquid effluent, even clean hot water. Abstraction of water from site boreholes can change the water table remote from the site and damage the ecology. Effects on vegetation by fume or dust can be severe and visible immediately around the site but windborne pollution can cause harm a long distance away. Piles of wastes awaiting disposal can give rise to leaching of pollutants into groundwater. Chapter 13 considers the problems of effluent disposal.

The interaction of processes in an emergency with a neighbour's operation has to be considered (section 3.2). When positioning plants and buildings, early consultation with the local authority is advisable and usually mandatory. High stacks can cause aerial hazards and may need warning lights.

Maintaining a pleasing appearance of the site, particularly from outside, is an important factor in preserving good local community relations. Tree-felling and levelling should be kept to a minimum, as apart from cost, they create barren, straight-line appearances which may conflict with natural surroundings. Plants and storage tanks can often be hidden from, or blended with, the view by using trees and by using existing natural contours as shown for a gas compressor station in Figs 3.3 and 7.13, (p. 88) and for a fire-fighting reservoir in Fig. 14.5, (p. 212). Building artificial earthworks (Fig. 11.1, p. 163) or semi-burying tanks can produce a similar result. This type of work may be vital if there are points near the site which are established as natural viewpoints, and from which the site would appear as an intrusion. Tasteful colour schemes can help in blending large buildings into their background. Occasional splashes of colour carefully chosen will break-up large, monotonous slabs and give visual relief and interest. Plant appearance need not, and should not, be offensive (either through poor siting, poor design or bad housekeeping) and the engineer should be proud to let a well-designed plant, which balances form and function, be seen. The works of man have a place in our environment and can be interesting complementary features of landscape so long as they do not seek to dominate both environment and mankind. This is the ultimate challenge of all engineers associated with the overall design.

Fig. 3.3   Blending a gas compressor station with surroundings (Courtesy: British Gas)

3.9   GEOGRAPHICAL FACTORS

The basic site selection process takes account of the economic geography of the area in which the site is placed, as described in section 3.10 but some aspects of physical geography have important site layout implications.

Plants should be more widely spaced if the site is in an area prone to seismic disturbance. Positions of plants may be influenced by variations in the ground load-bearing capacity or the water table level. Ground contours may encourage interplant gravity flow.

Extremes of weather must be known and allowed for in layout. Tropical monsoon conditions can produce up to 10 cm rainfall/hr and will consequently demand extensive surface water drainage, particularly from pump pits and tank farms. In some areas, severe electrical storms can occur one day in three and all plants must be within the 120 ° cone of protection of a lightning conductor. Severe cold conditions may require deep-burying of liquid pipes and special insulation of those above ground. In hot conditions, frost protection may be unnecessary. Personnel and vehicle movement inside and between plants is more difficult in extremes of temperature, high wind or heavy rain. These points must be reflected in planning of site emergency routes and choice of vehicles.

In hot sunny countries, solar heat gain on plants may be considerable and lead to uprating tank maximum pressures. Refrigerated plant should be sited in the shade if possible. This demands a knowledge of the sun's direction and elevation. In this connection, it should be noted that, south of the equator, the sun is in the *north* at noon.

The direction of the prevailing wind varies in different parts of the world and this affects the position of, say, cooling towers which have to be to the leeside of the plant. Sites at high elevations – say 600 m and more above sea level – will have lower barometric pressure and so design parameters, such as fan ratings, boiling points or evaporation rates, may need correcting before design commences.

All these factors may affect whether buildings are enclosed or open-structured (section 17.2, p. 242).

3.10   SITE SELECTION

In looking at possible sites,[1-6] account must be taken *inter alia*, of layout factors such as:

(a)   Desired layout of the proposed complex.
(b)   Cost, size, shape and contours of the land.
(c)   Degree of levelling and filling needed.
(d)   Load-bearing qualities and acidity of the soil.
(e)   Natural drainage.
(f)   Natural water table and flooding history.
(g)   Direction of prevailing winds and aspect.
(h)   Maximum wind velocity history.
(i)   Seismic activity.
(j)   Existence of old mineshafts and workings, culverts, pipelines or old chemical dumps.
(k)   Ease of obtaining planning permission.
(l)   Nature of adjacent land and activities (Table 3.1).

(m) Any future developments being considered by other bodies adjacent to the proposed site which could have beneficial or harmful interaction effects.

However, some important criteria for selection of a site are independent of the layout. They reflect instead the relationship of the site to its surroundings. A major consideration is the proximity of the site to raw material supplies and product markets and means of transport open to the site – road, rail, docks and (for small-volume, high-value goods), airports[2]. The optimum

Table 3.1 Possible adjacent types of land and activities

Airfield	Open storage
Ancient monument	Park
Barren land	Place of religion
Camping/caravans/chalets	Power station
Canal	Quarry
Cemetery	Quays
Cereals/ungrazed grass	Railway/station
Chemical manufacture	River
Cliff – above/below	Road
Docks	School
Effluent treatment	Scrubland
Emergency services	Sea – estuary
Footpath	Sea – open
Grazing land	Shops
Horticulture/vegetables	Sports ground
Hospital	Tank farm
Hotel	Theatre
Housing – low rise	Tip
Housing – high rise	Tunnels/caves
Incinerator	Warehouse
Lake/reservoir	Woodland/orchard
Leisure centre	
Manufacturing	
Marsh	
Mine	
Motorway	
Museum/gallery	
Offices	

combination of distance/difficulty/reliability of transport should be sought, bearing in mind any legal requirements for transport of materials, particularly hazardous or flammable ones. The availability of local labour, of suitable skills, and quality and motivation, and the existence of local subcontractors should be examined. The community and local infrastructure of schools, housing, social and cultural life may influence the willingness of key staff to move to the new site. Attitudes of local authorities and pressure groups can affect the responses both of the local community and society at large to the new site.

The most immediately measurable impacts of the site will probably be its consumption of public utilities (water, gas, electricity) and its generation of effluents (process, sewage, fumes and noise). The local services must be checked for their capacity to cope with the new demand and for their standards of quality and tolerance. Inevitably, fears of major plant accidents impinging on the community will arise and a level of risk certainly no greater, and preferably a lot less, than that imposed on members of the community by the existing industry should be the basis of design. Thus, due regard to adjacent fire hazards (buildings, factories, plants, tips and vegetation) should be made. Adequacy of local or regional firefighting and hospital services must be checked against the foreseeable major accidents. If the site development requires a large temporary influx of construction workers, their impact on a community can be considerable, particularly if the manners and customs of the two groups differ. In such cases, the setting up of a temporary 'construction community' must be considered. The nuisance effects of construction (noise, dust, etc.) should be minimized.

Economic factors also enter the choice of site through government incentives for investment and employment, assistance with buildings and developments of the social infrastructures such as roads, airports, railways, etc. These incentives are attractive. They are directly calculable cost benefits in the site investment decisions which might be allowed to override intangible negative considerations (e.g. the motivation of local labour). They can lead to wrong site-selection decisions.

It is very important to have, as early as possible, consultation with planning and other local authorities and with interested parties. The hiring of professional advisers knowledgeable in, for example, the local legal, safety, environmental and social practices, can also be advantageous. Such discussions should not only establish what effects the site may have on the locality but also should reveal if the planning authorities have any other plants or proposals that may affect, or even inhibit, future expansion. Indeed, a plant owner should continue to maintain contact with the planning authorities after the initial development in order to keep open the option to expand his process. There is the possibility that the planning authority will subsequently allow mixed – not selective – development up to the site perimeter so that the undeveloped part of the site becomes, for the owner, useless land.

REFERENCES

1. Townroe, P., *Planning Industrial Location*. Leonard–Hill, London, 1976.
2. Balemans, A. W M., *et al.*, 'Check-list. Guidelines for the safe design of process plants', in Buschmann, C. H., (ed.), *Loss Prevention and Safety Promotion in the Process Industries*. Elsevier Scientific Publishing Co., Amsterdam, 1974.
3. Maclean, W. C., 'Construction site selection – a US viewpoint', *Hydrocarbon Processing*, 110, June 1977.
4. Herzet, G. L., 'Construction site selection – a European viewpoint', *Hydrocarbon Processing*, 116, June 1977.
5. Cleveland, J. A. *et al.*, 'Using computers for site selection', *Environmental Science and Technology*, **13**, 792, 1979.
6. Conway Publications, Inc. 'Plant Location', in: Kirk-Othmer, *Encyclopedia of Chemical Technology*, 3rd edn., **18**, 44, 1982.

PLOT LAYOUT PRINCIPLES

4.1 PLOT LAYOUT OBJECTIVES

The theoretical minimum space a process plant can occupy is the total volume of its various components. Various constraints prevent the attainment of this minimum. These include allowing adequate clearances for access during operation, maintenance and construction and to prevent and isolate disaster. Subject to these constraints, the most economical plot layout is that in which the spacing of the main equipment items is such that interconnecting pipework and structural steelwork are minimized. Figures 4.1 and 4.2 show two typical plots and a further example may be found in Fig. 7.6 (p. 75).

Normally, equipment should be first laid out in a sequence to suit the process flow, but exceptions to this arise from the desirability to group certain items such as tanks or pumps, or perhaps to isolate hazardous operations. Equipment may also be grouped when a crane or trolley beam is needed for removing equipment and for materials handling. The choice between single or multiple stream flow patterns will affect layout as will the need to duplicate equipment (as standbys, etc.).

As a general rule, equipment should be at ground level, and high elevation should only be considered when ground space is limited or where gravity flow of materials is safe, economic, and gives greater reliability.

Plot buildings (section 4.11) should be kept to a minimum because most items of equipment may be safely installed in the open. The advantages afforded by buildings include protection of people and equipment against the weather and greater reliability of sprinkler installations. These have to be balanced against the disadvantages, such as impeding fire-fighting and the tendency to collect rather than disperse toxic or explosive vapours.

4.2 PROCESS CONSIDERATIONS

Where convenient the layout is based on the order in which items appear on the process flowsheet. The process design will have some particular layout specifications referring to, *inter alia*:

Fig. 4.1 A plot for a small additives plant (Courtesy: Burmah-Castrol (UK))

(a) Gravity flow.
(b) Limitations of pressure or temperature drop in transfer lines and heat exchangers.
(c) Sufficient head for orifices, reflux returns, control valves and pump suctions, particularly for liquids near their boiling points.
(d) Positioning of orifice meters.
(e) Length of instrument transmission lines.
(f) Requirements for operation particularly for manual materials handling operations.

Fig. 4.2 A plot for a large ethylene plant (Courtesy: ICI Petrochemicals and Plastics Division)

These requirements should not be overlooked during the layout procedure and the process designers should frequently review the layout during development of the project.

4.3   ECONOMIC CONSIDERATIONS

Economy in plant layout is concerned mainly with steelwork, concrete, piping and electric cables.[1]

Structures and, therefore, deep foundations can be greatly reduced by having most equipment on the ground. Where structures have to be used they should support more than one item. To minimize piling costs, the heaviest equipment should be located over the best load-bearing soil.

Equipment should be located to avoid excessive pipe and cable runs. Long runs increase the amount of conduit and insulation, have more fittings and greater energy losses.

Computer programs created for the economic optimization of layouts are discussed in section 6.2 (p. 54) and further ways of making economies are given in section 17.3 (p. 243).

It must be remembered, though, that economic considerations must not cause the constraints of safety, operation, etc. to be overlooked or ignored.

30

4.4 OPERATIONAL CONSIDERATIONS

Operational convenience is very important in achieving safe and reliable operation. It reduces the chances of making mistakes and increases the probability of malfunction being detected early.

Equipment, particularly that requiring frequent attendance, should be reached by the shortest and most direct routes from the control room, which itself must be in a safe location. Valves, instrument dials, etc. should be at a suitable height so that they can be easily used or read. Batch processes require more attention by the operator than continuous ones and so greater consideration has to be given to the ergonomics of the layout. Section 17.5 (p. 249) looks at the details of these requirements.

All plots require emergency escape routes and appropriate firefighting measures (section 17.7, p. 255).

4.5 MAINTENANCE CONSIDERATIONS

As with operation considerations, the layout designer should be looking for equipment arrangements which assist safe maintenance. Maintenance which is made safe and easy is more reliable, is often quicker and saves downtime which in the long run provides ample repayment for the thought and care given at the layout stage. Too often, space requirements for maintenance are outlined at the initial stages of design, and then forgotten during detailed design.

For equipment that is to be maintained *in situ*, space must be left for:

(a) Both men and their tools, to reach the equipment for inspection, and repair.
(b) The lifting gear to withdraw/lift/return internal or external fittings.
(c) Laying down new and used parts.

Where equipment is to be repaired in the workshop, space is required for:

(a) Both men and their tools to reach the equipment for inspection, disconnection and reconnection of pipework, electrics, etc.
(b) Removal and replacement of equipment.
(c) Loading/unloading of equipment on to workshop transport.

In both cases, the layout should provide a safe place of work with regard to access, lifting equipment, entry into vessels, electrical and mechanical isolation, draining, washing, etc.

Where precision equipment has to be used on open plants, weather protection may be needed or the equipment removed to the workshop.

Detailed maintenance aspects are given in section 17.6 (p. 253).

4.6 SAFETY AND EMERGENCY CONSIDERATIONS

The layout of a plot can have a number of important impacts on plant safety. These include:

(a) Protecting operators from such hazards as tripping, bumping the head, or coming into contact with hot surfaces (section 17.4.1, p. 246).
(b) Containing and channelling liquid spillages to safe recovery points, directing vents to safe locations and installing adequate ventilation (section 17.4.2, p. 246).
(c) Allowing vessels and pipework to be completely and safely drained.
(d) Reducing pipe and vessel fractures due to vibration, heat stress, impact, etc. (sections 17.4.3–17.4.5, pp. 248–9).
(e) Separating flammable and ignition sources and adopting electrical classification schemes (Ch. 6).
(f) Protecting equipment and adjacent plots from fires by separation, insulation, water screens, etc. (section 8.6.3, p. 124).
(g) Planning appropriate firefighting and emergency escape procedures (section 17.7, p. 255).

4.7 CONSTRUCTION CONSIDERATIONS

Construction factors must be considered in layout. A building may be needed for process and operating reasons but it will reduce construction access although it will provide weather protection during construction. Likewise, a multi-level plant on an open structure may be more difficult to erect than a ground-level plant.

High or heavy equipment should be located near the construction accesses which are allocated to the plot at the site-layout stage. Heavy equipment in a structure should be near main stanchions. The plot and, in particular, plot buildings should be so designed that adequate access is available to lift large items of equipment or columns into place. Access space planning for operation and maintenance, should be utilized during construction but additional space may be needed for scaffolding, rigging, etc. and for dismantling and removing construction equipment.

On extensions, care should be taken to ensure that constructional work interferes as little as possible with the running of the existing plants.

A recent development in construction techniques which should be considered is 'modular construction' in which the object is to have as much construction and erection work as possible done in the workshop and as little as possible on site.

Construction is considered in further detail in Chapter 16.

4.8 APPEARANCE

As a rule, an attractively laid out plot with equipment in rows is also economically laid out. Buildings, structures and groups of equipment should form a neat, balanced layout, consistent with keeping piperuns to a minimum, and allowing proper access for maintenance. Maintenance roads are provided parallel to the pipebridge and process equipment.

Preference should be given to having a single pipebridge with a minimum number of side branches. This pipebridge may be in the form of a ring main as this can obviate a complete shut down for repairing leaks, etc. Piping should be run orthogonally and, as far as practical, piping in different directions should run at the different elevations. It should change elevation when changing direction.

Where there are duplicated streams these, as far as possible, should be made identical. Such arrangements gain economies in design work, construction, operation and in reducing the amount of standby equipment.

Consultation with architects often leads to a better working environment such as a more pleasing-looking plant with softer lines and more thought about the relationship between the worker, his work and his surroundings. This improves morale and hence means better operation. However, changes in appearance must not conflict with requirements of operation, maintenance, safety, etc.

Further details on appearance are given in section 17.8 (p. 256).

4.9 FUTURE EXPANSION

Thought should be given to likely future expansion of structures, equipment and pipework so that the additions can be erected and tested with the minimum interference to plant operation. One approach to this is to draft the likely conditions on a permit to work and then see if the layout can be altered so that the conditions will be less restrictive on both operators and the construction team. A reasonable distance (depending on the degree of hazard) from plant areas is needed for safe welding where no special precautions are taken. The distance is fixed using the hazard assessment methods discussed in Chapter 8. On the other hand, the probable positioning of extensions should not involve excessive runs of pipework to link up with the existing plant. The distance is fixed by balancing the cost of extra piping against the cost of taking precautions during erection including the possibility of shutting down and then draining, purging, etc.

On the main piperuns it is desirable to leave room for future extensions.

4.10 CONSIDERATIONS FOR SOLIDS HANDLING PLANT

Whilst most of the preceding principles will apply to the layout of solids handling plant, there are other special needs which must also be taken into account.

This type of plant normally involves the use of heavy moving machinery and layouts should always aim to contain noise and dust within acceptable and permissible limits to provide for safe working conditions for operators and maintenance personnel. A vacuum ring main system or other equipment can be installed to clean dust from ledges, filling and emptying areas, and from spillages.

Sensitive equipment such as switchgear, controls and instrumentation should be housed separately for protection against contamination and vibration. Control rooms need protection against dust contamination, explosion and vibration. Steam, compressed air, and power generation plant, should also be housed separately, and isolated from the solids handling systems.

Layout is largely dictated by the process flow requirements. Simple processes requiring elevators can often be accommodated by one single lift elevator and the utilization of gravity flow. More complicated processes, with several recycles may need intermediate elevators to keep the height within practical limits. To avoid expensive structures, heavy equipment such as rotary driers, ball mills, and large crushers are best sited on the ground floor, with ancillary plant arranged on upper floors in a manner which allows easy access for operators and maintenance workers.

Sampling requirements should be considered at the design stage, with chutes arranged to convey samples to convenient collecting points.

Chapter 26 describes solids handling plant in more detail.

4.11 PLOT BUILDINGS

When layout is considered, an important decision affecting operations is the choice between an open plot and a building. Generally, a building is required for processes (or sections of a process) needing:

(a) Continuous operator and frequent maintenance attention.
(b) Extreme cleanliness, or special environments.
(c) Protection from the weather.
(d) Secrecy.
(e) Actual and psychological protection from high elevation.
(f) In some countries plants in buildings may attract less tax than those in the open.

Sometimes a building is not needed but individual items require weather protection.

A building is not desirable when the process uses flammable or toxic materials which must be dispersed quickly in the event of leakage. Since buildings are costly, their use must be carefully justified. The layout, whether in a building or not, should aim for plant items to be arranged at ground level with structures limited to those needed to maintain the essential elevation relationships required by the process.

Following the assessment of the need for a building, the choice of a control room or a set of local control stations must be considered. Local control stations are only practicable inside a well-lit and heated building (Fig. 17.7, p. 255) and the control room solution is most commonly adopted. Frequently mess rooms, amenities and plant laboratories are grouped together for operator and layout convenience, but their positions should minimize the numbers of people exposed to risk, consistent with safety. The location of the control room should be as near as possible to the plant.

The choice of plant structure is further considered in section 17.2 (p. 242) and control rooms in section 17.9 (p. 257).

REFERENCE

1. Kern, R., 'How to arrange the plot plan for process plants', *Chem. Engng*, May 8, 191, 1978.

5

PLANNING OF LAYOUT ACTIVITIES

5.1   THE ORGANIZATION OF STAGE ONE LAYOUT

This activity may also be known as 'proposal', 'front end', 'definition' or 'conceptual' layout.

The main object of the Stage One design is to provide sufficient information so that the feasibility, i.e. cost, economics, hazard, risk and the environmental and social impact of the proposed project can be estimated with sufficient accuracy for approval in principle by the sponsor and for new sites, by the regulatory authorities. The sponsor then allocates funds for Stage Two design and, if applicable, site purchase.

Stage One layout involves determining the relative positions and separation distances of the principal process units and buildings. Information for Stage One layout is obtained from the preliminary process design that takes the form of process line diagrams and data sheets giving the approximate sizes of process units. Other information is taken from design codes and standards, and manufacturers' preliminary information. Many site details will be known if the site is developed, but may be vague for a 'greenfield' project. The layout engineer has to estimate areas necessary for construction, maintenance and operation of the plant. This will involve consultation with construction departments and operators of similar plant.

Hazard control is the consideration of highest priority to which the requirements of operation, maintenance, and administration of the plant have to be related.

A typical sequence for Stage One layout is given in Fig. 5.1. In the iterative phase of design, the process design and engineering design specifications are improved until a satisfactory layout is obtained. Generally, the better the design standards, the more compact a safe layout can become. Stage One layout usually follows the order of equipment first, then plot and finally site layout as discussed in section 2.3 (p. 8), although an existing site may constrain the plot layout.

The review part of the iteration (shown as the diamond in Fig. 5.1) involves close co-operation between the process and layout engineers, particularly about hazard assessment. Operation, maintenance and construction staff

Fig. 5.1 Sequence for Stage One layout

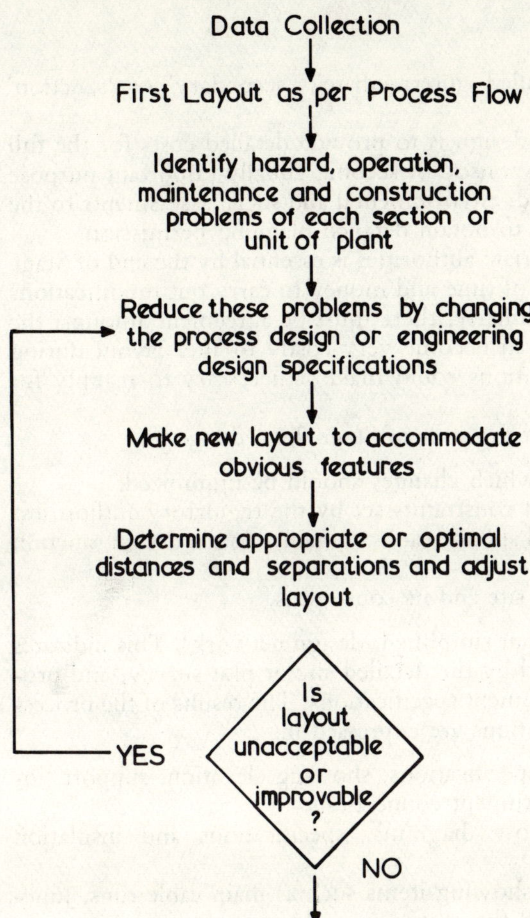

are also needed to participate in the review. Many cost-saving ideas (and costly mistakes) occur where two or more activities overlap such as between layout, process design and operation: or between layout, construction, design and maintenance.[1,2]

The engineers and others involved in the review should use a systematic method to scrutinize a proposed layout because *ad hoc* perusal is inadequate. As differing layouts can vary widely in construction costs, a selection of preliminary layouts should be produced, and the more promising ones refined to find the best one. In the case of a new site, site selection is carried out at the end of Stage One design and the reviews will involve reconciling the Stage One layout to the candidate sites.

It is important that a correct Stage One layout should be determined because it establishes the whole concept of the finished plant. If the conceptual layout is faulty, the plant is unlikely to be successful.

5.2 THE ORGANIZATION OF STAGE TWO LAYOUT

This activity may also be called 'intermediate', 'secondary' or 'sanction' layout.

One purpose of State Two design is to provide detailed costs for the full sanction of the project by the sponsor. A second, equally important purpose is to give comprehensive hazard, environmental and social assessments to the regulatory authorities in order to obtain detailed planning permission.

Agreement with the appropriate authorities is essential by the end of Stage Two as this avoids later waste of time and money to carry out modifications to meet legal requirements. Similarly, there must be agreement amongst the various disciplines because it can become very costly to alter layout during detailed design and such alterations could make it necessary to reapply for planning permission.

The information available at the start of Stage Two design is:

(a) The Stage One design to which changes should be minimized.
(b) The general conditions and constraints set by the regulatory authorities.
(c) Constraints, conditions and specifications contained in the design sanction or contract.
(d) The location of the chosen site and its constraints.

Figure 5.2 shows a typical, but simplified, design network[3]. This indicates that design sanction is followed by the detailed site or plot survey, and process design is followed by equipment specifications. The results of the process design and equipment specifications are expressed in:

(a) Equipment drawings and specifications, showing elevation, support, insulation, and pipe and instrument connections.
(b) Piping and instrument flow diagrams, specifications and insulation requirements.
(c) Electrical circuit diagrams showing items such as main cable runs, junction boxes and starters, and the motor control centre.
(d) Control room and control panel requirements, instrument cable runs.
(e) Building requirements for operation, maintenance, laboratory, warehousing, etc.

The site survey will reveal the:

(a) Load-bearing properties of the ground.
(b) Positions of services external to the site or plot.
(c) Details of the environment and neighbourhood.

The standards and policy of a project need to be established for:

(a) Road, rail and service layout.
(b) Buildings and construction requirements (including standards of protection against fire and explosion).
(c) Heights of pipebridges, overhead and buried piping, are decided partly from information provided by the site survey and partly on the basis of generally accepted and proven standards.

The main engineering design commences from node 3 as shown in

Fig. 5.2 Simplified Stage Two and Final Stage design network

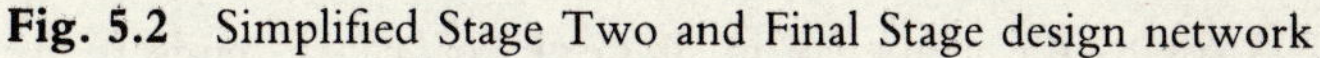

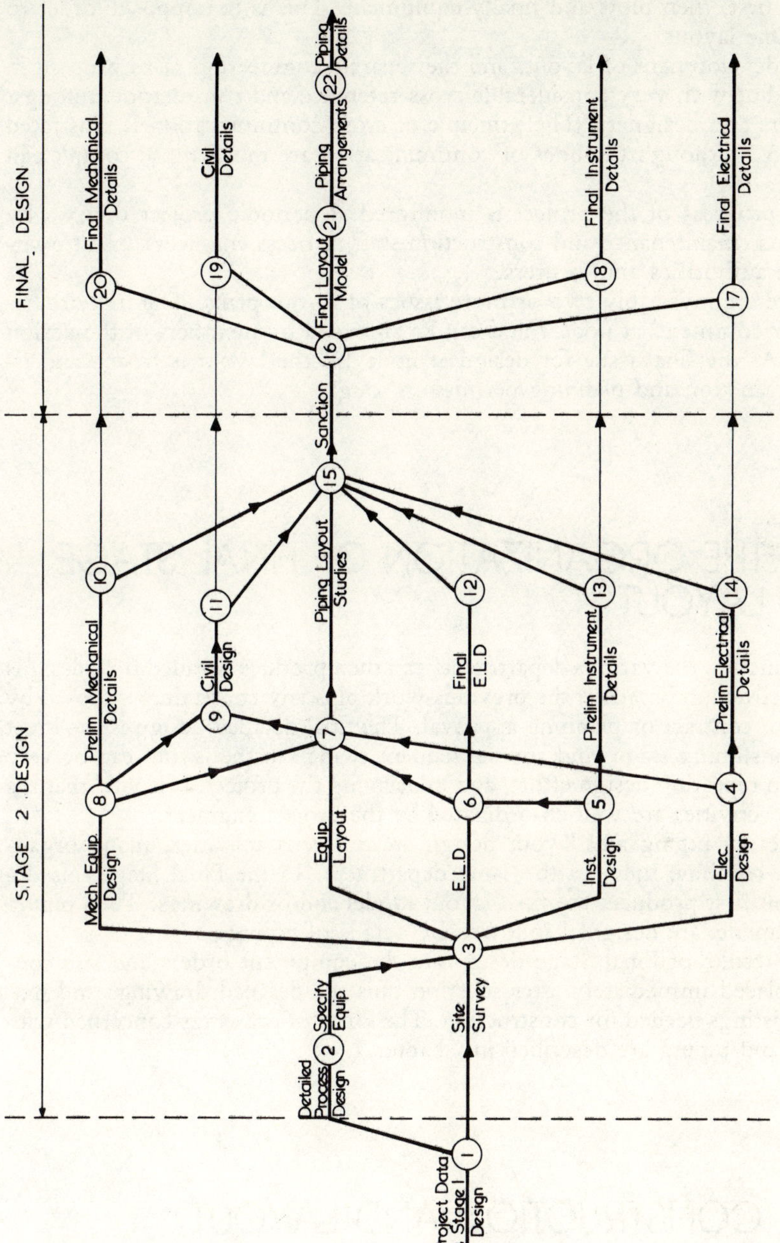

Fig. 5.2. The instrumentation, the electrical and the mechanical design follow separate paths in the network in parallel with detailed equipment layout. After the equipment loads and support positions have been determined by the mechanical design, civil engineering design can start from node 9. Piping studies begin when the equipment layout is complete at node 7. As men-

tioned in section 2.3 (p. 8) the sequence in Stage Two layout tends to be the site first, then plots and finally equipment. This is the opposite order to Stage One layout.

The development of layout, and the separate engineering tasks proceed in parallel but with very considerable cross-reference and consultation amongst engineers and designers. The principle of cross-communication is illustrated in Fig. 5.2 although the lines of communication are much more complex in practice.

The progress of the project is monitored at periodic project reviews by operation, maintenance and construction staff, process engineers, plant owners, fire authorities and insurers.

There are invariably two or more issues of layout plans. The first are 'issued for comment' at node 7 and can be amended by members of the design teams. At the final issue for design at node 15, the layout is frozen and financial sanction and planning permission sought.

5.3 THE ORGANIZATION OF FINAL STAGE LAYOUT

After sanction the various departments can then produce detailed final designs for construction based on the previous work plus any constraints imposed by sanction, contract or planning approval. The final detailed design is the most time-consuming stage, and any subsequent change to the layout can be very costly in repeating design effort, and in delaying the project. It is vital that the various activities are well co-ordinated by the project engineer.

Although piping and layout design are different activities, many organizations combine them in the same department. In the Final Stage this department first produces the final layout model and/or drawings. Then piping arrangements are designed followed by details of piping.

The results of Final Stage design are the equipment orders and site contracts placed immediately after sanction plus the detailed drawings and materials listings needed for construction. The kinds of drawings concerned with layout and piping are described in Chapter 7.

5.4 CONSTRUCTION AND LAYOUT

The network of activities which covers layout and engineering design also includes construction and commissioning. Construction starts before design is complete and similarly commissioning before completion of construction. Construction can start immediately after sanction when plot plans and structural drawings are available (nodes 16–19 of Fig. 5.2) and proceed while the final design details are drawn. Clearly the work prior to these nodes has to be frozen if construction is to start early. Otherwise, some construction work will have to be redone due to design changes.

Kern[2] points out that if there is clement weather, good planning and co-operation between the design and construction teams then calendar time can be saved by having on-site design (particularly of details such as pipework) proceeding alongside field fabrication and construction. This is especially true of plant extensions, repeat plants or rebuilt plants. However, for many projects it has been found that the above factors have not materialized so a trend has been in the direction of modularization with as little on-site fabrication and construction as possible. Where this is adopted, the detailed layout and design has to be much more complete before orders and subcontracts are placed. Thus more detailed design is included in Stage Two which means that the decision to use modular construction has to be taken prior to design sanction. So Stage One layouts will reflect the intention to modularize. Modular construction is described further in Chapter 16.

5.5 LAYOUT AND PROJECT PLANNING CONTROL

Because of budget and time constraints, a project has to be realistically programmed. However, often at commencement of Stage Two layout only a rough programme of milestone dates is available. So, during this stage, a detailed programme is developed in parallel with layout progress. Unless the project is either a near repeat or is very limited in layout alternatives, layout activities must be identified, allocated times and an overall timetable constructed. Also, the succeeding activities of plant details design, procurement, manufacture, inspection, test, construction and commissioning must be organized to satisfy the cash flow and time available, or likely to be available, for the entire project.

For small plants, e.g. of a chemical process already well proved in operation, and with less than, say, fifty plant items, simple bar chart programming can be used effectively. Fig. 5.3 shows a typical bar chart for this application.

For large complexes, however, and particularly for plants handling hazardous materials and for prototype processes, the planner must arrange for a plant development procedure which allows several reviews of the layout by other specialists, so that snags are quickly identified and remedied. In particular, to satisfy legal responsibilities under health and safety regulations of both the plant owner and of the contractor, the plant has to be designed according to procedures which consider safety of paramount importance. Formal recording of the results of each review is an important part of the discharge of these responsibilities.

Networks as typified in Fig. 5.2 have become commonplace as a tool in programming and progress reporting of projects from sanction through to construction and commissioning phases. In large projects as well as overall networks, detailed planning networks have to be developed for each separate

Fig. 5.3 Typical bar chart (Courtesy: British Nuclear Fuels)

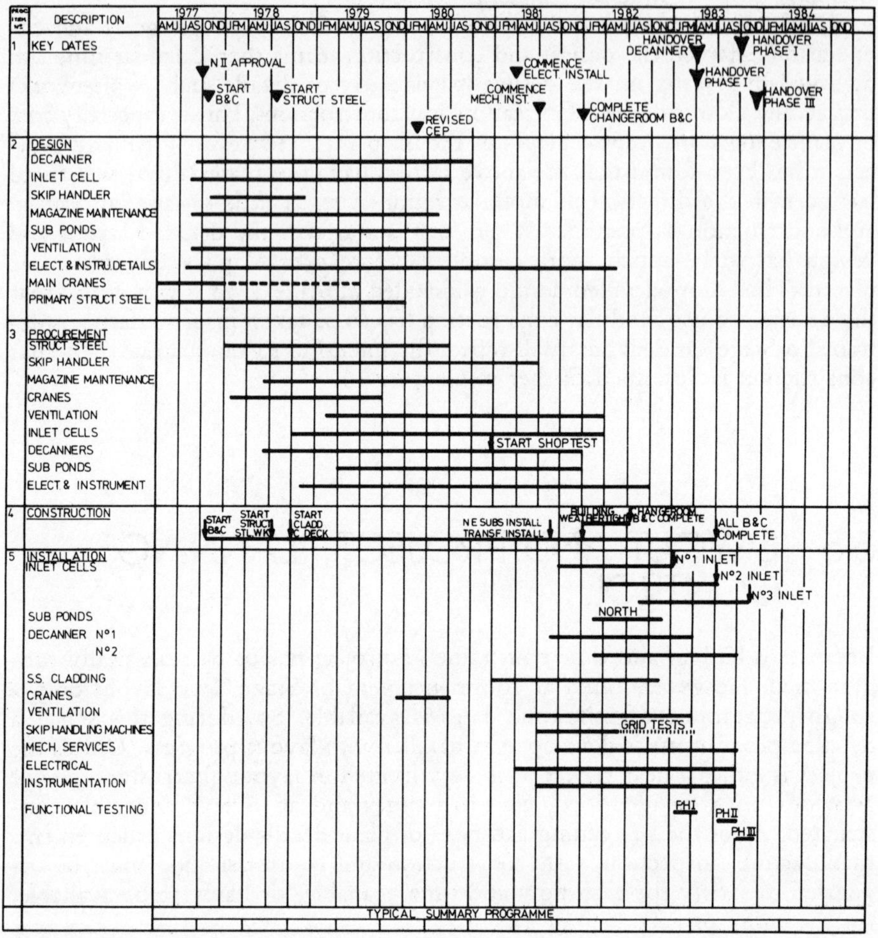

design engineering activity and also for each of the other specialist departments in the organization such as procurement, inspection, construction, etc.

Each of these department networks shows the sequencing of the department's work together with (a) the dependence on upstream departments for commencement, and (b) the flow of finished designs, etc. to downstream departments. The linking activities (a) and (b) are known as 'Interfaces In' and 'Interfaces Out' respectively.

Planning engineers use basic standard networks, previously prepared by each department, to produce a multi-department or overall network for the project. In large complexes a separate overall network is produced for each plant within the complex or for areas within a plant, depending on the amount of work to be programmed and the amount of concurrent working possible between constituent plants or constituent areas.

Computer programs have been developed to process the information established on the network, and diagnose the paths through the network which are critical in their timing and duration to the planned completion date. This procedure has been developed under a number of titles, the best-known being 'critical path analysis'[4] (CPA), 'programme evaluation and review technique' (PERT),[5] or simply 'network analysis'.[4]

The usefulness of a computer in programming varies considerably from project to project. Plant layout engineers usually start their work simultaneously with the overall project programmers (planning engineers). At that stage they are working to completion dates for producing the initial layout drawings and/or models from preliminary process and equipment information. Because the planning information is processed by computer, activity descriptions, durations, overlap and sequencing with other activities, etc. are required in accurate detail. Therefore the plant layout engineer must spend a considerable time compiling these details with the planning engineer, which can delay his own progress. This conflict can be resolved by using the following guidelines to gain the maximum advantage in the application of computers to plant design programmes:

(a) The number of activities processed and the computer program used, must be compatible with a total turnaround time of less than, say 25 per cent of the reporting period (i.e. less than one week if reports are required at monthly intervals).

(b) Within the turnaround time, the computer program must be capable of redistributing space (float) time realistically, within existing constraints, so that no 'impossible' activities emerge (e.g. involving increased manning levels on activities already maximized on manning).

(c) Everyone concerned with the project, particularly project managers, plant layout engineers and planning engineers, must be constantly aware that the design programme developed is an *estimate* of progress against time. Inevitably, the actual progress at certain stages or particular events in the design programme will not be achieved (Fig. 5.4) because of imperfections in the estimate; often the apparent lack of progress is blamed on the workforce simply because progress is not exactly to the estimate.

5.6   LIAISON WITHIN THE DESIGN OFFICE

Work proceeds in parallel in the various departments of the design office as outlined in Fig. 5.2. It is one of the tasks of the project engineer to co-ordinate these activities assisted by the layout engineer. It is very easy for one department to be working to a layout which has been already changed by another. It is also easy early in the project, to forget the layout requirements for departments whose work comes later in the programme.

There should be a series of project meetings for each layout stage attended by all the departments, until the frozen status is reached.

It is helpful for the layout engineer to co-ordinate each department's proposals and comments before each meeting and to present them in an easily understandable way with models, drawings, photos, etc. In this way the meeting can resolve any incompatibility without wasting time.

It is also important that between meetings the project engineer with the

Fig. 5.4 Typical process chart (Courtesy: British Nuclear Fuels)

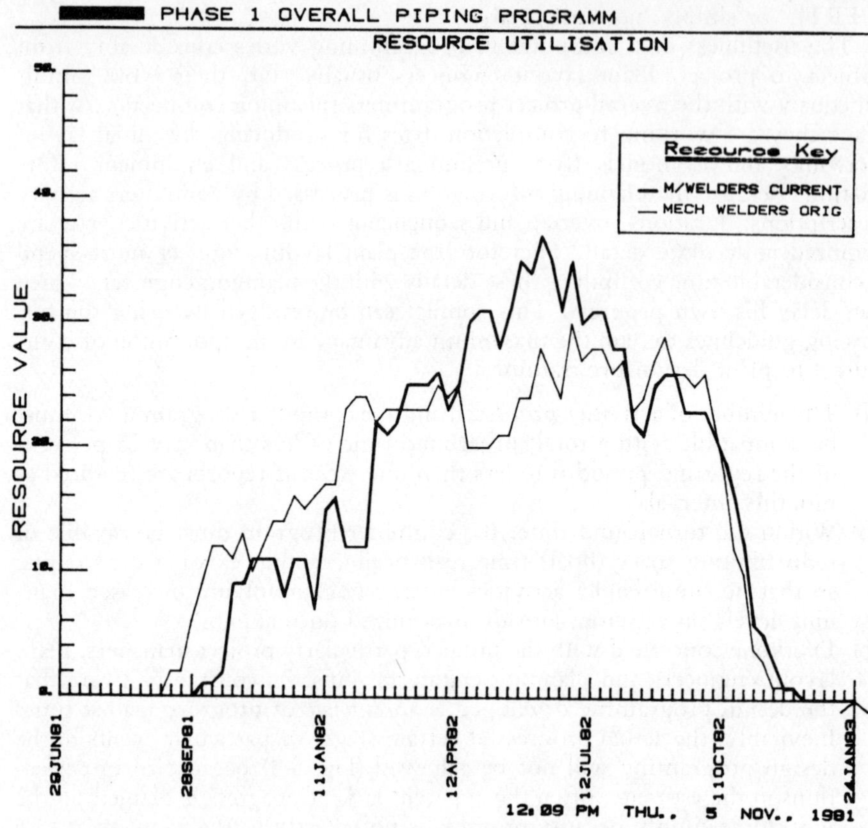

layout engineer, liaises informally between departments so that potentially incompatible plans are not allowed to proceed. There will be direct contacts between departments, but the project engineer must be aware of the decisions taken so that the layout design is kept up to date.

5.6.1 MECHANICAL ENGINEERING

The mechanical department receives:

(a) The process design including flowsheets and details of the process engineering design and the function of the equipment items.
(b) The preliminary layout.

This department undertakes:

(a) The detailed mechanical engineering design and/or specification of equipment items including ancillaries such as boilers, refrigerators, inert gas generators, etc. and vehicles, such as fire-engines and trucks.

(b) The planning of maintenance procedures (e.g. off-plant or *in situ*) and specifying additional items for this, e.g. manholes, lifting beams and access for maintenance. If need be, a fitting shop may have to be designed. Note that poor-quality equipment may require more maintenance access than good-quality items.
(c) The monitoring of the procedures for the installation of items.

5.6.2   PIPING ENGINEERING

The piping department receives:

(a) The process design including process flowsheets with main pipe-sizes, valve and control requirements, materials of construction and lagging needs.
(b) The preliminary layout.
(c) The detailed mechanical design of equipment, especially pipe nozzle positions.

This department undertakes the:

(a) Piping arrangement and isometric drawings; detailed design of the pipework and its supports including pipe stress analysis.
(b) Planning pipework erection procedures.

As already mentioned, the layout engineer is often a member of the piping department so liaison between piping and layout is generally good.

Further details of piping layout are given in Chapter 33.

5.6.3   ELECTRICAL ENGINEERING

The electrical department receives:

(a) The process design including information to determine the power requirements of the items.
(b) A preliminary layout indicating lighting requirements and electrical area classification.
(c) The detailed mechanical design showing the positions of the motors, drive characteristics, etc. and of ancillary equipment, e.g. cranes, not given on the flowsheet.

This department undertakes the:

(a) Detailed specification of electrical equipment, notably substations, switchrooms, transformers, major cables.
(b) Cable specification and routing.
(c) Planning and production of procedures for installing and commissioning electrical items, including earthing.
(d) Production of maintenance procedures, determining power requirements of lifting equipment and defining electrical workshop needs.

5.6.4 INSTRUMENT ENGINEERING

The instrument department receives:

(a) The process design including the instrumentation requirements for plant control in normal operation, startup, shutdown, emergencies.
(b) The preliminary layout, including control room location.
(c) The detailed pipe routing drawings showing proposed instrument positions.

This department undertakes the:

(a) Detailed specification of the instruments and approval of instrument positions.
(b) Instrument cable and piping specification and routing.
(c) Design of control panel and associated equipment, including specification of any microprocessors.
(d) Planning the instrument installation and commissioning procedures.
(e) Planning instrument maintenance procedures, specifying lifting equipment for instrumentation and defining workshop needs.

5.6.5 CIVIL AND STRUCTURAL ENGINEERING

The civil engineering department carries out a site survey and also receives:

(a) The process design including equipment needs of laboratories, capacities for canteens, rest rooms, etc. and drainage requirements.
(b) The detailed mechanical, electrical, instrument and piping designs and loadings.
(c) A fairly detailed layout.

The department then undertakes, preferably in consultation with architects, where they have been involved in the definition of the building requirements:

(a) The detailed structural design of all buildings, roads, railways and drainage.
(b) The detailed internal design of offices, amenities, laboratories, workshops, fire station, etc.
(c) The planning of building and site maintenance and specification of the necessary workshop and stores buildings, compounds, etc.

5.6.6 CONSTRUCTION DEPARTMENT

This department receives the complete detailed engineering design and layout and undertakes:

(a) The planning of the construction programme including the specification of construction equipment and buildings.
(b) Inspection and progress assessment of the actual construction.
(c) Provision of temporary accommodation for construction crew.
(d) Receipt and storage of materials and equipment delivered to site.

5.6.7 PROCUREMENT AND INSPECTION DEPARTMENTS

These specialist functions provide assistance to all engineering disciplines but do not usually impact directly on plant layout. However, one of their important functions that should be briefly mentioned is the requirement under health and safety laws to ensure that purchased equipment items or systems are, as far as is reasonably practicable, designed to be safe when used for the purpose for which they are intended. The reliability of equipment needs to be known for hazard assessment of layouts (see Ch. 8).

5.7 LIAISON OUTSIDE THE DESIGN OFFICE

5.7.1 PROCESS ENGINEERING

As far as the design office is concerned, the process engineer provides the main scientific and technical input, having brought together the work of the various scientific disciplines in the chemical engineering design. The documents of this design include flowsheets, process and instrument data equipment specifications, operating procedures and safety analysis. The design will impose process, safety and other constraints on the layout.

The process engineer, in conjunction with the safety, operating and commissioning personnel, will monitor the engineering design and the layout to see that process, operating and safety considerations are not violated. The layout engineer should keep the process engineer informed of progress by means of easily understood models, drawings, etc. There should be formal meetings where agreements and decisions are recorded, but there should also be plenty of informal contact between meetings.

However, many potential conflicts between layout and process design can be avoided if the layout engineer is kept informed of progress during the process design stage. In practice to save time, the layout and engineering design stage starts before the process design stage is completed. Liaison between the process engineer and layout engineer thus becomes even more important to ensure that the engineering design is not proceeding based on an out-of-date process design. To achieve this the successive stages of the process design must become 'frozen' after a certain point.

Because the detailed engineering design stage is very complex there is always the danger that process considerations, which should be uppermost, will be overlooked. There is thus an important but informal role for the process engineer. Formally it may appear that the chemical engineer only provides a safety and process design input whereas his influence should be on a much broader basis. He should be able to develop good relations with engineers and designers in the design office so that he can wander in and chat with them without upsetting anyone and without appearing to usurp the project engineer's role. He will then be aware of the various liaisons within and without the design office described in this section and section 5.8 (p.51) and will be able to spot, and quickly remedy, situations where process considerations are being overlooked or where process information is deficient or being misinterpreted. On the other hand, if the process engineer does not get on well with the design office, there is a great danger of creating a plant that operates badly and is unsafe and therefore likely to be contravening the law. It helps, of course, to produce in the first place a good process design that is easily understood and does not have to be subsequently corrected.

Layout is a team function with contributions from many disciplines. The project manager plays a key part in welding together and leading this team.

5.7.2   REGULATORY AUTHORITIES

Contact with these authorities should start on an informal basis during the preliminary layout stage before the site is selected or the project go ahead is given. This should encourage good working relations during design.

The various activities concerned are likely to be as described below.

Planning

Usually the local elected body is concerned with general planning such as the compatibility of the installation with others in the area, its effect on the road transport system and its desirability in relation to employment and generation of tax revenue. There may be further planning authorities having an interest such as national park boards (who have strict rules regarding industry) or government development agencies (who are briefed to encourage new industries). The layout will be affected mainly by appearance and road transport considerations.

Section 3.10 (p. 24) mentioned the importance of maintaining contact with the planning authorities after construction in order to keep open the option of using undeveloped parts of the site for future expansion of hazardous process.

Health and safety

This is usually the concern of a government-appointed authority with local branches who have to be satisfied that the process is acceptable for employees and public. The local branches can deal with general safety matters but major hazard investigation is usually considered centrally by a specialist unit. The health and safety and planning authorities usually maintain close liaison and, in some countries, are legally required to do so. Layout is greatly influenced both by general safety and major hazard considerations.

Pollution

In the UK, the planning authorities regulate solids and noise pollution, specialist air inspectorates deal with air pollution and smell problems and water pollution is controlled by the water authorities. In the US, most pollution matters are controlled by the Environmental Protection Agency.

Generally, pollution control affects the process design more than the layout though noisy, smelly or dusty plants should be kept away from site boundaries.

Emergency

The authorities concerned are the police, fire and ambulance departments and hospitals. If the public is likely to be evacuated then the planning authorities are also involved. If the site is near a waterway then the appropriate water transport authority is concerned. The safe handling of emergencies is an important factor in layout. Provision is needed for dealing with the Press and enquiries from the public during an emergency.

Transport

Roads, as already mentioned, are the domain of the planning authorities. The layout may be affected by the requirements of the rail company, dock and river boards. With high chimneys and buildings the aviation authorities may have to be consulted.

Quality assurance

In the food, drug and agricultural chemical industries the layout and design may have to satisfy the regulations of the appropriate quality assurance authorities of the countries in which the product will be *sold*.

Publicity and the Press

It smoothes dealings with the regulatory authorities if it is known that public reaction to a project is favourable. The layout engineer may be called upon to prepare visual aids, etc. for public meetings or press conferences arranged by his company.

Insurance

Unlike the health and safety authorities, the insurers are concerned with the protection of assets as well as people. Consequently they may require additional safeguards to minimize the spread of damage in a mishap. This may lead to additional separation requirements and management systems and it is necessary to consult the insurers throughout the project.

5.7.3   SUPPLIERS AND CUSTOMERS

Equipment suppliers

Where equipment is standard, it is usually economic for the layout to be adapted to suit. The layout engineer should ensure that he obtains all the pertinent dimensions from the manufacturer.

If a piece of equipment is being specially made for the project then it, and the layout, will be mutually accommodated. It is the layout engineer's task to agree *all* the relevant dimensions with the supplier and then see that they are incorporated in the layout.

Raw materials and products

The vehicles (road, rail or water) transporting materials have to be compatible with the loading and unloading facilities at both ends of the journey. The layout engineer should consult with the supplier or customer and the transport company to ensure that they are satisfied with the loading or unloading facilities, the site roads and the safety thereof.

Utilities

The provision and siting of substations, transformers, main cables, etc. will have to be agreed with the electricity suppliers. The piping and storage arrangements for taking potable, borehole, river or sea water will have to be made with the appropriate authorities. Gas and oil connections need to be discussed with the suppliers.

Effluents

The site layout will be affected by the point where liquid effluent is discharged to the sewer. The loading and transport arrangements for solid and special liquid effluents will have to be discussed with the company removing them.

5.7.4 CONSTRUCTORS

The construction planning department will agree the construction programme with the construction team. This team could be from the same company as the designers or could be from subcontractors.

The construction team have to be given all the details of the layout and during construction the layout engineer should ensure the layout is being maintained. Unplanned situations are bound to arise during construction and the layout engineer should ensure that resolution of these difficulties does not conflict with other requirements of the layout.

The construction engineer should attend the formal project meetings, certainly in the later stages.

5.7.5 COMMISSIONING TEAM

The commissioning team will be given the process design including the flowsheets and operating and safety instructions. However, to plan their programme they will need the plant layout, particularly the model. As they will probably find snags in the layout it is important that the process and layout engineers consult with the commissioning engineer about the layout throughout the engineering design.

If commissioning starts before the end of construction then the commissioning team and construction departments will have to liaise.

The commissioning engineer should attend formal project meetings.

5.7.6 OPERATING AND MAINTENANCE PERSONNEL

The operating team will adopt a similar approach to the commissioning team though the emphasis will be on operating the plant for several years. The prospective plant manager should be consulted both during process and engineering design so that results of operating experience can be built into the design and layout. He should attend the formal project meetings. In particular, he will recommend appropriate spare equipment.

The prospective works engineer should liaise with the various engineering design departments when they are planning maintenance procedures. He also should attend the formal project meetings.

5.8 OWNER, CONTRACTOR AND CONSULTANT RELATIONSHIPS

In the preceding sections of this chapter, it was implied that the various design activities were carried out in one organization. In practice they will be split between the plant owner and contractor and also in some cases between contractor and consultant acting for the owner. The allocation of work between client and contractor will depend on, *inter alia*:

(a) Ownership of the technology.
(b) Size and expertise of the plant owner's organization.
(c) Size of project.

Some examples are given as follows.

(a) A large operating company which also owns the technology of the process will probably carry out the preliminary design and feasibility work necessary for design sanction purposes. The contractor will then take over for detailed design and construction. Commissioning could be done by either owner or contractor. Similarly, the site work could be split. In many cases the plant owner will provide detailed layout specifications and will monitor the contractor's design work closely. In particular, the contractor's plant layout engineer will have many informal discussions with the owner's personnel such as those from process, safety, operation, maintenance, etc. in addition to formal meetings.

(b) A large operating organization has to have a fairly extensive engineering design team in order to monitor the contractor in detail. Consequently, this team often acts as an 'in house' contractor for smallish plants or extensions.

(c) Where the technology is owned by the contractor or where the contractor acts as an agent for the licensor then the contractor will also carry out the preliminary design work before sanction. Nevertheless, a large operating company will still follow the design work closely, in particular with regard to safety.

(d) A small operating company will not have resources to monitor the contractor's work. It will leave all the design, construction and commissioning, etc. to the contractor. The expertise that the contractor would obtain from a client with a large organization, e.g. on operating and maintenance requirements of a layout, would have to be found from within the contractor's own organization, e.g. from the commissioning and construction departments. Should a small plant operator want to monitor the design work, he would employ a consultant. In that case the consultant would have probably carried out a preliminary design and then advised on a choice of contractor.

Where there are two parties involved in the design and construction of a plant then the key people are the plant owner's project manager and the contractor's project manager. All formal contacts that vary the contract must be between the two project managers. The contractor's plant layout engineer will have many informal discussions with the owner's personnel from various disciplines. However, all decisions reached in such meetings must be ratified and recorded between the two project managers. Where there is a third party such as a specialist plant subcontractor, it is even more important (and difficult!) that all decisions should be formally recorded.

REFERENCES

1. Kern, R., 'How to manage plant design to obtain minimum cost', *Chem. Engng*, 23 May, 130, 1977.
2. Kern, R., 'Controlling the cost factors in plant design', *Chem. Engng*, 14 Aug., 141, 1978.
3. Mackenzie, G., 'The time and resource aspects of project management in the construction of chemical plants', *Chem. Engr, Lond.* **209**, CE118, June 1967.
4. Pemberton, A. W., *Plant Layout and Materials Handling*. Macmillan, London, 1974.
5. Daellenbach, H. G. and George, J. A., *Introduction to Operations Research Techniques*. Allyn and Bacon, Boston, 1978.

METHODS FOR LAYOUT CONCEPTION AND DEVELOPMENT

This chapter reviews mainly the various methods used to conceive, develop and examine preliminary layouts. Some of the methods are also applicable in later stages of layout.

6.1 SURVEY OF APPROACHES

One can identify five approaches to layout: intuition based on experience, plus four formal ways, namely, economic optimization, critical examination, rating and mathematical modelling. A formal technique is any logical method that provides definitive information on relationships between items or numerical data on spacing distances. It must be based on a procedure which is adequately defined and recorded and can be examined and criticized.

Before starting layout the relevant information should be assembled, such as process and site data, regulatory and contract requirements, company and other recognized codes of practice, etc. Sometimes not all such data are available at the start of a project and to avoid delay and to provide a starting point it is useful to have information on typical spacings such as that given in Appendix C. However, it is emphasized that such spacings are only to be used as temporary expedients and must be confirmed or replaced later on by the proper project data.

A first layout is almost always based on process flow as intuition drawn from experience indicates that such a layout is basically a good one and that it can be altered successfully to accommodate the requirements of operation, maintenance, safety, etc. Intuition in addition indicates immediately what these principal alterations should be. Thereafter formalized methods should be used since the intuitive approach alone misses the less obvious factors thus allowing a potentially good layout to become a bad one.

The impetus for developing formal layout methods is being generated by the changing attitude of society to the process industries and to the consequences of accidents in those industries. Hitherto, formal layout methods, mainly developed within the industry, had tended towards optimization for minimum capital cost. Although safety always was a major constraint in layout, its most visible effect on the layout was related to relatively simple rules for spacing and electrical zoning in accordance with codes of practice. The adoption of more severe processes, the larger scale of plants and associated storage and the shortage of skilled staff, coupled with greater public concern, however, require companies to be able to justify the reasons for selection of a given layout to a far greater extent than was necessary only a few years ago. Records of problems foreseen, alternatives considered and supporting data for decisions made will be required to an increasing extent to satisfy legislation.

At present, very few formal layout techniques are available to the designer and none can replace completely the designer's abilities either to conceive new solutions or to evaluate alternatives. Techniques reported appear to aim at one or more of three main objectives:

(a) Generation of spatial relationships between items.
(b) Specification of distances between items.
(c) Comparison of alternative layouts by numerate rational examination.

When used to supplement or verify the designer's experience present techniques go some way to improve the layout and provide rational justification of layout decisions. Some reported techniques and their operation are outlined in this chapter but very little information has been published on their large-scale industrial use and benefits.

Similarly, computer methods have been slow in being applied successfully commercially to process plant layout. In the future, formal techniques may be computerized to advantage so that results can be obtained quickly enough to give the engineer time to consider and make changes to the layout without disrupting the project time-scale.

6.2   ECONOMIC OPTIMIZATION

Conventionally-generated layouts can sometimes be improved by techniques which evaluate different factors selected by the designer. Alternative layouts can be compared and optimal layouts found.

Section A.3 (p. 478) outlines some methods correlation charts, travel charts and sequencing techniques which can be used. They were evolved for factory layout and therefore are most applicable for multi-storey enclosed building layout. However, they do not seem to have been used much for process plant layout. Similarly there are some computer programs (e.g. Computerized Relative Allocation of Facilities Technique (CRAFT),[1] Automated Layout Design Program (ALDEP)[2] and Computerized Relationship Layout Planning (CORELAP)[3] which tackle the conception of factory layout but have not been widely applied to process plant layout. One reason is that they fit plant into a known building size whereas, for process plant layout the building, if any, should fit the process plant requirements.

A number of workers including Newell,[4] Leesley and Newell,[5] Madden and Taylor,[6] and Cambridge CAD,[7]; Bush and Well[8] and Shuqair[9]; Gunn[10] and Al-Asadi[11]; Fine[12] and Mustacchi[13] have investigated various aspects of computer-aided process plant layout. A short review of these works is given in Appendix A and a number of conclusions can be drawn from their studies as follows:

(a) It is impractical at the present for a computer by itself to conceive a layout or undertake major changes of relational positions of items.[4,13] Layout conception and rearrangement should be left to the designer's intuition. Consequently, some authors advise that the computer system should be an interactive one with rapid visual and graphic output facilities.[6-8]
This is discussed further in section 7.6 (p. 95).

(b) With the layout details within the computer it is possible to calculate the cost in terms of space, piping, cabling and other connections. However, only limited facilities of this kind appear to be commercially available. With the display of cost trends it is possible, with interactive computer systems, for the designer himself to suggest new spacings and thus undertake optimization. The technique is more powerful if automatic pipe routing is available (see (d)).

(c) It is practical and rewarding for the computer itself to adjust the spacings between items and to optimize the piping costs providing the amount of computation[14] is reduced by the designer dividing the plant up into sections. Optimization is done independently within each individual section and overall between sections. However, the optimization is improved if the designer orientates sensibly the exit/entry nozzles of each section.[8,9,11] Automatic pipe routing is a necessary requisite for this (see (d)). Unfortunately, no commercial system is as yet available for optimization.

(d) Automatic pipe routing is feasible[9] (and commercially available[7]). The computer can arrange for the pipe to avoid items and regions set aside for access or other purposes. It is an advantage when major pipe routes (such as piperacks) can be specified by the designer so that the computer then takes the shortest orthogonal route between rack and item. Valves and other pipe fittings, and sizes of pipes and lagging should be taken into account in the routing operation.

(e) In order to avoid impossible or impracticable solutions there should be a routine to detect clashes of items and violations of forbidden regions. The designer should be able to call the routine at request. Time is saved, especially with a skilled designer, if there is the option of omitting the clash routine during say, pipe routing. The routine is then called when an apparently satisfactory arrangement has been achieved.

(f) It is possible to include features other than piping as optimizing criteria.[7,9,11]

6.3 CRITICAL EXAMINATION AND REVIEW

The layout has to be reviewed for operation, maintenance, construction, safety, emergency, insurance and regulatory purposes at frequent intervals during design. Many of the reviews can be of an intuitive nature, but there

Table 6.1 Critical examination sheet

The present facts		Alternatives	Selection for development
What is achieved?	**Why**	**What else** could be achieved?	**What should** be achieved?
How is it achieved?	**Why that way?**	**How else** could it be achieved?	**How should** it be achieved?
When is it achieved?	**Why then?**	**When else** could it be achieved?	**When should** it be achieved?
Where is it achieved?	**Why there?**	**Where else** could it be achieved?	**Where should** it be achieved?
Who achieves it?	**Why that person?**	**Who else** could achieve it?	**Who should** achieve it?

should be at least one formal review, near the finish of the Stage One layout and one formal review of the detailed layout just before sanction is sought (node 15 of Fig. 5.2, p. 39).

One method of formal review is the critical examination technique.

The principles of critical examination as supplied to process design have been known for some time.[15]

For a full examination of the process the full technique is used and questions are asked sequentially as indicated in Table 6.1. However, for layout interest is focused on 'place' (i.e. where it is to go). This is questioned with 'can we do our job regarding it?' (e.g. safely operate, maintain, construct, evacuate, firefight, insure the item or plant). Then preferred alternatives 'where else could it go?' are explored. When all this has been done, the next step is to apply them to the problem, to determine what arrangements or groups of arrangements are ruled out and which remain as possibilities, and of these to select for further development any that may compare favourably with the others.

Experience has shown that application of the general technique to any particular problem may be effective but it is very tedious and costly. Instead, before application, the method has to be adapted and developed for a particular type of problem, i.e. layout and maintenance, or layout and firefighting. In addition, a person should be trained in its use, so that he can act as examination leader and chair and conduct the critical examination. In order to be completely objective this person should not be the project manager, the process or project engineer, the layout designer nor the representative of the appropriate activity under examination (e.g. maintenance engineer or fire chief) nor any other member of the project team. The examination leader is responsible for seeing that all relevant information is available before the meeting (or more likely before each of a series of meetings), that each meeting proceeds quickly and constructively and its decisions are recorded and followed up. A series of meetings may be necessary because further design and investigation may be needed before conclusions can be reached.

So far, specific critical examination techniques for layout have not been developed. The most advanced development of critical examination for process engineering has been the HAZOP method for hazard and operability studies of flowsheets.[16,17] In the absence of comparable methods of layout, a check-list approach is useful (see Balemans[18], Rose[19], Anderson[20] and the general index). Appropriate manuals should be drafted before the main review covering operation, for maintenance and for emergency evacuation and fire-fighting. The relevant rating and mathematical modelling calculations and assumptions should also be available. The construction and commissioning programme should be available, even though these require preparation at an earlier project stage than is usual. In this way many of the deficiencies of the layout will come to light at the review though if a formalized critical examination method were available it would probably identify additional items requiring rectification.

6.4   RATING METHODS

The feature of rating methods is that the individual components and/or the layout are assigned classification or index ratings usually reflecting the degree of hazard. They are used either to assist in generating layout or to compare different layouts. The basic philosophy is to find the most economic arrangement that satisfies the rating scheme.

The actual values assigned to classes are mainly derived from experience though some are increasingly based on mathematical models or physical properties. Since each plant is unique there is a risk that it will prove to be the 'exception to the rule' and the use of generalized hazard ratings may generate a dangerous layout. However, rating methods are often easier to use in setting initial spacings in layout than mathematical model calculations. The ideal is therefore that these methods are used to generate layouts which are then checked using calculations based on mathematical models.

6.4.1   AREA CLASSIFICATION (ELECTRICAL)

Area classification recognizes the differing degrees of probability with which potentially flammable concentrations of gas, vapour or mist may arise in normal operation of plants in terms of the frequency of occurrence, the probable duration of existence on each occasion, and the physical extent of the hazardous concentrations produced.

The classification procedure results in an installation being regarded as having hazardous areas and non-hazardous areas. Non-hazardous areas (i.e. those considered 'safe') are those in which potential concentrations of flammable gas or vapour are not expected to be present in quantities such as to require special precautions in the choice, siting and use of electrical apparatus.

Hazardous areas will consist of one or more of three zones, defined in BS 5345,[21] in sequence of decreasing probability of flammable concentrations being produced, as follows:

Zone 0 A zone in which an explosive gas–air mixture is continuously present or present for long periods.

Zone 1 A zone in which an explosive gas–air mixture is likely to occur for short periods in normal operation.

Zone 2 A zone in which an explosive gas–air mixture is not likely to occur in normal operation, but if it occurs it will exist only for a short time.

Figure 8.2 (p. 132) shows a typical set of zones superimposed on a plot plan to indicate the use and extent of zoning.

To aid the determination of these zones the scheme in BS 5345 requires a classification of omissions or leaks from a plant as:

(a) *Continuous grade*: release is continuous or nearly so.
(b) *Primary grade*: release is likely to happen either regularly or at random times during normal operation.
(c) *Secondary grade*: release is unlikely to happen in normal operation and in any event will be of limited duration.

In general, these sources give zonal classifications, namely:

Continuous → *Zone 0*
Primary → *Zone 1*
Secondary → *Zone 2*

Certain situations can arise where a source is both primary and secondary under different circumstances with different leak rates. Thus a small Zone 1 could be surrounded by a larger Zone 2. However, Zones 1 and 2 can exist individually. In areas of restricted ventilation, secondary sources may be considered to give Zone 1 because of the persistence of the flammable condition.

Area classification can have a significant influence on plot layout and should be considered at Stage One layout studies. Opportunity should be taken to locate concentrations of electrical apparatus, such as the substation, switchroom, motor control centre, control room, outside a classified area and to minimize, and if possible group, primary sources and hence Zone 1 areas.

Formally, the area classification procedure does not consider non-electrical ignition sources such as furnaces, neither does it refer to the toxic risks associated with most flammable materials in concentrations which are usually very much less than the lower flammable limit (LFL). Although the area classification procedure does consider some 'abnormal' operating conditions, it does not take account of what are termed 'catastrophic abnormalities', such as the rupture of a process vessel or large pipeline. However, the calculation methods applied to determine the rates, duration and extent of release are common to major hazard situations, toxic release and non-electrical ignition as well as to area classification for electrical purposes. Thus, electrical area classification is considered in more detail in section 8.7.6 (p. 131) along with hazard assessment in general (Ch. 8).

6.4.2 ACCESS ZONES

For safety, reasons it will be necessary to restrict access to dangerous plants and it may be desirable to protect commercially sensitive processes. Security of the site was discussed in section 3.7 (p. 21) but within the site it may be necessary to fence-off particular areas with manned entrances. It also helps if the site is laid out in such a way that both vehicular and pedestrian site traffic avoid plant areas until reaching their destination (section 9.2, p. 142).

6.4.3   FLUID STORAGE

For many years the spacing of tanks from other equipment was determined on the basis of classifying liquid according to their flash points. Chapter 10 discusses this and other methods further.

6.4.4   CLASSIFICATION OF FIRE FIGHTING EQUIPMENT

The type of firefighting equipment depends on the process materials.[22] As it is sensible not to have a mix of firefighting equipment, it may be possible to rearrange the layout in order to have areas of similar firefighting equipment.[18]

6.4.5   MOND INDEX

This is a method[23] of generating inter-item spacings based on factors whose value is affected by considerations of, *inter alia*, process properties and amounts, operating temperatures and pressures, types of equipment, types of unit operation, etc. It yields spacing distances for related items but not positional relationships. The method may be used before layout has been formulated to enable objective spacing distances to be taken into account at all stages of layout development. Historically, the Mond Index is largely based on concepts originally developed in the Dow Index[24] which made use of material properties, etc. to assess relative hazards of plant units in a design, but which did not generate spacings.

6.5   MATHEMATICAL MODELLING

For layout in production engineering ergonomic models have been employed successfully. Except, perhaps, for the layout of packaging lines and ware-housing, it is unlikely that such numerical approaches will be beneficial in process plant layout. The solution of the ergonomic problem of operating, maintaining and constructing process plant items are very complicated, and it is usual to assume an adequate solution is provided by allowing space around each item. Space allowances are based on experience and are tabulated in Appendix C. Critical examination techniques are therefore more applicable than calculation methods.

Mathematical modelling is an aid to hazard assessment applicable to process plant layout. This is an increasingly comprehensive subject and so is discussed separately in Chapter 8.

6.6 APPLICATION OF LAYOUT METHODS

This section considers the way the methods described above, together with the hazard assessment methods of Chapter 8, the layout analogues discussed in Chapter 7 and the organization outlined in Chapter 5, can be used to formulate and develop layouts. It is convenient to assume a new site is to be built and so the development follows the sequence: Stage One preliminary plot layout, steps (1)–(9); Stage One preliminary site layout steps (10)–(15); site purchase and design sanction; Stage Two detailed site layout, steps (16) and (17); Stage Two detailed plot layout, steps (18) and (19). However, existing sites will impose particular constraints and some of the following site layout steps may not be needed.

6.6.1 STAGE ONE PLOT LAYOUT

1 Initial plot data
The data include preliminary flowsheets (which must show size of major pipework and suggested elevations of major equipment), process engineering design of equipment (e.g. size and shape), the results of preliminary hazard assessment of the flowsheet and the codes of practice to be followed in the plant design.

2 First plot layout
The first layout is made with the above data in the sequence of the process flow using the experience of the engineer to recognize constraints such as major piping and cabling. Typical layout spacings, such as given in Appendix C, are a help at this stage. Simple drawings, cutouts and block models are employed.

3 Elevation
The elevation assumptions in the flowsheet are questioned (e.g. by critical examination). This enables the process objectives and constraints on elevation to be defined. Various alternative elevation arrangements are generated, possibly by using formal techniques such as travel and correlation charts.

The cost of each elevation alternative is examined, primarily for differences between, for example, the number of plant items needed to achieve the objective or differences in the material transfer costs such as piping, pumping to elevating items and power consumption.

Simple elevation drawings can be prepared showing only heights and relative positions of items, but structure and floor levels are not introduced at this point.

4   Plot plan

Plant items, buildings and principal pipe and cable runs are laid out in plan ensuring that the obvious layout constraints (operation, maintenance, construction, environmental, safety, drainage areas, etc.) are accommodated. Cutouts are helpful at this stage. A costing is made of each alternative arrangement. The more promising arrangements may be optimized (by computer) to produce even more economical layouts.

5   Plot buildings

Housing plants in buildings is more expensive than having plants in the open, even for plants on elevated structures. The need for enclosed buildings specified in the process design should therefore be examined critically.

6   Second plot layout

The selected plan and elevation layouts together with building studies are now combined to determine possible positions of support and access structures and to study civil requirements (e.g. foundations). These may force relaxation of earlier constraints. The layout alternatives are presented by block models (either physical or on visual display unit (VDU)). These models will help both the layout designer and other disciplines to visualize functional and safety aspects. Consequently, it is useful at this stage to have brief and mainly intuitive reviews of the layout by the various disciplines.

The acceptable layouts are re-costed and a short-list (of one, ideally) of particular layout arrangements selected and recorded as plot plans (section 7.3.2, p. 73).

7   Hazard assessment of plot layout

Sources within the plot where loss of containment can occur must be identified and the amount of material lost predicted. The consequences of each loss from explosion, fire or toxicity are calculated as indicated in Chapter 8.

Within the plant these calculations will indicate separation distances between sources of ignition and sources of leaks and will specify the various hazard zones for electrical equipment and fired equipment. The safe positioning and/or protection of control rooms will also be calculated. Outside the plot, the danger to people, equipment and buildings from fire, explosion and toxicity at various distances from the plot arising from the predicted losses will be assessed. The layout may well have to be adjusted.

The Mond Index method may be used prior to (but not instead of) the above assessments.

8   Layout of piping and other connections

The principal piping and/or pipe routes are confirmed. Also checked are principal electrical mains routes. Various connecting arrangements are considered and the most promising one further optimized. Piping models are used as aids and computerized versions are particularly good for optimization. The best layout arrangement should now be selected and recorded (section 7.3.3, p. 77).

9   Critical examination of plot layout

The proposed arrangement should satisfy all the obvious requirements in the light of all the information available. It should be examined formally by the various disciplines to make sure less obvious features have not been omitted. The aspects to examine include:

(a) Ease of operation.
(b) Ease of maintenance.
(c) Ease of construction.
(d) Ease of commissioning.
(e) Ease of escape and firefighting.
(f) Safety of operators and other personnel during construction, commissioning and full operation.
(g) Environmental impact.
(h) Future expansion.

Models are good aids to the review process. Check-lists are provided under the heading 'Plot Layout' in the general index.

The results of the critical examinations by the various disciplines and of the hazard assessment have to be reconciled. Thus multi-disciplinary consultation is essential. In some cases it may be found that the layout is impractical or even impossible. Then it will be necessary to rethink the process design or even undertake further laboratory and other development work. In most cases, though, the results, of the critical examination will mean adjusting the layout, albeit by a further iteration from step (2) onwards.

6.6.2   STAGE ONE SITE LAYOUT

10   Initial site data

Steps (1)–(9) will be carried out for each separate plant and storage area within the proposed site. This will provide:

(a) The size and shape of each plot.
(b) The access requirements for vehicles and people during construction, operation, maintenance, emergencies, etc.
(c) An approximate evaluation of the separation needed around each plant for hazard containment.

From the process data of the various plants, the following should be compiled:

(a) Site materials and utilities flowsheets. This should include pipe, conveyor and vehicular traffic capacities for both internal and external movements.

(b) Size and shape of the plots, buildings, utilities, central services, amenities, etc.

(c) Pedestrian traffic movements.

11   First site layout

The materials flowsheet for the site allows the various processes to be positioned relative to one another. The flow pattern may be distorted in order to isolate hazardous processes and to accommodate the proposed rail and road entry points or wharf positions. Next, the services (e.g. boiler house, effluent plant, etc.) are added in the most convenient positions subject to the provision that they are not likely to be put out of action by a disaster. The central buildings (administration, canteens, amenities, medical centre, fire station, stores, central workshops, laboratories) are placed so that the distances travelled by personnel who use them are minimized, providing that these buildings are in safe places. After this the road and rail systems are marked in more detail, trying to keep the various types of traffic segregated as far as is possible and desirable. There should be access from at least two directions to all parts of the site to allow for emergencies.

The size of the site is found from the area of individual plants, storage areas and central buildings plus the clearances between the plants. It is also necessary to allow ample space for parking, loading and unloading, stores, fire-fighting water storage, etc. Typical clearances, size and areas are given in Appendix C. During the development of this preliminary layout, allowance must be made for future plant expansion and for general construction and other access considerations.

It is essential to establish, within the overall layout, all the important positional relationships between elements of the layout which must be maintained. This consideration is crucial because available sites may not conform in shape or topography to the preliminary layout and some compromises and amendments to the layout are almost certain to be needed. In this event, it must be known which relationships can be relaxed to fit the plants in the available space.

Cutouts are very useful for site layout development together with simple drawings. Block models may also be used especially to gauge the visual impact.

12   Hazard assessment of site layout

The plots on the site where loss of containment can possibly occur are noted. Vulnerable parts of the site are listed, such as offices, central utilities, key commercial plants and the site boundary (representing the start of the public domain). The consequences of each loss through fire, explosion or toxicity on the vulnerable items are calculated. The layout is adjusted so that the consequences become acceptable; in particular, the chance of escalation of an occurrence throughout the site ('domino' effect) is made less likely. Chapter 8 goes into further detail on this topic.

13 Site layout optimization

When there are feasible alternative arrangements they should be costed, usually with respect to transport and piping connections between the various plots, etc. The most economical layout so far found can then be subjected to further optimization of the plot spacings, subject to the hazard constraints.

14 Critical examination of site layout

The proposed site plan should be examined critically by the various disciplines first separately and then together. Points to review include:

(a) Containment of hazards and safety of employees and public.
(b) Emergencies.
(c) Transport and piping systems.
(d) Access for construction and maintenance.
(e) Environmental impact including drift of airborne effluents and discharge of liquid effluents.
(f) Future expansion.

Other points are included in the check-list under 'Site layout considerations' in the general index. In the worst case it may become obvious during the review that one or more of the proposed plants is unacceptable but in most cases the review will result in adjustments to the site layout, i.e. iterating from step (11) onwards.

15 Site selection

The results of the preliminary site layout will be:

(a) The size and shape of the site.
(b) The pipeline, road, rail and water access needed to the site.
(c) Necessary hazard separation distances around the site.
(d) Position of the various structures and their foundation loading.

These factors, plus the others outlined in section 3.10 (p. 24), will guide the selection of a suitable site. In Fig. 5.2 (p. 39) step (15) is executed at node 1.

6.6.3 STAGE TWO SITE LAYOUT

16 Stage Two site data

No site will be ideal, however much care is taken in its selection. So, after site purchase, the engineer has to adjust the layout to the constraints of the site and it is important that these are clearly established. They could include:

(a) Site topographical details referring to:
    (i) the load-bearing ability of the soil and subsurface conditions;
    (ii) site grading and drainage features.
(b) The atmospheric conditions with regard to:
    (i) extremes of weather which may make it desirable to provide special shelter or protection for equipment or operators;

(ii) prevailing wind direction for consideration when locating intake or exhaust stacks, or furnaces up- or downwind in relation to the remainder of the plant. Also sand, sea-spray and leaves can be blown by the wind on to a plant.

(c) Environment conditions relating adjacent properties, e.g. residential property or public places, neighbours' hazardous or vibratory operations, roads, railways, airfields, rivers (see Table 3.1, p. 25).

(d) Site boundary and service parameters for normal and emergency conditions, e.g. access from public roads, waterways and rail systems, sewers, water supplies, power supplies, pipe trenches, drains, public paths and rights of way.

(e) Legal requirements, e.g. planning and building laws and by-laws, requirements for dealing with effluent pollution and noise, traffic regulations, fire, insurance and other safety requirements.

Site standards should also be established such as:

(a) Road width, radii, gradients, etc.
(d) Service corridors.
(c) Pipebridges (e.g. height over roads, railways) and pipetrenches.
(d) Building lines.
(e) Architectural finish to buildings, etc.

These should be based on the owner's and national standards and codes of practice.

It is likely that while the site is being selected and purchased, further process and engineering design and market research work has been undertaken on the individual plants and their products. The plot layouts could have been updated and the further information, relevant to the site layout, made available.

17 Stage Two site layout

Steps (11)–(14) should be repeated but in greater detail and subject to the constraints of the selected site. Possible layout changes to the original plan could be caused by:

(a) The desirability of placing heavy plants on good load-bearing soil.
(b) The position of road, rail and service access points.
(c) The need to put hazardous plants away from public places such as schools, etc. and to take note of neighbouring hazards.
(d) The desirability to have a good environmental impact as described in section 3.8 (p. 22).
(e) Planning restrictions.

The hazard assessment can now take account of known vulnerable features outside the site boundary. The critical examination will, in addition to the items given in step (14), also check that site constraints and standards have not been violated. Extensive consultation will be made with the various regulatory and emergency authorities during the detailed layout stage.

The final site plan will show the roads, railways, site pipetracks, sewers, central buildings, services, etc. It will be produced in the form of, and with the aid of, detailed drawings and models, both computer and manual (section

7.3.1, p. 72). With reference to Fig. 5.2 (p. 39), steps 16 and 17 occur between nodes 1 and 3.

6.6.4 STAGE TWO PLOT LAYOUT

18 Stage Two plot layout data

The detailed plot layout information includes:

(a) Standards, etc. in particular:
  (i) owner's basic practices and standards;
  (ii) national and/or international codes of practice, standards, specifications and regulations;
  (iii) contractor's standards where the above are not available.
(b) The detailed site information given in step (16) which could impinge on plot layout.
(c) Site plans and details giving the features that might influence the plot layout.
  (i) location and relationship of roads and railways surrounding the plot and estimates of traffic that might interact with the plot's loading and unloading facilities;
  (ii) steam, water, sewage disposal, and other services, particularly the terminal points relating to the plot;
  (iii) raw material and product pipeline terminals;
  (iv) sources of atmospheric pollution that might affect the process operators or maintenance staff.
(d) The detailed process engineering design which contains:
  (i) flowsheets indicating (with identification codes) the process equipment and instrument requirements and showing the pipeline connections;
  (ii) flow diagrams showing the flows and composition of each stream;
  (iii) line lists giving for each pipe its size, specification and the temperature and pressure conditions;
  (iv) equipment lists and drawings providing the specification of each item together with its plant size, register number, critical dimensions, process power requirements, process and utility nozzle connections and flows, materials of construction, process conditions, operation and maintenance requirements;
  (v) process design data sheets containing the process design data, philosophy and calculations and indicating any process layout requirements;
  (vi) the results of the hazard and operability studies of the process design.
(e) The Stage One plot layout conceived in steps (1)–(9).

In Fig. 5.2 (p. 39), this step (18) occurs at node 3.

19 Stage Two plot layout

Most of the initial steps, particularly (4), (6)–(9) are repeated in greater detail and subjected to the site constraints given in step (18), steps (2)–(5) occur between nodes 3 and 7 of Fig. 5.2 (p. 39), steps (6) and (7) at node 7, step (8) between nodes 7 and 15 and step (9) at node 15.

It is important that in repeating step (6) (node 7) there is good co-ordination between the layout, process, operating, piping, civil, structural and mechanical departments, etc. The piping arrangement studies (repeat of step (8)) are discussed in section 7.3.4 (p. 77). The hazard reassessment (step (9)) will be mainly concerned with internal plant spacings such as area hazard classification zones and control room and other plant building positions. Inter-plot spacings are considered in hazard assessment of the site layout. The repeat of the critical examination step (9) should, as well as considering the aspects given in step (9), also see that the standards regulations, etc. and site constraints on the plot have not been violated.

Aids to Stage Two layout are detailed physical models, drawings and VDUs for computer modelling.

6.6.5 OVERALL SITE AND PLOT LAYOUT HAZARD ASSESSMENT

Increasingly, the regulatory authorities will require a combined hazard assessment of the site and plot layouts after both have been tentatively finalized. Some existing activities will also be required to submit hazard assessments.

(a) Interactions between items within the plot.
(b) Interactions between plots within the site.
(c) Interactions between the site and its surroundings.[25]

REFERENCES

1. Armour, G. C., Buffa, E. S. and Vollman, T. E. 'Allocating facilities with CRAFT', *Harvard Business Review*, **42**, (2), 130, 1964.
2. Evans, W. O. *Automated Layout Design Program*. IBM Corp.
3. Lee, R. C. and Moore, J. M. 'CORELAP – Computerized Relationship Layout Planning', *Journal of Industrial Engineering*, **18**, (3), 195, 1967.
4. Newell, R. G. *Algorithms for the Design of Chemical Plant Layout and Pipe Routing*, Ph.D. thesis, London, 1974.
5. Leesley, M. E. and Newell, R. G. 'The determination of plant layout by interactive computer methods', *I. Chem. E. Sym. Ser.*, **35**, 2:20, 1972.
6. Madden, J. and Taylor, V. T. 'Design data requirements in the process industries', *Computer Aided Design*, **11**, (3) 142, 1979.
7. Computer Aided Design Centre Publications, Maddingley Road, Cambridge, England.
8. Bush, M. J. and Wells, G. L. 'The computer aided production of unit plot plans for chemical plant', *I. Chem. E. Sym. Ser.*, **35**, 2:15, 1972.
9. Shuqair, M. M. *Studies on Plant Layout*, Ph.D. thesis, Sheffield, 1978.
10. Gunn, D. J. 'The optimized layout of a chemical plant by digital computer', *Computer Aided Design*, **2**, (3). 111, 1970.
11. Al-Asadi, H. *Computer Aided Layout of Chemical Plant*. Ph.D. thesis, University College, Swansea, 1980.

12. Fine, B. 'Piping design in chemical plant', *I. Chem. E./A.I.Ch.E. Sym. Ser.* **4**, 107, 1965.

13. Mustacchi, C. 'Optimal process layout by a branch and bound technique', *Ing. Chim. Ital.*, **10**, (12), 203, 1974.

14. Reed, R. *Plant Location, Layout and Maintenance.* Richard, D. Irwin, Inc., Homewood, USA, 1967.

15. Elliott, D. M. and Owen, J. M. 'Critical examination in process design', *Chem. Engr. Lond.* **223**, CE 377, Nov. 1968.

16. CIA *A Guide to Hazard and Operability Studies.* Chemical Industries Association, London, 1977.

17. Kletz, T. A. *HAZOP and HAZAN – Notes on the identification and assessment of hazards.* I. Chem. E., London, 1983.

18. Balemans, A. W. M., *et al.* 'Check-list guidelines for safe design of process plants', in: Buschmann, C. H. (ed.), *Loss Prevention and Safety Promotion in the Process Industries.* Elsevier Scientific Publishing Co., Amsterdam, 1974.

19. Rose, J. C., Wells, G. L. and Yeats, B. M. *A Guide to Project Procedure.* George Godwin/I. Chem. E., 1978.

20. Anderson, F. V. 'Plant layout', in: Kirk-Othmer, *Encyclopedia of Chemical Technology*, (3rd edn.) **18**, 23, 1982.

21. BS 5345, Part 1, (1976), 'Basic requirements for all parts of the code'; Part 2, (1983), 'Classification of hazardous areas; ' "Code of practice for the selection, installation and maintenance of electrical apparatus for use in potentially explosive atmospheres (other than mining applications or explosive processing and manufacture)" ' British Standards Institution. Superseding BS CP 1003 'Electrical apparatus in explosive atmospheres', British Standards Institution, 1963–67.

22. BS 4547 'Classification of fires (EN2)', London, British Standards Institution, London, 1972.

23. Lewis, D. J. 'The Mond, Fire Explosion and Toxicity Index, Applied to Plant Layout and Spacing', A.I.Ch.E., *Loss Prevention*, **13,** 20, 1980.

24. Dow, *Dow Safety and Loss Prevention Guide.* A.I.Ch.E., New York, 1973.

25. EEC Directive 82/501/EEC, 'Major accident hazards of certain industrial activities', *Official Journal of the EEC, L230*, 5 Aug. 1982.

LAYOUT ANALOGUES AND VISUAL AIDS

7.1 GENERAL

Layout analogues and aids can serve at least three functions related to the recording or conveying information about the layout in:

(a) Development.
(b) Communication.
(c) Documentation.

Three-dimensional scale models are most useful as aids to development and for communicating with people not familiar with reading engineering drawings. Drawings are good for conveying quantitative layout information and as documentation of the layout for record purposes.

Computers have the potential advantage of both systems in that three-dimensional visual displays can be used for development and then the computer can be asked to plot the conventional record drawings. The computer's database holds up-to-date information as the layout is developed and so can produce excellent documentation for a permanent record.

7.2 CO-ORDINATE DIMENSIONING

In discussing layout aids, reference is made to the term 'co-ordinate dimensioning'. This system is essential if models are employed and it can be applied as an alternative to conventional methods of dimensioning on drawings. It is used on site plans, plot plans and elevations, piping arrangements and isometrics. It is also used on certain drawings produced by other design sections particularly those of civil and electrical origin.

Co-ordinate dimensioning has several advantages namely:

(a) Interrelated parts of a plant can easily be checked to show up interference and clashes, in particular between drawings prepared by other disciplines.
(b) Certain types of drafting are simplified and made easier to read.

(c) Locations and dimensions can easily be transmitted by letter or telex without ambiguity.

(d) It can assist programming and application of the input procedure where a company employs a computerized system.

The basic principle is that all dimensions are related to a fixed origin. Either Cartesian or polar co-ordinates may be used or a mixture of both. The origin is usually and preferably located outside the plant area so that all dimensions are positive. A fixed reference origin at N 100.00/E 100.00 is usually selected to the South West of the plot preferably where it is possible to set up a permanent point physically in the field (Fig. 7.1). Ideally, all plan drawings within a project are prepared with the plant North at the same edge of the drawings, conventionally at the top. Plant North is not necessarily true North but is often chosen by relating the longest straight feature (road, building, etc.) to the nearest cardinal compass point. The high point of drainage surfaces is established and given a notional elevation of 100.000 so that the origin is notionally 100 units below the plant. On a large site, subsidiary datum points are often used, for convenience, on individual plant units but these are always related back to site datum.

Fig. 7.1 Co-ordinate dimensioning

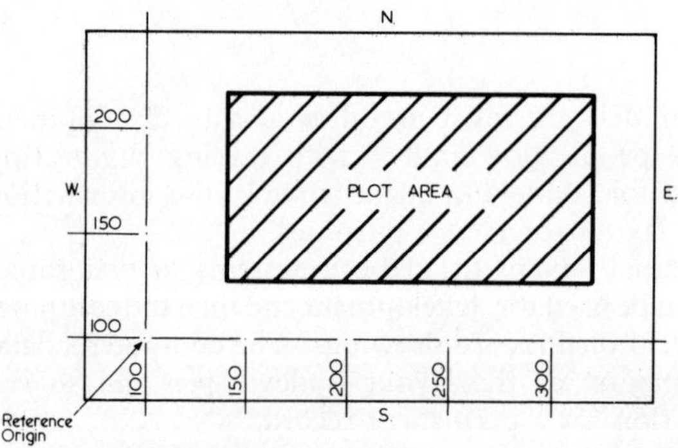

Locations are always specified in the same order so as to avoid confusion. Northings are usually first, followed by Eastings and elevations, e.g. N 105.645/E 134.250/H 110.875. Units can be expressed in either metric or imperial forms. If the scope of work changes such that the plant is extended in a westerly or southerly direction beyond the fixed reference origin, locations can be given, for example, as N 075.000/E 050.000. Should this occur the figures selected for an origin reference permit locations of up to 100 units in a westerly or southerly direction.

Any point is therefore uniquely located by quoting a maximum of three numbers. It is not always necessary to quote all three at each change of direction or height. Examples are given in Figs 7.2 and 7.3. In Fig. 7.2(a) the pipe, although dimensioned, is not absolutely located like Fig. 7.2(b). To fix the pipe in Fig. 7.2(a), two additional dimensions are required tying into other items. The line in Fig. 7.2(b) is completely defined both as to dimensions and location with only the ends located fully and the changed co-

Fig. 7.2 Defining a piperun

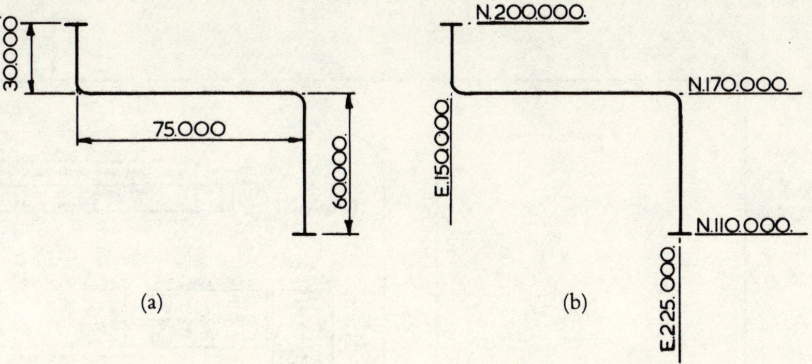

Fig. 7.3 Defining a nozzle position

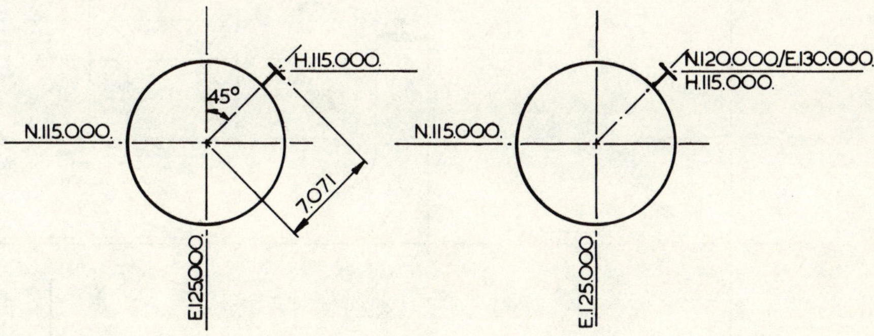

ordinates given. In this way, dimension lines are minimized on the drawing, making it easier to read.

A similar example is shown for a nozzle in Fig. 7.3.

For clarity, nozzle and connection details and locations for all equipment can be tabulated on the general arrangement drawing quoting equipment number, nozzle number, size, rating and facing, angular orientation, and the N, E and H co-ordinates.

7.3 DRAWINGS

The success of producing a safe, efficient and economically constructed plant depends upon good communication between the different disciplines and departments throughout the various stages of design, erection, commissioning and operation. The traditional means of conveying the numerical parameters by which the overall project design should proceed is provided by approved drawings.

Fig. 7.4   Typical site plant (Courtesy: APV Hall International)

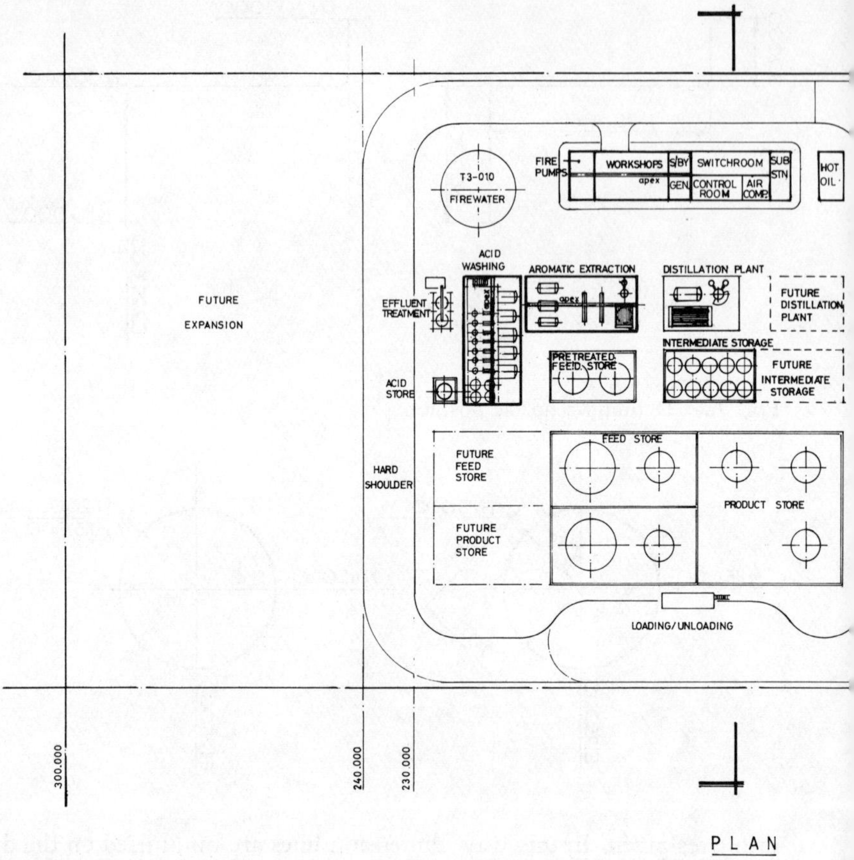

7.3.1   SITE PLANS

The main site plan (as in Fig. 7.4) indicates the overall area of an existing process plant complex or a new 'greenfield' area. The drawing shows the dimensions defining property boundaries, service roads and railways; plots containing process units; storage areas including tank farms and warehouses; administration buildings, car parking facilities, control rooms, electricity sub-stations and transformer bays, utility plants, pipe tracks and racks, power supplies, etc. A special edition of the plan should show safety and emergency features such as escape routes, firefighting access, emergency control points, safety separations and hazard area classification zones. Other drawings developed from the main site plan are used to study and establish maintenance, access, underground drainage and so on (see section 7.3.5).

Landscaping should be indicated where important and contours shown where the site is uneven. If the site is reasonably level around the plant areas it is not necessary to show an elevation through the site; it will be sufficient to indicate spot heights of key points above site datum level. On large sites

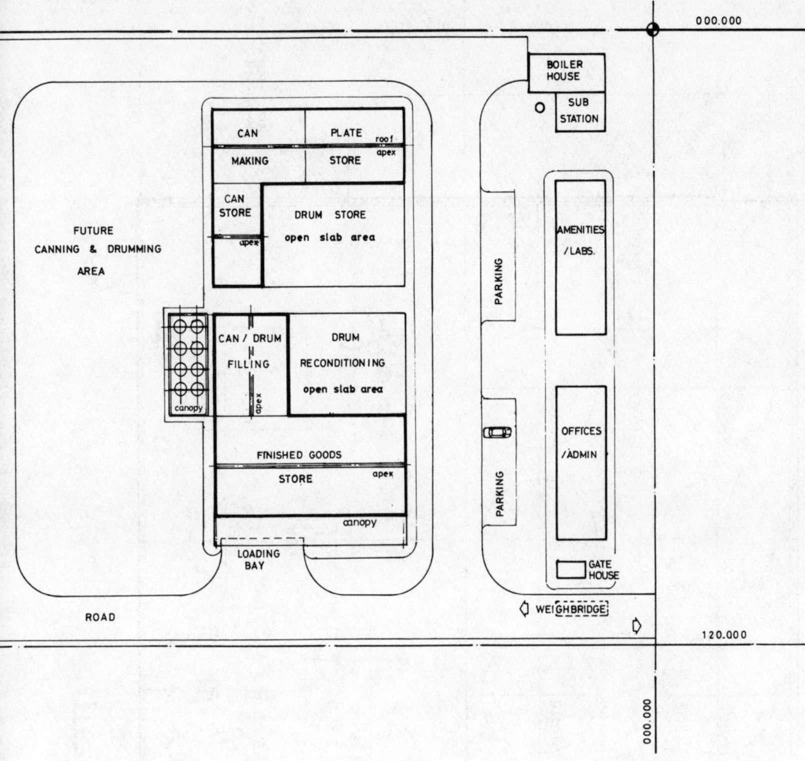

the site plans can be divided into a grid system so as to identify more easily individual plant areas.

The site plan is an essential reference source and study tool for the complete project from design initiation through construction to commissioning of a plant.

7.3.2 PLOT PLANS AND LAYOUT ELEVATIONS

These drawings show to a larger scale, and in more detail, the arrangement of a plot (Fig. 7.5). Similar but separate drawings show the elevation (Fig. 7.6), although these may not be required for single-storey plants. Other examples are given in Figs 19.5 (p. 289) and 21.2 (p. 315). It is one of the most important functions on a project to prepare plot plans and elevation plans as quickly as possible to enable other sections to proceed with their work.

Fig. 7.5 Typical plot plan (of a dextrose plant) (Courtesy: APV Hall International)

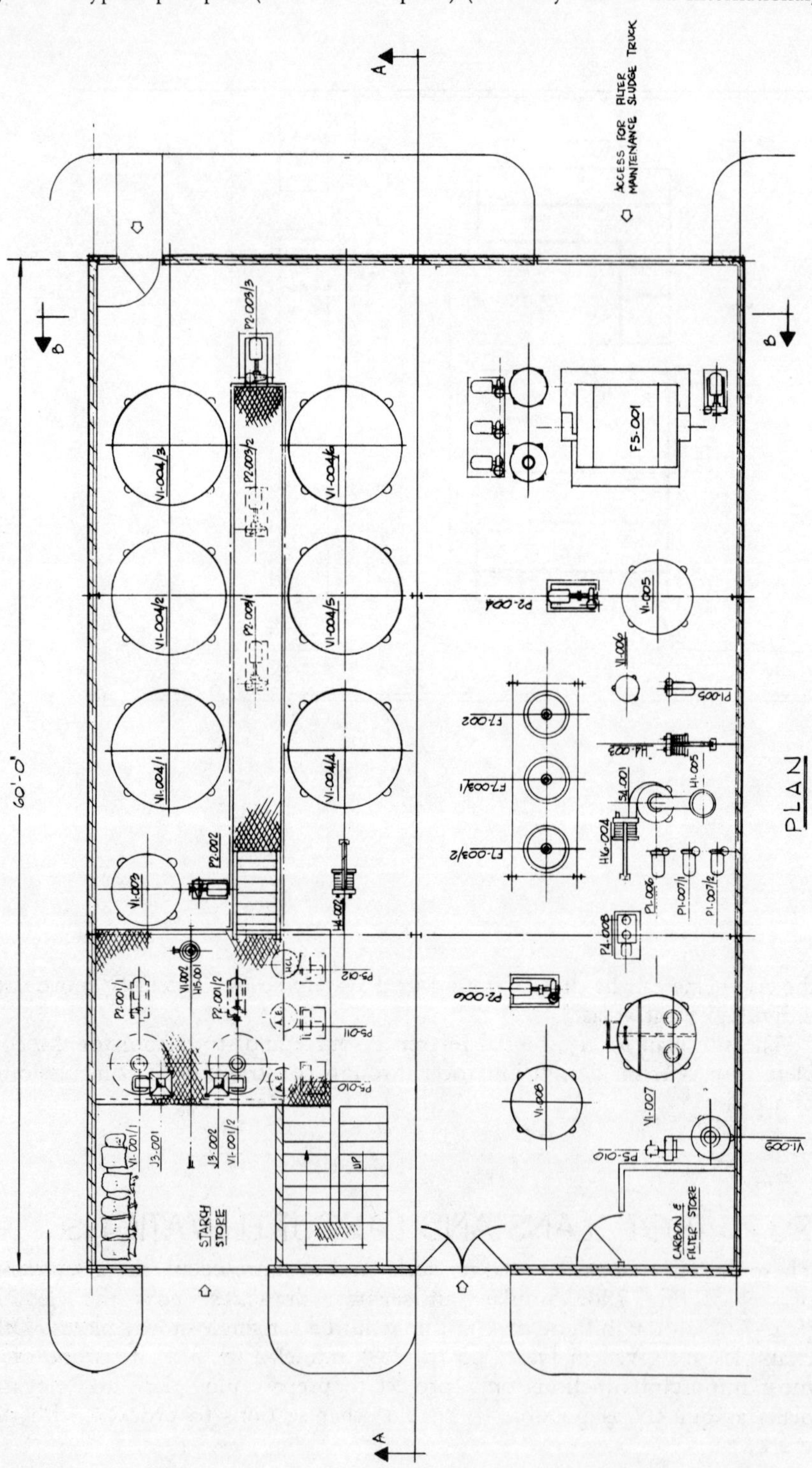

Fig. 7.6 Typical layout elevations (of a dextrose plant with associated perspective) (Courtesy: APV Hall International)

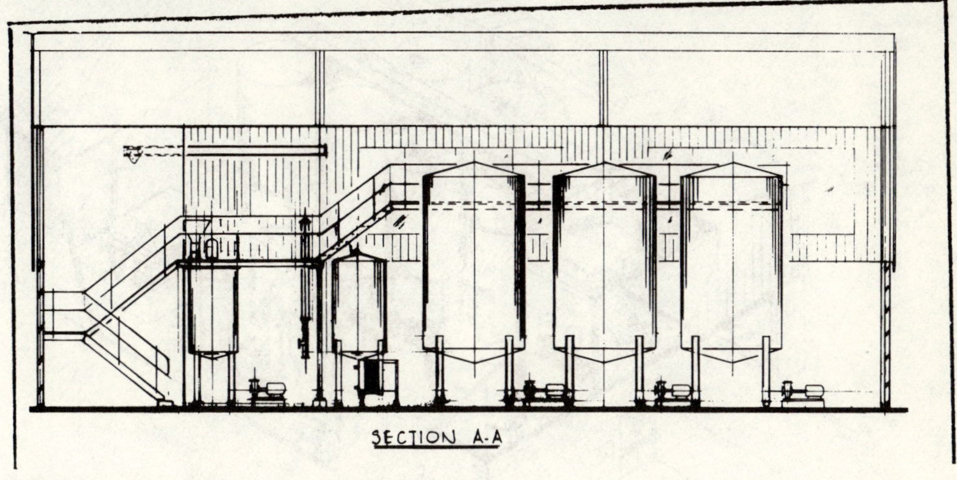

(a)

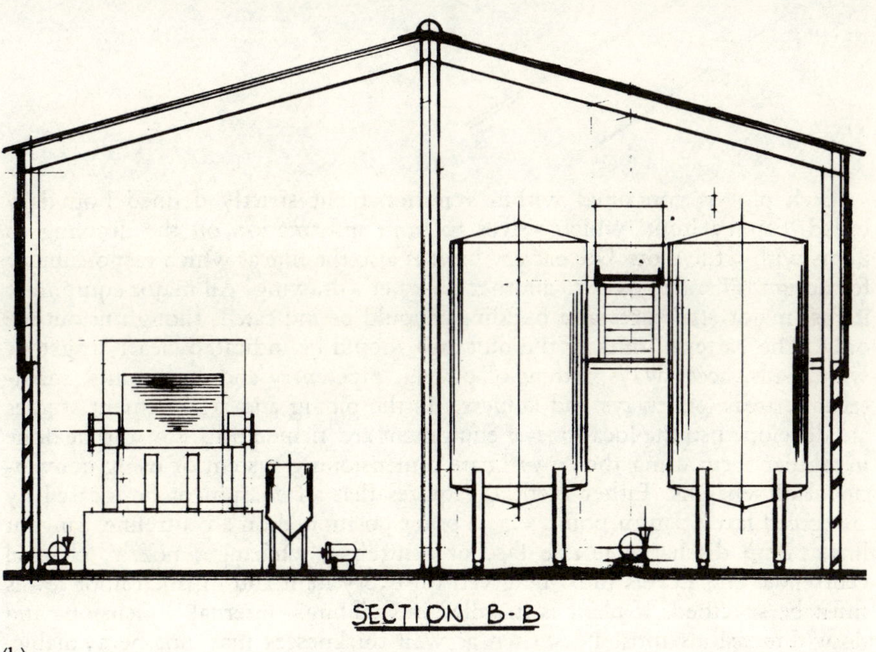

(b)

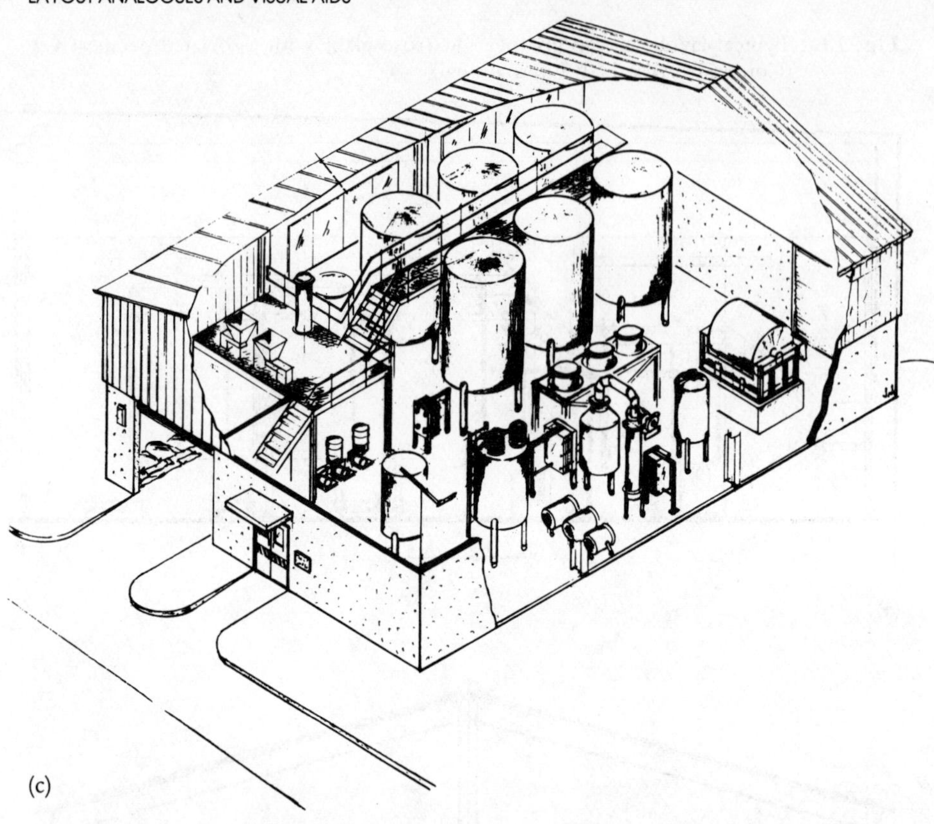

(c)

Each plot is contained within a notional but strictly defined boundary called 'battery limit'. which serves to limit information on the drawing to items within the plot. The battery limit is also the line at which responsibility for design is handed over to another designer's drawing. All major equipment items, major structures and buildings should be indicated, though in outline only. The battery limits of the plot area should be indicated clearly together with roads, accessways, extent of paving, pipe entry and exit points, maintenance areas, stairways and ladders. As the piping and arrangement studies are developed so the locations of equipment are 'firmed up'. This can be done in tabular form using the co-ordinate dimensioning system or using conventional dimensions. Either method requires that all equipment be located by an agreed fixed datum point, e.g. a point positioned on a centreline, tangent line, pump discharge nozzle face or centreline, exchanger nozzle, channel centreline, etc. Levels must be given for every item and finished floor levels must be specified. If plant is installed in buildings, internal dimensions and door dimensions must be shown as wall thicknesses may not be available. The plot plan is a key dimensional drawing corresponding in importance to the engineering line diagrams. It requires careful review and checking at the end of Stage Two before client approval and release for detail design. If a detailed plant and piping model is to be made for later design stages the plot plans and elevations can be left as simple, accurate specification drawings, but if no model is made, more detail must be shown to supplement the outline dimensions and pictorial data.

76

7.3.3 PIPING AND ARRANGEMENT STUDY DRAWINGS

The purpose of these drawings is to develop an adequately accurate representation of the equipment and piping arrangement as quickly as possible and with minimum use of man-hours. Difficult problems or likely trouble spots can be foreseen and highlighted during the early stages of a project, when their solution can easily be accommodated, without necessitating extensive redrafting or affecting the work of other sections. The study drawing provides the basic data which enables other sections to proceed with equipment requisitioning and initiate their own detailed design. It has therefore, an important temporary function and must not be considered a final drawing, but used to establish the following information:

(a) Positions, including elevations, of all major equipment, pipeways, buildings, foundation plinths, structures, platforms, access ladders and stairways.

(b) Equipment nozzle orientations, including manholes, manhole davits, handholes and instruments.

(c) Main routeing for instrument and electric cable trays and ducts.

(d) Accessways and clearances in high-activity areas used for erection, maintenance and operational reasons, e.g. catalyst loading, batch reactor operation, filter opening, discharging and cleaning, tanker loading/unloading, bagging and drumming-off points, packaging and warehousing, conveyors, davit swing, trolley beam travel, internals dropping and withdrawal areas.

(e) Runs of piping, generally limited to DN 80 and larger unless alloy or expensive material is used or the piping is lined, or has components that are long-delivery items.

(d) Lines having a high design temperature or pressure are considered critical and require stress analysis. Anchors, guides, special supports, expansion loops, etc. should be shown.

(g) The location of major valves, control equipment, orifice assemblies, special piping components, all in-line and vessel-related instruments, effluent drain points, drain gullies and underground piping.

The study drawing should be prepared with sufficient detail to establish the basic data given above. Drawing representation should be in its simplest form and unnecessary or repetitive details avoided provided the overall criteria are met. Figure 7.7 shows a typical piping study, Fig. 19.3 (p. 286) gives a study for a batch reactor and Fig. 26.4 (p. 366) illustrates a study drawing for a solids handling plant.

7.3.4 PIPING GENERAL ARRANGEMENTS

These drawings should adhere to the principles and battery limits set out for the study drawings. They represent the final piping design for a given area and should terminate at a battery limit match-line which can be checked against an equivalent match-line of the adjacent drawing. By comparison with studies these drawings are more complete and drafting should follow

Fig. 7.7 Part of typical piping and arrangement study (Courtesy: Petrocarbon Developments)

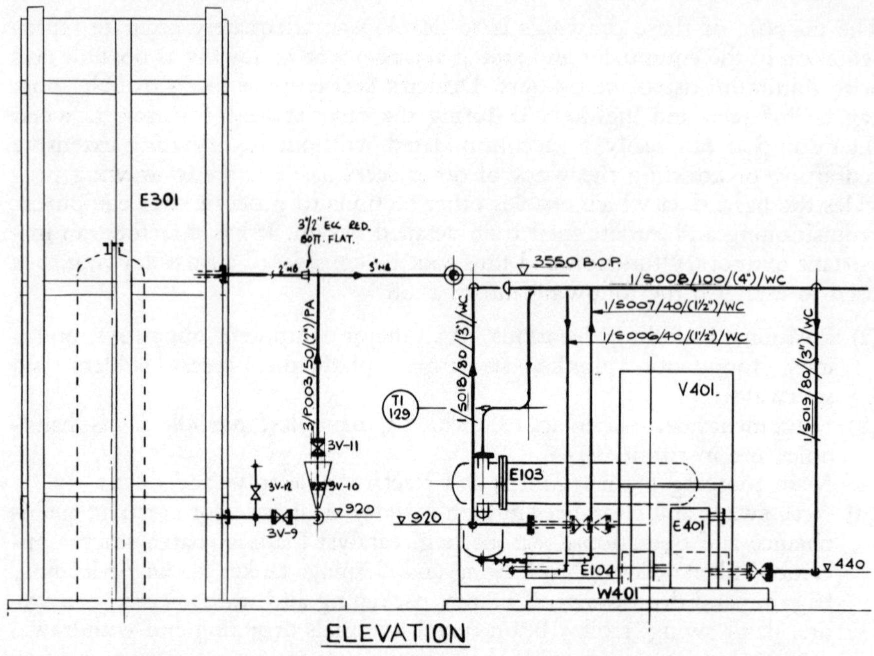

a company standard procedure for symbols, presentation and dimensioning technique (Fig. 7.8). Whether a co-ordinate dimensioning system or linear dimensions is used will depend upon the standards adopted but it is important that they are kept to the minimum which will convey the information necessary for piping isometric production. Clarity is essential and all detailed dimensions for individual pipes will be transferred to the appropriate isometric for that pipe. Piping is shown in heavy single line using conventional symbols and abbreviations, and where arrangements are complex or larger-diameter pipes are to be shown a double line portrayal can be used. Equipment is drawn in light line and shown in significant detail whereas only necessary details of walls, steelwork, etc. are given to establish and indicate clearances. In general it should be sufficient to prepare a plan drawing to indicate all the piping in a given area. Elevation drawings should be avoided, relying on the isometric to define each pipeline fully, but where specific detail is required in complex areas, localized views can be shown as small insert sketches on the plan drawing. Nozzle locations for each item of equipment are indicated in tabular form on the drawing along with size, connection type and rating data.

Computer methods for piping general arrangements are becoming available (see section 7.6).

7.3.5 PIPING ISOMETRICS

Detailing of piping is necessary both to provide pipe fabricators and construction with precise instructions and to act as a final materials check. Piping

Fig. 7.8 Part of typical piping drawing (Courtesy: Humphreys & Glasgow)

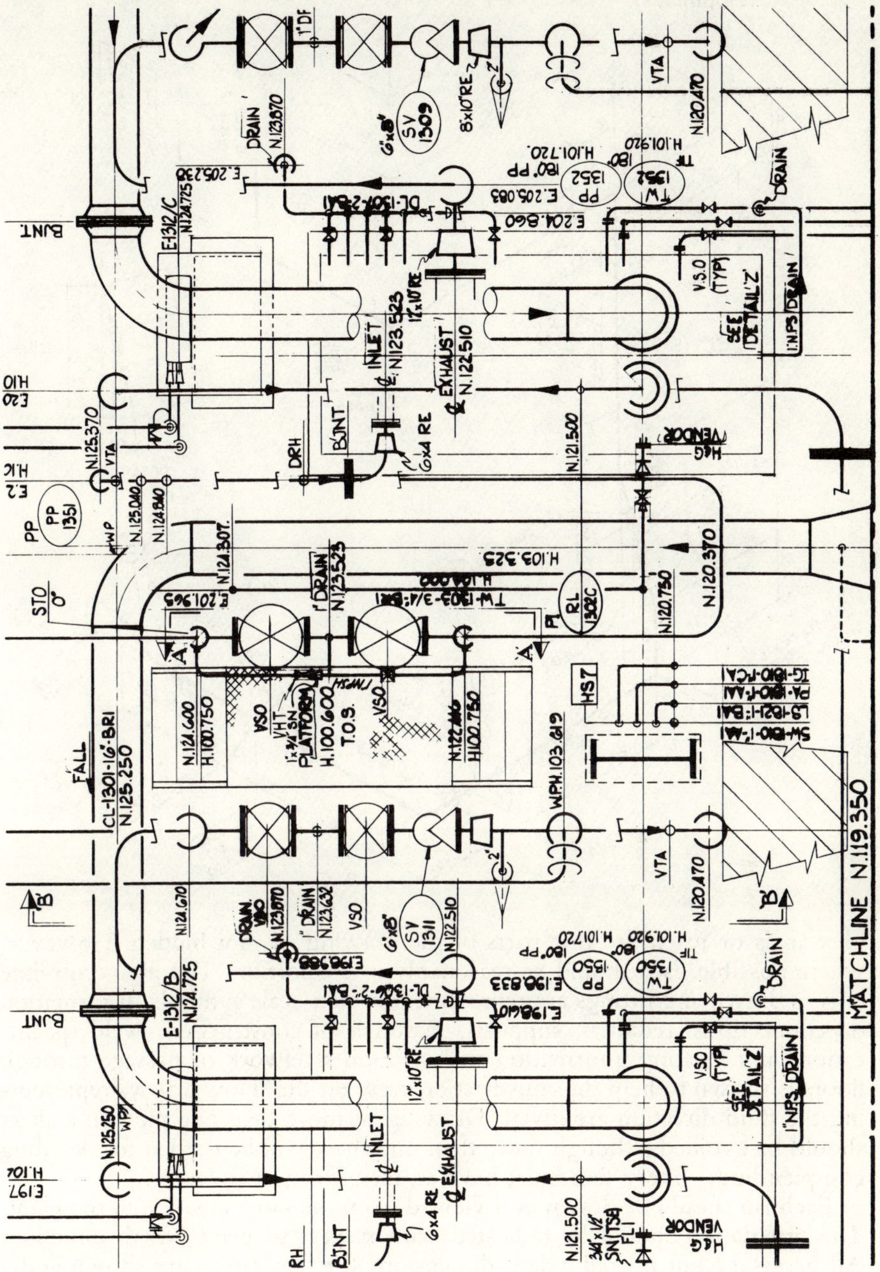

isometrics are now the almost universal technique employed for this purpose. An isometric drawing (or 'iso') is a pictorial view of one pipe, using isometric projection conventions to present a two-dimensional view of a three-dimensional pipe (Fig. 7.9). The isos are not, though, truly isometric since they are not to scale and draughtsman's license is employed freely to highlight com-

Fig. 7.9   A manually produced piping isometric drawing (Courtesy: Petrocarbon Developments)

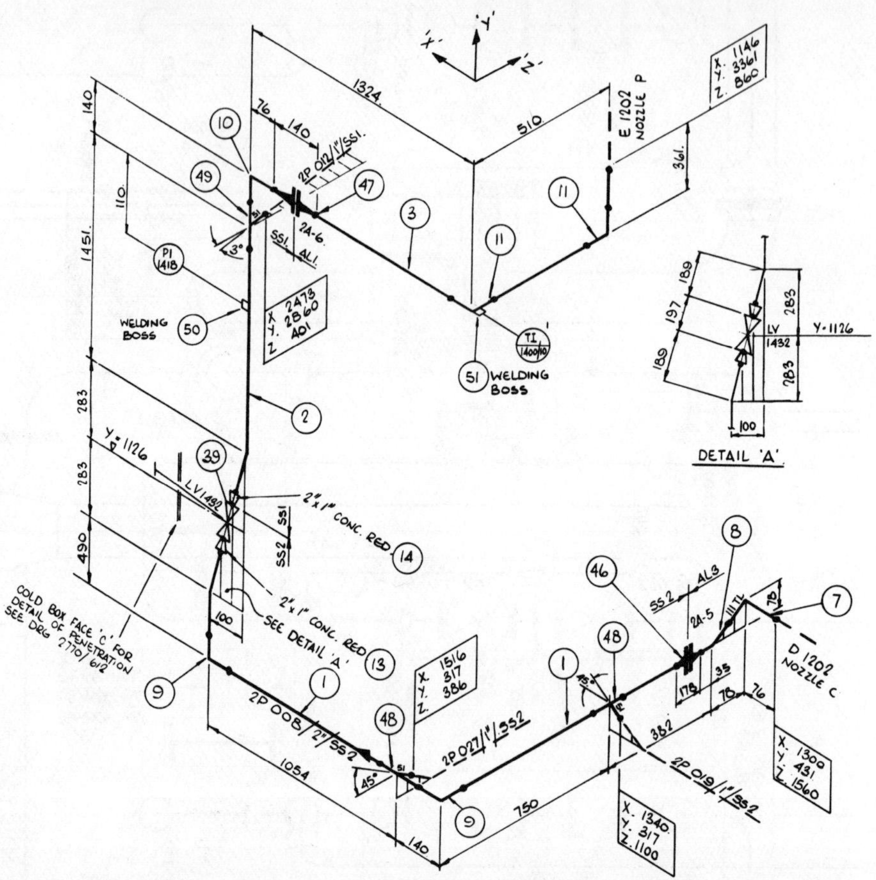

plex areas or to ensure that parts of the drawing are not hidden. However, where possible, isos should be reasonably in proportion. The pipe centreline only is drawn and fittings are represented by non-scale symbols. Information on connections, reducers, supports, materials of construction, weld specifications and locating information such as local steelwork or passage through floors is shown to help the construction crew on site. Flow arrows representing the fluid direction are given. Drawing of more than one pipe on a sheet should be avoided although more than one sheet will be needed for detailing complex lines. Every line must have an identification code.

Each iso should be drawn as if viewed from the same position in the plant. The orientation should be indicated with respect to plant North direction. All necessary but no redundant dimensions for fabrication are shown and a bill of material is included on the iso. Small lines (i.e. less than DN 40 for mild steel, DN 25 other) which are run and fabricated on site require only an isometric view with leading dimensions so that these can be proved in the field. Larger lines fabricated in a shop will require isos showing all dimensions, orientations and details including those of the fittings, such as piping offsets, eccentricity of reducers, direction of valve stems, etc.

Fig. 7.9 (*cont.*)

	QTY	SIZE	P.D. STANDARD REF.	REMARKS	
PIPING	2·5 M	2" NB	SS2		1
	1·75 M	2" NB	SS1		2
	2·0 M	2" NB	AL1		3
					4
					5
					6
BENDS	1	2" NB	AL3	90° LRWE	7
	1	2" NB	AL3	45° LRWE	8
	2	2" NB	SS2	90° LRWE	9
	1	2" NB	SS1	90° LRWE	10
	2	2" NB	AL1	90° LRWE	11
					12
REDUCERS	1	2" × 1"	SS2	CONCENTRIC	13
	1	2" × 1"	SS1.	CONCENTRIC	14
					15
					16
					17
					18
					19
FLANGES					20
					21
					22
					23
					24
					25
GASKETS					26
					27
					28
					29
					30
					31
BOLTS					32
					33
					34
					35
					36
					37
					38
VALVES	1	1" NB	SS2	TAG LV 1432.	39
					40
					41
					42
					43
					44
					45
	1.	2" NB	AL3 / SS2	TP 2A-5	46
	1.	2" NB	SS1 / AL1	TP 2A-6	47
	2.	2"×2"×1"	SS2	REDUCING TEE.	48
	1.	2"×2"×1"	SS1	REDUCING TEE	49
	1.		SS1	DRG DO·5 DETAIL 2	50
	1.		AL1	DRG DO-1. TYPE 2A	51
					52

LINE WORKING PRESSURE	KG/CM2. G / P.S.I. G	PIPE SHOP HYDRAULIC TEST	KG/CM2 G / P.S.I. G
LINE WORKING TEMPERTURE	°C / °F	LINE HYDRAULIC TEST	KG/CM2 G / P.S.I. G

The maximum size of each shop-fabricated line is fixed by checking it will be convenient to make, transport and erect (and to remove for maintenance if this is likely). Normally this is so if it fits into a region of dimensions 2.5 × 2.5 × 7.5 m but occasionally the length may be up to 12 m if essential. Connections between lines are denoted on the iso as either flanges or field welds together with the identification of the contiguous line. Piping assemblies requiring stress-relieving, galvanizing, or lining processes applied should have regions of smaller dimensions imposed in accordance with the 'after fabrication' process.

Production of isos usually takes place towards the end of a design and is recognized to be tedious and rather error-prone when done by hand. It is also time-consuming; an average iso requires 10–12 manhours for drafting, checking, material listing and final material takeoff.

Fig. 7.10 A computer-produced isometric drawing (Courtesy: Isopipe)

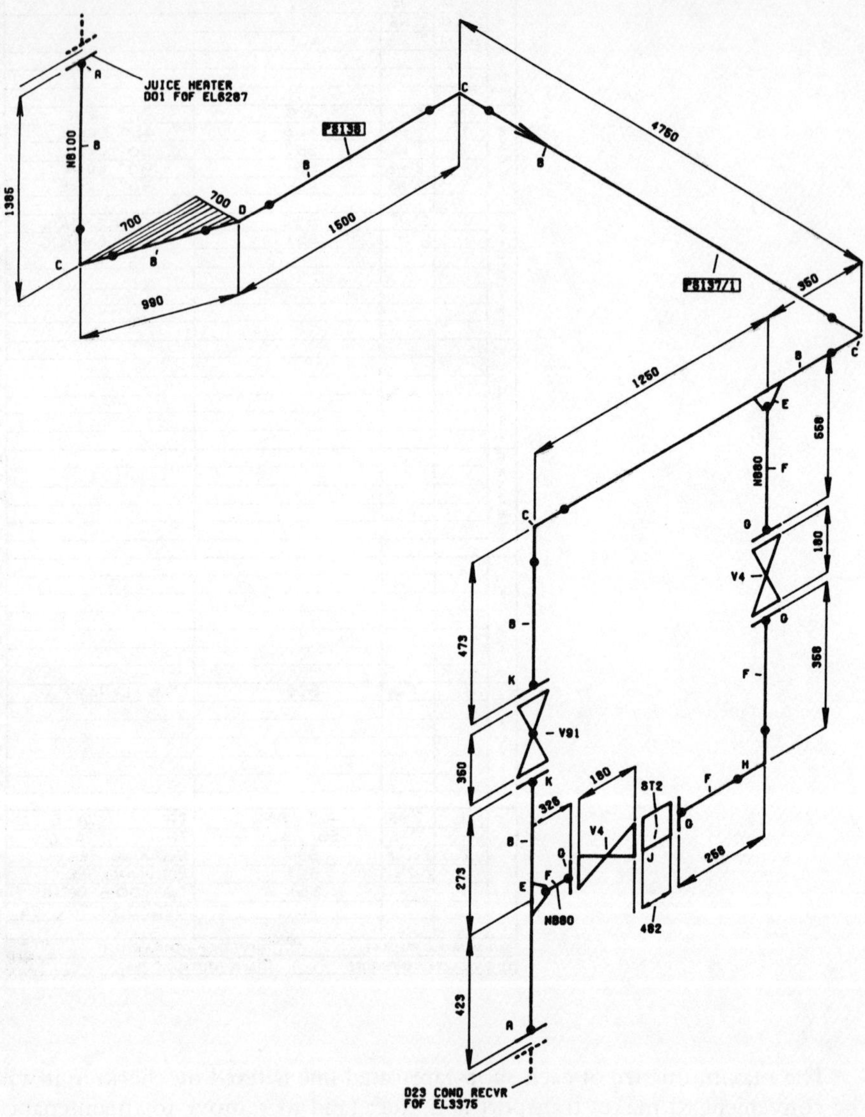

In consequence, there are several well-proven computer systems which produce isos almost identical to manually produced drawings in function, format, and content (Fig. 7.10). Thus they can be used by all grades of staff without special training. Advantages of producing isometrics by computer include:

(a) Fewer drafting errors.
(b) Consistent drawing standards.
(c) More accurate material listing.

Fig. 7.10 (Cont.)

ITEM NO	PART NUMBER	QUANT	DESCRIPTION	
			FIELD	1
				2
B	21EA-NJOOE250	10.1M	100MM PIPE.MEDIUM.BVLD ENDS.TO BS1387.UNIT N50	3
F	21EA-LJOOE250	1.1M	80MM PIPE.MEDIUM.BVLD ENDS.TO BS1387.UNIT N50	4
C	23DB-NJOOA250	4	100MM ELBOW.90 DEO LR.SCH 40.BW.TO BS1640 ORD WPB.UNIT N50	5
D	23DB-NJ0OC250	1	100MM ELBOW.45 DEO.SCH 40.BW.TO BS1640 ORD WPB UNIT N50	6
H	23DB-LJOOA250	1	80MM ELBOW.90 DEO LR.SCH 40.BW.TO BS1640 ORD WPB.UNIT N50	7
A	22FD-NOMML070	2	100MM FLANOE.SO PLATE TYPE.BS4504 NP6.UNIT N70	8
O	22FD-LOMML070	4	80MM FLANOE.SO PLATE TYPE.BS4504 NP6.UNIT N70	9
K	22FD-NOPML070	2	100MM FLANOE.SO PLT. TYPE.BS4504 NP16.UNIT N70	10
	251A-NCMOOV70	2	100MM=1.5MM OASKET.CAF.INSIDE BOLT CIRCLE.BS4504 NP6.UNIT N70	11
	251A-LCMOOV70	5	80MM=1.5MM OASKET.CAF.INSIDE BOLT CIRCLE.BS4504 NP6.UNIT N70	12
	251A-NCPOON70	2	100MM=1.5MM OASKET.INSERTION RUBBER.INSIDE BOLT CIRCLE.BS4504	13
			NP16.UNIT N70	14
	25FA-O840AA70	44	16MM=60MM BOLT.BLACK.C/W NUT.BS4882 BS1506-111.UNIT N70	15
E	23D2-NJLJ2250	2	100MM=80MM BRANCH BEND.REDUCINO.BW.UNIT N50	16
J	2700-ST2	1		17
	26AA-LMFV4	2	80MM=18OFF VALVE OATE FLANOED NP6 FF CI BODY TAO V4	18
	26AC-NPFV91	1	100MM=35OFF VALVE OLOBE FLANOED NP16 FF CI BODY TAO	19
			V91	20

REV	DATE	DRN BY	CHKD BY	STRESS BY	MTO BY	APPD BY	DESCRIPTION

DAR

BN

EVAPORATOR HOUSE

PROJECT	6000	AREA RF	SPEC J2

ISOPIPE LTD
NOTTINGHAM

DRO.NO.	LINE NO. C1/12	SHEET 1OF1	REV 0

(d) Less time spent on an iso (about 5 manhours).
(e) Possibility of linking with material and cost control (e.g. purchasing, expediting, receiving, storing and using; material, fabrication, erection and modification costs).

Fig. 7.11   Schematic of isometric production

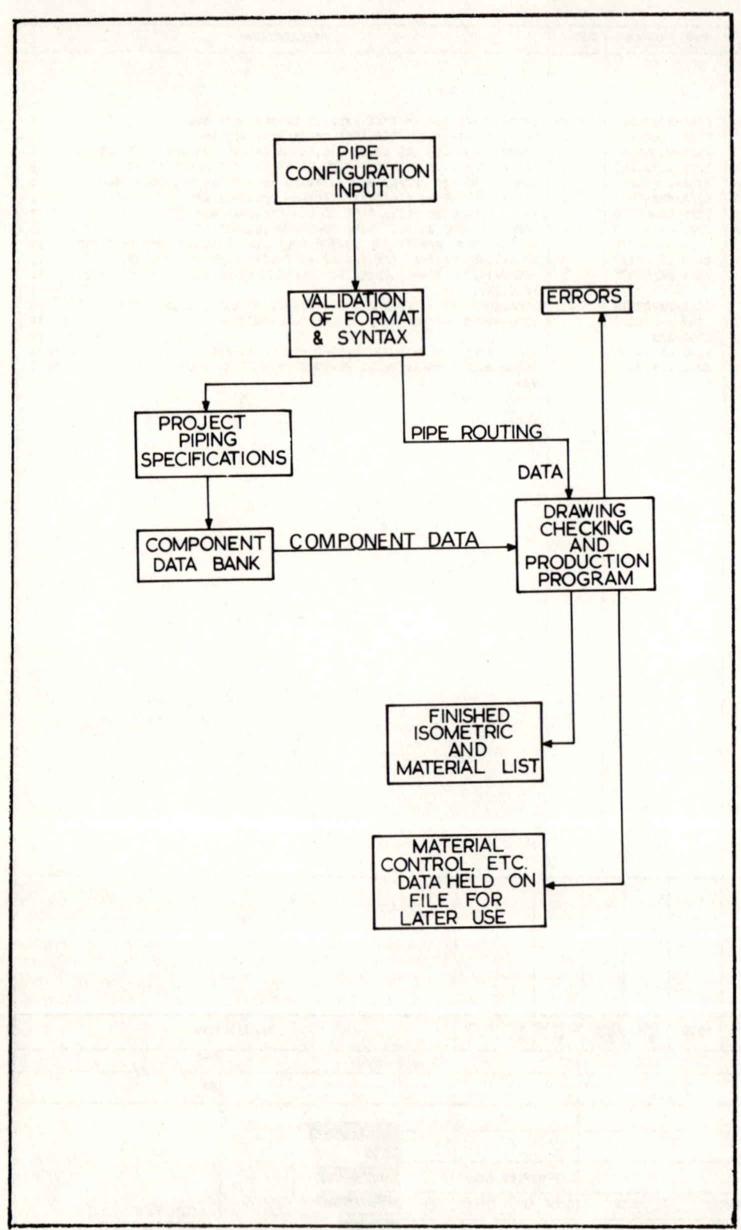

Most computer isometric systems operate as shown schematically in Fig. 7.11 and have the following characteristics.

Component data bank

This consists of drawing symbols, dimensions, connection type and description for a wide range of piping components, selected according to the project piping specification. Common examples are elbows, flanges, etc. but items like bellows, sightglasses, stream traps, supports, flow arrows, pipe falls, sets and rolling sets can be included or handled.

Piping specifications

These are codified methods of selecting the standard components from the databank for each pipe to suit the pipe operating conditions.

Pipe configuration input

This is a means of describing to the computer the route, leading dimensions and components of each pipe for which an isometric is to be drawn. The information is obtained from a conventional layout drawing or model or from a computer model. Routeing is described as N–S, E–W and U–D directions and dimensions are given directly or by co-ordinate data. Components are called-off in their correct order and positioned along the pipe. To aid the draughtsman, standardized pipe configuration coding sheets are provided.

Error checking

Mistakes such as incorrect components, inconsistent dimensions and connection mismatches (e.g. flanged to non-flanged) can be revealed by inbuilt logical checks in the computer system.

Drawing production

The iso with material list is produced for each pipe (Fig. 7.10).

Management control

Data on pipe material quantities, fabrication and erection work, etc. are retained on computer file for cost and control purposes.

7.3.6   MISCELLANEOUS DRAWINGS

A number of other drawings will be produced on a project depending on the scope of work. These include the following.

Battery limit drawings

These define the piping location, component, and support details at the agreed boundaries of the plant where perhaps the owner or another contractor might connect their piping systems.

Drainage drawings

These provide the detail and location of the underground drainage systems. They are of particular interest to the civil department and the owner. The drainage points are located in conjunction with the engineering and utility flowsheets, plot plans and studies, showing service type, flow rate and discharge temperatures. Invariably the drainage drawing is prepared by superimposing on the site or plot plans as appropriate.

Electrical area classification drawings

These are prepared to ensure that electrical equipment and cabling is selected correctly within potentially dangerous areas (see section 6.4.1 (p. 57) and Fig. 8.2 (p. 132)). These drawings are usually in the form of additions to copies of the plot plans and elevations. They will show the sources of hazard together with the classification and extent of the dangerous area as derived from the applicable regulations, codes of practice and internal reviews. This information can be expanded by a table listing the source of hazards, the vapour or fluid involved, the gas grouping and an indication of whether the gas is lighter or heavier than air.

Hazard area and separation drawings

These are similar but serve to locate any equipment which constitutes a non-electrical source of ignition (e.g. fired heater).

Emergency provision drawings

These show escape routes, assembly points, firefighting access, positions of hydrants and other firefighting equipment, emergency control points and stores. These will be developed in consultation with the emergency authorities.

Perspective sketches

These are sometimes used to enable persons not used to engineering drawings, to appreciate and understand proposed layouts (Figs 7.6 and 7.16).

7.4 MODELS

7.4.1 CUTOUTS

This is a quick and very effective means of developing a two-dimensional plant or site layout, usually in plan but sometimes in elevation. Shapes are cut out of sheets of paper, cardboard or plastic sheet to the nominated scale representing plan views on both equipment items and buildings. The site or plot area, or the floor-by-floor arrangement of a chemical process housed in a building, is then drawn on a sheet; divided into grids of suitable units. The scale templates or cutouts are shifted about the sheet until a reasonable layout is found. The cutouts can then be attached temporarily to the sheet and a copy taken of the arrangement for record purposes. The designer is then in a position to reconsider the parameters set down and examine fresh ideas. When the activity is completed, engineers and designers can then review the various alternatives and approve the most feasible layout.

It is necessary for the cutout to represent not only its equipment size, but also the amount of space required for general access around the item and any maintenance considerations. For example, if the unit is a heat exchanger with 4.8 m-long tubes, the clearance at the channel end of the exchanger required for the removal of the tube bundle would be 6.3 m with 1.5 m extra manoeuvring space. In addition, 1.5 m should be left in front of the shell cover and 1.0 m each side of the shell. Therefore, the plot area required for this unit is 12.6 m × 2.8 m for an exchanger shell of 0.8 m diameter and the cutout should be as shown in Fig. 7.12(a).

Where an equipment item is located adjacent to another where the access or maintenance area can be shared for economy of space the cutouts can be overlapped as can be seen in Fig. 7.12(b). This technique is also useful in laying-out areas for vehicle movements, e.g. loading/unloading bays and fork-lift truck movements for drum and pallet handling.

Once the layout, in principle, has received approval, the arrangements can be drawn with all the equipment, buildings, steelwork, etc. located to enable other design disciplines to proceed with their work.

Fig. 7.12 Cut outs showing access space (a) outline of exchanger (b) outline of adjacent exchangers

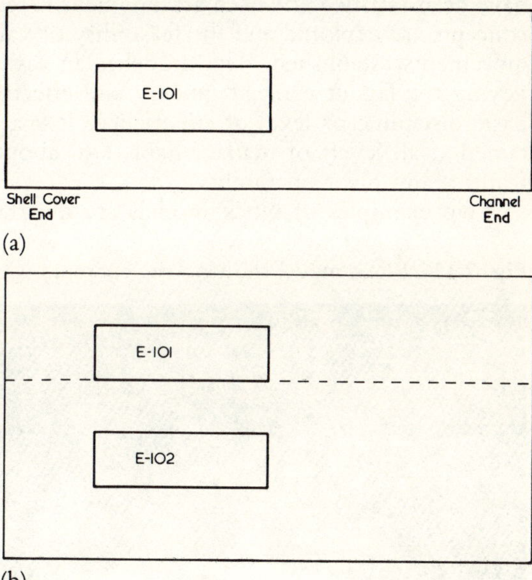

7.4.2 BLOCK MODELS

Block models can be utilized as a three-dimensional design aid for establishing preliminary plant layouts where process or space utilization requires a multi-level layout. Alternative arrangements can be explored rapidly and the best general overall layout can be discovered even before final sizes of equipment have been established. Block models are low-cost aids which show only the major items and structures in very simple form and in correct relationship to each other. Buildings, control rooms, switchrooms, piperacks, roads, etc. and critical piperuns should also be indicated.

The block model is developed generally to a scale of 1 in 50 where accuracy is not absolutely necessary. Simple equipment models can be constructed from wood, cardboard or polystyrene and located on a baseboard which may have a scale grid surface. Main structures can be indicated in outline using similar materials. One simple and effective system, particularly where multi-floor structures are envisaged, is to use clear Perspex sheets supported on aluminium threaded rods. The rods can be used to represent the vertical columns and the Perspex sheets the floor levels. The threaded rod allows the floor level to be adjusted to suit the layout as it is developed. The overall extent of a floor or a removable floor area needed for lifting equipment through the structure can be shown by using adhesive tape. Polystyrene is a material to be recommended with this system. It is easily shaped or cut with a hot-wire machine and equipment which passes through floor levels can easily be split for attachment to the upper and lower surfaces of the Perspex floors.

The initial layout is made to suit the requirements of the process and then recorded, preferably using photographs. As further information becomes available, alternative arrangements are prepared ensuring that all constraints have been satisfied and each arrangement is recorded. In this way, the design concepts are explored and the feasibility of various process and operating requirements established. The model is an easily constructed method of conveying the layout concept quickly and effectively to all staff whatever their basic discipline or level of experience. It enables basic agreement to be obtained at all levels of management, e.g. about construction methods, safety features and plant operability.

Two examples of block models are illustrated in Figs 7.13 and 7.14 and

Fig. 7.13 Block model showing site contours (Courtesy: British Gas)

Fig. 7.14 Block layout of plot equipment (Courtesy: Babcock Woodall-Duckham)

others are shown in Figs 10.5 (p.158), 14.5(b) (p. 212), 16.7(a) (p. 234), 18.1 (p. 266), 18.4 (p. 269).

7.4.3 PIPING MODELS

Once approval has been given to the overall equipment layout of the block model, a piping model can be started (examples are shown in Figs 7.15, 16.7(b) (p. 234), 16.7(c) (p. 235) and 26.5 (p. 368). The information contained in the block model is combined with data from engineering together with reasonably firm data and drawings in respect of equipment, foundations, structural steelwork, etc. In some cases it may be necessary first to develop piping study drawings of the regions of the model, particularly if it is congested or contains critical piping. The piping model must be accurate as it is built for design, development and detailed production purposes. It must be well prepared and presented, and consequently the scale must be increased from the 1 in 50 of the block model to a minimum of 1 in 33. When it is finished and the design is complete, detailed arrangement drawings may be made. Alternatively a 'follow on' model is prepared in much greater detail than the 'design' model.

The number of piping model baseboards is established by reference to the overall plot plan in much the same way as used for piping layout drawings. Care must be taken to ensure that the baseboard size and the vertical height of the model allows for clearance through doorways and exits of the model shop or office where it is being constructed. This also applies to its final site location if it is to be used for construction and training. If flammable, and

Fig. 7.15 Piping model examples (a) petrochemical plant (Courtesy: BP Chemicals) (b) evaporator system (Courtesy: Babcock Woodall-Duckham)

(a)

possibly toxic, model paint sprays are used then there must be a separate paint-spray room, which should be made fire resistant and be well ventilated.

Model baseboards are usually built of 12 mm-thick high-density chipboard with the upper surface covered with 0.5 m-thick plastic sheet, preferably in a white matt finish suitable for drawing on or taking sticky tape. A printed grid is an advantage to assist correct spacing. It is essential that there is also a 0.5 mm-thick plastic sheet of similar consistency on the underside to prevent warping. It is then suitably framed and strengthened for support legs and transportation. All edges are protected with plastic strip except where they butt up to an adjoining baseboard; each connecting baseboard is provided with brass location dowels for alignment of adjacent sections. Equipment is usually made in plastic, although wood or metal might also be used where suitable. All items should be accurate in scale but need only be approximate in shape, showing relevant spatial requirements and avoiding unnecessary, time-consuming detail. Construction of these items is eased by making use of manufactured systems of snap-in plastic parts. These systems provide ready-made models of tube, dished ends, hemispherical ends, exchanger flanges, pump bodies, motors, etc. which, by selection to the adopted scale, can be assembled very quickly. Orientation of nozzles is helped by pre-scored lines at 90 ° intervals of the tube. Manholes with correct hinge

90

(b)

pins, davits and foundation plinths are also available to ensure access space is provided. Platforms and walkways around tanks and vessels can be modelled from standard parts. If equipment is lagged this can be shown in localized areas to scale thickness and be coloured white.

Piping can be represented by a 'wire and disc' system or by a full-bore system. The original wire and disc method consisted of 0.5 mm diameter brass wire representing the centreline of the pipe and fibre discs spaced at intervals along the wire, indicating the outside diameter of the pipeline including lagging where relevant. This system is now very rarely used because of its soldered construction and the need to paint the wire of a colour code. It has now been generally replaced with the plastic-coated wire. This is known as the 'centreline' system where the 0.5 mm-diameter wire is used to indicate the centreline of the pipe as it would appear on an isometric drawing. A variety of colours is available, allowing service identification to be recognized quickly with a colour code. Sleeves are used to indicate actual and insulated diameters with snap-on plastic valves and components.

However, the most widely accepted method currently in use is the full-bore plastic system. Pipelines are represented by plastic tube to a colour code and size to scale. Overall systems including valves, controls and steam tracing are assembled with snap-in and snap-on components which permit quick and efficient construction. Once the designer is satisfied with the arrangement, all joints can be secured with the use of a solvent. The limit on pipe size to be modelled is usually set by the purpose of the model: DN 80 for a 'design' model and DN 25 for a 'follow-on' model.

Identification labels are attached to, or near, all items of equipment and pipelines are also suitably labelled. Pipe labels are stick-on arrow line tags indicating flow direction and designate line number, line size and line specification. Lagging is indicated by the use of white insulating sleeves clipped at intervals to the pipework.

Instrumentation is shown both for 'in line' and equipment instruments. It is identified by labels indicating whether the instruments are local or remote/panel mounted and are positioned for convenient viewing of the model.

Electrical and instrument trays are shown in green and white plastic sheet respectively. Lighting points can also be shown on the model.

Underground routes for process pipes and firefighting mains are shown chain-dotted on the baseboard. If trenches are used for piping, electrical and instrument cables, black plastic sheet is used showing the trench width to scale.

Piperacks, pipebridges and structures are constructed in plastic or brass material section, additionally using tube or rod for correct appearance where necessary. Concrete structures and fireproofed steelwork can be represented in square or rectangular section in either plastic or wood. Standard components are used for stairways and ladders using the nearest scale size available. Platforms are cut from grey plastic sheet. Buildings are shown in outline only with faces broken away at the corners where necessary to show model detail within. They are generally made in either Perspex or wood with window locations shown distinct from surface finish. All roads are indicated in black plastic sheet.

Each baseboard is completed with a datum point defining reference coordinate locations and elevations for equipment, steelwork and piping and a North arrow. One baseboard for the project should be selected to carry the project title, company nameplate and a code box. The code box is a key to model conventions showing pipe and components and the colour code used for the project.

Summarizing, a suggested colour code scheme might be:

Baseboards	Grey
Roadways and access	Black
Special access for firefighting	Red
Columns, heat exchangers, tanks and vessels: insulated	White
uninsulated	Grey
Equipment skirts and saddles	Grey
Foundations and concrete	Cream
Pumps, compressors, machinery	Grey
Electric motors	Green
Buildings: concrete	Cream
brick	Terracotta
Steelwork, piperacks, pipebridges	Grey
Pipe tracks – concrete sleepers	Cream
Instrument trays	White
Electrical trays	Green
Trenches	Black
Piping: process lines	Orange
steam	Red
condensate	Yellow

water   Green

water	Green
inert or purge gas	White
instrument and plant air	White

The most economical approach to using a model as a design tool is to design piping direct onto the model. The process and instrumentation diagram becomes the major document required and use of sketches and studies should be avoided. This is, however, a technique which is only likely to be successfully achieved with experienced designers.

Once the model is under construction and equipment and piping start to be shown, other design groups can begin to co-ordinate their activities. Structural designers can plan their supporting structures and particularly locations of bracing and tie-in steel. Civil engineers can establish drainage routes. Electrical engineers can locate lighting, cable and wiring routes, and switchboxes. Process engineers can examine pipeline routeing for flow patterns and pressure drop, while the instruments department can comment on orifice, pressure and temperature instrument locations. The stress engineer can commence his analysis.

Throughout the design phase of a project a number of model reviews are held. Key staff from all the disciplines involved with a project are brought together to examine the overall design at a particular stage using the model as the focal point. There might initially be internal reviews where possible changes are debated and noted. Once project approval is received then the owner's representatives are invited to review. All agreed comments are incorporated on the model and the procedure repeated at each review. Only when each area or model base has received its final review should it be photographed for record purposes. These photographs should be taken from as many angle points as possible to give maximum coverage of the plant.

If a computerized isometric system is employed piping fabrication detail drawings can be made direct from the model either via sketches or by coding up suitable input forms. The three-dimensional view obtained from the model allows this activity to be carried out rapidly and accurately.

A piping model can prove beneficial to the construction department in the planning of their work. It enables:

(a) The size of the site team to be established.
(b) Heavy lifts and crane requirements to be planned.
(c) Rigging problems to be solved.
(d) Equipment erection and installation to be scheduled.
(e) Scaffolding, pipe construction, painting and insulation requirements to be considered.

An owner usually requires a model for training his operating personnel and to satisfy himself that the plant being built conforms with his requirements with regard to layout, access, operation and maintenance. It has particular value in enabling operator training to commence before plant erection is complete and for maintenance schedules to be considered carefully at an early stage.

Models are not cheap to construct but the cost can often be outweighed by the advantages in which they:

(a) Save in drafting time, depending on the type and complexity of the plant.
(b) Provide an effective form of communication during design and construc-

tion and co-ordinate the requirements of the contractor's process, design, construction and project departments with those of the owner.

(c) Save costs and time by routeing piping more economically and meeting process, operating and maintenance requirements.

(d) Ease the engineer's approach to piping flexibility analysis, pipe-support selection and location, and to planning fabrication schedule and isometric details.

(e) Allow equipment clashes to be highlighted and designs to be criticized by all levels of management.

7.5 PHOTOGRAPHY AND PHOTO-GRAMMETRY

Photography may be used both in the early and later stages of layout development. Photographs of different plant arrangements prepared as cutouts may be produced as enlarged prints, or as transparencies in a fraction of the time required to produce drawings of the arrangements. Models of plants on several floors may be constructed by surmounting floors including block models of equipment on lower floors; floors complete with equipment blocks may then be removed in sequence and photographed in plan and elevation. Different floor arrangements may be documented as a set of photographs quickly and economically.[1]

Photographs of models are useful in producing operating manuals, in operator training and for publicity purposes.

Applications of photography in the later stages of layout development have been concerned principally with the estimation of dimensions from scale models of vessel and pipe arrangements.[2-4]

Two disadvantages have been found. The first is that any inaccuracy in the model is scaled-up. The second is that it is difficult to photograph the centre of the model without dismantling it. There are two possible solutions to this: (a) the use of modelscopes (Fig. 18.4, p. 269); and (b) to employ laser beams. However, it is not thought that either of these techniques have been used for process plant photogrammetry.

Recent work[5] has shown that photographic and photogrammetry methods are very useful for extensions. A simple scheme is to use a combination of photographs and piping isometric drawings which largely eliminates the need for piping arrangement drawings. A photograph of the existing plant is reproduced with the isometric with the start of the new piping marked on the photograph. To be successful the new pipework should not be prefabricated completely but spool pieces should be inserted on site to allow for inaccurate measurements. The position of these spools should be given in the isometrics.

More advanced schemes exploit the benefits of photogrammetry which is a means of extracting dimensional data from stereo photos taken for that purpose. Until recently it has been applied by specialist photographers almost exclusively to aerial surveys for the preparation of maps and the planning of civil works such as road and railway systems. *Close range* photogrammetry as the term implies is the same principle applied on the ground, usually to buildings and industrial plant. 'As built' co-ordinate dimensioning can be established as and when required *after* the stereo photography exercise providing that photographic interpretation, equipment and skills are available.

The various stages of the close-range photogrammetry procedure are:

(a) A civil survey establishes datum points on the subject plan.
(b) Photogenic markers are positioned strategically to appear in the photographs.
(c) Pairs of stereo photos are taken by design engineers/photographers who previously have planned the precise number of photographs and contents of each photograph.
(d) Each day's filming is developed on site.
(e) In the design office enlarged referenced prints are used to define the dimensional information required.
(f) Co-ordinate dimensioning is established by a special comparator which enables the photographs to be viewed in three dimensions by the designer. Electronics are used to register and file the results.
(g) The co-ordinate dimensional data may be used directly in the design office to aid design work on plant extensions.
(h) Alternatively, the data can be transferred automatically into a computer modelling system so that a range of three- and two-dimensional drawings can be produced as described in section 7.6.

7.6 COMPUTER MODELS

Computer models can be graded in order of increasing sophistication.

(a) Nozzle positions and piperack routes specified, with piping computer generated.
(b) Vessel dimensions and positions are added to (a).
(c) Detailed pipe routes and components added to (a) and (b).
(d) As (c) but with full details of structures, cables, ducting and obstructions given.
(e) As (d) but with process conditions contained in the computer model.

All are three-dimensional models of increasing complexity with (e) being complete. A test for clashing of items of pipework etc. is a valuable enhancement to any such system (e.g. Plant Design and Management System (PDMS)[6]).

All models can be used to produce initial and final layout drawings, piping isometrics, material takeoffs and costs, and data for pipework, purchase and control of erection. With model (a) piping layout and piperack drawings can be produced. Equipment layout drawings can be made with model (b). Model (d) allows the computer to output a complete engineering specification of the plant, e.g. layout and engineering drawings, equipment schedules, purchase and erection control documents, etc. Current (1983) advanced development is at this stage.[7] Model (e) is for the future but would allow, *inter alia*, hazard assessment and simulation of plant operation to be done. It will probably link up with computer-aided process design. The less sophisticated models tend to be for recording so there is the tedium of feeding much data to the computer. The more complex models are likely to be used with computer-aided layout development and the resultant generation of information will reduce the input task.

7.6.1 COMPUTER HARDWARE

Equipment used is fairly standard. A minicomputer is physically small and relatively inexpensive but can be very powerful, enough to handle the computer model of plant in the £150 million (1981 value) range, and can be installed near the design office or located remotely and accessed by telephone links. Access is frequently (but not exclusively) interaction from 'work-stations' in the design office. A 'work-station' is usually one or a combination of:

(a) Console:

On which commands can be sent to the computer through a keyboard.

(b) Visual display unit (VDU):

A device which can produce pictures on a precision cathode ray tube (CRT). Most VDUs are linked to a keyboard and can send commands and receive responses. Many VDUs can produce a paper copy of the picture on the CRT for drawing office use or record.

(c) Printer:

The normal way of making 'hard' copy of both text and drawings.

(d) Digitizer:

A flat board (up to 1.5 m × 1 m) flat board equipped with a movable cursor whose xy position on the board can be detected very accurately (to 50 μm) and communicated to the computer. Special-purpose computer software decodes the digitizer data, creates a drawing image and can then modify and store the drawing for future use. The device is used mainly to capture and modify drawings by locating relevant lines, symbols and points with the cursor. With appropriate software, digitizing can be very rapid.

7.6.2 COMPUTER SOFTWARE

Several computer systems, intended for design office use, are commercially available. They fall into three distinct classes as follows.

1 Turnkey drafting systems

These incorporate all hardware and proprietary software designed for rapid creation and modification of two-dimensional drawings. They operate by providing a set of standard symbols which can be scaled, positioned, orientated and connected on a drawing very rapidly. Some systems can handle limited amounts of non-graphical data associated with the drawing, e.g. properties and relationships of the real objects represented by the drawing symbols. Few systems can handle the complex three-dimensional geometry of a process plant. This type of system is good for diagrams and similar symbolic drawings and for any drawing where extensive modification is expected. They offer a means of drawing and changing layouts very quickly but most systems of this nature are essentially only very fast, reliable tools which assist the layout designer by responding to his design ideas, without significant feedback to assist decision-making.

2 'Standalone' drafting systems

These are identical in function to (a) but are provided as software only to run on the client's or any computer. Some of these systems have more extensive data-handling features and may in the future be developed into more versatile and flexible layout aids but at some sacrifice of speed of response.

3 Three-dimensional design systems

Unlike (a) and (b), these systems build up a three-dimensional model of the actual design, rather than a drawing of it. Some systems also have very extensive data-handling features to store and use data associated with the real objects in the design. Interactive response is usually slower than the drafting system because of the much greater volume of data to be processed during interaction. Since layout is a three-dimensional operation involving a large amount of non-graphical data, the three-dimensional design systems probably offer the best long-term prospect of providing useful layout aids. One three-dimensional system in commercial use[6] (Figs. 7.16 and 7.17) operates from data derived from the process flowsheet and from vessel positions defined by the designer. When a layout has been agreed after study of system-produced drawings, all the data (equipment positions, sizes, dimensions, etc.) are passed directly to the design data base without any transcription of data by the designer, ready for development into the final plant and piping layout. The system has additional techniques to check the proposed layout for clashes and can use the computer systems data transcription facilities.

7.6.3   FUTURE DEVELOPMENTS

The above illustration of one available three-dimensional design system highlights the greater potential of this type of software. As algorithms for, for example, spacing calculations, hazard ratings and other layout parameters are developed, they can be made compatible with three-dimensional design systems so that the designer can easily and quickly validate his intuitive layout ideas without holding up progress of the design. The combination of three-

Fig. 7.16 Examples of three-dimensional computer output (a) plot (b) heat exchanger grouping (Both drawn by PDMS; courtesy: Babcock Woodall-Duckham)

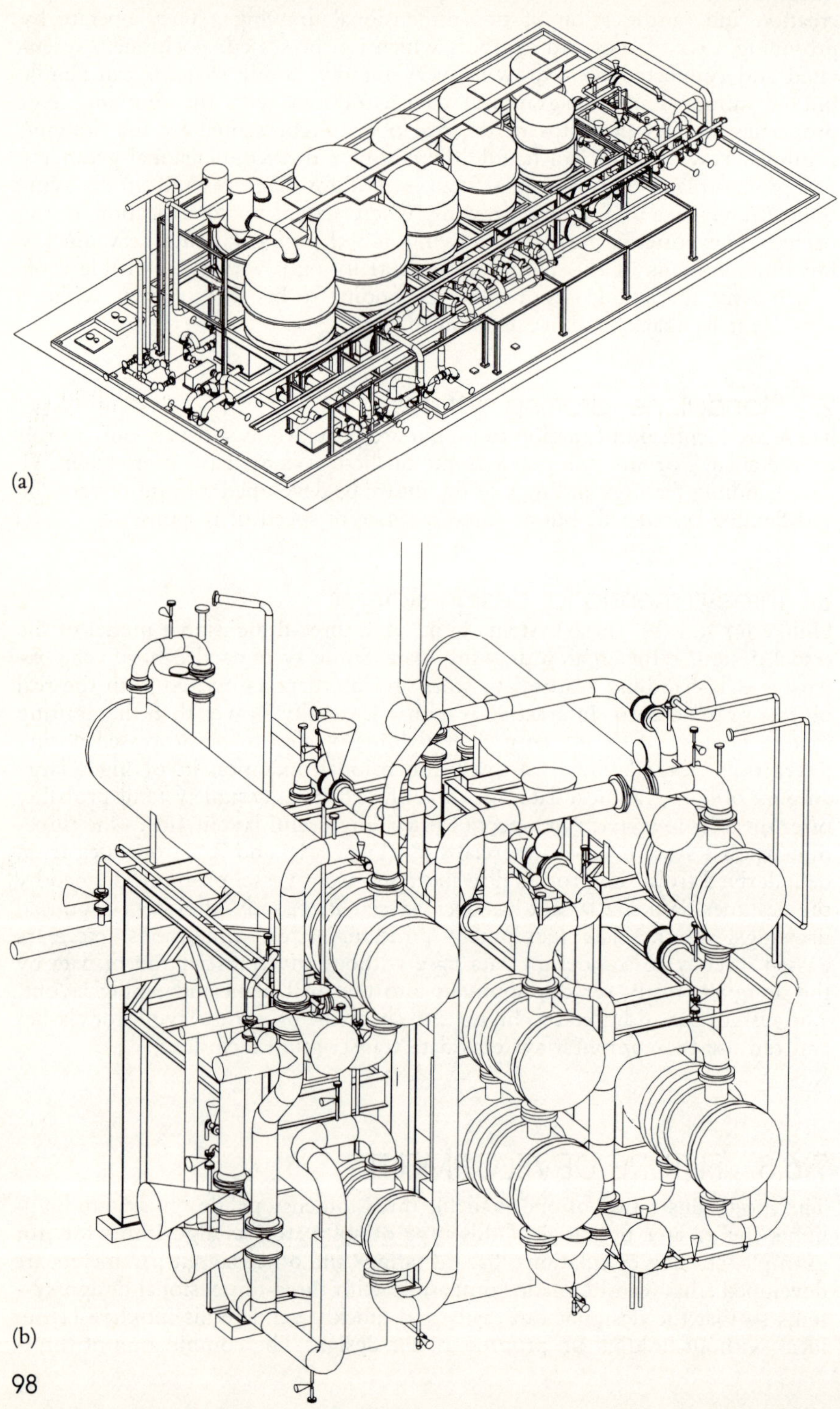

(a)

(b)

Fig. 7.17 Section of plant drawn by computer PDMS (Courtesy: Babcock Woodall-Duckham)

dimensional and formal technique is, as yet, in an early stage but it promises to lead to genuine advances in layout procedures.

However, progress has been slow (see section A.2.2, p. 467) because although the hazard assessment programs are fundamentally straightforward, they and their associated data bases are large. This has meant many manhours of development and the use of large mainframe computers leading to high costs. Also problems of transcribing data from drawings or models to the computer have meant incurring unacceptable manpower and elapsed-time penalties. The data to be transcribed defines such topics as inter-item relationships and properties of items such as dimensions, process materials contained, temperature and pressure, etc., all of which affect the spacings and relationships of items in the layout.

Recent progress[7] in computing methods, however, can reduce these problems to manageable proportions. Current mini-computers are sufficiently economic to be used as dedicated design office tools and are becoming powerful enough to handle major computation PDMS work. Terminal hardware (VDUs, digitizers, consoles, etc.) which can provide accurate, design-office-quality drawings is also becoming available and increasingly economic. Finally, the software available for 2-dimensional graphics and 3-dimensional modelling can be coupled to powerful databases able to handle real plant-scale problems in an interactive working environment. These latter developments allow designers to express layout ideas in engineering language at the drawing office terminals and have the computer create and store the layout drawings and, with some systems, also store in a database the data associated with the layout.

In the process of layout creation and evaluation, the crucial impacts of the design office mini-computer lie in its ability, firstly to respond quickly to the designer's needs; secondly, to produce drawings; and, thirdly, to derive large amounts of data quickly, reliably and in formats suitable for interfacing with the other programs.

REFERENCES

1. Kern, R. 'How to manage plant design to obtain minimum cost', *Chem. Engng*, 23 May, 130, 1977.
2. Farrand, R. 'Photogrammetry applied to pipe systems of chemical plant', *Photogrammetric Record*, **5**, 100, Oct. 1965.
3. Farrand, R., 'Drafting photography' in *Photography in Engineering*, I. Mech.E., London, 1970.
4. Ari Elo, *Making an accurate orthographic projection from a model*, Amer. Engr Model Soc. (Meeting, San Francisco, May 1979).
5. Klement, U. R. and Bracewell, P. A. 'Easier routes to "as built" records', *Process Engineering*, 45, Mar. 1982.
6. Anon. 'Computer aided engineering', *Computer Systems*, July/Aug., 49, 1981.
7. Parker, D., 'Can CAD untangle the pipework?', *Engineering Today*, 21 Oct., 16, 1980.

HAZARD ASSESSMENT IN PLANT LAYOUT

8.1   INTRODUCTION

The safety aspects of the siting and layout of process plant are related to the probability, size and consequences of loss of containment of the process materials.

The size of leakage ranges from small but frequently occurring flows from drains, vents, pump seals, etc. to large and almost instantaneous but less likely releases from pressure vessels or large-bore pipes. Damage to plant and people ('targets') results from the release through fire, explosion, toxicity and chemical or physical attack such as embrittlement or corrosion. The amount of damage will depend on the size of release, the properties of the released material, the proximity of potential targets to the source of release and the strength and resistance of such targets against damage. It will also depend on the response to the emergency by the personnel, etc.

The technique by which these phenomena are identified, evaluated and quantified is called 'hazard assessment'. Its role and position in the procedure of layout conception and development was indicated in section 6.6 (p. 60). The systematic approach follows the sequence for general Stage One layout development given in Fig. 5.1 (p. 37), having the following steps:

(a) Establish the process design (flowsheets, equipment, etc.) plus a preliminary layout.
(b) Identify sources of failure and the vulnerable targets, and, if possible, eliminate them by better design.
(c) Identify parts of the plant containing dangerous materials, which if released, could cause undesirable consequences even though no particular point of loss can be specified.
(d) Estimate the frequency of loss and the amounts and rates of leakage.
(e) Evaluate the consequences of the discharge of hazardous materials on the targets.
(f) Adjust the layout and/or design and repeat the assessment until the consequences are acceptable.
(g) Plan the emergency response procedures which may mean further adjustments of the layout.

It should be emphasized that although a good design and layout may make the consequences of particular losses 'acceptable', this does not alleviate the obligation of those responsible for the operation, maintenance and modification during the lifetime of the plant, to take all reasonable measures to prevent such losses and if they occur to respond efficiently.

The nature of the hazard assessment will vary between plot and site layout and site selection.

Plot layout is concerned with the hazards presented by more numerous smaller-scale events and should be arranged in order to reduce the risk that these events might escalate to give major damage in that plot, or initiate 'knock-on' events which could involve adjacent plots and buildings. The assessment will determine area classification for electrical equipment, the spacing to prevent the spread of fire between items of equipment and the distance the control room and other occupied buildings should be from hazardous parts of the plant. The hazard of explosion may sometimes influence the layout of a large plant but it will most probably be found that the available area makes it impossible to achieve a meaningful separation of equipment. The release of toxic materials will seldom influence plot layout as the separation distances required to achieve a useful dilution are too large and protection has to be sought by other means.

With site layout, the distance between the various plots will be chosen to stop the spread of fire and to reduce the consequences of explosion. The siting of central offices, services and utilities will be governed by the need to reduce the dangers of explosion, fire and toxicity.

Site location with respect to its environs is concerned with those effects, such as explosion and toxic release, which can cause damage up to a considerable distance outside the site boundary and fire, which effects the environment adjacent to the property. Siting decisions should take cognizance of the density and susceptibility of the population in the adjacent public domain in relation to the hazard. Different criteria may be expected to be used when the site is adjacent to suburban, rural or coastal areas or if there are schools, hospitals or centres of high population density in the vicinity.

The approach to hazard assessment will also vary between Stage One and Stage Two layout. In preliminary plot layout, adjacent plots are not known and so a system of hazards contours should 'surround' the proposed plot layout. This will aid the preliminary spacing of plots on the site layout. Similarly, before site purchase the surrounding community is not known and so the proposed site layout should be 'encircled' by hazard contours to aid site selection. After site purchase, the Stage Two site layout hazard assessment can take account of the consequences of hazardous release on particular community items. Likewise, within the site the effect on a particular vulnerable item, like an office, from a release on a specified plot can be calculated.

The methods and data available for hazard assessment are parts of a rapidly developing subject. The layout designer should be aware of the need to keep abreast of current thinking and practice contained in the literature and legislation and presented at learned society meetings (e.g. I. Chem. E.[1,2] and EFCE[3]), at private meetings with consultants, or with co-operative organizations such as the UK Systems Reliability Service.[4]

8.2 RELEVANT HAZARDS

Hazard assessment looks at four possible hazards resulting from loss of containment.

(a) Overpressure from unconfined vapour cloud explosions (UVCEs).
(b) Thermal radiation from fires.
(c) Toxicity effects of a vapour cloud.
(d) Flammable concentration of a vapour cloud.

In the first three, the consequences lead to fatalities, injuries and damage. The fourth item is not strictly a hazard although it is convenient to treat it as such. It leads to the first and second hazards and flammable limits are used, as a precaution, to define electrical classification zones and the separation of sources of ignition from flammable leaks. The hazards of chemical and physical attack are usually covered indirectly as items (b), (c) and (d) require the determination of the size and position of liquid pools and jets.

The potential to generate hazardous conditions is related to amount released, the behaviour of the substance after release and the flammable and toxic properties of the material. The amount released from a vessel or pipe is governed by the geometry of the leak, the process condition and inventory of the item. The immediate behaviour of the material on release, and in particular the manner in which vapour clouds are formed is a function of the initial state of the material. The subsequent behaviour over extended time and distance is determined by the weather conditions, topography, and how material is released.

8.2.1 RELEASE OF MATERIAL

The amount released and the rate of release, depend on:

(a) The inventory of material available to supply the release.
(b) The duration of the release.
(c) The physical state of the material on release, e.g. gas liquid, or mixture.
(d) The process conditions (temperature, pressure) prior to and during the release.
(e) The release geometry, e.g. whether by vessel failure, pipe failure, flange leak, etc.

The combination of these factors will result in two extreme forms of release: instantaneous release and continuous release.

Instantaneous release

As an example, the opening of a valve to atmosphere on a gas system or the massive failure of a pressure system can approximate to instantaneous release. Provided the energy is available, the amount released will generally be the total isolatable inventory of the system.

In assessing what is the extent of the isolatable inventory, some credit should be taken for the following (provided the equipment is scheduled for regular inspection and/or testing as appropriate):

(a) Non-return valves operating.

(b) Manually operated isolation valves operating within a few minutes, provided the operators are informed by signals from gas detectors, or in other ways, that a leak has occurred. If the valves are not remotely controlled but on the plant, their location must permit operation in the event of release.

(c) Valves which close automatically, by fall in pressure or in other ways.

Continuous release

In this case the release rate approaches a steady state and process conditions are substantially maintained.

Between these extremes of instantaneous and continuous releases there is a range of possibilities and the process condition may alter during, and because of, the release so that release rate varies during the time-span of the release. Fortunately, in practice, one of the two limiting extremes, instantaneous or steady continuous, can be chosen.

8.2.2   BEHAVIOUR OF MATERIAL AT RELEASE

The following conditions need to be considered.

A.   Liquid at atmospheric pressure

A1. Liquid with a boiling point above ambient temperature which is processed/stored at a temperature at or below its normal boiling point.

A2. Liquid with a boiling point below ambient temperature which is processed/stored at low temperatures and atmospheric pressure (e.g. cryogenics).

B.   Gas

C.   Flashing liquid under pressure

C1. Liquid with a normal boiling point above ambient temperature which is processed/stored under pressure at a temperature above its normal boiling point. Upon release any residue liquid is in condition (A1).

C2. Liquid with a normal boiling point below ambient temperature which is processed/stored under pressure at a temperature above its normal boiling point. Upon release any residue liquid is in condition (A2).

For release with condition (C), the decrease of pressure causes flash vaporization. If the resulting volume of vapour is significantly larger than that of the residual liquid, there will be a violent expansion which will atomize the liquid into small droplets. Initially these will probably be small enough not to precipitate but coalescence may cause fallout. On the other hand, as the cloud or plume dilutes with air, the droplets will evaporate and diminish and the cloud will become less dense. The presence of mist or aerosol may be ignored in calculating the initial volume of the cloud but should be allowed for in the assessment of the flammable mass and later dispersion. The release of a gas (condition (B)) gives a similar cloud or plume but without the complications of liquid entrainment although release of a supercritical vapour may produce a small amount of liquid.

In a release under condition (A), there is no flash vaporization accompanying the release, although there may be subsequent evaporation.

With instantaneous failure the whole inventory can be released to atmosphere. For a less catastrophic failure, account has to be taken of the rate of release and this requires the postulation of possible modes of failure and dimensions of the leak. Also with conditions (B) and (C) consideration has to be given to choked flow and for (C), the vaporization during the leakage.

There may be enough residual liquid from release of condition (C) fluids and there certainly will be for condition (A) to form a liquid pool on the ground or in the remains of the vessel. The temperature of such pools will be at or below the liquid boiling point and the subsequent rate of evaporation depends on the size of the pool, the vapour pressure, ground temperature and the windspeed. With condition (A1), the rate of evaporation depends primarily on windspeed. For a cryogenic liquid (condition (A2)), contact with surroundings at ambient temperature can produce initially high rates of heat transfer to the liquid and consequent boil-off. These rates diminish as the surroundings cool down and eventually the evaporation rates become governed by the same rules as for normal liquid pools. Boil-off from cryogenic liquids is unlikely to be rapid enough to give liquid entrainment but there will be a visible cloud from water condensed out of the air. Because of the low temperature, these clouds will usually be denser than the surrounding air and this has a significant influence on the subsequent dispersion.[5]

The position and rate of spreading of the pool will depend on the topography of the surrounding area. In addition, when the material issues as a free jet, it will be necessary to calculate the trajectory of a liquid stream in order to determine the point at which a liquid pool will form. The engineer should minimize the possible surface area (or surface/volume ratio) of a liquid pool in order to reduce the size of the vapour cloud caused by the evaporation.

Methods for calculating release rates and amounts of vapour flashing are given in section B.3.1, p. 497.

8.2.3   VAPOUR DISPERSION IN THE OPEN

Three consecutive modes of dispersion in the open are usually considered:

(a) Momentum dispersion.
(b) Dense-phase dispersion.
(c) Neutral buoyancy dispersion.

For a gas or vapour issuing as a free jet, there will be entrainment of air into the jet and dilution to below the lower flammable limit is possible before the velocity has fallen sufficiently for the jet to have lost identity and become a drifting plume under the influence of atmospheric turbulence. However, dilution by jet entrainment works best in preventing a hazard when the jet is deliberately directed vertically upwards so that there is no chance of impingement on any other object. For a jet resulting from a random leak, it may be prudent to assume that the jet momentum will be dissipated immediately by contact with the ground or other equipment and the leak disperses totally under the influence of atmospheric turbulence.

With an instantaneous release, the amount of entrainment due to momentum is uncertain. Consequently, momentum dispersion is often ignored for instantaneous clouds which may lead to conservative separation distances.

After loss of residual momentum, the cloud or plume travels downwind and becomes mixed by turbulence (modes (a) and (b)). This is a process where molecular diffusion is so slow that it may be ignored.

Most materials have a molecular weight greater than air and many could have an initial temperature below ambient after flashing from condition (C), particularly (C2). Both features give rise to denser-than-air clouds and plumes which tend to hug the ground.[6] Because of this, special attention is needed to determine where vapours might collect in hollows, bunds and trenches or might run down inclines to lower levels. Materials with low molecular weight or high initial temperature will, however, be lighter than air and they will tend to rise and therefore be less of a hazard. When sufficient dilution has occurred both negative and positive gravitational buoyancy effects will be overcome as the cloud will achieve approximately the same density as air.

A release does not necessarily go through all three modes of dispersion. A release from a low-pressure gasholder will not have much momentum, whereas a jet from a high-pressure tank may entrain so much air that the dense phase stage is omitted. Lower flammable limits are typically 1–3 per cent and are attained in the jet or dense-phase modes. Toxic limits which are often of the order 10 p.p.m. will be usually found in the neutral buoyancy mode. The classical treatment of Sutton[7] and Pasquill[8] for dispersion was based on the neutral buoyancy mode. However, this is now considered to be insufficient even though the mathematics of dense phase dispersion is much more intractable.[6]

There is a critical windspeed below which mixing of air and gas occurs only slowly by molecular diffusion and it is probable that the material from an escape would spread out round the point of release. The critical windspeed is not well defined and appears to be in the region of 1.3 m/s. For the calculation of safe distances a minimum windspeed of 2 m/s is used to validate the method.[9] Windspeeds of less than 2 m/s occur for less than 15 per cent of the time in the British Isles. It should be noted also that, within plant areas, the obstruction to air movement by equipment tends to maintain turbulence.

Atmospheric temperature gradients significantly affect atmospheric turbulence and, hence, the dispersion of windborne material. Normally, the air temperature falls with increasing height at a rate of about 1 °C per 100 m in the neutral condition. In the unstable lapse condition, the fall in temperature exceeds 1 °C/100 m and in the stable inversion condition the air temperature rises with increasing height until at some altitude it begins to fall with further height. Pasquill[8] has defined the stability categories and others[10] have given the typical distribution in order of increasing stability shown in Table 8.1.

It is normal practice to use stability conditions B, D and F given in Table 8.1 to categorize the three atmospheric temperature profiles. It will be noted that the neutral condition occurs for about 75 per cent of the time in the UK and it is customary to use this in initial estimates of dispersion. Note, however, that the worst case for ground level releases would be the inversion condition (F) where vertical dispersion is restricted and the exceptional stability promotes little turbulence to assist horizontal diffusion. The lapse con-

Table 8.1 Weather stabilities

Pasquill stability category		Typical wind speed (m s^{-1})	Description of weather	Probability in UK
Unstable lapse	A	1	Very sunny summer	0.013
	B	2	Sunny and warm	0.064
Neutral	C	5	Partial cloud during day	0.150
			Overcast day or night	0.620
	D	5		
Stable inversion	E	3	Partial cloud during night	0.067
			Clear night	0.084
	F	2		

dition is unstable and promotes dispersion at or near ground level but a release from a high point can give the highest ground level concentrations as the plume spreads to the ground.

It is probable that, within a plant structure, the turbulence produced by the obstruction to airflow and the thermal uplift produced by heat released from the plant cause sufficient instability to give rise to permanent lapse conditions. It is certainly safe to assume that in plant areas the atmospheric stability does not exceed normal, i.e. there is never an inversion.[11]

Methods of calculating the dispersion from instantaneous or continuous sources under various weather conditions are given in sections B.2 (p. 487) and B.3 (p. 497).

8.2.4 VAPOUR DISPERSION IN BUILDINGS

Dispersion mechanisms within buildings do not seem to have been widely reported in the literature. In section B.8 (p. 529) we have proposed a scheme whereby initially jet action occurs until the vapour velocity reduces to the room ventilating velocity or the jet hits equipment, etc. This is then followed by molecular diffusion if the airflow is laminar or by complete mixing if the flow is turbulent. Often laminar flow occurs in natural ventilation and turbulent flow with forced ventilation.

Complete mixing is postulated in the case of turbulence because the turbulent action is reflected off adjacent walls, equipment, etc. However, there can be stagnant areas where this mixing will not occur.

Gravity will have some effect, but in the simple treatment given in section B.8 (p. 529) it is assumed that for small leaks its effect is not important.

8.2.5 FIRE AND EXPLOSION HAZARDS

The types of ignition can be classified as follows:

(a) Liquid pool fire.
(b) Flammable jet burning at point of escape.
(c) Fireball or firestorm
(d) Boiling liquid expanding vapour explosion (BLEVE).
(e) UVCE or aerial deflagration.
(f) Aerial detonation.

Although it is by no means certain that ignition will occur in every case, and there have been many cases of vapour clouds reported that did not ignite, in hazard studies the assumption is made that ignition will occur. The fundamental properties of flame propagation are far from understood and so the following picture is only approximate.

On release, a vapour cloud will be initially over-rich and capable of burning only relatively slowly as a diffusion flame around its periphery. If there is an ignition source close to the leak which ignites the escaping vapour the result is a fireball or jet flame, both without explosion, and the hazard is due to heat radiation. The term 'firestorm' is associated with large burning clouds which can create strong convection currents in the surrounding atmosphere.

If ignition is delayed, so that a substantial part of the vapour cloud is in a flammable condition, the possibility of a UVCE or aerial deflagration is considerably enhanced. A deflagration has large but subsonic flame speeds and moderate overpressures caused by the blast wave.[12-15]

Exceptionally, the flame speed can become supersonic with very high pressures giving rise to a detonation. However, detonations appear to occur in confined spaces like long pipelines and not, generally, in the open, except at localized points around dust particles, etc.

Empirically, it seems that the size of the cloud has some influence on probability of combustion occurring as a UVCE since in larger clouds there will be more room for the flame front to accelerate. Thus, the concept has developed of a minimum size of vapour cloud, expressed as a 'threshold weight' of fuel, which must be present before the probability of explosion is significant. There is at present no theoretical justification for this concept, but it is supported by surveys of past incidents, such as those of Strehlow[16] and Davenport.[17]

The ignited vapour from spills can flash back to the pool, open tank or issuing jet causing a fire. If other intact containers of flammable liquids are exposed to the fire, the metal can be so overheated as to lose its strength and fail. This type of failure is a BLEVE and can eject parts of the vessel and contents for considerable distances. There will always be some overpressure effect due to the expansion of the vessel's contents and there may be a subsequent UVCE although a fireball is more likely as ignition should be immediate. A well-known example of a BLEVE is that which occurred at Feyzin in France in 1966.[18] In general, it is the thermal radiation effects from a BLEVE which cause the most damage because, although fragments of the vessel can be deposited over a wide area, the resulting damage is localized.

8.2.6 THE FLAMMABLE AND TOXIC HAZARDS

There are important practical differences, in respect of concentration and range, between the two hazards.

The toxic hazard is concerned with low concentrations (a few p.p.m.) persisting over relatively long periods of time is a problem which extends to large distances. By contrast, the lower flammable limit of most substances is typically 1–3 per cent 10,000–30,000 p.p.m. and the flammability hazard is essentially of short range and duration.

The distinction between the two hazards with respect to the time-scale of concentration measurement or prediction is particularly important. In general, with a normal respiration rate, it takes two minutes for air in the lungs to come into equilibrium with the atmosphere and significantly longer for equilibrium with the whole body to be obtained. Therefore, the dosage (concentration × time) and not the instantaneous concentration is likely to be of most value in layout toxicity considerations. Exceptions to this statement might be very highly toxic substances.

Unfortunately, the available toxicity data are incomplete as the understanding of the body's reaction to toxicity is far from complete. This means that toxic calculations can only lead to approximate conclusions.

With flammability considerations, instantaneous concentrations are the most useful. The data on flammability of materials in air at atmospheric temperature and pressure is more comprehensive (e.g. CRC[19]). Thus flammable calculations can be more accurate, though there are still the uncertainties in the weather conditions and release geometry.

8.3   IMPLICATIONS FOR LAYOUT

8.3.1   IDEAL APPROACH

The word 'target' is often used to denote the possible victims or casualties (both human and equipment) of a potential incident even though there is no deliberate 'aiming' as implied in the conventional use of 'target'.

The probability that a loss of containment will cause a given amount of damage, fatalities, etc. to a particular target can be split into three separate probabilities: loss of containment, transmission and damage.

Probability of loss of containment
This depends on the process conditions (e.g. temperature, pressure, corrosiveness), the quality of engineering (e.g. vessel thickness, control equipment, the possibility of collapse of one item on another) and the quality of operation and maintenance.

Probability of transmission
For fires and explosions this includes the probability of ignition, and for toxic and flammable cloud drift the probability of the wind having a certain direction and speed.

Probability of damage

This is the probability that the hazard having reached the target will cause damage, etc. This probability is a function of the intensity (overpressure, thermal radiation flux, or concentration) of the hazard, the duration of the incident and the robustness of the target.

The intensity of hazard can depend on:
(a) The properties of the materials released (e.g. heat of combustion, toxic or flammable limits, density).
(b) The size of released inventory.
(c) Weather conditions (e.g. lapse or inversion conditions).
(d) Distance between source and target.

The total risk to a target is a combination of the risks from each source of loss of containment. A series of risk contours can be developed.[20] A number of actions can be taken if a target is in an area of unacceptable risk. If there is room the target can be shifted. Alternatively, it may be given better protection. However, the most economical remedy is usually to reduce the areas of high risk by such measures as lower inventories, less severe process conditions and chemicals or by better engineering.

Computer programs to deal with this procedure are being developed[20] and have been used to assess risk to communities from a number of adjacent sites. The possibility exists of combining such a description of risk with computer-generated layout and piping models so that a full description of the geometry of the plant is available and the effects of layout change can be quickly assessed. However, there is one major snag in applying the risk contour procedure: there is a dearth of data on probability of failure of particular items of process equipment. Most failure data available are based on nuclear and aerospace equipment and are heavily biased towards instruments and instrument systems. The poor situation for the process industries is being remedied and in the UK the Systems Reliability Service (SRS) collects and correlates failure rate data.[4]

8.3.2   CURRENT APPROACH

For the present, to accommodate this lack of reliability data a slightly different approach has to be used. This approach has the incidental advantage in that it is amenable to manual or simple machine calculation. It is in two stages which use respectively intensity criteria and risk criteria.

The first part is based on the idea of selecting all likely sources of loss of containment and then using critical intensities or *criteria*. If the intensity at the target is less than the criterion then the target is considered acceptably safe irrespective of the risk of loss of containment. The criterion reflects the degree of protection given to the target.

The target intensity can be reduced by:

(a) Shifting the target away from the source.
(b) Reducing inventories.
(c) Having less severe process conditions.
(d) Reducing toxicity and flammability through having safer chemicals.

Many layout situations can thus be resolved by arranging the intensity at the target to be below the criteria. Probability is merely considered subjectively in choosing or rejecting the sources for investigation. Only in the few cases not so resolved in the second stage employed in which finite probabilities have to be used.

The approach for the second stage is to take the acceptable risk at the target and ask, probably subjectively, if the risk of loss of containment is consistent with this or can be made so by improving engineering and operational and maintenance standards and by training operators in the correct disaster reactions. It is assumed that immediately a criterion is violated then casualities change from zero to 100 per cent. Clearly, such a sharp division does not occur in practice but the variation of casualties with intensity is only crudely known (see section B.6, p. 519). Similarly, the probability of transmission is assumed to be 100 per cent even though for fires and explosions, not all releases are ignited and for toxic releases the wind direction varies.

These simplifying assumptions lead to the proposition that if the critical intensity is violated then the risk at the target is the same as the risk of loss of containment. The design procedure for the second stage is to reduce this risk to acceptable levels called 'risk criteria'.

One advantage of the first stage in not using finite probabilities but intensity criteria, is that the concept of the maximum credible accident is applicable. After preliminary examination of the likely sources of release, only the ones giving the greatest damage are used for further assessment. The basis for this is the axiom that if it is safe for the big events it must be safe for the little occurrences.

However, when finite probabilities and risk criteria are used in the second stage, the contributions from all the credible sources violating the criteria should be considered in assessing the risk at the target. This involves more calculation effort. It is advisable that when finite probabilities such as risk criteria and plant reliability have to be considered, the general layout engineer should consult with specialists.

8.3.3   ACCIDENT MODELLING

This is a closely allied topic to hazard assessment in that it uses the mechanisms of fluid escape, cloud dispersion, explosion overpressure, thermal radiation, etc. to model the accident after the event. However, some uncertainties are removed, e.g. the source of the loss of containment and the weather conditions are known. Thus the use of more accurate representations of the above mechanisms are justified and probably needed. In particular the generation, trajectory and impact of missiles[21] will have to be considered if aerial fragmentation occurred during the incident. However, since the chance that an item is struck by a missile is so small, the consideration of missiles in hazard assessment is unlikely to be justified.

The accuracy of the equations and models given in Appendix B has been geared to hazard assessment of layouts and not to the higher demands of accident modelling.

8.4 APPROPRIATE CRITERIA

The governing principles in selecting critical intensities or criteria are:

(a) The effect on plant and property adjacent to a hazardous plant should avoid damage sufficient to cause the plant or property to become involved in the incident and so escalate the risk.
(b) People should only suffer minor injuries.

8.4.1 CRITERIA FOR BLAST PRESSURE DAMAGE

Damage to equipment and people depends on the magnitude and duration of overpressure, with the magnitude being taken as the more important, certainly with our current state of knowledge.

The possible overpressure caused by a UVCE can be estimated at various distances from the epicentre of the explosion by the methods given in section B.2.3 (p. 492). Containment failure which could release enough material to form a vapour cloud capable of explosion will probably be classified as instantaneous and it is customary to consider the epicentre of such a cloud as the point of release. Using the scaling laws for the attenuation of overpressure, contours can be drawn on layout diagrams to show the possible extent of various levels of overpressure. Layout decisions based on the level of damage to personnel and various types of property can then be made. To this end Table B.3 (p. 493) lists some possible overpressure criteria for the separation of vulnerable equipment from sources of hazard. Section B.7.2 (p. 525) also gives the likely damage to plant items for various overpressures.[22]

The human body can be harmed by an explosion in a number of ways, e.g. being:

(a) Hit by the blast pressure.
(b) Thrown onto and tumbled along the ground and possibly then hitting an obstruction.
(c) Hit and buried by falling debris.
(d) Struck by missiles such as glass fragments.

Expected levels of injuries and fatalities are given in section B.7.1 (p. 525).[22] Man's tolerance of blast pressure in the open is surprisingly large, especially if the rate of pressure increase is moderate and the duration short. For example, a person lying down can survive a 2 bar overpressure although lung and ear damage will be received. On the other hand, glass fragments can be lethal at overpressures as low as 0.1 bar.

It should be noted that special treatment[23,24] is needed to deal with the problem of the design and location of control rooms which, for operational reasons, have to be sited close to the plant. This means that they could be subjected to far higher blast and heat effects than other buildings and may even be inside the deflagrating cloud (see section B.7.3, p. 525).

8.4.2 CRITERIA FOR FLAMMABLE LIMITS

The two critical concentrations are:

(a) *Lower flammable limit (LFL)*: the concentration of the vapour in air *below* which combustion will not take place.

(b) *Upper flammable limit (UFL)*: the concentration of the vapour in air *above* which combustion will not take place.

The LFL is the more useful criterion for layout assessment being mainly used to determine the separation between ignition sources and points of loss of containment (see sections B.2.2 (p. 488) and B.3.2 (p. 502)). Lower flammable limits are available in the literature for many compounds. The amount of combustible material in a cloud is better represented as that contained above the LFL and not the amount released. However, the inaccuracy of hazard assessment does not justify such precision and the combustible amount should be taken as the total amount.

In the context of hazard assessment a flammable material is defined as one which is stored at or above its flash point, i.e. the vapour pressure is greater than the LFL $\times$ atmospheric pressure.

8.4.3 CRITERIA FOR TOXIC LIMITS

A toxic material could be defined similarly as one which is stored at a temperature such that its vapour pressure is greater than the toxic limit $\times$ the atmospheric pressure. However, the concept of dosage criteria for the assessment of toxic risk was introduced earlier in this chapter (section 8.2.6). The available data are in the form of fixed time dosages known as time-weighted concentrations. They include:

(a) *Long-term exposure limit (LTEL)*   The time weighted average (TWA) concentration for a normal 8 hr working day, 40 hr week, to which nearly all workers may be exposed repeatedly, day after day, without adverse effect.[25]

(b) *Short-term exposure limit (STEL)* The maximum concentration to which workers can be exposed for periods up to 10 min. continuously without suffering from:

  (i) irritation;
  (ii) chronic or irreversible tissue change;
  (iii) narcosis of sufficient degree to increase accident-proneness, impair rescue or materially reduce work efficiency;

  provided that no more than four excursions per day (or shift) are permitted with at least 60 min. between each exposure and provided the LTEL for that day is not exceeded.[25,26]

(c) *Emergency exposure limit (EEL)* Emergency exposure limits are intended as guides for use in advance planning for dealing with emergencies only and they refer to concentrations without permanent impairment to health but not necessarily without acute discomfort or other evidence of irritation or intoxication. Emergency exposure limits should only be exceeded in circumstances where impairment to health is justifiable in order to prevent a still more serious event.[27,28]

A similar quantity is the 'immediately dangerous to life or health' (IDLH) concentration which is defined as the maximum level from which one could tolerate for 30 min. without any escape impairing symptoms or any irreversible health effects.[29]

Limit values are reviewed annually by the American Conference of Government Industrial Hygienists (ACGIH) and in the UK by the Health and Safety Executive (HSE).[25] These limit values are the only comprehensive[19] and widely recognized standards for toxic exposure but they are of limited value in layout planning as they relate to continuous exposure, whereas the hazard assessment aspects of layout are usually concerned with acute but infrequent exposure.

The IDLH is a more useful concept for hazard assessment and layout considerations as it relates specifically to severe infrequent emergency situations. However, IDLHs along with EELs and STELs are available for only a few substances. A short-list is given in section B.9.4 (p. 546).

It will be recognized that values for LTEL, STEL, IDLH and EEL are merely points on a spectrum of exposure conditions and physiological response. Provided sufficient data are available, it is possible to construct a diagram showing the effects of exposure to a wide range of dosage in time/concentration terms and this allows a more flexible and useful approach to emergency release problems. In principle, this can be applied to any gas but Sellers's[30] data for chlorine is one of the few examples available.

An alternative approach would be to have a model of the body's reaction to toxic materials so that realistic dosages for varying times could be calculated from the LTEL, STEL or EELs. However, knowledge in this field is not advanced enough for this. Thus, in most cases it will be necessary for limits to be assessed individually by professional industrial hygienists or occupational health physicians.

It is not expected that the internal layout of a plot will be significantly affected by consideration of toxic release. The control of a small continuous release, such as that from a pump seal, to limit the area subject to concentration above the LTEL is a matter of design for ventilation hygiene rather than layout.

An important consideration in toxic release is the protection provided by a building with close-fitting windows and doors. The internal concentration takes considerably longer to build up than the rise outside the building. This point is discussed further in sections B.2.5 (p. 495) and B.3.4 (p. 506).

An extensive treatment of a design and engineering approach to the reduction of toxic hazard and of layout and other strategies for the reduction of risk has been given by Green.[31]

8.4.4 CRITERIA FOR EXPOSURE TO THERMAL RADIATION

Assessment of the effects of thermal radiation falling on targets of various types is usually considered in terms of energy flux at the incident surface. Methods for estimating the emission flux of flames together with the distance and view factors for calculating incident flux are given in sections B.3.3 (p. 503) and B.4.2 (p. 513).

Criteria for the effect of thermal radiation on buildings may be obtained from the requirements of various national standards for building construction and materials, e.g. BS 476.[32]

Such standards usually state that the building should withstand a given flux for a given time (e.g. after 1 hr) before failure. For normal buildings having exposed wood and glass the flux[33–35] is about 14 kW m^{-2}. For special buildings having flameproof doors and no windows the criterion is 25 kW m^{-2} for 1 hr. Structural steelwork should be insulated (e.g. concrete cladding) so that it takes some time (e.g. 1–2 hr) to heat up to its yield temperature. Unprotected, the time taken is less than 15 min.

In general, for plant equipment, a criterion based on limiting the rise of surface temperature can be used to determine the maximum incident flux to which the equipment should be exposed and, hence, its layout and spacing from hazardous areas (see section B.5, p. 516). The temperature limit may be met with regard to structural integrity or to the auto-ignition temperature of flammable materials in contact with the irradiated surface. In considering the structural integrity of a pressure vessel, for example, one would determine the temperature at which the yield or ultimate tensile stress of the material became equal to the actual stress at the relieving condition due to the internal pressure. This pressure may rise with the increase in temperature if the fluid is a gas or a liquid on the boil and there is no pressure relief. Metal in contact with a boiling liquid is partially protected in that it is approximately at the boiling point of the liquid and not at the temperature the incident flux would give alone. On the other hand, it is usually assumed that metal in contact with a gas or with a liquid below its boiling point has the incident flux temperature because heat transfer to a non-boiling fluid is slow.

Equipment, especially storage tanks, are often protected by water-drench systems. The temperature rise and evaporation of the water keep the surface temperature at 100 °C and allows quite high incident fluxes (see section B.5, p. 516).

The most vulnerable items that are placed at risk by thermal radiation and hot-combustion products are electrical and instrument cables. Those with plastic insulation and cladding will deform and become damaged above 120–140 °C, but special cables can be used up to 1,000 °C. The incident flux required to damage plastic insulated cables is around 2 kW m^{-2} and consideration must be given to protection when these cables are run in areas likely to be exposed to fire. The protection can be in the form of shielding when the direction of radiation can be predicted (as from a flare), by enclosure in fireproof trunking, or by fireproofing individual cables.

Work[36–41] on the effects of exposure of persons to thermal radiation suggest that significantly high flux (6 kW m^{-2}) can be tolerated for short times, e.g. by escaping personnel. Even light clothing provides significant protection; a person can be cooled by airflow; he may move so that the radiation is not always incident on the same patch of skin; he may protect himself by seeking shelter and he may escape to a greater distance from the fire.

The safe limit of exposure for stationary personnel and for members of the public is usually taken as 1.5 kW m^{-2} which is the level of mild sunburn. Higher levels, say 3 kW m^{-2}, may be allowable in infrequent emergency situations for up to 30 min.

The average level of flux at which vegetation of various kinds ignites is usually taken at 10–12 kW m^{-2}.

8.4.5   RISK CRITERIA

In deciding levels of risk (in terms of health and safety, and also financial loss) there is inevitably some choice, and consequently the views and policies of the owner and the regulatory authorities are most relevant. Until these aspects have been fully discussed and an appropriate policy formulated, the following criteria can be used on the initial stages of a project.

The principles used as a guide for the selection of risk criteria are:

(a) The increase in risk, caused by the presence of the hazardous plant, of the local community (i.e. neighbouring public) should be negligible in comparison to the risk they already face in everyday life.

(b) The workforce on the hazardous plant should be expected to accept a potentially greater risk than the members of the local community, since the workforce (unlike the local community) have elected to work in plant and more importantly can be protected against the hazards and trained to avoid them, thereby reducing the actual risk to themselves.

Various authorities in the UK[42–44] and elsewhere have made policy statements concerning acceptable or non-acceptable levels of risk to which individuals in the community may be exposed. Risk in this context is usually expressed as the probability of death for an individual in a year (8,760 hr) of exposure. The acceptable annual risk rates range as high as 10^{-2} for voluntary activities such as heavy cigarette smokers but the risk from most other normal mishaps (transport, fire, falls, etc.) range from 10^{-4} to 10^{-6}. The annual risk of death from natural causes (excluding accidents) for people in the prime of life is also about 10^{-3} and at no time does it drop below 10^{-4}. The risks from events over which people feel they have little control (involuntary risks) tend in general to be lower than risk from voluntary activities. It thus is concluded that a risk to the individual of more than 10^{-4} per year is not acceptable as a result of serious hazards incidents from any proposed process plant development. However, a risk of 1 in 10^{6} could be regarded as acceptable as a criterion applicable to individuals. For risks of death to individuals within the community near the plant, between 10^{-4} and 10^{-6} per year the plant should be examined with the view of reducing the risk.

For the individual worker; the generally recognized form of presentation for fatal accident statistics is the fatal accident rate (FAR) defined as the average number of fatalities in 10^{8} working hours. Gibson[45] has given values of FAR for various industries in the UK and the chemical industry has a FAR of 4 which is equivalent to 8×10^{-5} deaths per exposed person per year and is in agreement with Dutch experience. This is considerably better than other industries (e.g. modern coalmining FAR = 14). It is apparent that this rate has found general acceptability but includes a variety of conventional accidents of conventional kinds such as falling or being hit by falling objects. A risk assessment of process hazards would not consider such accidents. For this reason a FAR for all process malfunctions of 2 is a generally accepted target. This gives an annual risk to an employee of 4×10^{-5} from all major incidents which, in the interests of progress, we will call 10^{-5}. It should be noted that this agrees with the fact that the average worker in the chemical industry tolerates a higher risk at work than when he is in the community by a factor of at least 10. So, keeping this factor of 10, the unacceptable level for employees compared with the community can be put as above 10^{-3} per year and the acceptable level at 10^{-5} or below. Improvements should be considered where the risk is between 10^{-3} and 10^{-5} per year.

For risk of multiple fatalities there have been several studies, e.g. by the US Atomic Energy Commission[46] and Provincial Waterstraat.[47] The former plots the frequency of a given number of deaths being exceeded against the number of deaths for a number of cases of both natural and man-made disasters in the US. Such curves depend on the size of the population at risk and so, to develop criteria it is better to discuss the shape of the curve. It appears that for most disasters the frequency is inversely proportional to the number of deaths.

However, account must also be taken of society's reaction to multiple fatalities. Western society begins to be worried at 10 deaths in a single incident and finds intolerable more than 1,000 deaths in one incident, although tolerance of large numbers of single-figure fatalities is much higher than simple proportioning would indicate.

Basic statistical theory can be applied (see section B.6, p. 519) to the above in order to relate the risk of multiple fatality to the risk to the individual. Like the latter, a range of multiple risk from the acceptable to the unacceptable can be found. If an actual risk falls between the two levels, then more assessment and evaluation of the design is needed to see if the risk can be decreased towards the acceptable level.

The various values of risk to individuals ($10^{-3} - 10^{-6}$), and the critical population levels of 10 and 1,000 can be altered to accommodate the customs, practices and values of the country in which the site is located.

8.5   ASSESSMENT PROCEDURE

The layout procedure in general was outlined in section 6.6 and this section amplifies the steps concerned with hazard assessment. As with section 6.6, a new site situation is assumed and the procedure will have to be changed as appropriate for fitting new plots in existing sites.

8.5.1   STAGE ONE PLOT LAYOUT

At this stage information concerning plant design may be restricted to process flowsheets which give quantity, quality and operating conditions in equipment together with process data sheets giving basic sizing and design data for major items of equipment. A preliminary hazard and operability study of the flowsheet may have been undertaken.

As indicated in section 6.6, (p. 60) a layout incorporating process and economic considerations will be available for the hazard assessment which takes the following steps.

1. Data

All relevant data such as physical and chemical properties, flammable limits, combustion properties, toxic limits, physiological effects, etc. should be collected, collated, and recorded. As an option, Mond Index[48] or Institute of Petroleum type code[49-50] assessments can be carried out before the following steps.

2. Minor leaks and area classification

All sources of small but likely losses in reasonably normal operation should be identified. For toxic materials this study will determine ventilation requirements. For flammable fluids it will define the electrical area classification zones and the hazard areas for non-electrical ignition sources such as furnaces.

This is a very important aspect of hazard assessment and so is discussed in further detail in section 8.7.

3. Major sources of leak

Those sections which are likely to lead to major loss of containment by vessel, pipeline failure, etc. are ascertained by a review of the operating conditions and procedures. The condition of the release as listed in section 8.2.2 should be noted.

The plant should be divided into sections that are separated or can be isolated from each other by valves which can be rapidly closed in an emergency. By this means the inventories of the major sections of the plant and hence the maximum amount of material which could escape may be determined. This amount should be made as small as possible.

4. Catastrophic failure of a pressure or gas source

This can occur for conditions (B) and (C) of section 8.2.2 and would give an instantaneous vapour cloud.

The size of the cloud should be calculated and then the following sets of circles determined (see calculations in Appendix B).

(a) Distance cloud disperses to LFL in order to help determine position of distant ignition sources such as offices, housing.
(b) Overpressures contours.
(c) Radius of fireball within which vents and liquid pools will ignite.
(d) The flux from fireball in which people are at risk.
(e) Isopleths of toxicity in the open and the likely penetration into buildings of various degrees of airtightness. Reasonably airtight buildings can sustain life for a surprisingly long time.

(f) Toxic dosage at various distances.

If there is liquid residue after the failure, it should be treated as in step (6) below.

5.   Major steady leakage from a pressure or gas source

Like step (4) this can occur for conditions (B) and (C) of section 8.2.2 but gives rise to a jet which decelerates into a plume.

The rate of escape should be determined and then the following calculated, using the methods given in Appendix B.

(a) The distance the jet (or possibly the plume) takes to reach the LFL to find the position of the nearest major ignition source.
(b) Thermal flux contours for the jet fire.
(c) Isopleths of toxicity in the open and the likely penetration into buildings of various degrees of airtightness.

If there is liquid escaping which can form a pool, it should be considered as the next step (6).

6.   Failure of liquid source

This happens with condition (A) of section 8.2.2.

If the top of the container fails, the liquid will evaporate or burn from the vessel. With other failures the liquid will run out and form a pool which will then either evaporate or burn. The engineer should fix the position of the pool away from other vessels and make its area as small as possible.

From the rate of evaporation or burning, can be found:

(a) The distance the plume travels before being diluted to the LFL which marks the distance of the nearest major ignition source.
(b) Thermal flux contours for the pool fire.
(c) Isopleths of toxicity in the open and the likely penetration into buildings of varying airtightness.

The calculation methods are outlined in Appendix B.

7.   Internal plot layout

This is mainly determined, for hazards, by the zone 1 and 2 hazard area classification discussed in step (2) and by the need to locate permanent ignition sources outside zone 2.

Except in very large plots there is insufficient room to mitigate the effects of overpressure on equipment and the toxic effects. However, it will be possible to site vent discharges in order to prevent the ignition of emergency releases. It may be possible to space items to stop the spread of fire and avoid equipment collapsing on to other equipment.

The results of the studies in steps (4), (5) and (6) should be used to position the control room and other plot buildings containing personnel. Such buildings should be situated and protected to resist the expected overpressures and fire radiation and allow escape. They should not be in classification zones 0. 1 or 2 (see BS 5345[51]) and should be capable of being made airtight and have internal air supplies when there is the risk of toxic release. If location inside a classified zone is unavoidable, protection by pressurization may be used (see BS 5345).[52]

8. External plot separations

In order to prepare for site assessment, the effects of the various losses of containment within the plot should be combined. One useful way of doing this, at this level of assessment, is to think in terms of the maximum credible incident. So for overpressure, flammable and toxicity effects one should take the respective incidents that give the biggest extent of contours. However, it is possible that the whole or large parts of the plot will be on fire. This is because additional losses of containment may be caused by UVCEs or because equipment is close enough to allow spread of fire. The plot thermal radiation contours must thus be based on the flux of all the relevant plot items on fire.

It may be obvious at this stage that the contours are too extensive, i.e. the plot too hazardous. Various remedial measures described in section 8.6 may therefore be considered before proceeding to site assessment.

8.5.2 STAGE ONE SITE LAYOUT

9. Data

The data available includes a first site layout (see step (11) of section 6.6.2 (p. 62), plus the hazard assessment of the various plots in the form of contours.

10. Vulnerable plots

These should be identified and could include:

(a) Central offices and amenities.
(b) Workshops and laboratories.
(c) Central utilities.
(d) Emergency services.
(e) Main site roads etc.
(f) Key commercial plants.

Items listed in (a)–(d) could themselves be hazardous to a certain extent, especially with regard to fire and therefore their thermal radiation contours should be found. Items in (f) are treated as normal plots as already described.

11. Internal site layout

The size and arrangement of the proposed site are adjusted so that the relevant criteria of overpressure, flammability, toxicity and thermal flux are not violated at the vulnerable items. In particular, there should not be any escalation of an incident from one plot to the next (domino effect).

12. External site separations

To aid site selection hazard contours should be 'drawn' around the site.

As flammability and thermal flux are fairly local in effect, the contours will be based on the various plots placed near the edge of the site. Overpressure contours can probably be based on the maximum-sized UVCE occurring on the site, though two sources may be considered for very large sites. Similarly, the toxicity contours can probably be those of the worse case unless there are possible releases of materials with very different physiological effects.

13. Site selection

It may be found from steps (11) and (12) that the site and its surrounding 'sterile' zone are going to be too large irrespective of the location chosen. In this case, remedial actions given in section 8.6 should be applied to the appropriate plots and then the site reassessed.

The hazard assessment of the site will indicate the kind of location needed, e.g. hazardous sites cannot be put in densely populated areas.

8.5.3 STAGE TWO SITE LAYOUT

14. Data

As mentioned in section 6.6.3 (p. 64) the site layout has to be accommodated to the site purchased. This means that the plots may be in a different arrangement to that on which the Stage One site assessment was based.

In addition, the vulnerable and hazardous installations outside the site are now known.

15. External vulnerable installations

As well as the internal items listed in step (10), vulnerable items external to the site should be identified and marked on a map of the area around the site.

These could include (see Table 3.1, p. 25):

(a) Public roads, railways.
(b) Housing.
(c) Large concentrations of people such as schools, hospitals, theatres, shops, stadia, etc.
(d) Factories and public utilities.
(e) Vegetation.

16. External hazardous installations

Ideally, existing adjacent factories should provide a site hazard assessment for the use of the new site owners. How much information will be given will depend on the goodwill of the adjacent owner and on the legal requirements of the country in which the site is located. Certainly, some degree of assessment of adjacent sites must be available.

17.   Internal site layout

Step (11) is now repeated but with the additional hazard information on adjacent sites. With a given size of site it will probably be found that not all the criteria for overpressure, flammability, toxicity and thermal flux can be satisfied. At this stage these violations should be noted but not resolved.

18.   External site spacing

Hazard contours are produced as in step (12) but the contours can be laid on the map around the site giving the position of the vulnerable installations. Adjustments are made to the site layout so that most situations obey the various criteria. As in step (17) the violations should be noted.

8.5.4   STAGE TWO PLOT LAYOUT

19.   Data

A much more detailed layout is available for Stage Two assessment compared with Stage One (section 8.5.1). This detailed layout is based on more comprehensive process and project engineering designs. Deficiencies found in the required hazard properties when Stage One was undertaken, should have been largely remedied.

20.   Calculations

Step 2 on area classification, etc. and the major leak calculation steps (3)–(6) are repeated with the more reliable data.

21.   Internal plot layout

The layout should be adjusted so that the spread of fire from one item to another and the collapse of one item on to the next are limited as far as the size of the plot allows. Escape, firefighting and other emergency procedures should be accommodated in the layout.

Emergency vents should be spaced sufficiently far from ignition sources such as furnaces and electrical equipment.

The position and degree of protection of the control room and other plot personnel buildings must be consistent with the appropriate overpressure, thermal flux and toxicity criteria. If not, the probability of a loss of containment causing casualties in the control room has to be within acceptable limits. This will mean improving the engineering standards of the plant and emergency escape facilities and procedures from the buildings.

22.   External plot separations

The combination of the effects of the various losses of containment should be carried out as in step (8). However, unlike Stage One layout, the positions of adjacent plots are now known and it may be found that the forecast intensities, particularly flammability and thermal flux in the adjacent plots, are greater than the appropriate criteria. Rearrangement of the items within the plot may remedy this. Similarly, it may be found that intensities from the adjacent plots are unacceptable but that rearrangement within the plot improves the situation. For example, a vapour source and ignition source may face each other across a road separating plots. Either could be moved to the other side of their respective plots. Another example is the moving of items within the plot so that electrical classification zones do not impinge on roads or site rail spurs.

8.5.5 FINAL ASSESSMENT

23. Overall site and plot layout

With the comprehensive plot layout assessments, steps (17) and (18) should be repeated. Hopefully, most situations can be resolved within the appropriate intensity criteria. Those that are not have to be reconciled with the relevant criteria of risk. As pointed out in section 8.3.2 this can be quite lengthy as several sources of loss have to be considered, instead of the ones giving the greatest damage.

The results of the final overall assessment may well have to be submitted to the regulatory authorities, in which case expert advice should be sought in its preparation.

8.6 APPLICATION TO LAYOUT DEVELOPMENT

The preceding assessment identifies the sources of hazard and the vulnerable targets and quantifies the effects of release from the sources on the targets. It may well then be necessary to alter the process and the layout to reduce the effects on the targets to acceptable levels. The first three methods given below (sections 8.6.1–8.6.3) are concerned with reducing the intensities at the target to below the appropriate criteria. The fourth (section 8.6.4) deals with containment of the hazard to reduce the risk to the acceptable level when violation of the intensity criteria cannot be avoided.

8.6.1 ELIMINATION AND REDUCTION OF THE HAZARD

The purpose is to reduce or even eliminate the potential hazard of a loss of containment at source. Methods include:

(a) Use of less hazardous process materials, e.g. ones with lower flammability, toxicity or corrosiveness.

(b) Use of lower maximum attainable temperatures and pressures so that less material is lost. In particular, change from pressure to cryogenic storage.
(c) Reduction of inventories by having smaller vessels and pipes or installing isolating valves (see section 8.2.1) so that there is less material to escape.

8.6.2   SEPARATION OF SOURCE AND TARGET

If the hazard cannot be removed the first thought is to isolate the hazard from targets.

All spaceborne hazardous effects diminish with distance in accordance with various natural laws of which the inverse square law is typical. It follows that separation of hazard sources and vulnerable targets is of significant benefit. However, distances cannot be extended indefinitely because of the cost of efficient land utilization and because of the increased chance of loss of containment in long connecting pipelines. Consequently, it is necessary to determine distances at which the effects can be tolerated, in some cases after incorporating appropriate protection of targets.

The qualitative aspects of separation and segregation include:

(a) Large concentrations of people both on and off the site must be separated from hazardous plants.
(b) Separation of ignition sources from leakage sources is necessary.
(c) Firebreaks are wanted and they are often provided by a grid–iron road plan.
(d) Tall equipment should not fall on other equipment or buildings.
(e) Drains should not spread hazards.
(f) Large storage areas should be separated from plants.
(g) Central and emergency services should be in safe areas.

Generally, toxic and explosion effects determine site and community distances, while thermal flux and flammability limits fix site and plot distances.

As already mentioned, in order to be able to make hazard calculations, the philosophy is adopted such that if the intensity of the hazard at the target is below a certain criterion, then there is complete safety. If the intensity is above the criterion, there is complete jeopardy. In practice, such a sharp division does not exist, so calculated separation distances must not be treated as rigid but as a guide.

8.6.3   PROTECTION OF TARGET

When it is not possible to place targets sufficiently far from the hazard, protection of these vulnerable targets should be considered so that the consequences of the event may be minimized and the danger of escalation reduced.

Explosion

Methods are now available for the structural design of buildings to resist the shock loading imposed by a UVCE and a typical application in areas close to hazardous plants would be for the design of control buildings.[23] At greater distances from the sources of hazard, strengthening the design of more conventional buildings may be considered, especially those buildings containing high numbers of personnel.

Fire

Fire protection may be applied to structures still in the areas of possible pool fires. The principle is to provide a delay time, typically 1 hr, so that the structure will outlast the fire and not collapse, bringing more fuel, etc. into, and thus escalating, the incident. Conventional thermal insulation provides some fire protection. Uninsulated equipment may be protected by the provision of fixed sprays as is often done in the case of storage tanks.

The fire resistance of buildings can be increased by the elimination of wood and plastics and, in some cases, windows.

Fire protection is applied to instrument and power cables in high-risk areas because they are vulnerable to short-term fire exposure. Their integrity is often essential to achieve an ordered shutdown and their replacement is time-consuming and expensive.

Calculation methods for fire protection are given in Appendix B.

Flammability

Electrical equipment and instruments that are within possible areas of flammable cloud and plumes must be of a design appropriate to the area classification[51] (section 8.7).

Fig. 8.1 Steam curtain for dispersing flammable gas (Courtesy: ICI Petro-chemicals and Plastics Division)

Dispersion aids such as steam curtains (see Fig. 8.1 and Anon.[53]) and water walls may be considered near furnaces, etc. or leak sources to prevent the ignition of flammable clouds.

Toxicity

Control buildings may, where necessary, be designed as places of refuge from the effects of toxic leakage. This requires that the building can be sealed to a reasonable standard and has an uncontaminated air supply, either from a remote source or more usually from individual breathing apparatuses.

A normal building with window, etc. closed can give a great deal of protection. As clouds drift relatively slowly there is often time to warn the public to shut windows or to evacuate the area.

The philosophy of using criteria implies that if the criteria are met by the protection installed, then complete safety is obtained. However, the note of caution given at the end of section 8.6.2 on separation also applies to protection.

8.6.4 CONTAINMENT OF HAZARD AT SOURCE

There are various ways of reducing the chances of a hazard occurring as opposed to preventing it or reducing its size. As the hazard can still occur it is preferable not to rely completely on containment but also use the previous methods for the protection of the target and separation of target and source as well.

Operating conditions

Even if maximum attainable pressures and temperatures cannot be reduced (see 8.6.1(b)) the operating values may be lowered.

Crash shutdown and isolation

A system should be devised where potentially dangerous aberrations and leaks are quickly detected so that there can be either manual or automatic operation of valves and switches which close off heating, start emergency cooling, isolate the inventory into smaller amounts, turn off pumps, activate blowdowns, and so on.

Non-return valves

These can also help to isolate sections of the plant, but should not be considered reliable. Roughly, they can fail to act in one case in ten.

Vents and drains

Vents and drains on equipment handling hazardous materials should be designed so that the discharge is contained and directed to the appropriate disposal facility at a predetermined rate.

Ventilation

Good ventilation should be provided, particularly in buildings, where dangerous vapours can collect.

Blast walls

Unstable materials may be surrounded by blast walls as for explosives. However, the walls should not collapse onto vulnerable equipment, nor should any escaping shockwave be capable of causing damage.

Engineering

The appropriate mechanical design standards should be used for all pressure systems and the requirements for fabrication, inspection and testing should be applied rigorously. The correct materials of construction should be used, allowing for stress changes, corrosion, erosion, etc. over the whole range of operating conditions – normal, startup, shutdown and emergency. These can be identified using techniques of an operability study.[54,55] Pump and compressor seals should be engineered so that the rate of escape of process material on seal failure is restricted and, if possible, dispersed in a safe manner.

Maintenance

Regular maintenance inspections and programmes should be followed. This is especially true if corrosion is anticipated. On-line monitoring of equipment conditions should be considered.

Operation

Good-quality management and well-trained operators should be used. Good housekeeping is essential.

Modification

Plant modifications should be engineered to the same high standards as the original construction.

8.7 MINOR LEAKS AND AREA CLASSIFICATION

8.7.1 MINOR AND MAJOR HAZARD ASSESSMENT

Fundamentally, there is no difference in the basic treatment of major or minor losses of containment. Both cases consider the consequent threat of ignition, fire and toxicity. However, the difference in size means that the approach to assessment can be different in detail. In addition, most plants of all sizes will have to consider minor leaks but only large plants will involve major hazards. Consequently, it is convenient to differentiate between the two aspects.

As already discussed in this chapter (section 8.6.1), major hazards usually result from fracture of equipment or pipework. There may be a catastrophic instantaneous loss or steady leakage. The area of risk extends from the plot to the site and its environs. The probability of a major loss happening should be small.

Minor losses do not usually involve damage other than wear of packing materials or seals, etc. Examples of minor sources are given in section 8.7.3. These may be steady but are often intermittent and the probability of them happening can be quite high. The area at risk is mostly confined to the plot on which the plant stands.

However, the distinction between major and minor hazards is blurred. For example, a badly worn seal may give a bigger discharge than a pinhole leak in a pipe. Emergency relief valves have to be classified carefully. One producing a large release would give rise to a major hazard and should discharge to a scrubber, flare, etc. But one producing small intermittent releases might discharge to atmosphere and would determine area classification and ventilation requirements. Consequently, the distinction between major and minor hazards and losses should be made for each plot.

8.7.2   MINOR HAZARDS

With minor leaks it is unlikely that there will be explosions with any force, except in confined spaces in buildings. The hazards considered are thus threat of ignition, fire and toxicity.

The threat of ignition cannot be countered completely such that fire will never occur. Therefore, it is advisable to consider the amount and consequences of fire damage. The toxicity hazard may partially be eliminated by the measures for the reduction of the ignition hazard, but as acceptable toxic concentrations are nearly always much lower than the lower flammable limit extra ventilation and precautions may be needed.

Thus, in minor hazard assessment the ignition hazard is the primary consideration, with fire and toxicity important secondary factors.

8.7.3   SOURCES OF MINOR LOSS

As for major losses the leak may be gaseous, flashing liquid, or just liquid. For the first two the source is the initial point of loss but liquids may collect away from the release point to form a secondary source by evaporation. However, with minor losses the height of release and whether the release is inside a building are important factors. With a major release the volume of the building or the height may be insignificant compared with the size of the release.

Initial release points include:

(a) Seals on moving machinery.
(b) Flanges on permanent pipe connections.

(c) Temporary connections (e.g to tanker).
(d) Vents allowing breathing of tanks.
(e) Sample points.
(f) (Small) relief valves.

Evaporation from liquid spills may occur in:

(a) Open areas near the initial release.
(b) Drains.
(c) Collection pits.

Methods for calculating the discharge and evaporation rates and amount of dispersion of minor losses are given in Appendix B. For these calculations it is necessary to know the size of the leak. This information is derived from both process data (e.g. for venting) and from mechanical details such as the leak sizes of seals and flanges. Prior to calculation typical values (given in Appendix C) of zone sizes may be used for preliminary layouts.

However, before deciding which leaks merit investigation, estimates of the frequency and duration of the leak are needed. These can be found from reliability data but this may be imprecise. British Standard 5345[51] on electrical area classification recognizes this imprecision by requiring just a three-part classification of emissions or leaks as follows:

(a) *Continuous grade*: release is continuous or nearly so.
(b) *Primary grade*: release is likely to happen either regularly or at random times during normal operation.
(c) *Secondary grade*: release is unlikely to happen in normal operation and in any event will be of limited duration.

8.7.4   TARGETS

A cloud drifts harmlessly until it reacts with someone or something, which is called in general (in absence of a better word) a 'target'.

The targets for the ignition hazard are sources of ignition which can include:

(a) Sparks from electrical equipment.
(b) Hot surfaces.
(c) Naked flames.
(d) Electrostatic and other random sparking.

In addition, the targets may be fixed (plant items) or moving (vehicles, people).

The targets for toxicity are the plant personnel and for fire are personnel and equipment.

The criterion for flammability is the LFL and that for fire, thermal flux, both as for major hazards. However, for toxicity whilst the criterion for intermittent releases will be IDLH value as for major releases, for more frequent and persistent releases the LTEL may be more appropriate. However, these values are usually so low and the distances so short in a plot or building that hardly any continuous leak is acceptable.

8.7.5 DESIGN STEPS

Basically the strategy follows that given in sections 8.3.2 and 8.6 for major hazards, i.e. either by means of prevention, separation or target protection, ensure that any loss has no dangerous consequences, or by means of containment of the source reduce the risk of damage to an acceptable level. However, as the consequences of minor loss are not necessarily unacceptable, it pays to consider containment as a possible solution even though the former approach may be feasible. With major hazards, as was pointed out in section 8.3.2 the policy of containment with some finite acceptable risk is only to be considered if the approach of having no dangerous consequences proves unfeasible.

Thus step (2) of section 8.5.1 – the treatment of minor leaks and area classification in Stage One plot layout – may be amplified to give the following procedure. (It will be noted that it is an important part in the conception of the layout, as it occurs in Stage One.)

(a) Identify all likely minor sources of loss of containment such as seals, glands, flanges, temperature connections, vents and relief valves, sample points, etc. Also, identify where liquid escapes are likely to run and collect. Conditions under emergencies (loss of utilities or control, equipment failure), startup, shutdown and changing load should be considered as well as steady running.

(b) Consider the use of less dangerous materials, the lowering of operating pressures and the reduction of inventories. Consider how sources of loss can be eliminated, e.g. by replacing flanges in high-pressure/temperature or hydrogen usage with welded joints or using magnetic stirrers. Decide which relief valves and vents should discharge to a closed relief system.

(c) Using a set of typical spacings (see Appendix C) establish the areas around sources of leak which should not contain sources of ignition, both permanent and fleeting.

(d) Examine whether sources of leak can be grouped together in order to reduce overall extent of ignition-free areas.

(e) Identify all likely sources of ignition, e.g. electric equipment, furnaces, hot surfaces, sources of static, vehicles, etc.

(f) Examine whether any ignition source too near a source of loss can be moved, eliminated, or replaced by lower energy/voltage equipment.

(g) In cases where ignition and leakage sources cannot be kept apart and where the ignition source cannot be eliminated establish the size, duration and frequency of the leak and the ventilation conditions nearby. From this establish the type of protection needed around the ignition source, e.g. continuous purge or enclosures with appropriate maximum aperture size to suppress vapour ingress and flame egress. For electrical equipment the provisions of BS 5345[52] should be followed (see section 8.7.6).

(h) Calculate the rate of leakage and the dispersion to the lower flammable limit in order to check the separation distances used in step (d) or the degree of protection used in step (g) (see section B.8, p. 529).

(i) Calculate the dispersion distances to the acceptable toxic levels (section B.8, p. 529). For the steady leaks this could be the LTEL, and for intermittent ones IDLH value. If these distances mean that personnel are at risk it may be necessary to increase ventilation or improve the engin-

130

eering of the potential leak in order to reduce the leakage rates, frequencies, etc.

(j)   Examine the effect of the leaks and pools catching fire using thermal radiation calculations (section B.8, p. 529). These calculations may show that to prevent the spread of fire the separation distances are too small or that equipment needs protecting by insulation or with water sprays.

(k)   Where sufficient distance or protection cannot be achieved the engineering design of the equipment has to be improved so that the probability of loss of containment and therefore of damage by fire is consistent with losses that can be tolerated.

(l)   Even where sufficient distance or protection can be achieved it is sensible to check whether, providing there is no threat to personnel, it is more economical to carry the risk of having to replace equipment and lost production after a fire rather than to install expensive preventative measures.

(m) Hazards originating in adjacent plants must also be considered. However, it may not be possible to identify fully all sources, either on or off the plant until Stage Two plot layout (section 8.5.4). In these circumstances appropriate locations should be given so as to give guidance, e.g. for electrical installations, and 'firmed up' at a later date.

8.7.6   EXTENSION FROM ELECTRICAL AREA CLASSIFICATION

Traditionally, consideration of minor leaks was mostly confined to electrical matters and ignored non-electrical ignition sources such as furnaces and also did not consider toxicity and fire. Consequently, area classification was developed[56,57] as a mechanism allowing the selection of the appropriate types of electrical apparatus (and their correct use and maintenance) in areas where flammable materials are generated, prepared, processed, handled, stored or otherwise encountered. Until recently this subject has been the province of the electrical engineer but it is now recognized[58] that considerable process knowledge is needed in order to predict plant behaviour under various conditions and therefore to define classification areas satisfactorily. Thus current practice is that area classification should be carried out by a process engineer assisted by other engineers (such as safety, control, maintenance, electrical) who are familiar with all the likely risks, both electrical and non-electrical. The process engineer will ensure that area classification agreed will be formally recorded (e.g. on a plot plan showing zones and non-hazardous areas (see Fig. 8.2)). When the range extends across plot and site boundaries, this should also be formally recorded.

The classification scheme was outlined in section 6.4.1 and full details are given in BS 5345.[51] Its use means that neither the probability nor the duration or release have to be precisely calculated but put instead into three grades, continuous, primary and secondary (section 8.7.3). This is an advantage with our present lack of knowledge of reliability of plant. Thus, it can be useful to extend informally the classification schemes to toxicity and thermal flux (steps (h)–(l) of section 8.7.5). For example, a high toxic risk zone ($\equiv$ Zone 0) would cover regions above the LTEL inside tanks or around continu-

Fig. 8.2 Part of an area classification drawing (a) plan (b) elevations (Both courtesy: Humphreys & Glasgow)

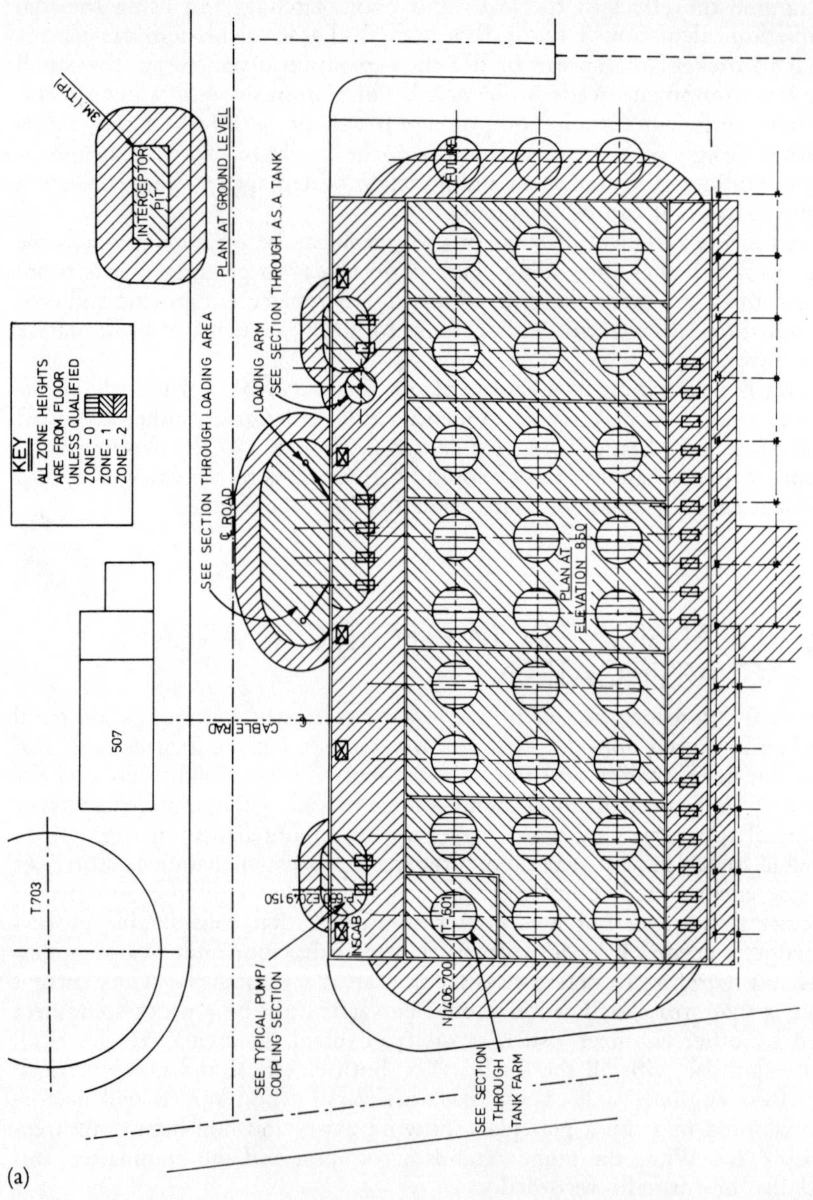

(a)

ous small leaks whilst medium- and low-toxic-risk zones ($\equiv$ Zones 1 and 2) would encompass regions above the IDLH value emanating respectively from intermittent small Gland infrequent large releases.

Thermal zones can be related to particular equipment and the heat they can withstand. For example, a high thermal risk zone ($\equiv$ Zone 0) would cover regions near where flammable fluids are always present which if ignited would burn with sufficient flux and sufficient time to destroy or badly dam-

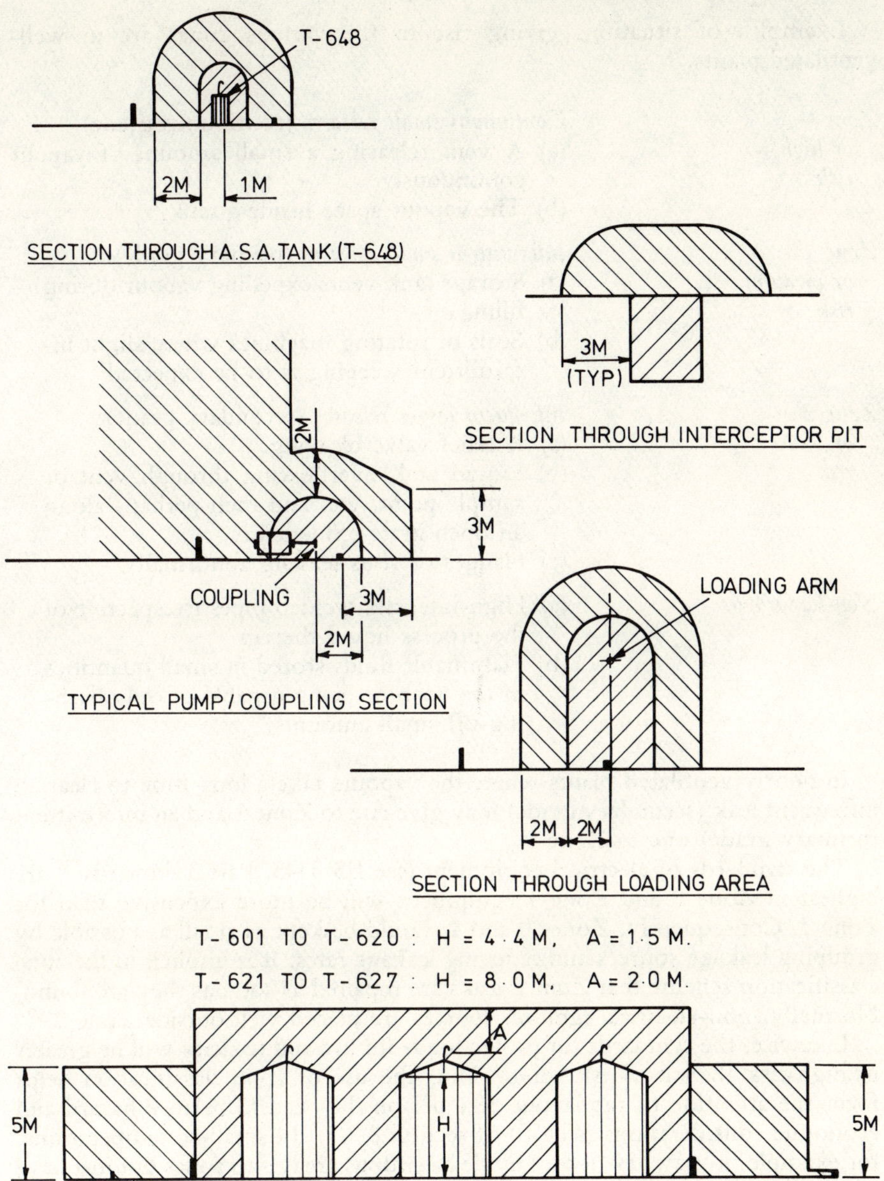

(b)

age the equipment. Medium- and low-fire-risk zones ($\equiv$ Zones 1 and 2) would cover regions with similar fluxes and times but result from the ignition of intermittent small or infrequent large releases respectively.

Thus the informal extension of the zone classification idea means that a particular zone relating to a particular leak or group of leaks has different boundaries depending on whether flammability, toxicity or thermal flux is considered.

Examples of situations giving rise to the various zones are in well-ventilated plants.[51]

Zone 0	*Continuous small releases* (continuous grade)
or high	(a) A vent releasing a small amount of vapour
risk	continuously.
	(b) The vapour space inside a tank.

Zone 0
  or high
    risk

Continuous small releases (continuous grade)
(a) A vent releasing a small amount of vapour continuously.
(b) The vapour space inside a tank.

Zone 1
  or medium
    risk

Intermittent small releases (primary grade)
(a) Storage tank vents expelling vapour during filling.
(b) Seals of rotating machines where slight intermittent weeping is to be expected.

Zone 2
  or low
    risk

Infrequent larger releases (secondary grade)
(a) A relief valve blowing.
(b) Liquid tank overflowing through vent or sample point, etc. and then perhaps along an open grate effluentway.
(c) Flanges or seals leaking abnormally.

Non-hazardous

(a) High-integrity welded pipes irrespective of the process liquid therein.
(b) Flammable fluids stored in small quantities in the open or being capable of only flashing-off small amounts.

In poorly ventilated plants where the vapours take a long time to clear an infrequent leak (secondary grade) may give rise to Zone 1 and an intermittent (primary grade) one to Zone 0.

The standards of electrical equipment (see BS 5345, Part 3 onwards[52]) are highest in Zone 0 and Zone 1 equipment will be more expensive than for Zone 2. Consequently, Zones 0 and 1 should be kept as small as possible by grouping leakage sources and reducing leakage rates. It is implicit in the zone classification scheme that Zone 2 leaks are repaired as soon as they are found. Normally, non-electrical ignition sources are placed well outside Zone 2.[59]

Likewise, the standards for protection from fire and toxicity will be greater in high-risk than low-risk zones. Thus the use of zone classification helps focus the attention of the layout designer on the need to balance hazard and economic considerations (step (l) of section 8.7.5). In small-scale operations, for example, it may pay not to divide hazardous areas into zones but just have a simple split between hazardous and non-hazardous areas. All ignition sources and any equipment that must not be damaged would be in the non-hazardous areas.

For minor leaks explosions are unlikely in the open, and internally explosions are so destructive[60,61] that the idea of zoning is not applicable except that barricades[62] may be erected in the latter case.

The subject of area classification[58] and, indeed, hazard assessment in general is constantly developing and engineers will need to be aware of the current state of progress.

REFERENCES

1. I.Chem.E. 'Chemical process hazards VII', *I.Chem.E. Sym. Ser.* **58**, 1980.
2. I.Chem.E. 'The assessment of major hazards', *I.Chem.E. Sym. Ser.* **71**, 1982.
3. EFCE 'Fourth international symposium on loss prevention, Harrogate', *I.Chem.E. Sym. Ser.* **80**, 1983.
4. Systems Reliability Directorate, 'Unclassified Reports', United Kingdom Atomic Energy Authority, Wigshaw Lane, Culcheth, Warrington, WA3 4NE.
5. Kaiser, G. D. 'The accidental release of anhydrous ammonia to the atmosphere. Evidence on the occurrence of denser-than-air vapours', I.Chem.E., *Loss Prevention Bulletin*, **38**, 1, 1981.
6. Britter, R. E. and Griffiths, R. F. (eds) *Dense Phase Dispersion*. Elsevier Scientific Publishing Co., Amsterdam, 1982.
7. Sutton, O. G. *Micrometeorology*. McGraw-Hill, 1953.
8. Pasquill, F. *Atmospheric Diffusion*. Ellis-Horwood, 1974.
9. Cude, A. L. 'Dispersion of gases vented to atmosphere by relief valves', *Chem. Engr. Lond.*, **290**, 629, 1974.
10. NRPB–R91 'A model for the short and medium range dispersion of radio nuclides released to the atmosphere', National Radiological Protection Board, Harwell, 1977.
11. Simpson, H. G. 'Design for loss prevention – plant layout', *I.Chem.E. Sym. Ser.* **34**, 105, 1971.
12. Clancey, V. J. 'The development of hazardous clouds', Second International Symposium on Loss Prevention and Safety Promotion in the Process Industries, Heidelberg, Dechema, 323, 1977.
13. Lee, J. H., Guirao, C. M., Chiu, J. W. and Bach, G. G. 'Blast effects from vapour cloud explosions', A.I.Ch.E., *Loss Prevention* **11**, 59, 1977.
14. Clancey, V. J. 'Dangerous clouds, their growth and explosive properties', *I.Chem.E. Sym. Ser.* **49**, 111, 1977.
15. Brasie, W. C. and Simpson, D. W. 'Guidelines for estimating damage from chemical explosions, A.I.Ch.E. *Loss Prevention* **2**, 91, 1968.
16. Strehlow, R. A. 'Unconfined vapour cloud explosions – an overview', 14th International Symposium on Combustion, The Combustion Institute, Pittsburgh, Pa., 1189, 1973.
17. Davenport, J. A. 'A survey of vapor cloud incidents', A.I.Ch.E. *Loss Prevention* **11**, 39, 1977.
18. Anon. 'Protecting liquid petroleum gas vessels from fire', *The Engineer*, **221**, 475, 1966.
19. CRC *Handbook of Chemistry and Physics*. CRC Press, revised annually.
20. Comer, P. J., Cox, R. A. and Sylvester-Evans, R. 'A formalized procedure for the analysis of risks imposed by complex petrochemical developments', *Bull. of Inst. Maths. and its Applic.*, **16**, 121, 1980.
21. Symposium 'Design of chemical and nuclear installations against impacts from plant generated missiles', *Nuclear Energy*, **19**, 149, 1980.
22. Walker, F. E. *Estimating Production and Repair Effort in Blast Damaged Petroleum Refineries*. Stanford Research Institute SRI Project No. 6300–620, July 1969.

23. CIA, *An Approach to the Categorization of Process Plant Hazards and Control Building Design*. Chemical Industries Association, 1979.
24. HSC, *Second Report of the Advisory Committee on Major Hazards*, HMSO, London, 1980.
25. HSE Guidance Note EH *Occupational Exposure Limits*. HMSO, London, 1984; also ACGIH POB 1937, 'Documentation of threshold limit values for substances in workroom air', American Conference of Government Industrial Hygienists, Cincinnati, 1982.
26. Pennsylvania Dept of Health, 'Short-term limits for exposure to airborne contaminants', Penna. Dept. of Health, Chapter 4, Art. 432, Jan. 1968.
27. Frawley, J. P. 'Emergency exposure limits', *Am. Ind. Hyg. Ass. J.*, **25**, 578, 1964.
28. Ferguson, D. M. 'Short term exposure limits', *Ann. Occup. Hyg.*, **19**, 275, 1976.
29. Mackinson, F. W. (ed.), *Pocket Guide to Chemical Hazards*. US Depts of Health and Human Services/Labor, 1980.
30. Sellers, J. G. 'Quantification of toxic gas emission hazards', *I.Chem.E. Sym. Ser.* **47**, 127, 1976.
31. Green, S. D. 'The engineering response to health hazards', *I.Chem.E. Sym. Ser.* **58**, 137, 1980.
32. BS 476 'Fire tests on building materials and structures', Parts 2–8, British Standards Institution, London, 1970–81.
33. Lawson, D. I. *Fire Research Bulletin No. 1*. HMSO, London, 1954.
34. Lawson, D. I. and Simms, D. L. 'The ignition of wood by radiation', *British J. of Applied Physics*, **3**, 288, 1952.
35. May, W. G. and McQueen, W. 'Radiation from large liquefied natural gas fires', *Combustion Science Technology*, **7**, 51, 1973.
36. Buettner, K. 'Effects of extreme heat and cold on human skin: 1. Analysis of temperature changes caused by different kinds of heat application', *J. Applied Physiology*, **3**, 691, 1951.
37. Stoll, A. M. and Greene, L. C. 'Relationship between pain and tissue damage due to thermal radiation', *J. of Applied Physiology*, **14**, 373, 1959.
38. Stoll, A. M., Chianta, M. A. and Piergallini, J. R. 'Thermal conduction effects in human skin', *Aviation Space and Environmental Medicine*, **50** (8), 778, 1979.
39. API *Guide for Pressure Relief and Depressurizing Systems*. American Petroleum Institute RP–521, 1980.
40. Brzustowski, T. A. and Sommer, E. C. *Predicting Radiant Heating from Flares*. American Petroleum Institute's 38th mid-year meeting, Philadelphia, API reprint 64–73, 17 May 1973.
41. Oenbring, P. R. and Sifferman, T. R. 'Flare design – are current methods too conservative?', *Hydrocarbon Processing*, 124, May 1980.
42. HSC *First Report of the Advisory Committee on Major Hazards*, HMSO, London, 1976.
43. Webb, G. A. M. and Maclean, A. S. *Insignificant Levels of Dose: a practical suggestion for decision making*. United Kingdom National Radiological Protection Board, Report NRPB R62, Harwell, 1977.
44. United Kingdom Royal Commission on Environmental Pollution, 7 Reports, HMSO, 1971–79.
45. Gibson, S. B. 'The design of new chemical plants using hazard analysis', *I.Chem.E. Sym. Ser.* **47**, 135, 1976.

46. WASH 1400, *Reactor Safety Study. An Assessment of Accident Risks in US Commercial Nuclear Power Plants.* US Atomic Energy Commission, Washington D.C., 1974.

47. *Pollution Control and the Use of Norms in Groningen.* Provinciale Waterstraat, Groningen, 1979.

48. Lewis, D. J. 'The Mond, fire, explosion and toxicity index applied to plant layout and spacing', A.I.Ch.E. *Loss Prevention* **13**, 20, 1980.

49. Institute of Petroleum, *Model Codes for Safe Practice in the Petroleum Industry.* Applied Science Publishers, London, 1956 onwards.

50. Home Office, *Code of Practice for the Storage of Liquefied Petroleum Gas at Fixed Installations.* HMSO, London, 1971.

51. BS 5345, Part 2, 'Classification of Hazardous Areas', 'Code of practice for the selection, installation and maintenance of electrical apparatus for use in potentially explosive atmospheres (other than mining applications or explosive processing and manufacture)', British Standards Institution, 1983.

52. BS 5345 (see ref. 51), Part 3 onwards. 'Installation and maintenance requirements for electrical apparatus with various types of protection', British Standards Institution, 1983.

53. Anon. 'The containment and dispersion of gases by water sprays', *Chem. Engr. Lond.*, **376**, 25, 1982.

54. CIA, *A Guide to Hazard and Operability Studies.* Chemical Industries Association, 1977.

55. Kletz, T. A. *HAZOP & HAZAN – Notes on the identification and assessment of hazards.* I.Chem.E. Lond., 1983.

56. RoSPA/ICI, *Electrical Installations in Flammable Atmospheres*, Royal Society for the Prevention of Accidents, Report IS91, 1972.

57. IEC *Electrical Apparatus for Explosive Gas Atmospheres, Part 10: Classification of Hazard Areas.* International Electrotechnical Commission, Report 79–10, Geneva, 1972.

58. Peck, G. C. and Palles Clark, P. C. 'Flammable atmospheres and area classification', *Chem. Engr, Lond.* **377**, 39, 1982.

59. Kletz, T. A. 'Plant layout and location: some methods for taking hazardous occurrences into account', A.I.Ch.E. *Loss Prevention Bulletin* **13**, 147, 1980.

60. Maidstone, R. J. *et al. Structural Damage in Buildings Caused by Gaseous Explosions and Other Accidental Loadings 1971–1977.* Building Research Establishment Report, HMSO, 1978.

61. Baker, W. E. *et al. Explosion Hazards and Evaluation.* Elsevier Scientific Publishing Co., Amsterdam, 1983.

62. Tunkel, S. J. 'Barricade design criteria', *Chemical Engineering Progress*, **50**, Sept. 1983.

PART II

DETAILED SITE AND PLOT LAYOUT

TRANSPORTATION

9.1 GENERAL

An essential consideration in site layout is the transportation network for the movement into, out of, and around the site of materials, personnel and emergency traffic.[1,2] A good network optimizes these movements, taking into account the types of materials, traffic volumes, site operations, economics and safety.

A rectangular grid is often the most economical arrangement in a chemical plant complex[1] (Fig. 2.1, p. 7). Overhead piperacks, sewer systems, underground piping systems, trenches, electric cables and instrument lines usually follow the road configuration. Thus, curved or angular roads (creating triangular or odd-shaped plots) and dead-ends should be avoided when possible.

Materials movements to, from and within the site are commonly by road because of the flexibility and convenience of road vehicle operation. For large material flows to and from the site, however, rail transport can be more economic and is often used for coal, minerals, liquefied gases, etc. Railways within the site will impose some constraints in layout because of the large turning radii and shallow gradients needed for trains.

Bulk materials may occasionally be transported by canal or river. Such traffic and its loading arrangements will be a fixed element in planning site layout and its transport system. Rail and, perhaps, water transport to and from site, though less flexible, are likely to become increasingly important as energy costs rise and thus increase road transport costs.

Within the site itself, bulk material transport (for process materials and services) is best handled in pipes arranged in conventional piperacks. Large volumes of solids can be piped if they can be handled pneumatically or in fluid suspension. The economies of this technique may justify a size reduction stage. If piped transport of solids is impossible, conveyors will probably be necessary despite their disadvantages of inflexible layout, high capital cost and high operating costs. As they will be very inflexible in layout they must be planned at the earliest possible stage (see Ch. 29).

For small or intermittent flows of liquids or solids, batch transport by road can be economic. The site transportation network, then, will almost certainly include a road system, a piperack system and some footpaths. Rail traffic may be handled at an internal spur connected to the main line or a rail system may be needed on site.

9.2   TRANSPORT PLANNING

In most process industry site planning, materials transport influences site layout far more than considerations of personnel and emergency traffic. Thus, planning the site transportation network starts with a clear flow diagram, showing flows of all site materials and services. This is studied in conjunction with initial proposals for layout of the various plots.

Considerations of plant location, material flow and transportation methods will enable an initial site layout to be drawn up. Successive iterations will be necessary to refine the initial ideas into an acceptable layout and during these iterations, transportation considerations such as outlined below must be examined.

(a) A good site layout will minimize the distances over which materials (including services such as water, steam, etc.) have to flow, either in processing or into and out of the site.

(b) It should keep to a minimum transhipment and double handling of materials in transit from one system or storage area to another to minimize cost, delay and damage. Thus, unless there is punitive demurrage, materials are left in the delivery vehicle until offloaded at the store.

(c) A major constraint is that only traffic authorized to serve a plant should go to that plant. All other traffic should use site routes which bypass the plant with spur, layby or preferably loop connections, to the plant.

(d) The incoming/outgoing movements of raw material/product should not interact with each other, with the internal interplant transportation nor with the site personnel and services traffic. For heavy densities it may be necessary to separate the four systems. Some gain in safety will be made here because external transport drivers, who may not know the site, will be kept separate from internal traffic. However, with lower densities, segregation may not be justified but the roads will have to be sufficient in number and width to cope with the demand expected and one way traffic systems are sometimes used.

(e) Some isolated hazardous plants may have their own secondary security fences and will, therefore, need extra gate control, restricted access to routes serving these plants and facilities for controlling vehicle access. These will increase the difficulty of planning an economic site transportation network, the capital costs of road and pipe track construction and the operating costs of transport and pumping. This aspect of layout and transport planning highlights the need to ensure that plant segregation policies are based on rational, quantitative assessment of hazards so as to minimize transportation cost without prejudicing safety standards.

(f) Although materials flow is the dominant factor, the final design must give adequate emergency and pedestrian access. These points must be checked before the design is agreed.

9.3   SITE EMERGENCIES

Roads, or other access over firm ground, should be provided to allow appliances to approach within reasonable distance of each plant. The route must be kept free of obstruction and fire appliances should not have to cross railway lines. The actual distance of approach will depend on the circumstances and risk, but will be in the range 18–45 m. Adequate water supplies should be available at these places, in accordance with the potential requirement of the processes involved (see section 14.3, p. 210).

There should be a peripheral road with at least two connections to the public road system. The site roads should permit two approaches to all major fire risks, to allow for possible closure due to debris, heat, fumes or leakage.

Roadways for fire appliances, if not two lanes wide, should have passing spaces. Cul-de-sacs should be avoided but, if used, there should be adequate turning areas to preclude the need to reverse.

Each large storage of flammable material or major plant process unit should be accessible from at least two sides which should preferably be the longest opposite sides.

Besides the road dimensions given in Appendix C for operational requirements, typical dimensions required for firefighting purposes are given in Table 9.1. The dimensions given may have to be increased to suit large appliances.

Table 9.1   Typical fire appliance dimensions

Dimension	Appliance		
	Pump	Turntable	Hydraulic platform
Road width(m)	3.7	3.7	3.7
Turning circle(m)	17.0	21.0	21.0
Clearance(m)	3.7	3.7	4.0
Gate width(m)	3.0	3.0	3.0
Laden weight(t)	10.5	14.0	18.5
Standing width(m)	>3.0	4.3	5.0 (+2.1 free boom space)
From face of building (m)	—	5–11	1.9–7.5

One or more hardstandings should be provided beside each open water source to enable fire appliances to be positioned at strategic points, without blocking roadways (see section 14.3.2, p. 211). A waiting area should be allocated near each main entrance to the site as a rendezvous point for appliances, where this is warranted by the size or nature of the installation. Emergency schemes must be considered in consultation with the fire authority (see section 15.5, p. 220, and BS 5908[3]).

9.4   STORAGE LOCATION

Decisions on centralized or dispersed storage of raw materials and finished products affects transportation planning. Centralized storage may need excessive transport access and decentralizing can be a means of spreading transport more evenly around the road network and thereby increase its capacity. Storage areas must take into account the characteristics of the means of transport, the variety of materials, their potential hazards and the possibility of being loaded incorrectly or incoming material being transferred to the wrong tank, pile, etc. Storage and transportation layout must also take into account the interchange of materials between different plants on the site.

The raw material unloading facilities and the product loading areas should not interact, though it is unnecessary to have only one unloading area, one storage area, etc. The numbers depend, amongst other things, on the range, throughputs and hazard classifications of the raw materials and products and on the various manufacturing processes. The collection, storage and removal of waste products must also be accommodated. Ideally, all storage loading and unloading areas are best on the site boundary near the main gate. When hazardous or unpleasant materials are used, this will not be allowed and these facilities must then be suitably located within the site or be placed on the boundary where they will not inconvenience or put neighbours and the public at risk.

Storage areas should be adjacent to loading and unloading areas to assist in controlling these operations correctly. Storage local to plant should have an adjacent loading area and the whole storage complex should be near the plant it serves. The storage area itself should be separated from the plant by a service road or shipment area. However, closeness to plant is not feasible with hazardous materials and the unloading/loading area should be fenced-off from less hazardous areas. (Detailed discussion of fluid, bulk solids and warehouse storage and of effluent disposal is given in Chapters 10–13.)

9.5   ROADS AND PARKING AREAS

When laying out the road system it is necessary to analyse both the immediate and future traffic requirements in order to fix the routes. Important variables are widths, kerb radii and gradients which are determined by traffic density, type of vehicles and their turning circles. Data on these matters should be gathered at an early stage of planning. The advice of traffic engineering consultants is often advantageous.

Main site traffic should not go through process areas other than when at their destinations and, even then, area classifications must be heeded. Entry of vehicles, including tankers and maintenance trucks into hazardous areas should be governed by a suitable permit or control system. Roads through plant areas should therefore be laid out only for access to the plant. Ideally, the outside of a plant area should be accessible on all four sides by road. It should not be surrounded by railway tracks. Adequate access must be provided to areas where it is known that equipment or material must be brought in for maintenance purposes, e.g. the compressor house. Similar access is needed for reactors or converters where it is known that catalyst removal and replacement must be affected. Likewise, access and space for delivery and removal of drums, bags, etc. is equally important. These roads should be wide enough to permit easy manoeuvring of vehicles and mobile cranes. The entry of handling equipment into the plants from roads should not be obstructed by kerbstones, drainage ditches, pipeways at ground level, etc. Access for firefighting equipment should be accommodated (see section 9.3). Adequate turning space is needed for tankers and other supply vehicles. It is important that reversing of vehicles is avoided as many accidents, often fatal, are caused by this.

Single-carriageway roads should allow two vehicles to pass comfortably. Corner radii should suit both the inside and outside turning circles of the largest vehicles and any special loads. Blind spots, due to dips or humps in roads or to buildings, should be avoided. Pipebridges over roads should be kept to the minimum. Railway crossings, crossroads, junctions, dead-ends, right-angle bends and ramps should be as few as possible. Roads downwind of cooling towers should not be liable to water drift and freezing. Draining of roads must prevent pools forming and in cold climates there should be space at the sides for snow removed from the roads. Road widths must allow for adjacent drainage channels. Footpaths adjacent to roads should be allowed for in areas of high personnel concentration and traffic movement. Road lighting levels should be determined by identifying safety and security needs at night time.

There should be adequate parking space for vehicles waiting to load or unload, to be weighed or to receive clearance to enter or leave the site. Car and bus parks for personnel and visitors (Figs 2.1 (p. 7) and 4.2 (p. 30)) and their access roads should be in a safe area way from wind-blown dusts, etc. and outside security zones. The parking area for night-shift employees should be well lit and under observation by the gatekeeper. Factory gates should be sited so that the effect on the surrounding road system of personnel coming off duty is small. The parks should be adequately sized to avoid congestion at shift changeover. Employees should be able to reach their workplace from their entrance gate without crossing process areas.

Standard road signs should be used to assist visiting drivers. Speed control ramps may be needed for long straight sections. These must have adequate warning signs and lighting. Travel routes and regulations should be clearly defined, understood and enforced. The gate/guardhouse should afford a full view of the main road, along the fence and down the road into the plant.

9.6 RAIL TRACKS

Rail tracks within works and rail links with the national system should be laid out in consultation with the railway and regulatory authorities to meet the works' and safety requirements for raw material reception and product dispatch. The rail layout should be considered very early on in the layout of a site in conjunction with the road plan, e.g. whether whole trains go to one plant on a 'merry-go-round' system or whether trains have to be shunted for breaking and making up in sidings.

A rail system is relatively inflexible in layout and will require large radii for tracks, shallow or no gradients and substantial areas for sidings and loop lines. The railbed will need good surface or other drainage and sufficient area should be allowed for this. If a railway has to pass near a hazardous area, special ignition-free locomotives will be needed and the possibility of sparks from brakes and from wheels on the rails must be considered.

Account should be taken of any special requirements for rail tanker loading and unloading, with adequate radii at curves for the longest tanker or railcar.

Rail tracks can be an obstruction to operation and maintenance, so spurs and rail/road crossings should be kept to a minimum and it is undesirable that rail tracks cross the main entrance road at grade. Plants needing rail transport should be placed near the railway and others away from it. A plant should not be boxed-in by branch lines.

9.7   DOCKS AND WHARVES

The location of water transport terminals will depend on the size and draught of the ships, the depth of water and the nature of the local tides and currents. These factors and not those of layout convenience determine the terminal's location. These are therefore key factors in site selection. If a satisfactory layout cannot be arranged on the site near the water terminal then another site ought to be chosen.

Layout on the docks, wharves and jetties should follow normal layout principles, e.g. for fluid and bulk solids storage, pumps, conveyors and piping (Chs. 10 and 11, 23, 29 and 31 respectively). Extra facilities may be needed to provide ships with ballast, cleaning and effluent treatment facilities.[4,5]

Care must be taken to see that a major incident, e.g. fire or explosion, etc. starting on the terminal does not spread to, nor affect, the remainder of the site.

REFERENCES

1. Simpson, H. G. 'Design for loss prevention – plant layout', *I. Chem. E. Sym. Ser.* **34**, 105, 1971.
2. Balemans, A. W. M. *et al.* 'Check-list guide lines for the safe design of process plant', in: Buschmann, C. H. (ed.), *Loss Prevention and Safety Promotion in the Process Industries*. Elsevier Scientific Publishing Co., Amsterdam, 1974.

3. BS 5908 'Code of practice for fire precautions in chemical plant', British Standards Institution, London, 1980.
4. Rutherford, D. *Tanker Cargo Handling*. Charles Griffin and Co., London, 1980.
5. Various authors, 'Shipping of dangerous materials', *Chem. Engr. Lond.*, **382**, 275, 1982.

10
FLUID STORAGE

10.1   APPROACH TO DESIGN

The principal aims of the layout of fluid storage areas are, as far as is reasonably practical, to ensure that the areas are in a safe position, away from process and public areas and to contain any spillage from the vessels. In this way the risk to personnel and the public is minimised.

The practical objective is to produce the most economical layout consistent with safety, operational and maintenance requirements. If land is at a premium then the objective reduces to one of finding the minimum area, subject to the same constraints. When land is cheap, the objective becomes one of optimizing hardware (tanks, pipes, etc.) costs subject to these constraints.

A systematic approach to fluid storage is as follows:

(a) Identify the operating requirements and the engineering and site constraints. With these make a preliminary layout based on sound economics.
(b) Identify the latest relevant codes of practice and ensure compliance of the layout with them as a minimum standard.
(c) Confirm the layout by the latest available hazard assessment methods (Ch. 8).
(d) Satisfy the local planning authorities and the community to ensure planning permission.
(e) Satisfy the statutory regulatory authorities for permission to operate.

Note: If points (d) and (e) are ignored, experience has shown that unacceptable delays and costs can occur.

There are many codes of practice relevant to layout and separation of fluid storage complexes. However, up to now (1984) most separations recommended in the codes have been given either as actual distances or distances as ratios of tank diameters. In many cases the reasons for these recommendations are not given. Thus it has not been possible to check if they are suitable, or relevant, to a particular installation. In the light of current attitudes towards safety and the greater hazard potential of modern plants, future codes may contain either justification for their recommended distances, or suggested methods for justifying the distance and separations for particular installations.

Nevertheless, the present codes represent accummulated good design practices and rules from different industries. They can, indeed, be shown to have shortcomings. For example, Robertson[1] has pointed out that for layout based on UK Government codes, an incident involving fire on one storage tank of a multiple tank group must inevitably envelop other tanks in the group when both thermal radiation and wind effects are taken into account. He also showed that spacings could be reduced by optimizing tank designs, thus saving precious space. However, there has not been a large number of serious incidents in storage areas when the codes have been followed. We therefore recommend in this book the use of a consistent set of codes as a starting point for discussing storage layout in terms of hazard. Critical areas can then be strengthened by the use of hazard assessment (Ch. 8).

Probably the best and most widely used codes are the National Fire Protection Association (NFPA) codes in America[2] and the Health and Safety Executive (HSE) guidance notes in Britain.[3] Most of the major operating companies have their own in-house codes as well, built up over years of experience. The codes for storage of hydrocarbons are the most widely available.

In this book a summary of typical separations for preliminary layout is given in Appendix C. More detailed distances as well as non-layout considerations such as materials of construction will be found in the code of practice appropriate for the particular application. A check-list of storage requirements is given by Balemans.[4]

10.2 LOCATION

Tank farms should preferably be placed in the open and on one side or not more than two sides of the process plant area. This arrangement allows adequate segregation and gives the possibility to expand either the tank farm area or process plant areas at any time in the future (Fig. 10.1).

The tank farm should be placed on good load-bearing soil, otherwise considerable piling for the foundations may be needed. The tanks should preferably be above ground unless space is restricted. Underground tanks should not be under process areas.

Materials should not be stored adjacent to urban developments when there is a possibility of either an unconfined vapour cloud explosion (UVCE), a fireball, or a toxic cloud drift over the population. Such installations should be on the rural side of the site if such an aspect is available. A tank farm containing flammable materials should be surrounded by an area in which no ignition sources are allowed (plant vehicles, matches, etc.). This could correspond to an electrical classification zone[5] determined, for example, by the expelling of vapour during filling. A fire starting in a process area or, indeed, from elsewhere within and without the site should not spread to the tank farm. Similarly, a fire in a tank, or from a spill inside a bund, should not spread to adjacent plant or buildings, ignite vegetation nor injure employees or the public. These criteria will help determine the location of tank farms within the site.

For smaller plants,[6] storage tanks can be sited to suit the flow arrangements and be individually located. However, the general principles hold for separ-

Fig. 10.1 Typical tank farm location (Courtesy: BP Chemicals)

ation to accommodate ignition and fire hazards, cloud drift, etc.

The tank farm should be made secure against unauthorized entry by fencing etc.

10.3 TANK SIZE

The total quantity to be stored is set initially by process, commercial and political considerations though subsequently hazard considerations may reduce the quantity.

The minimum number of tanks is determined by operational methods, in particular whether fluids are quarantined after unloading (or before loading) while analysis is carried out. The size and frequency of transportation will be a factor here. Other factors will tend to produce a farm with a larger number of smaller tanks than the first layout. For example, on large tank farms, it would be impractical to have a large number of smaller tanks, not only because of cost but because the increased number of pipe connections would itself present an increased chance of loss of containment. However, on small installations it is often worth considering a reduction in storage inventory. The regulatory authorities are likely to require prior notification of storage installations of certain dangerous fluids, and before giving approval they would probably inspect the proposed designs.

A further limitation on tank size is the need for firefighters to be able to approach near enough to a burning tank. They either try to put out the fire or let it burn and cool adjacent tanks to prevent the fire spreading. The maximum distance a hose can throw water is about 60 m. So the firefighter should be able to stand within 60 m of the *far* side of a tank without being injured. If this is not possible then fixed unmanned foam, sprinkler, drench, steam curtain, etc. systems must be installed to such an engineering quality that they will not fail in an emergency. Ideally, these devices should be installed even with good firefighting access. It is good practice to have a prominent label on each tank stating the name of its contents, the appropriate safety label to assist firefighters and its plant reference number, e.g. T 123 (see Fig. 31.6, p. 440).

10.4 TANK SPACINGS

Tanks should be spaced so that the intensity of thermal radiation from a tank on fire is low enough to give a temperature below either the ignition (or auto reaction/polymerization) temperature of the contents, or the stress failure temperature of adjacent tanks. In addition, the tank location should be such that the radiation from possible fires outside the farm does not cause damage to tanks and fires within the bund must not cause danger to the public, nor cause vegetation fires, etc. Separation distances between tank (un)loading points should be far enough to prevent fire spreading between each point and to the tank farm. The effect of wind on the flame angle should be included in these considerations.

Tank tolerances to radiation can be increased by having the tanks clad in reflecting colours (white and silver) as opposed to absorbent colours (black), by providing insulation, by installing cooling coils and by having automatic drenching systems or steam curtains (Fig. 8.1, p. 125). These provisions can lead to reductions in tank spacing.

Tanks that have been initially laid out according to simple rules of practice such as in Appendix C should be checked to see if there are unacceptable levels of thermal radiation. Discussion on determining the acceptable levels is given in Chapter 8 and Appendix B.

Tanks may be of fixed,[7] or floating roof design. The latter have greater intrinsic safety and, for that reason, tank spacing and dimensional limitations are generally lower than those for a fixed roof tank.

For certain liquids, protection from solar radiation by burial of sunshields is either necessary or desirable. Cooling facilities may be required on occasions.

10.5 BUND AREAS

A bunded storage area is needed for slightly or non-volatile liquids (e.g. asphalt, heavy fuel oil, some aqueous solutions) only for effluent control. Many aqueous liquids can be treated by the site effluent plant, unless there is toxic release or violent reaction on different fluids mixing (e.g. sodium cyanide and acid). They should be guided by curbs, etc. to the effluent treatment system and not flood over walkways, roads, railways, etc. nor flow into natural waterways. Spills which cannot go to the effluent plant should be safely collected in special areas or tanks for treatment or recovery. Liquid flows from vents when tanks are accidentally overfilled should also be channelled and contained in safe locations.

Bunded areas must be used for tanks containing volatile materials (Figs. 10.2 and 10.3). Tanks are put into groups that contain compatible liquids, i.e. they must not react on mixing and they must conform by composition so that firefighting requirements are compatible. A group of compatible tanks may be further divided into small groups in separate bunded areas. This is to give access to each tank for firefighting, construction and possibly maintenance. Insurers also like small groupings in order to minimize the property at risk. Usually either single or double lines of tanks are used within a bund and 2×2 or 2×3 arrangements are common.

Tanks must not be shielded from firefighters so, for example, 3×3 arrangements are unacceptable. Access must be allowed for firefighting on two sides of each tank bund area and all roads should be linked in such a way that access is still possible should any road be cut by fire.

The free volume of the bund area is sized to contain the contents of one tank (the largest if sizes vary) plus a margin – usually 10 per cent – to allow for slopping and boiling. A more precise margin can be determined by fluid dynamic calculation and investigation. In addition, the wall of the bund must be high enough consistent with its distance from the tanks to contain liquid jets coming from side gashes. Apart from this consideration walls should not be very high. Firefighters find there is less obstruction with low walls (lower than 1.5 m). This means larger bund areas. Such arrangements in addition promote good natural ventilation and dispersion of small spills. Low walls also make for easier means of escape in emergencies without having to resort to steps and ladders (see section 10.8). On the other hand, if a bund full of liquid ignites, one with a low area (and high walls) produces a lower-intensity flame. Thus process plants, roads, etc. can be placed that much nearer to the storage area. Usually considerations lead to the choice of low walls (< 1.5 m). Ideally with land available, when tanks or their connecting pipes fail, the released liquid should be contained and drained by means of catchment ditches, barriers, slopes, etc. to a bunded area which is well away from the original tank area. Here it can be allowed to evaporate (i.e. liquefied petroleum gas (LPG) or otherwise treated, and the tanks themselves will not be affected by any subsequent fire from the spill.

An alternative of bund walls is the double-walled tank to provide complete secondary containment (see Fig. C.8, p. 575). If evaporation is undesirable the cavity is topped.

Most gases on release, due either to high molecular weight or low temperature, are denser than air and flow along the ground like liquids. So, similarly, they should be channelled to safe areas. In particular, cryogenic spills should not envelop structures that could fail due to low-temperature embrittlement.

If space is at a premium and the spill rests near the tanks then special precautions should be taken in order to stop a pool fire heating up and bursting

Fig. 10.2 Part of typical tank farm layout

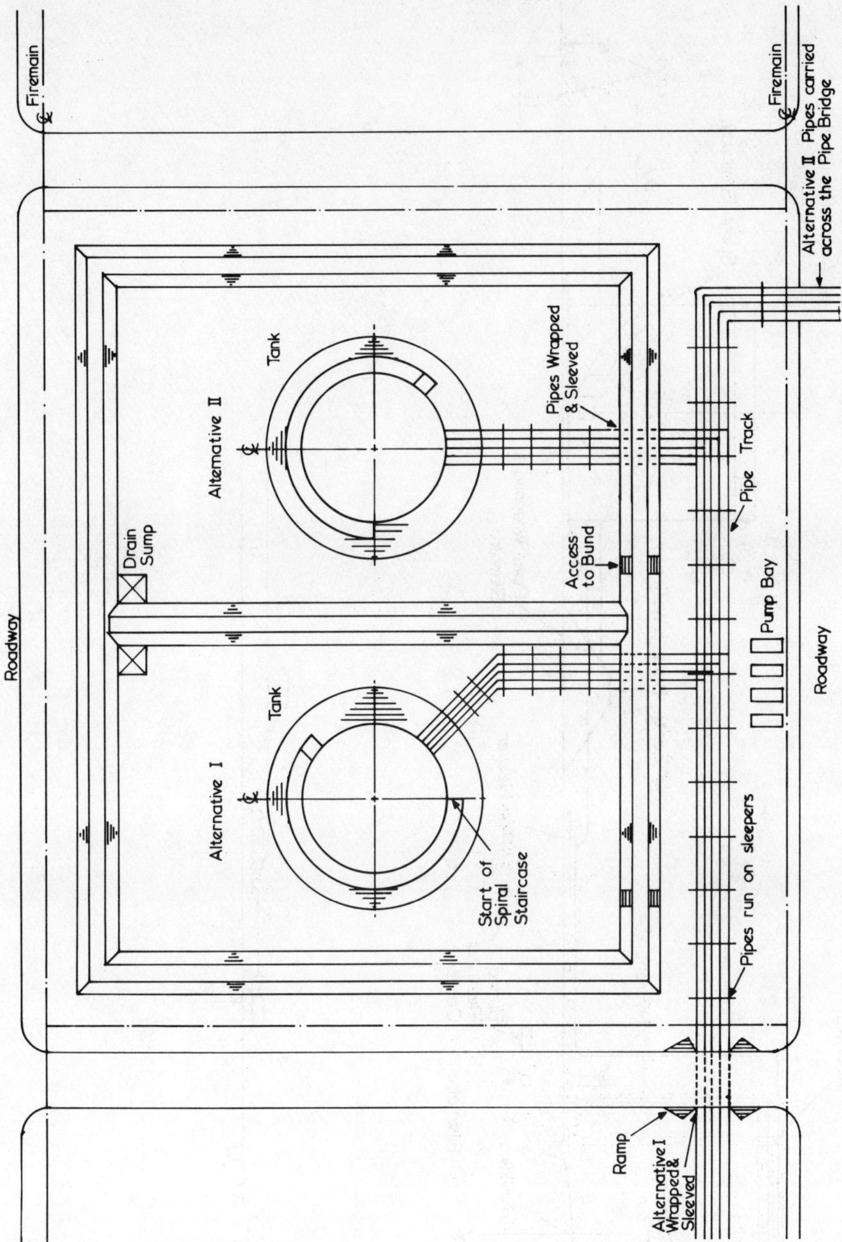

the tanks. The tanks should be vented adequately or have floating roofs to prevent build-up of vapour pressure. The relief valve should be set to take account of the fact that high temperatures may result in tank failure at pressures lower than the design value based on ambient temperature. If the end of the vent is significantly higher than the base of the tank then the tank must

Fig. 10.3 Typical cross-sections through tank farm

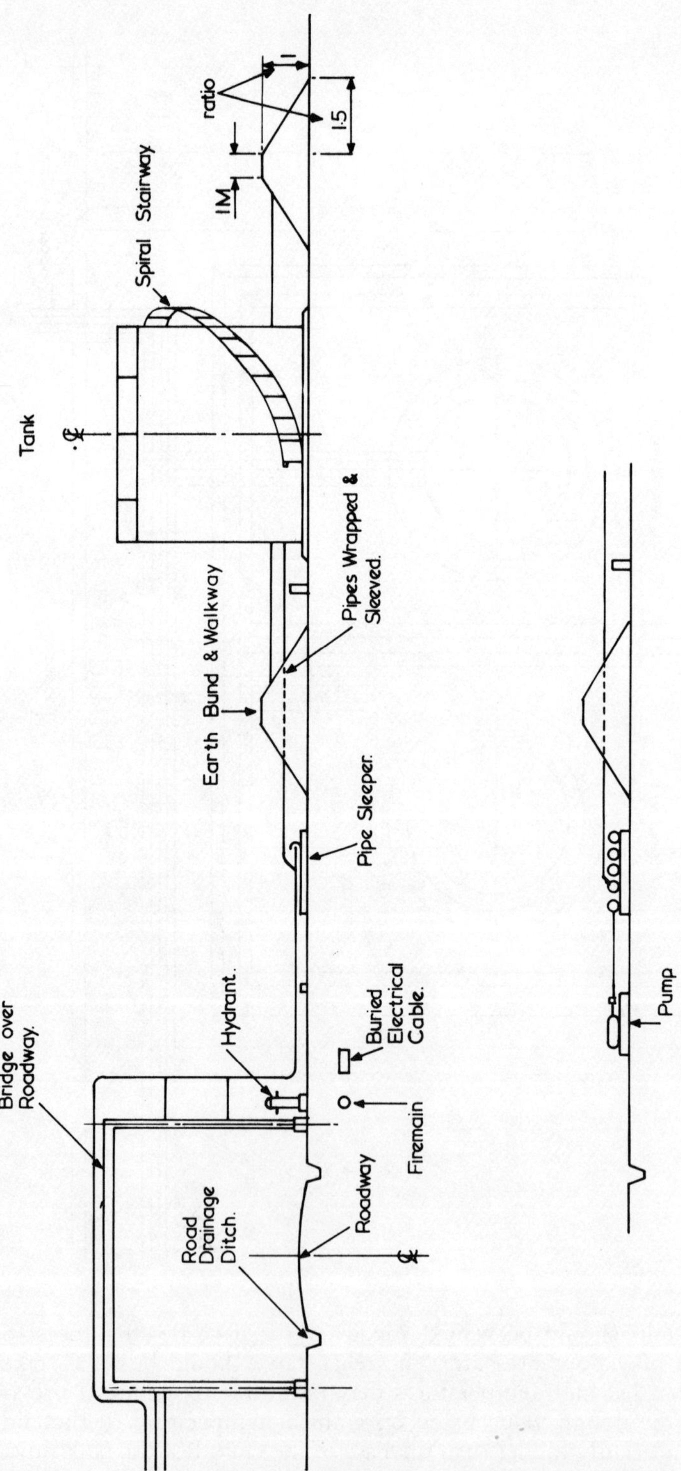

be designed to withstand the hydrostatic pressure if the vent is accidently filled with liquid.

All drains from bunds should be equipped with a value outside the bund regardless of whether the drainage goes to a ditch or sewer system. This prevents liquid spills from entering the sewer or being released from the bund area. These valves should be kept closed and blanked off except when withdrawing water. Better still, instead of a valve a pump powered by hand, steam, air or (flame proof if with a flammable substance) electricity should be used to raise the water over the bund wall. Whether valve or pump is used, surface water from areas storing flammable, water immiscible liquids should be routed via an interceptor before entering the drains (Fig. 13.2, p. 196).

10.6   BUND AND TANK CONSTRUCTION

The planning of the storage areas should take account of ground slopes and contours. It is sometimes possible to provide a bund wall on only three sides of the tank if the slope of the ground is sufficient. It can sometimes be more economical to use a bund wall circular in plan to follow the profile of the tank. Bund walls are constructed to withstand the hydrostatic head when full and can be made of concrete or earth. The latter are designed such that in cross-section (Fig. C.1, p. 566), the two sides have a 1 in 1.5 slope and a 1 m-wide walkway along the top. Aesthetically it is desirable to have all tanks requiring the same bund height in the same area, subject to compatibility, etc.

Given reasonable soil conditions it is normal practice to locate a tank on sealed mounds. A walkway should be allowed around the tank base. If soil conditions are poor or if settlement is critical, a piled concrete foundation may be necessary. In all foundations it is recommended that the tank is elevated a minimum of 0.3 m above the surrounding ground level to ensure it is clear of surface water. Support legs for spherical tanks, etc. should be insulated against fire. It is important that all tanks must be secured so that they can withstand the up-thrust, even if empty, when the bund is flooded, e.g. with storm or firefighting water.

10.7   PIPES AND PUMPS

The design of all piping connected to storage tanks must be arranged to allow for tank settlement and thermal and seismic movement. The first pipe support must be at a sufficient distance from the tank to allow for settlement or movement, without overstressing the pipe or tank nozzle. The vertical distance between tank outlet and pump inlet should ensure flooded suction and good drainage is maintained after the anticipated tank settlement. Tanks should not be raised to provide gravity discharge because of the difficulties of stopping flows under fire conditions.

Overhead piperacks should be kept to a minimum and pipes should be run in banks at grade on sleepers (Figs. 10.2 (p. 153) and 31.6 (p. 440)). The sleeper height is frequently determined by the requirements for draining and trapping lines but in no case should be lower than 300 mm. Trenched piping should be avoided unless there is no risk of flammable vapours collecting. Where pipes cross an access road, they should be culverted and steps should be provided where they cross walkways. Piping passing through bund walls should have sealed sleeves. Chapter 31 discusses piping further.

Pumps related to storage can be in groups or located individually to serve one or two tanks. Groups of pumps will facilitate centralized operation but may require costly long suction piperuns (Ch. 23). Lines carrying hot or flammable materials should be as short as possible, consistent with accommodating thermal stresses.

Pumps and manual emergency valves should be situated outside the bund area where they can be more easily operated and be in areas protected from fire and flood. For toxic or flammable liquids the pumps should preferably be in the open or at least in well-ventilated structures (section 23.2, p. 336). Emergency valves that have to be within bunds should be remotely controlled and quick-acting. As they, and the pipework, can be destroyed in a fire they should be insulated and also backed-up by valves outside the bund. Valves operating installed water and foam spray systems should always be in a safe location outside the bund.

For multiple product storage where the products vary according to seasonal or other changes in demand, layout is important to prevent accidental mixing of two products and to permit flushing and cleaning of tanks and pipes. In these cases individual tanks might not be hard-piped to the production plants nor to the tanker or container-filling points. A number of lines are run from the production area to the storage area and from the storage area to the filling area and then connected-up with flexible hoses according to production requirements (section 19.2.1, p. 289 on batch reactors). The filling pumps, whilst situated near the tanks should be controlled from the filling points. With such schemes care is needed if the fluid is hazardous. Also it is easy to get lines crossed so fixed piping with removable sections and/or swinging bends might be preferable in some cases.

10.8   ACCESS WITHIN BUNDS

There should be adequate accessways (Fig. 10.4), over the bund wall with at least two means of escape and lighting should be provided within the bund area. Many incidents with storage tanks occur during maintenance, i.e. while men are working inside the bunded area. The easiest and simplest route is out over the bund wall, if it is not too high. Steps and ladders have to be found – which is not always easy in an emergency situation.

Access is required to the roof of tanks for manholes, dip hatches, valves and instruments. On smaller tanks straight access up to 10.5 m in height is normal as shown in Fig. 10.4 but, for larger storage tanks spiral staircases

Fig. 10.4 Access over bund wall (Courtesy: The Boots Company)

shown in Figs. 10.2, 10.3 and 31.6 (p. 440) are used with connecting platforms for adjacent tanks.

If storage tanks have heating or cooling coils, maintenance space between bund wall and tank is needed for coil withdrawal and replacement.

10.9   LOADING AREAS

The layout of loading areas must encourage safe working practices. A tanker when in position for filling from, or discharging to, tanks must be in the open and not in a walled enclosure from which the escape of liquid or heavy vapour is restricted (Figs. 10.5 and 10.6).

The loading area required both for road and rail will be made up of the area for operating platform and equipment plus the area for the tankers. As an example for a loading island with, say two or three loading arms, the equipment area would be 7.5 m × 1.8 m with 3 m each side of the island for the vehicles to stand. The dimensions of the tankers to be used, both immediately and during the life of the installation must be ascertained. There should be adequate platform access to the top of tankers. The various pipelines must be identified clearly with material and destination.

Fig. 10.5 Model of loading area layout (Courtesy: Petrocarbon Developments)

Fig. 10.6 Example of moderate-size tanker unloading area (Courtesy: The Boots Company)

Spillages from both the loading and parking areas should be guided to safe bunded areas where a decision on disposal can be taken. Hosing-down facilities are usually required. Kerbs and barriers must be provided to prevent a tanker damaging any part of the installation or other tankers. Operating personnel should have unobstructed freedom of movement and a good overall view of the whole loading operation from the control point. Adequate lighting must be provided. The whole loading and parking areas should be considered with the total tank farm area in planning electrical area classification (see Appendix C), firefighting and emergency facilities. In addition, as mentioned in section 10.4, fire should not be able to spread between tanks and the (un)loading areas.

The materials of construction for loading and parking areas must be non-porous and resistant to chemical or solvent action by the fluids concerned.

10.10 OUTDOOR DRUM STORAGE

Drum storage should not be used if tank storage is practical.

10.10.1 STACK SIZE

A permissible fire load for outdoor drum storage has been defined as that which is equivalent to the limit acceptable for storing timber in the open at docks and timber yards, namely, 75×10^8 kJ. This is equivalent to 250 t of hydrocarbon.

For drums stored on end, a stack height of 4.5 m can be built with a forklift truck; this allows stacking of 0.18 m^3 drums five-high if there is drum-to-drum contact, and four-high if they are on pallets (Fig. 10.7).However, if the drums contain dangerous (i.e. flammable, toxic or corrosive) liquids non-palletized storage should be restricted to three-high in order to remove leaking drums easily.[7] It is suggested that drums on their sides be piled to a maximum of four-high (three-high for dangerous materials) since, unless they are carefully chocked, a 'runaway' situation may occur when drums are being removed from the stack. This would also reduce the possibility of bouncing drums over the others down the side of a pile, which could rupture them.

Fig. 10.7 Example of outside palletized storage

Aisle	4.0 m★
Storage	2.4 m
Storage	2.4 m
Aisle	4.0 m
Storage	2.4 m
Storage	2.4 m
Aisle	4.0 m

4 drums/pallet
4 pallets high
1.35 m between pallet centres along storage rack.

★Fire protection may require 5 m or more

The number of 0.18 m³ drums stored in a stack should not exceed 1,500. To allow for operational and firefighting access, no stack of drums should be larger in area than 230 m² and, in general, the maximum length of any side should be 18.5 m. If, however, it is desirable for the stack to be long and narrow, the length may be extended to 30 m. These dimensions should be reduced for particularly hazardous materials such as carbon disulphide or ether.

Demarcation lines showing the limits of the stacking area outside which drums should not be stacked should be marked on the ground.

10.10.2   STACK SEGREGATION

Highly flammable liquids (i.e. those with a flash point below 32 °C), must be stored in a properly constructed and locked drum storage area which should be segregated from sources of ignition, and other combustible materials. The area is generally classified as Zone 2 but becomes Zone 1 if drums are filled (see Fig. C.11, p. 577). Every drum should be labelled and checks should be made to ensure conformity with current legal requirements.[3,7,8] Drum storage areas, especially those containing highly flammable liquids, should be remote from any building or plant and stacks of drums should not be located beneath pipebridges or cable runs. Fire should not spread between stacks nor between stack and plants, buildings or the site boundary. Fires in both drums and in spillage areas should be considered. The effects of exploding drums must be examined. It should be possible for firefighting water jets to reach stacks from a safe distance to the benefit of the firefighters.

All stacking areas shall be sloped, drained and have impervious surfaces. The direction of slope will depend on how it is proposed to arrange the drums within the stack. Where there are adjacent stacks, the slope should be towards the central access between the stacks. When the boundary of the stack is adjacent to ground sloping away from the stack, the potential hazard of liquid escape in the direction of the slope must be estimated and measures (e.g. diversions, sills, etc.) taken to prevent them if necessary. Consideration should be given to see if it is desirable to put a bund around the storage area.

For preliminary layout purposes, a number of suggested separation distances are given as follows.

Stacks of drums containing highly flammable liquids should not be placed nearer than 7.5 m to any working building, amenity building or plant so that adequate firebreaks are provided. Similarly, stacks of drums containing combustible materials (having a flash point above 32 °C) should not be placed nearer than 4 m to any such building. Likewise, where a stack of drums is located near the boundary of the plant, there should be a clear space of 7.5 m between the stack and the boundary fence if the stack contains highly flammable liquids, or 4 m if the stack contains combustible liquids. For detailed layout the separation distances quoted above should be checked by thermal radiation calculations discussed in Appendix B.

If highly toxic materials are stored, the distance of drift of the vapour cloud until dispersion below the toxic limit should be estimated (again, see Ch. 8).

10.10.3 FIREFIGHTING

The above separation dimensions allow foam or water to be applied across the stack area from any peripheral point, allowing for the 'stand back' position of the firemen and a reasonable water pressure at the nozzle of the hose. Access for firefighting should be on at least three sides of the stack.

At least 5 m should be left clear between adjacent stacks of drums to provide access for firefighting. This distance should be increased for particularly hazardous materials.

No stack of drums should be situated so close to a fire hydrant that a fire would prevent its use. Clear, preferably straight-through, access must be available on two sides of the hydrant and room on the fourth side must be left clear for fire appliances and equipment. There should be an adequate number of fire hydrants around the drum storage area connected to the firefighting main. Depending upon the fire risk and the facilities available, consideration needs be given to the provision of fixed monitors for drenching the stacks with water.

Stacking areas should be arranged so as to leave space for possible hose runs in the event of a fire on any drum stack in the area.

Portable dry-powder fire extinguishers should be placed around the drum storage area. An absorbent material should be provided for soaking-up spillages.

REFERENCES

1. Robertson, R. B. 'Spacing in chemical plant design against loss of fire', *I. Chem. E. Sym. Ser.* **47**, 157, 1976.
2. NFPA National Fire Protection Association, Boston, Mass.
3. HSE *Safety in Stacking of Materials*. Health and Safety at Work Booklet No. 47, HMSO, London, 1975.
4. Balemans, A. W. M. *et al*. 'Check-list guidelines for the safe design of process plants', in: Buschmann, C. H. (ed.) *Loss Prevention and Safety Promotion in the Process Industries*. Elsevier Scientific Publishing Company, Amsterdam, 1974.
5. BS 5345, 'Code of practice for the selection, installation and maintenance of electrical apparatus for use in potentially explosive atmospheres (other than mining applications of explosive processing and manufacture)', (1983), Part 2, 'Classification of Hazardous Areas', British Standards Institution, London.
6. BS 2654, 'Vertical steel welded storage tanks with buttwelded shells for the petroleum industry', British Standards Institution, London, 1973.
7. HSE *The Storage of Highly Flammable Liquids*. Guidance Note CS2, HMSO, London 1978.
8. Home Office *Code of Practice for the Keeping of Liquefied Petroleum Gas in Cylinders and Similar Containers*. HMSO, London, 1973.

11

BULK SOLIDS STORAGE

The layout of bulk solids storage depends upon the type of transportation used for delivery to, and removal from, storage area, and the method adopted for store loading and unloading. An example is shown in Fig. 11.1.

The main storage systems are:

(a) Open stockpiles for materials unaffected by weather.
(b) Stockpiles within closed warehouses.
(c) Bunkers, including bins, hoppers and silos.

With all systems, the design requirements given in Chapter 26 should be followed.

11.1   BULK SOLIDS INTAKE

Bulk solids arrive at the factory by road, rail or sea, except for some mining- and quarrying-type operations, which can be direct linked by conveyor. Factory intake systems from these incoming routes have a number of common layout features which are essential for economic operation.

To minimize demurrage costs, space should always be available to receive incoming transport in truck parks, rail sidings, or at quaysides, and for parking empty transport after discharge. Space should also be available in bulk stores to receive incoming consignments at the maximum rate of discharge to avoid demurrage costs.

Buffer storage should be provided where unloading operations are interrupted, such as movement of trucks or rail wagons to and from their discharge station, or the intermittent operation of a grabbing crane. This allows for a maximum dumping rate, whilst economizing on the size of downstream handling equipment by regulating material flow into the bulk stores at a controlled rate.

Ideally, intake equipment should be capable of handling materials whose flow characteristics may change during transit by compaction, chemical reaction or the effects of weather. However, designing for this flexibility is

Fig. 11.1 Example of a layout of a bulk solids handling plant (Courtesy: Walker Engineering)

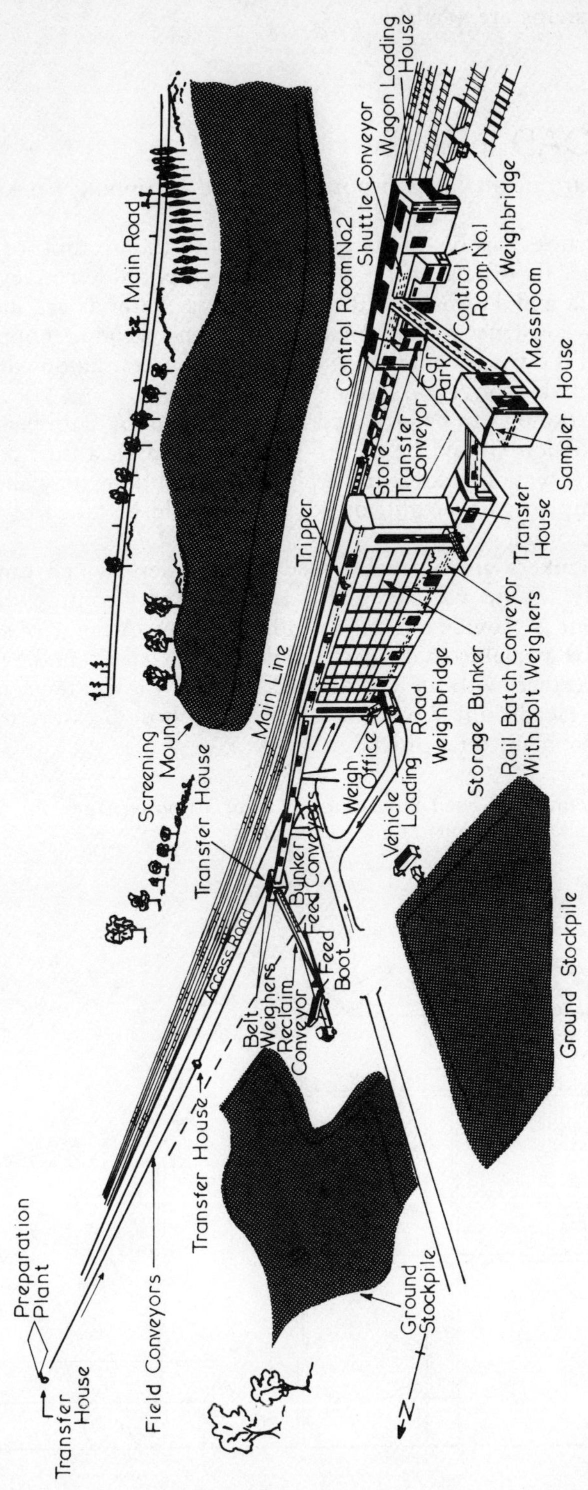

difficult in practice. Weighing the incoming material will facilitate mass balancing, purchasing and stock control, and both batch and continuous-belt weighing systems are available.

11.1.1 ROAD INTAKE

Bulk solids are normally transported by road in tipping trucks or pressure tankers.

Tipping trucks can be discharged direct onto the ground, or, by controlling the outlet flow gate on the vehicle, into a portable conveyor as shown in Figs. 11.2a and 11.5b (p. 170). Where these methods are unsuitable it is necessary to construct a pit containing a batch-receiving hopper, which is sized to hold the truck contents, and arranged for continuous discharge onto the factory intake conveyor system (Fig. 11.2b).

Weatherproofing can be provided by using a canopy with flexible curtains, and dust emission should be controlled by providing a dust-collecting unit (Fig. 11.2b). To avoid excavating a pit, local conditions may allow construction of a ramp to gain height, or advantage could be taken of site contours on sloping ground.

Pressure tankers are normally used for powders which can be handled pneumatically. Some vehicles each have their own compressor, otherwise compressed air is provided at the unloading station. A range of suitable adaptors should be available to connect the tanker to factory intake pipework via a suitable receiver with air filter, sized to hold the contents of the tanker (Fig. 11.2c). Headroom should be allowed for those pressure tankers which are tipped to obtain total discharge.

Fig. 11.2 Example of road unloading (a) from a tipping truck (b) from a tipping truck using a pit (c) pneumatic tanker

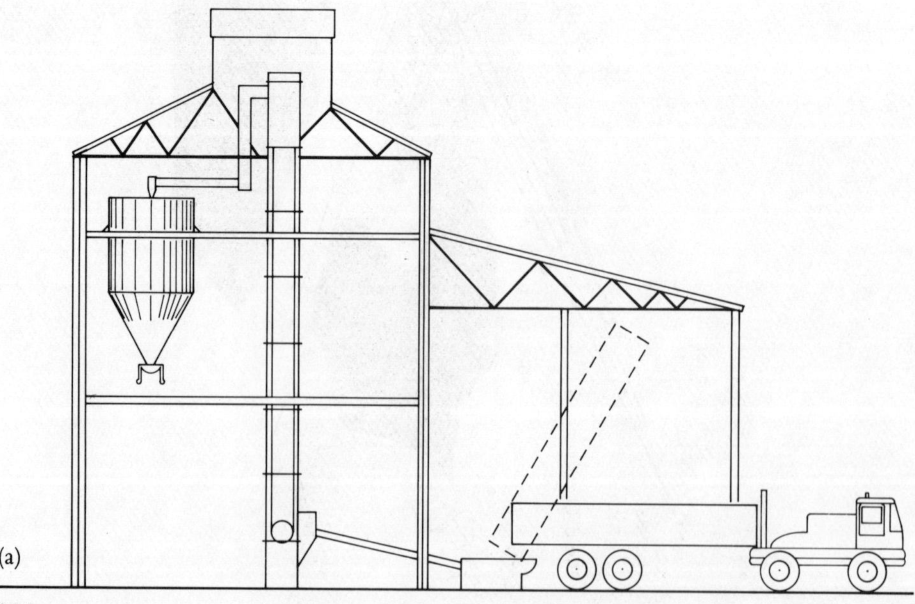

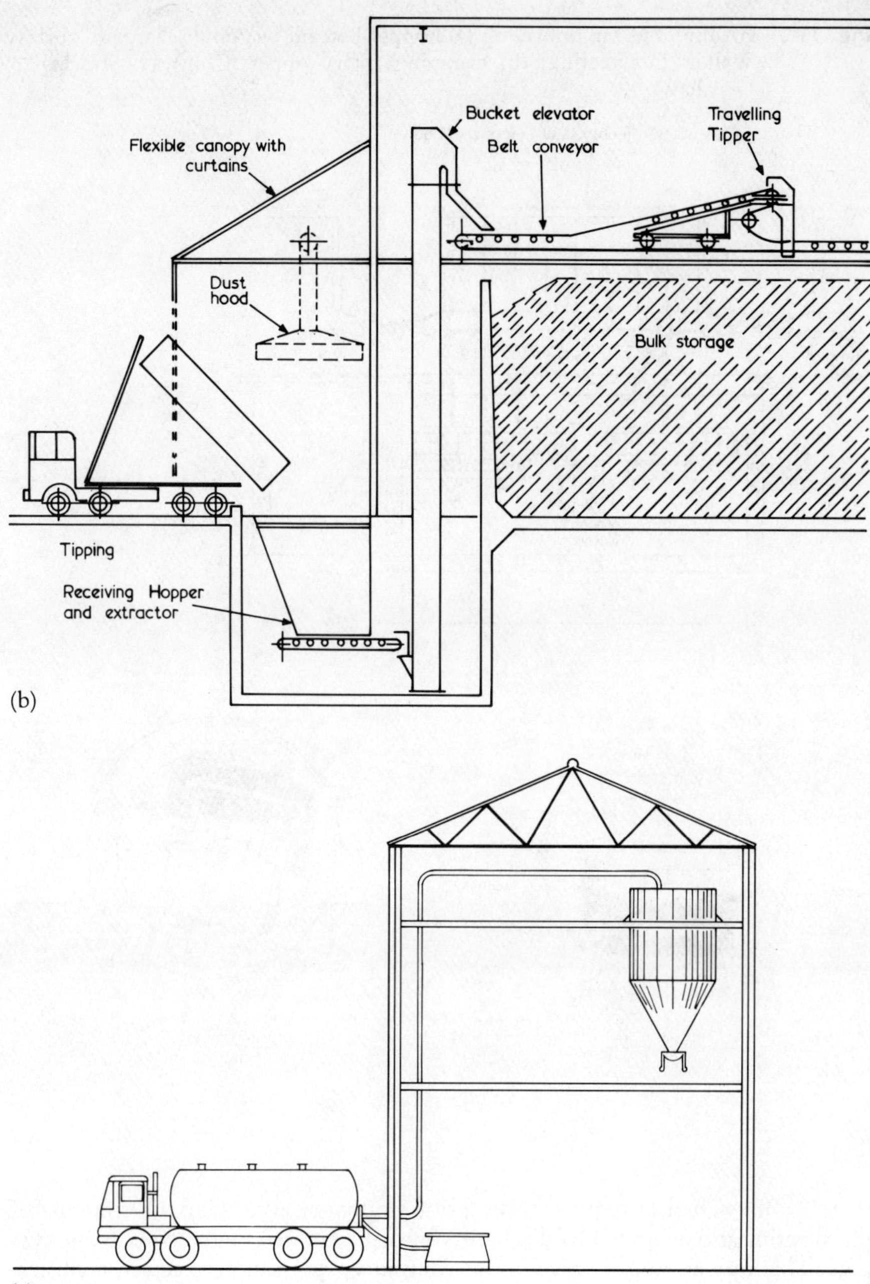

11.1.2 RAIL INTAKE

Bulk solids are normally transported by rail in open, hopper bottom discharge, or pressurized wagons.

Open wagons are usually discharged by manual clearing through side doors, by bottom discharge (Fig. 11.3a), or by inverting on a wagon tippler

Fig. 11.3 Examples of rail unloading (a) hopper bottom wagon discharge (Courtesy Walker Engineering) (b) clamping rotary tipper (Courtesy: Strachan & Henshaw)

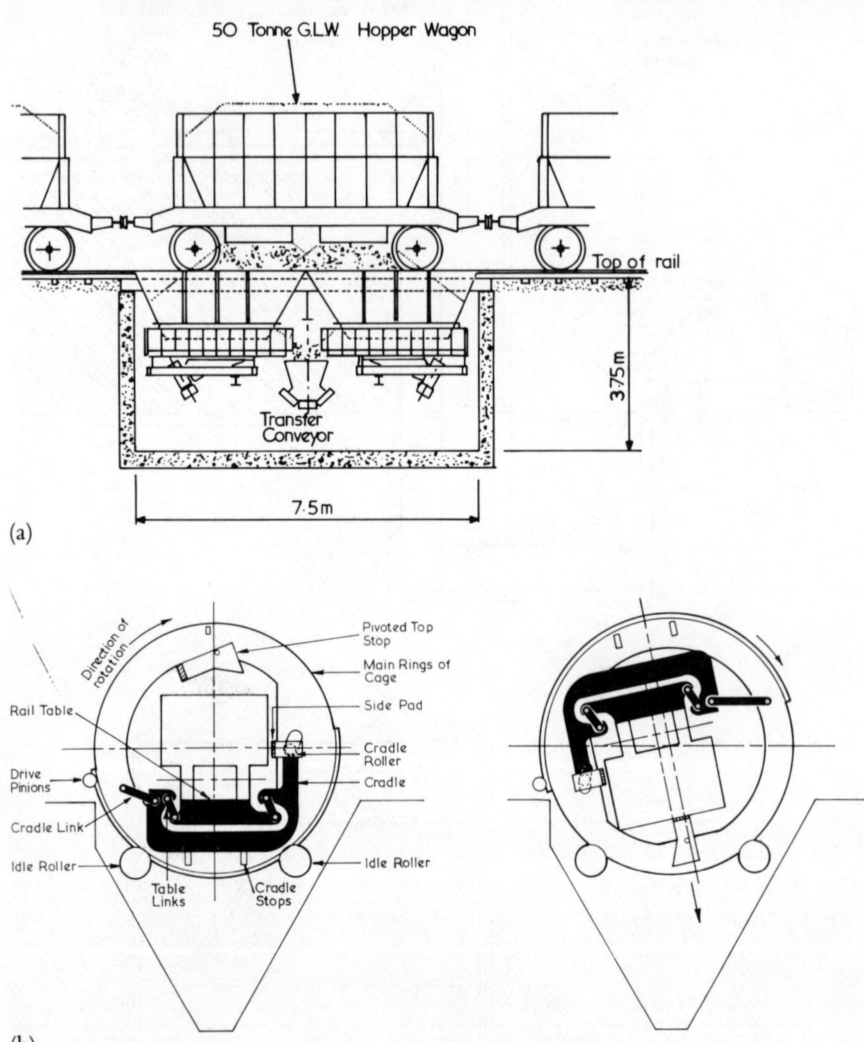

(Fig. 11.3b). The last requires a reception hopper sized to take the contents of the wagon, and arranged to discharge onto the factory intake conveying system. This operation can create a dust nuisance, requiring collection facilities. Hopper bottom wagons discharge into a reception hopper with a means of transfer to the factory intake system; a typical arrangement is shown in Fig. 11.3a. Space should be allowed for manual operation of discharge valves, although some purpose-built stocking-out systems are automatically operated to allow continuous discharge with the train in motion.

Pressurized wagons are used for powders which can be handled pneumatically, and require similar offtake facilities to road tankers. Discharge rates can be increased by using a series of takeoff points, spaced to suit the length of rail wagons. This reduces the number of movements of the train, but access is required for coupling, valve operation and maintenance.

11.1.3 SEA INTAKE

Fig. 11.4 Examples of ship unloading (a) grab unloader (Courtesy: Walker Engineering) (b) screw lift ship unloader (Courtesy: Siwer-Tell (UK))

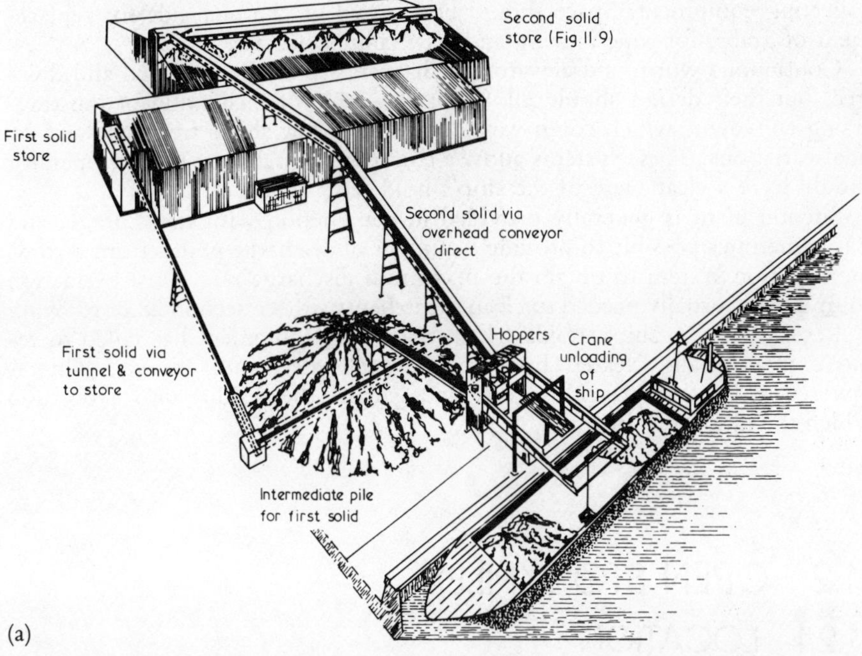

Ships are normally unloaded by grabbing crane (Fig. 11.4a), suction plant, continuous worm dischargers (Fig. 11.4b), or continuous elevator discharges. These unloading facilities are provided either on self-discharging ships, or on the quayside. Quayside facilities should be mobile, in order to travel to each hold of the ship, rather than move the ship along its berth, which is slow and costly.

Grabbing cranes should be provided with dumping hoppers which allows grabs to return and refill whilst the hopper is discharging at a continuous rate on to the factory intake system (Fig. 11.4a). Dust generated by this method can be minimized by shielding the dump hopper, and by providing dust-collecting equipment. Space should be allowed on the quayside for replacement of grabs, for rope reeving and for maintenance access.

Continuous worm and elevator dischargers are totally enclosed and dust-free, but their design should allow for the angle of inclination of the traversing conveyor, which could vary according to the ship's freeboard and by tidal variations. These systems allow a continuous discharge, but the operator should have a clear view of the ship's hold (Fig. 11.4b).

Suction plant is generally provided in conjunction with silo storage, and it is sometimes possible to provide a number of discharge points from a common suction system to obtain the maximum discharge rate. Dust extraction equipment is usually needed to clean the exhaust air (see section 28.3, p. 395).

Residues left in ships' holds which unloading equipment has failed to remove is normally cleared by using a small mechanical shovel. This is lowered-in by the quayside crane, and gathers the residue into a position which the unloader can reach.

11.2   OPEN STOCKPILES

11.2.1   LOCATION

Many variations of the large open pile system are used all with different types of stocking out and reclaiming methods. These are the cheapest ways of storing bulk materials, but usually because of dust, etc. have to be located in good open spaces away from the main usage point, other plants and the public. Dust nuisance can be minimized by considering the prevailing wind, by limiting the free-fall height of materials, and by using fine-water sprays at points of dust generation. Water sprays with binding agents added can also be used on the stockpile surface with some materials to prevent wind scouring. Apart from this dust problem, the stockpile location should be convenient to road, rail and sea intake systems, and to subsequent process plant and dispatch routes.

11.2.2   SIZE OF STOCKPILE

The size of the stockpile should be sufficient to safeguard continuity of processing and to accept incoming materials, taking into account lead-time, the size and frequency of consignment, reliability of transport, weather conditions, and industrial relations disputes.

The height of stockpile is limited by the load-bearing properties of the ground. Extra loading, and therefore height, can be obtained by casting a reinforced concrete base supported from a grid of piles. The height is also determined by the type of equipment needed to place the material on the pile (see section 11.2.4). The area of the stockpile is governed by the available space. However, the area and height are related by the angle of repose to the volume of the stockpile (see section C.8, p. 577). Thus, an economic balance has to be struck between the cost of height (foundations, conveyor length) and the cost of land.

11.2.3   SITE PREPARATION

Ground conditions and types of material to be stored determine the extent of preparations. Good draining ground may only require levelling and rolling, possibly a hardcore fill and top surface compacted by rolling-in the material to be stockpiled.

Cast *in situ* concrete floors are preferred for more sensitive materials, but precast concrete slabs can also be used, laid on a bed of sand. Adequate drainage must be allowed to avoid waterlogging. If it is required to segregate materials or prevent spread of the stockpile, walls can be constructed from reinforced concrete, designed for forces from either side, and cast *in situ*. Another form of retaining wall for outside use, particularly when changes in storage patterns are expected, is constructed from free-standing precast reinforced concrete wall units.

11.2.4   STOCKPILE EQUIPMENT

The choice of method of operation of a stockpile depends on its size, the rates of stocking and discharging methods of transport and the life of the installation.

The simplest way to load a pile is to discharge material from a dumper truck near the storage area, then use a front-end loader to move it into a neat pile. Reclaiming is carried out by a front-end loader which dumps the material into the feed hopper of a bucket elevator, conveyor, or possibly directly into process equipment. In plan, this type of storage requires an area equivalent to the largest pile required, plus space around it for manoeuvring the loading trucks and mechanical equipment (Fig. 11.5a). The space allocated will be connected to the unloading apron and to the road network coming from the material source. The loader's cab should be enclosed as protection against pile collapse.

A more sophisticated method is to use a portable conveyor which has a reception hopper designed for charging from tip vehicles or another conveyor (Fig. 11.5b). The pile height is dependent on the angle of repose of the material, on the inclination of the conveyor and on the outreach of its boom. The pile will be conical. The system is not suitable for friable materials due

Fig. 11.5 Examples of open conical stockpiling (a) single conical open stockpile fed by belt conveyor (b) stockpiling with portable conveyor (c) boom stacker for large stockpiles (Courtesy: Babcock-Moxey)

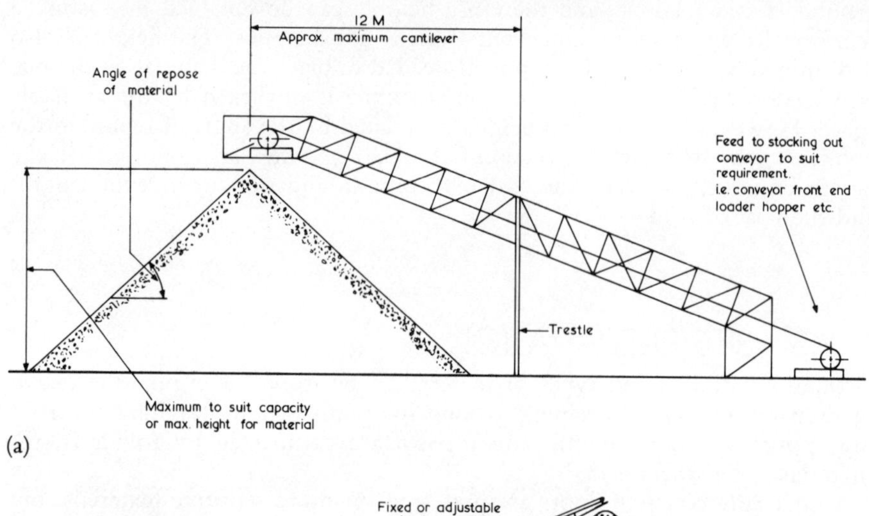

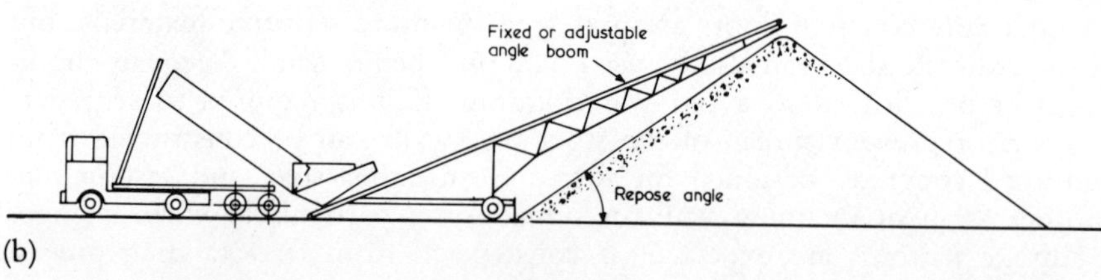

to the dropping height. This can be overcome partially by making the top section of the feeding conveyor hinged. Reclaiming can be by front-end loader but any intermediate conveyor supports which are normally buried must be reinforced against damage by shovel buckets.

For stockpiling larger quantities, distributor conveyors are used, having fixed or travelling trippers or ploughs. If located at ground level, a boom loader is used to reach the stockpile apex (Fig. 11.5c) but if at high level,

Fig. 11.6 Large open stockpile fed by travelling tripper conveyor (a) diagrammatic arrangement (b) overhead stockpile conveyor (Courtesy: Crone & Taylor)

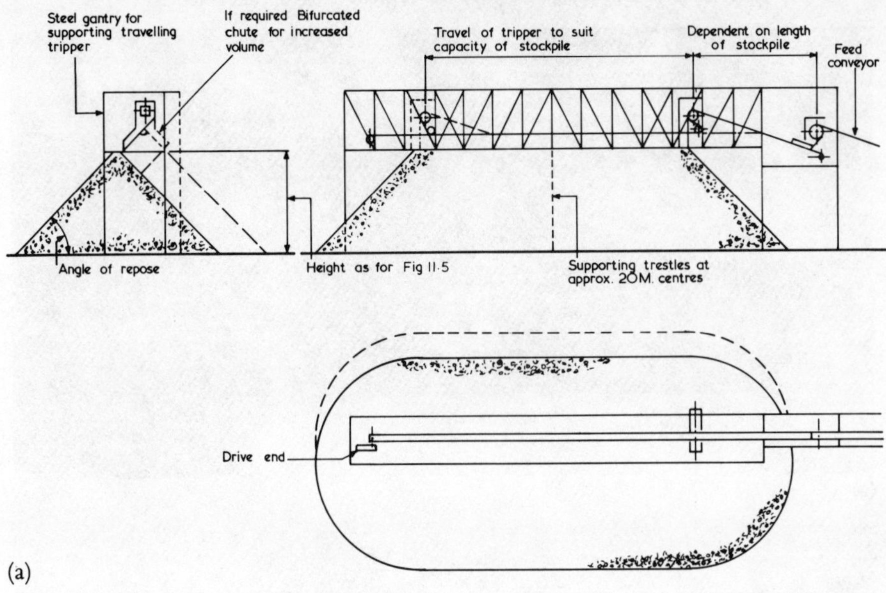

(a)

(b)

central location is necessary for maximum filling (Fig. 11.6). The piles will be rectangular in shape and can be of any length. Reclamation can again be by front-end loaders (Fig. 11.7a) or other types of reclaiming machine. For example, bucket wheel reclaimers (Fig. 11.7b) are used for arduous, high-capacity duties. They can be combined with the boom loaders as dual-

171

Fig. 11.7 Examples of reclamation (a) direct loading of bulk trucks by front-end loader (Courtesy: Norsk Hydro Fertilizers) (b) bucket wheel reclaimer (Courtesy: Babcock-Moxey) (c) typical drag scraper scheme

(a)

(b)

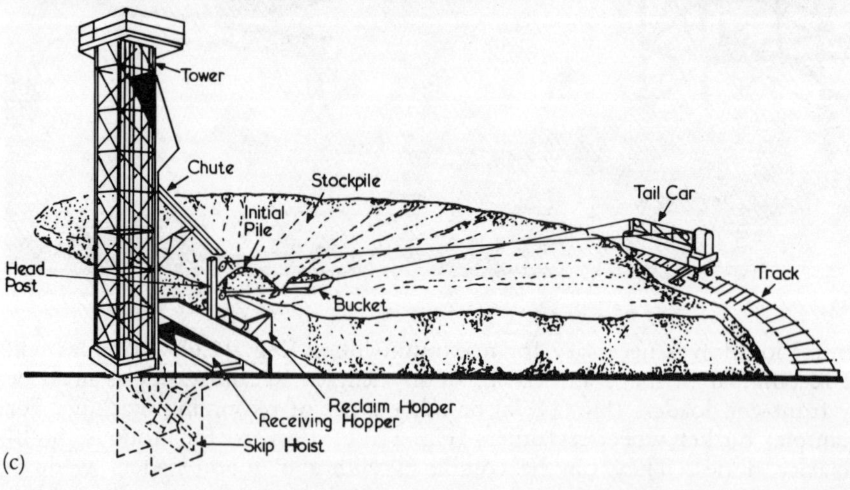

(c)

172

purpose machines when simultaneous loading and unloading is not required.

Drag scraper systems can also be used for large-capacity stockpiles. These have a centrally situated headpost to which a track-mounted tail car, travelling around the perimeter of the pile is connected by cable. The stockpile is built up by spreading delivered material with the scraper bucket, which also reclaims to a discharge point located adjacent to the headpost (Fig. 11.7c).

11.3   CLOSED WAREHOUSES

Where materials cannot be stored in the open air, bulk warehouses can be used. These are usually purpose-built, their design and shape depending on the type, reactivity, quantity and numbers of materials to be stored.

11.3.1   BUILDING LAYOUT

Floors, retaining walls, and dividing walls are normally of load-bearing reinforced concrete construction, with steel, concrete or timber-framed superstructures shaped to suit the angle of repose of material (Fig. 11.8). There should be a minimum number of ledges on which dust can settle and surfaces should be chemically resistant to the material stored. Cladding materials should be compatible with the internal and the external environment. Approximate space needs are given in section C.8, p. 577.

Ventilation should be provided, and should take into account diesel fumes

Fig. 11.8   Examples of closed warehouses for bulk storage (a) circular 'Beehive' bulk store building (b) 'A'-frame warehouse arrangement (c) 'A'-frame warehouse (Courtesy: Norsk Hydro Fertilizers)

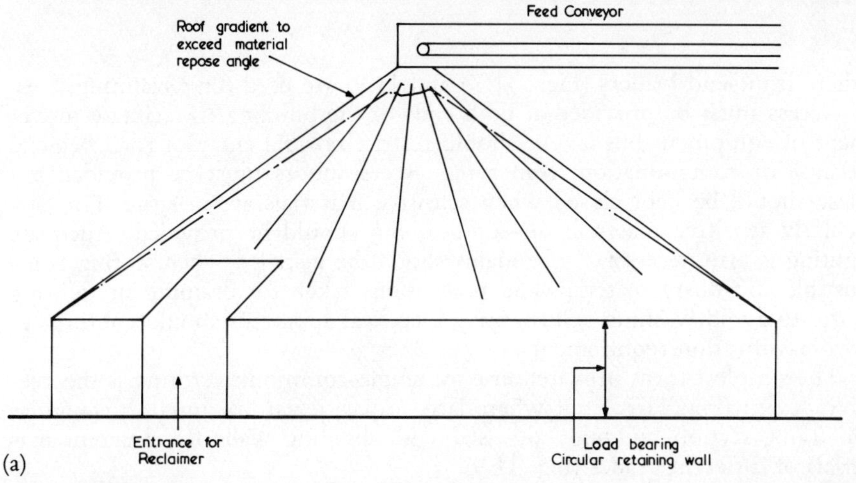

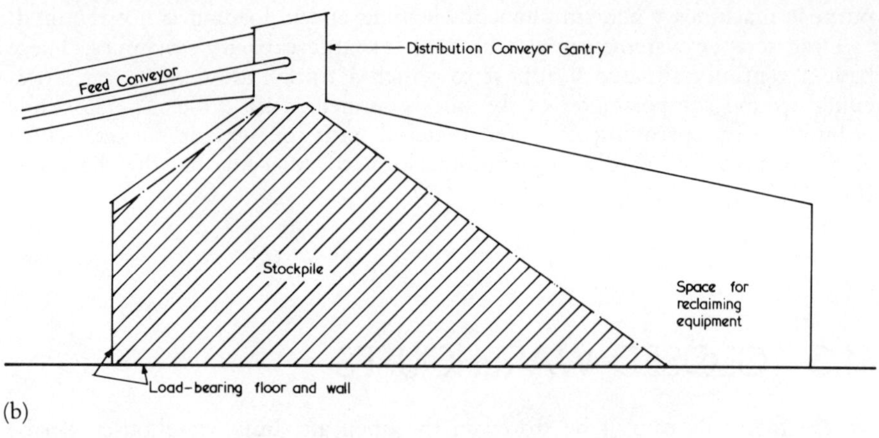

(b)

(c)

where front-end loaders (Figs. 11.7a and 11.9) are used for reclaiming.

Access must be provided at both ends of the building to facilitate movement of equipment but layout should restrict normal entry of road vehicles because of contamination from tyres. Access doors must be provided but these should be kept closed when sensitive materials are in store. For particularly sensitive materials air-conditioning should be provided. Adequate lighting is also necessary. Fire mains should be installed when storing combustible and dusty materials and precautions taken for draining firefighting water to avoid washing solids away. Electrical apparatus should conform to area classification requirements.

The simplest form of warehouse for single-commodity storing is the beehive design (Fig. 11.8a) but where large capacities are required, rectangular 'A'-frame section buildings are used with dividing walls for different materials or different grades (Fig. 11.9).

Fig. 11.9 Compartmentalized bulk warehouse

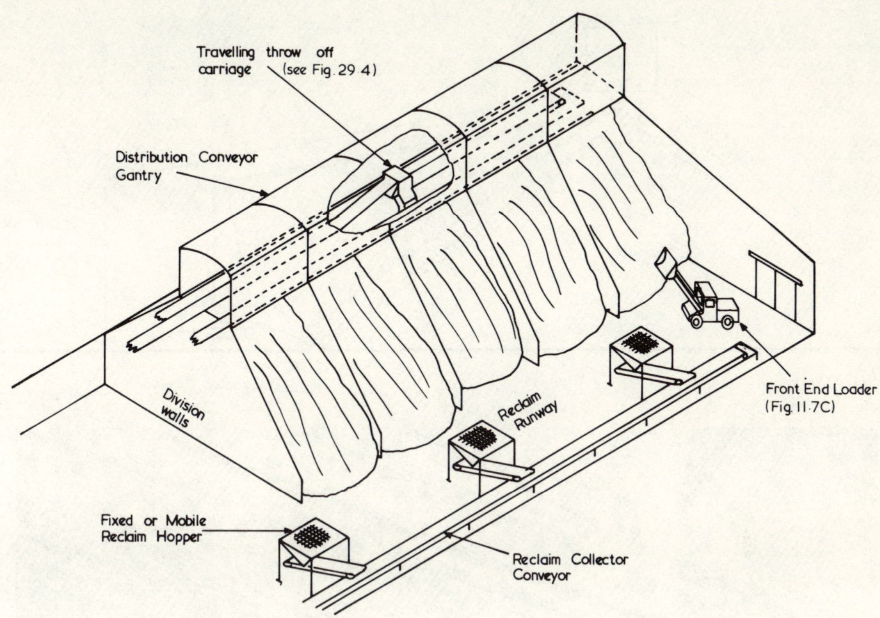

11.3.2   FILLING EQUIPMENT

All the methods used for open stockpiling can be used. For beehive stores, an end-conveyor discharge is sufficient, but for rectangular 'A'-frame buildings, conveyors with travelling throw-off carriages (Fig. 29.4, p. 409) are fitted, which can deliver along the whole length of the store (Fig. 11.9).

11.3.3   RECLAIMING EQUIPMENT

For complete flexibility, the front-end loader (Figs. 11.7a and 11.9) has much to offer if the engines can be made safe in the working conditions prevailing. As in open storage, closed cabs are necessary for safety and environmental reasons, and driver skill is necessary to avoid damage to machinery and buildings. Reclaim rates reduce as the store length increases; it may be preferred to minimize the length of the vehicle run by providing additional, or travelling charge hoppers, with collector conveyors running the length of the store building (Fig. 11.9).

Scraper conveyors or worm conveyors are an alternative to the front-end loader. These are normally mounted on a bridge over the stockpile, or a carriage alongside, and direct the materials into an underground or shielded collector conveyor (Fig. 11.10). Safe access is required for maintenance.

Fig. 11.10 Scraper reclaimer in warehouse (a) arrangement (b) example of reclaimer (with loader) (Courtesy: Babcock Minerals Engineering)

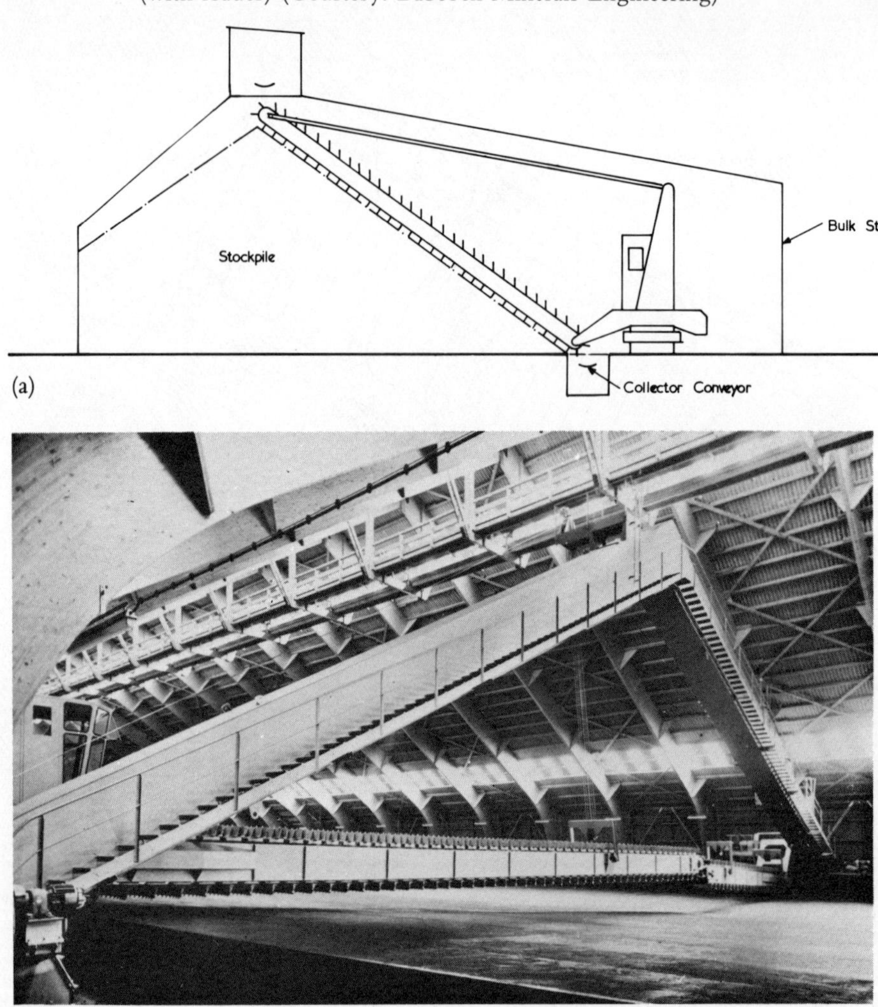

11.4 BUNKER STORAGE

Bunkers are used for both free-flowing powders or granules and for cohesive materials such as coal and iron ore. They are particularly useful for materials sensitive to the atmosphere and are widely employed for intermediate storage as well as on- and off-loading (Figs. 11.11 and 11.12).

11.4.1 BUNKER CONSTRUCTION

Bunkers (otherwise known as 'bins') consist of a vertically-sided section (the 'surcharge') beneath which is a converging section (the 'hopper'). The surcharge cross-section may be circular, square or rectangular and the hopper conical, pyramidal or plane. The term 'silo' is usually restricted to the combination of circular cross-section surcharge and conical hopper with a large height/diameter ratio as in Fig. 11.12 and to use with free-flowing materials. A rectangular bunker may have one planar hopper along its length or a series of conical or square hoppers as in Fig. 11.11.

Bunkers are usually built from steel or concrete (Fig. 11.11), although there are applications for flexible wall construction. Selection depends upon the type, quantity of corrosiveness of materials, method of filling and discharging, available construction period, capital cost, and life expectancy of the installation. Where bunkers contain combustible and dusty solids, their tops should be fitted with explosion-relief doors, capable of preventing wall failure due to explosion pressure. The doors should relieve to safety (see section 26.4 (p. 365) and Institution of Chemical Engineers[1,2]). There should be safe access for both operating and maintenance personnel especially from falling into the bunker.

11.4.2 FILLING EQUIPMENT

Bunkers are filled from mechanical or pneumatic conveying systems. Mechanical filling should be at the top centre, and voids caused by the angle of repose of materials can be filled by using spreader conveyors or rotating

Fig. 11.11 Bunker discharge (Courtesy: Babcock-Moxey)

Fig. 11.12 Typical layouts of vertical silos (a) external arrangement (Courtesy: Babcock-Moxey) (b) internal arrangement (c) flow and control

(a)

chutes. Pneumatic systems usually require tangential entry for disengagement (Fig. 11.12c). Air displaced when filling should be vented through filters, particularly when pneumatic filling systems are used.

11.4.3 DISCHARGE EQUIPMENT

Free-flowing materials will discharge by gravity through the bunker bottom outlet to downstream handling plant. The hopper angle and outlet dimensions are interrelated and depend on the friction between the particles themselves and between the particles and the wall. Unless the bunker has been designed specifically[2] physical assistance will be necessary to induce flow. This can be provided by aeration, by mechanical vibration of the hopper (see

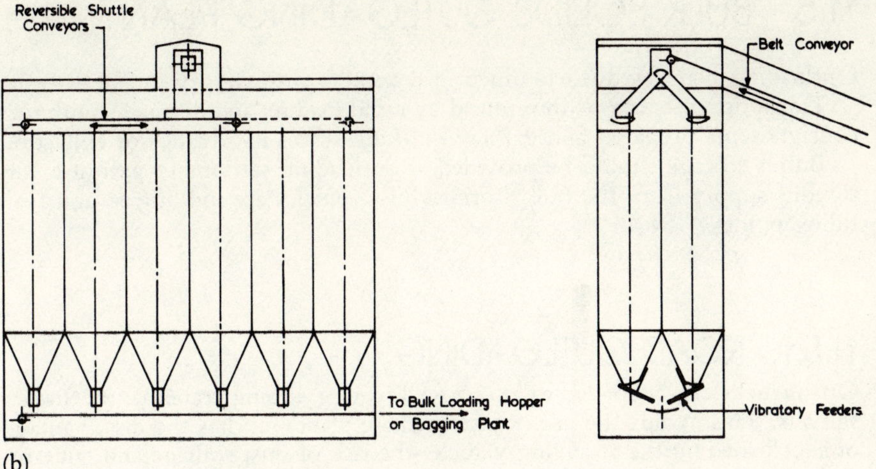

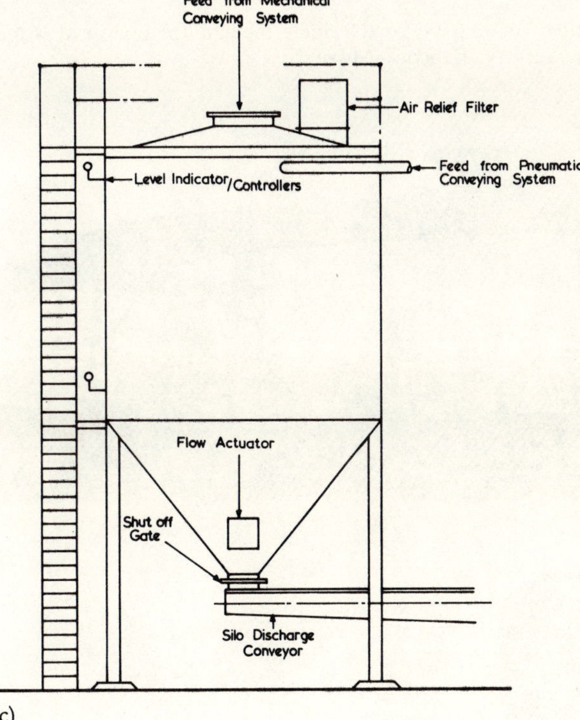

Figs. 11.11 and 11.12), or by a 'spider' – a rotating horizontal wheel with arms that rake material from the bottom of the bunker.

Flow control is achieved by use of worm or belt dischargers, rotary valves, and vibratory feeders. These should be protected by discharge gates to permit maintenance work whilst the bunker is full of material. Various forms of level indicator are available. These should be fitted to indicate the contents or high/low levels (Fig. 11.12c).

Safe access should be provided to all equipment for maintenance purposes.

179

11.5   BULK SOLIDS OUTLOADING PLANT

Outloading plant layout has much in common with factory intake systems.

Transport costs can be minimized by rapid loading and dispatch, with total consignments prepared and gathered in bulk stores in advance of collection.

Buffer storage should be provided at outloading stations to permit a continuous supply from the bulk stores whilst vehicles are moving to and from filling points.

11.5.1   ROAD OUTLOADING

Open trucks can be loaded by direct feed from grabbing cranes or mechanical shovels. This avoids the use of intermediate plant, but has the disadvantages of shock-loading the receiving vehicle, the risk of dust spillage and contamination, and uncertainty of weight and vehicle axle loadings (Fig. 11.7c).

Fig. 11.13   Overhead hopper loading (a) road vehicle loading (b) open rail wagon loading (Both courtesy: Babcock-Moxey)

(a)

(b)

180

These can be overcome by dispensing from an overhead hopper (Figs. 11.13 and 11.14), with weighments predetermined by load cells on the hopper or a continuous weighbelt, or by stationing the vehicle on a pre-tared weighbridge. An observation platform should be provided to ensure proper distribution in the vehicle. Dust emission can be contained by minimizing the free fall of material into the vehicle, by using a retracting discharge chute (Fig. 11.15), by providing dust hoods and collecting plant.

Fig. 11.14 Pneumatic road and rail loading arrangements (a) side elevation with siding removed showing typical equipment arrangement and method of loading (b) end view with siding removed showing air and dust withdrawal system (Both courtesy: Midwest International)

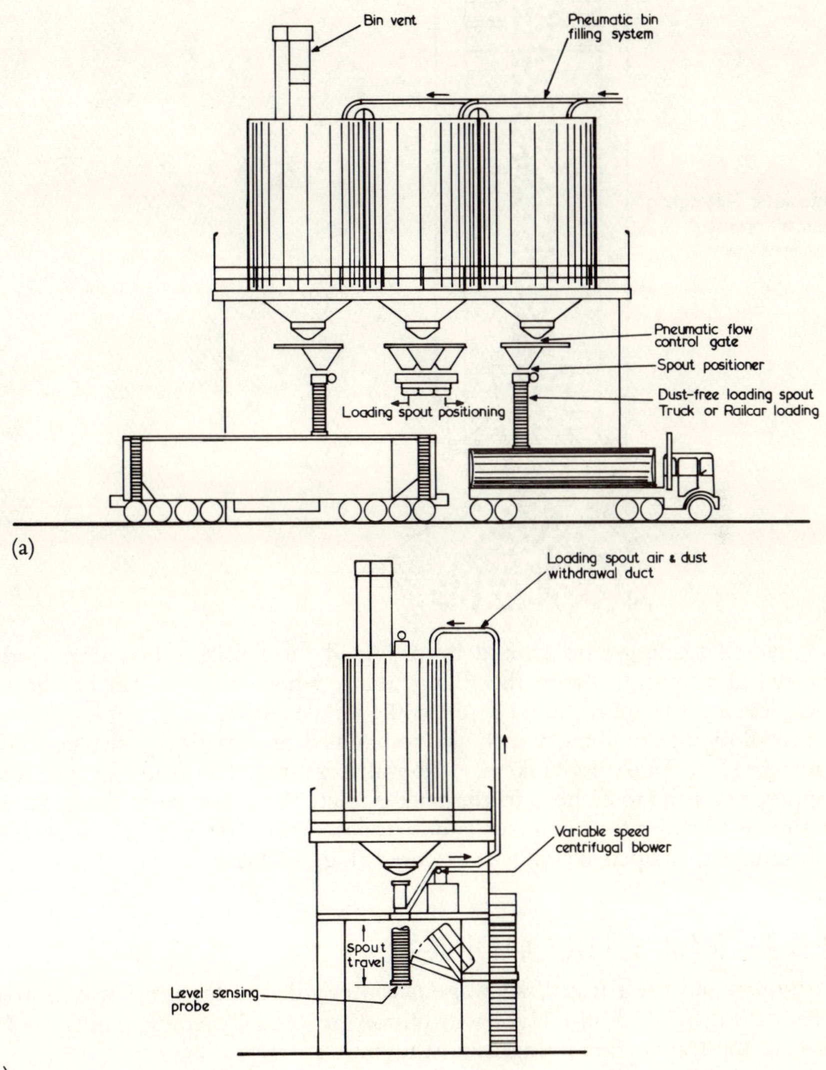

Fig. 11.15 Retractable chute (a) flow arrangement (Courtesy: Midwest International) (b) external arrangement (Courtesy: Babcock-Moxey)

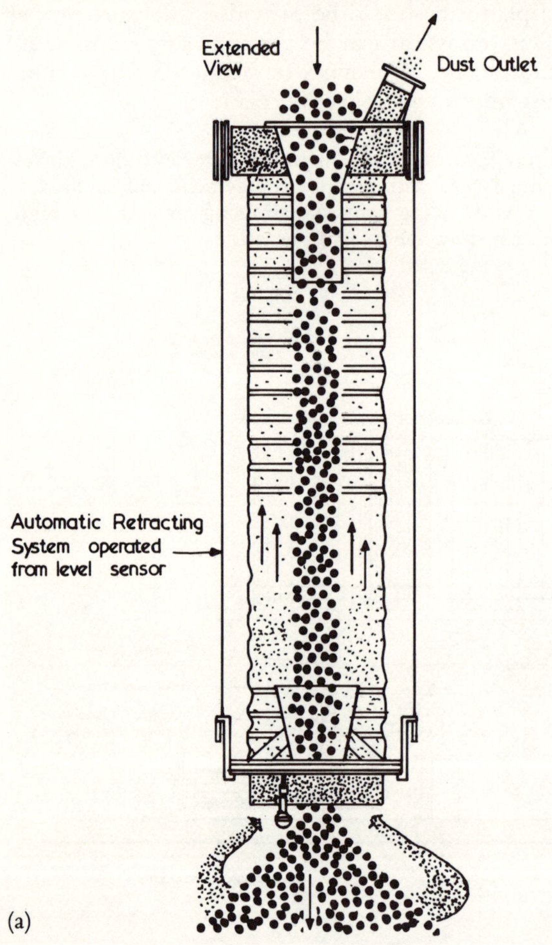

Covered loading-bays should be provided for sensitive materials, and a covered area, remote from the filling point, where vehicles can be sheeted over, leaving the filling point free for the next vehicle.

Free-flowing powders, which can be handled pneumatically, are normally transported in pressure tankers. Material is gravity-fed from an overhead hopper, via a flexible chute to the tanker inlet. Weighing procedures would be similar to those for open trucks but attention must be given to controlling dust emission if open weighbelts are used (Fig. 11.14).

11.5.2   RAIL OUTLOADING

Both open and pressurized rail wagons can be filled in the same way as road vehicles (Figs. 11.13 and 11.14). Provision of several loading points avoids moving the trains after filling each wagon.

(b)

11.5.3   SEA OUTLOADING

The choice of outloading equipment is influenced by the size of ship to be filled and type of material to be handled. If the ship is small or can easily be moved from its berth, a fixed outloader is normally adequate, but otherwise the ship should remain moored, with the outloader arranged to travel to each hold in the ship. The factory outloading system should be arranged to transfer material to the outloader at any position along the quay (Fig. 11.16).

To reach all parts of the ship's hold, the outloader boom can be designed to luff, slew or traverse and can be fitted with a rotating discharge chute. The boom height should allow for clearance of the ship's freeboard and for tidal variations.

Dust emission can be minimized by making the final discharge chute retractable, with the outlet rising automatically as the hold fills, thereby limiting the free fall of material after leaving the chute. Construction can be either telescopic tube or flexible sleeve. Either should be suspended vertically, and the latter can be arranged with air suction across the free area of the chute exit (Fig. 11.16a).

Fig. 11.16 Ship loading arrangements (a) for clay (Courtesy: Babcock–Moxey) (b) for coal (Courtesy: Humphreys & Glasgow)

(a)

(b)

REFERENCES

1. I.Chem.E. *A User Guide to Dust and Fume Control*. Institution of Chemical Engineers, 1981.
2. *Pow Tech. '83, I. Chem.E. Sym. Ser.* **69**, 1983.

WAREHOUSE STORAGE

12.1   GENERAL

On most sites using or producing packed items, storage of materials and of finished goods occupy more space than the production plants. It is important, therefore, that the supplies and finished goods warehouses are correctly and economically designed not only to store goods, but to receive, service and supply them. They should be considered as part of the total distribution systems, both external and internal. Activity in and around the warehouse is high, as well as where goods are stacked.

Solids can be stored in a variety of containers such as bottles, boxes, bags, barrels drums, tote-bins and intermediate bulk containers (IBCs). Similarly, liquid containers include barrels, kegs, drums, cans, bottles, and tote-bins whilst gases are stored under pressure in heavy-gauge cylinders. The layout of the storage area depends on the way the packages are to be stored (e.g. whether on pallets) and the method of handling them into and out of the store. Most designs are done on a modular basis around a unit load because this leads to a more flexible system. A unit load may be described as a group of items or bulk materials so arranged that the load can be picked up and transported as a single unit. Thus it is desirable to specify the kind of packaging as it is the size and type of package which determine the character of the unit load. This load, together with the quantities and distances involved, will indicate the type of transport to be used.

The size of the storage area is principally determined by:

(a) The required stockholding with some flexibility for changing demand patterns.
(b) Gangway space which is governed by the throughput and the need to recover packages quickly.
(c) Areas for handling the goods which depend on the throughput.
(d) Space for offices, amenities, analytical facilities, electrical truck recharging and maintenance (if unavailable elsewhere).

Technical details of design can be obtained from one of the standard textbooks.[1] Before any design can be started it is necessary first to establish the different functions which will occur within the warehouse and then to collect relevant data. Normally the three main operating functions are:

(a) Goods inwards.
(b) Storage and servicing, e.g. sampling, repacking, labelling, etc.
(c) Goods outwards.

Having obtained all the necessary data on space, movement within the warehouse and transport patterns to and from it, consideration has also to be given to points of access, lifts, staircases, conveyor openings and the siting of potential obstructions from heating, lighting, ventilation, drainage and building supports. There must be provision of housekeeping equipment, e.g. running water/drains for clearing up after spillages from, for example, ruptured or dropped containers.

Requirements of the local authority, particularly of the fire officer, must be satisfied. Security against theft is important and in addition the warehouse may be a bonded area for customs purposes.

12.2 GOODS INWARDS

This consists of unloading, checking, reformation of loads for storage, temporary storage, advice of receipt (including validity check) and redistribution.

Space must be provided for vehicle (or container) parking during loading and offloading. About 15 m × 2.4 m (depending on vehicle size) should be allowed for the vehicle with 4 m space between and round each vehicle for fork-lift truck access (see Fig. 12.1). The dock area must be designed so that

Fig. 12.1 Warehouse unloading bay (Courtesy: Boots Co.)

there is minimum congestion from vehicles or stock. Note must be taken of the mean frequency of vehicle arrivals, peak delivery requirements, types of delivery vehicles, load constitutions, etc.

Goods must be checked, unpacked and the loads organized into a suitable form for subsequent use. These activities could include unit load transformation, e.g. pallet loads reduced to individual cartons or loose cartons palletized. Control data for these activities includes validation of goods received and a quality-control check against incoming documents. Post unloading analysis could also be necessary to provide process information.

Temporary storage for sorting and smoothing of the workload needs to be provided near the dock area.

12.3 STORAGE

This involves the movement of goods to discrete storage locations, holding and retrieval on demand.

The warehouse should be arranged so that it is impossible to stack so high as to overload the floor. Items should be grouped so that they require similar firefighting equipment. Facilities for washing-away spillages safely are needed. Toxic materials should be stored in a well-ventilated area, especially if the containers are breakable. For hot or cold stores, the location of doors is important to prevent draughts. In many countries petroleum and petroleum products having a flash point below 23 °C must be stored in an approved building or in the open air in an area specially constructed to accommodate spillage, e.g. in bunded areas (see section 10.10, p. 159). Where flammable materials are stored, fork-lift trucks, cranes, lighting, etc. should conform to the appropriate hazard area classification standards (see Chs. 6 and 8).

Objectives for storage are:

(a) Maximum use of space.
(b) Minimum distance of movement.
(c) Accurate location and retrieval of stock.
(d) Maintenance of quality of the materials and their containers.

Method study techniques are used widely to assess movement, labour and equipment requirements.

Location systems are either fixed or random. The former are used when the range of products is small and quantities held per product line show little fluctuation between maximum and minimum demand. In such conditions a 'block storage' system is often used. This consists of 'unit loads' stacked directly on to one another. Access to stock is poor and control is difficult. It usually relies on the fork-lift truck driver's knowledge of the layout. Shelving or binning are also based on fixed-location techniques.

Random location systems are the most common with pallet racking. The racking is marked in such a way that each pallet position can be identified individually. It is usually operated by a 'two-ticket' system. The driver placing the pallet in storage seeks out the first suitable empty location. The location number is then recorded on one part of the ticket which is returned

Fig. 12.2 Palletized keg store (with pallet) (Courtesy: The Boots Company)

to the central office. Here the tickets are filed by product with the newest stock being filed at the back to facilitate good stock rotation.

Many chemicals have a limited shelf-life so a first in/first out policy has to be adopted. The layout and operation of the warehouse should aid this policy.

Pallet racking systems utilize floor space to a maximum whilst still allowing access to any individual item stored. A typical pallet size is 1.2 m × 1.2 m × 150 mm high (see Fig. 12.2). Pallets are handled principally by fork-lift truck and overhead cranes, though the latter are likely to result in a less flexible layout.

The limits of racking height are determined only by the height of the building and by the capability of the device used for placing and removing the unit loads. There is direct access to every single item in the store allowing the use of random storage with considerable savings in floor area requirements.

Racking and handling plant should always be considered as one integrated system. Up to heights of 6 m, normal fork-lift trucks can be used but aisles need to be as wide as their swept turning circle and the tolerance of the swinging load (see Fig. 12.3). A typical aisle width is between 2 m and 2.7 m. For a higher cubic capacity with the minimum of floor area free-standing racking can be built to over 12 m high. For this work free path turret trucks are available which require aisles only a little wider than the pallet length. Experience has shown that a figure of 80 per cent pallet utilization is reasonable for an efficiently run warehouse. An example of a conventional single racking system is shown in Fig. 12.4.

Fig. 12.3 Palletized racking systems (Courtesy: The Boots Company)

Fig. 12.4 Example of conventional (single) racking

Aisle	2.7 m
Storage	2.4 m
Aisle	2.7 m
Storage	2.4 m
Aisle	2.7 m

Pallets stored 8–high 1.4 m between pallet centres
along storage rack

Many products are received, handled and dispatched in large quantities. These are often stored in 'blocks', i.e. several pallets deep handled from a single aisle so that maximum use is made of the floor area. A system of 'live storage' may be used for this case. It uses sloped racks into one end of which the pallets are inserted and through which they flow either under power or by gravity to the other end for disposal (see Fig. 12.5).

Fig. 12.5 Example of live storage

Aisle	2.7 m
Storage 10 pallets deep	12.0 m
Aisle	2.7 m
Storage 11 Pallets deep	13.2 m
Aisle	2.7 m

Pallets stored 7-high 1.35 m between pallet centres along storage rack

Powered mobile racking has proved successful for varied stock with a medium selectivity rating but it is expensive and slow in operation.

Certain items can be stored outside which is cheaper than covered storage. Examples of these are drums (section 10.10, p. 159) and tote-bins. Items can be typically stored on pallets four high and handled by standard fork-lift trucks capable of lifting 3 t and operating in aisles of 4 m width (Fig. 10.7, p. 159).

Unpalletized stores are usually serviced by electrically hand-operated trucks for the smaller containers (see Fig. 30.7, p. 425). The aisle should be wide enough for trucks to manoeuvre, i.e. truck lengths (not widths). For large containers such as bags, overhead cranes may be used. In both cases the floor should be marked out into manageable lots for stocktaking purposes.

12.4   GOODS OUTWARDS

A programme of goods recovery is planned from the various customers' orders (see Fig. 12.6).

Ideally the goods are taken directly from the racks to the vehicle by fork-lift trucks or by retractable conveyor. However, space may be needed for collecting, checking, marking and packing of loads, prior to vehicle loading. The layout of the loading dock (e.g. in Fig. 12.7) is the same as for unloading (see section 12.2).

REFERENCE

1. Falconer, P. and Drury, J. *Building and Planning for Industrial Storage and Distribution*. Architectural Press/Halsted Press, London, 1975.

Fig. 12.6   Goods recovery and order make-up area (Courtesy: The Boots Company)

Fig. 12.7   Warehouse loader bay (a) internal (b) external (Both courtesy: The Boots Company)

EFFLUENT DISPOSAL AND NOISE REDUCTION

13.1   SOLIDS DISPOSAL

The two principal methods of solids disposal are incineration and tipping (often referred to as landfill), and the method adopted will depend upon the physical and chemical nature of the waste and its rate of production.

In both cases there will be intermediate storage at the plant and, in addition, for incineration, at the furnace. This is to allow for fluctuations in either production or disposal rates. These stores should be situated to prevent nuisance by way of dust, smell, fire risk or seepage. Storage in open piles is allowable provided the waste is not hazardous, does not deteriorate under the action of the weather or is not blown by the wind. Otherwise weather-proof, but well ventilated, storage has to be provided and arrangements for dust extraction and filtration may have to be made. Depending upon the condition, volume and hazardous nature of the materials, bins, overhead silos or piles with mobile loaders may be provided (see Ch. 11). Good security is needed for on-site storage of toxic or hazardous materials awaiting disposal.

Loading points for waste in plant areas should be sited in low-hazard areas and should be close to access roads so that the waste may be loaded and transported with a minimum of interference with process areas. Routes within the site should be as short as possible if a high density of traffic is required for disposal.

13.1.1   INCINERATION

Incineration (or thermal destruction or thermal oxidation) of solids may be considered for a wide range of throughputs. It may be followed by a scrubber to reduce the concentration of acidic or toxic gases released on combustion. The incinerator should be located conveniently near the plant effluent source, although a compromise site will be necessary if it is to serve two or more process areas or additional fuel has to be supplied. However, the incinerator should be sited remotely from high-hazard process areas as it is a potential source of ignition. It should be placed so that drift and fume neither cause difficulties in process areas nor create environmental problems (see section

Fig. 13.1 Reception area for a refuse incinerator (Courtesy: Babcock-Moxey)

13.3). Stack heights should be chosen so as to minimize difficulties of wind drift and the regulatory authority should be consulted if acidic or other toxic gases are likely to be emitted in stack gases.

The layout of an incinerator is usually specified by the manufacturer (Fig. 13.1) though a preliminary layout may be undertaken by treating it as a furnace (Ch. 20) with solid fuel feed (Chs. 11 and 29).

13.1.2 TIPPING OR LANDFILL

Disposal by tipping on sites controlled by the local authority may be possible if the material is no more noxious than normal household waste. However, if the material is toxic or dangerous it may be necessary to arrange tipping on specially licensed sites operated either by local authorities or by waste disposal contractors. In the latter case it is normal for the contractors to collect waste for tipping, or in some cases, for processing. Care must be taken to avoid mixing of incompatible substances, especially if leading to harmful by-products, and so accurate and complete records should be kept.

Dumping wastes off-site can give rise to leaching of pollutants into ground water and cause damage to people, wildlife and plants. All effluent disposal from site requires very careful planning and control to ensure that slow reactions, occurring over a long time-scale, do not damage the environment. Increasing onerous legislation means that the 'polluter' is likely to remain responsible indefinitely for the restitution of any such damage.

13.2   LIQUIDS DISPOSAL

Liquids to be disposed-of include storm water, plant effluent, storm water contaminated by process effluent or other chemicals and domestic sewages and spillages which may accidentally occur.

13.2.1   EFFLUENT SYSTEMS

Plant effluent and contaminated storm water are normally run in a separate system and not mixed with domestic sewage or storm water for treatment reasons and to avoid the risk of corroding sewage pipes. Thus usually three effluent systems need to be run on the site. Sewage and plant effluent are often put in pipes but harmless aqueous effluents and storm water may be run in open channels if practicable, and if there is no risk of freezing, or vapour collecting or accident to plant personnel or indigenous wildlife. Effluent ways should run alongside plant roadways and should take account of future road plans. Short cuts across undeveloped site areas should be avoided to ensure that future plant construction does not jeopardize the sewer network. If space is limited, effluent pipes may run under roadways, but this means that site traffic problems may arise when pipe maintenance is required. For this reason culverts should be used and in any case pipes should not run under major site emergency access roads. Open ditches should be avoided at access points for maintenance and emergency vehicles and, because of fire risk, under pipeways. If not so avoided, the ditches should be culverted at these locations. In chemical works it is advisable to run pipes containing potentially hazardous materials where they can be seen and to bury them only as a last resort.

Storm water from buildings such as offices, stores and gatehouses having no obnoxious materials can be drained to the public sewerage system, provided that it is not overloaded. On large sites, storm water is very seldom allowed in the public sewage system because of high peakloads. When uncontaminated it may be pumped into watercourses or to sea, if the authorities approve. Contaminated rain water from plant areas should be treated along with aqueous process effluent. Rain water from areas that are only occasionally contaminated may be stored in holding ponds and checked before release to a public sewer or watercourse or to the plant effluent; this method avoids the necessity of designing the effluent plant to deal with peak rates due to rain water, and thus reduces the required size of the plant.

Special provisions may be necessary for disposal of water used in firefighting operations. As a general guide, with large open-air plants, the volume of water for firefighting operations is about five times the volume of water allowed for as storm water. It is usually too expensive to design the drains to take the maximum amount of firefighting water that may be used. The excess must either be pumped-off or must be run-off into unused land. The mixing of fire effluents and normal effluents should be prevented if it produces dangerous fumes, etc. which may hamper the emergency services.

Consideration should be given to the problems associated with flammable effluents that are immiscible with water. There is always a possibility of burning liquid entering the system (particularly open trenches) and spreading fire over long distances. Thus, prior to runoff, the whole of the effluent should

Fig. 13.2 Effluent interceptor (Courtesy: The
Boots Company)

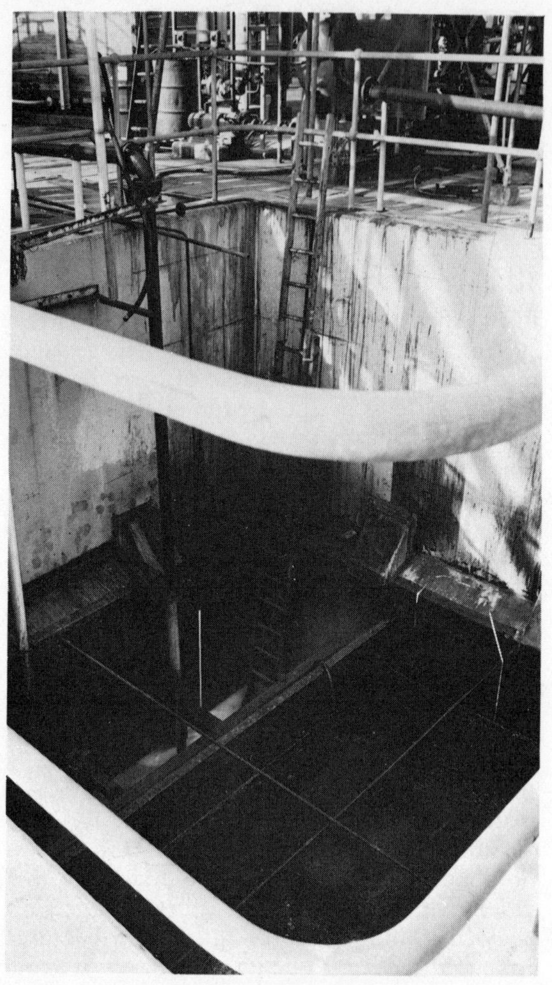

be passed through efficient interceptors, (Fig. 13.2 and Fig. C.10b, p. 576) located at the perimeter of fire-risk areas. To avoid overloading, primary interceptors may be used in the collecting areas to give preliminary separation. Precautions against sedimentation and freezing of the interceptors may be necessary. Similarly, suitable traps or lutes may also be needed to prevent spread of fire, unless a fully-flooded drainage system is used.

All site effluent ways, whether for sewage or for process effluents, must have adequate gradients to provide for flow of liquids at velocities which will prevent settling of entrained solids and grit. Connections from plant into main runs should be made at liquid-sealed boxes to prevent dangerous or flammable gases passing through the effluent system. Provision should be made for venting the boxes to a safe location, e.g. 3 m above grade and at least 4.5 m from plant platforms and 12 m from furnace walls.

The need to provide gradients in all effluent ways means that the height of paved areas on plants must be checked to ensure that plants can drain into the correct system (see section 17.4.2, p. 246). Areas allocated to future plants should also be checked so that plant paved levels are not too high and consequently require extra landfilling prior to construction.

Care should be taken to avoid flooding sensitive areas such as pump pits and bunded areas. If the latter become flooded the tanks in them may float. Liquid effluent must not be allowed to run off plants onto adjacent property, or vice versa. Extra precautions are necessary if the site slopes or contains natural watercourses.

13.2.2 EFFLUENT TREATMENT PLANTS

There is an economic balance to be made between constructing and operating a treatment plant on site or paying the water authority to treat the effluent. One disadvantage of the latter scheme is that the authority has to be informed before a new effluent is discharged or the composition of an existing one altered. A compromise is to partially treat effluents on site.

Treatment plants are usually placed near the perimeter of the site at a point suitable for the collection of the various effluent streams. They may provide for:

(a) The treatment of general organic wastes.
(b) Neutralization of acid or alkaline liquors.
(c) Removal of oil contamination.
(d) Elimination of heavy metal ions.
(e) Precipitation and removal of unwanted chemicals from aqueous process streams
(f) Removal of suspended solids.

Equipment used includes columns, filters, centrifuges, thickeners, heat exchangers and agitated vessels, plus storage for treatment chemicals. The layout principles described in Chapters 19–25 should be followed. If effluent streams are produced that require different treatments it is desirable to process the streams without mixing. In such a case, the entire system of collection has to be laid out for separate collection of the different streams. Where immiscible streams are already mixed, separators can be used (Fig. 13.3 and Fig. C.10a, p. 576).

Effluent treatment plants may require large site areas if oxidation ponds, aerobic digesters or lagoons for the conditioning of acqueous suspensions are required (Fig. 13.4). Suspensions of finely divided solids may be produced that may be toxic, give rise to high biological oxygen demand, or may otherwise contaminate the sewage system. If dumped in lagoons, storage may be necessary for a considerable period of time. Lagoon areas should be secured against the possibility of runoff of suspended solids into rivers, seaways or sewage runs. If solids, after conditioning in lagoons, are to be subsequently dumped off-site consideration should be given to minimize interference of truck routes with the rest of the site. Consideration of the routes of disposal tankers is also necessary when the solids are transported for off-site disposal as an aqueous suspension (see section 13.1).

There should be provision against the possibility of accidental process discharge. One method[1] is to pass all process effluent through effluent-holding basins with a capacity of about 1 hr's holding time at peak conditions. These

Fig. 13.3 Oil-water separator (Courtesy: ICI Petrochemicals and Plastics Division)

basins are placed at the inlet to the treatment plant. It is recommended that basins should have a minimum depth of 1 m. If the problem of accidental discharge is likely to be serious it may be necessary to provide additional ponds for containment.

The design of effluent systems should cater for using 24 hr monitoring equipment for both raw and treated streams.

13.3   GAS DISPOSAL

This concerns systems for venting, blowdown and pressure relief.

If a gas is inert or chemically harmless such as a stream, air or nitrogen it may be discharged to atmosphere. Other relatively harmless materials may

Fig. 13.4 A biological effluent treatment plant for pharmaceutical manufacturing (a) sludge clarifiers (b) neutralization tanks (Both courtesy: Humphreys & Glasgow)

(a)

(b)

be discharged to atmosphere if the discharge is infrequent. Toxic and flammable materials may be discharged to the air if the discharge point and velocity are high enough to prevent any public nuisance or danger. This will have to be agreed with the authorities charged with controlling air pollution.

Hot volatile liquids or suspensions may be collected in a horizontal drum fitted with a vent for quenching by water or steam. The vent should be sized to relieve steam formed in the quenching. Both drum and vent stack are normally located close to the relief or blow down valve.

Normally open vent systems discharge at least 3 m above the highest working platform within a radius of about 20 m. The actual heights and distances depend on the potential health risk from fire, toxicity or noise and it may be that a tall stack will be needed. For example, a high discharge velocity will give good dispersion but may also generate noise, so a high discharge point would be needed to obtain both acceptable dispersion and noise values at ground level. Advice should be sought from the regulatory authorities on stack heights and allowance should be made for the effects of prevailing or potentially troublesome winds and ingress of rain. A check must be made to see whether high stacks may cause aerial hazards and whether warning lights are needed. The authorities may require sample points to be installed for them to check discharge compositions, etc.

Venting systems for flammable and toxic fluids are connected by piping to a safe disposal facility such as a flare, burning pit or scrubber. Scrubbers are used to remove toxic materials prior to discharge to atmosphere. Gases containing large quantities of flammable liquid are burnt in a pit whereas, without liquid, the gases are burnt in a flare. Toxic gases may also be burnt in a flare if the combustion products are not a nuisance or potentially hazardous.[2]

Headers to flare stacks from relief and blowdown valves and vents are run first to a knockout drum (Fig. 13.5). It is desirable that lines should be run overhead to the top of the drum and they must slope at a minimum of 1 in 400 so that all condensate runs into the drum and does not collect in the lines or around the outlet side of the valve. It may be necessary to put the drum in a pit to achieve this, but not if significant quantities of liquid have to be removed from the drum. Occasionally, the header lines may be heated to prevent condensation. The knockout drums are located near the stack to avoid condensation of liquid between drum and the stack. It is advisable to have automatic discharge of drum contents so that the operator does not have to go into the high thermal radiation area. The required size of drum is difficult to estimate but 3 m diameter and 12 m long, is frequently considered necessary.[3] The drum should have adequate drainage lines and be fitted with heating coils if necessary.

Corrosive vapours should be led to a separate knockout drum at the base of the flare and a separate pipe should be run inside the stack.

Flares should be designed and sited so that:

(a) Nuisance is not caused on and around the site from combustion products and noise.
(b) Heat radiation or dropout of burning liquid does not cause ignition or damage to plant or people.
(c) Any releases of flammable vapour from other parts of the site disperse to below the lower flammable limit before reaching the flare.

200

Fig. 13.5 Flarestack showing knockout drum (Courtesy: Babcock Woodall-Duckham)

Stack heights normally vary from 5 to 90 m although shorter stacks or horizontal pipes may be enclosed in brickwork kilns, possibly for use in an emergency to avoid overloading the main flare. There should be provision for steam jets and associated equipment at the top of the stack if a smokeless flare is required by local regulations. As efficient dispersion into the atmosphere calls for high discharge velocities, a check should be made on the noise levels developed.

A sterile radius of at least 60 m should be specified around the flare, within which only flare access roads, flare knockout facilities and non-flammable storage are permitted. Drums and timber should not be permitted within a high radiation area and the area should be clearly signposted. Normally, people should not work within the sterile radius but, if they must, radiation should not exceed 1.5 kW m^{-2} for prolonged periods or 6 kW m^{-2} for 30 s of emergency work by persons in full protective clothing including gloves and faces cover.[1]

Flare stacks are normally located downwind from a process area and at least 100 m away to allow for dispersion of vapour releases (i.e. well outside zone 2 areas) (Figs 2.1 (p. 7) and 4.2 (p. 30)). The particular values of stack height, sterile radius and separation from plants should be checked using thermal radiation (see API RP 521,[1] Bland and Davidson[3] and Ch. 8) and dispersal calculations such as discussed in Chapter 8 and in texts on stack emissions.[4]

The flare stack may be self-supported if short, or supported by two-level guys spaced at 120° intervals if space allows (Fig. 13.5). Aviation warning lights and lightning protection may be needed on tall flares. In congested areas, derrick structures containing ladders and supporting platforms are economically viable for stack heights from 30 to 60 m. A platform should be provided at the top of the stack for access to flare pilot burners (which ensure ignition of the flame) and other equipment. The platform is reached by suitably protected vertical ladders either secured to the stack or supported in a derrick structure if used.

13.4   NOISE REDUCTION

It may be possible to reduce noise to meet levels acceptable to the community by siting plant which constitutes a potential noise problem as far away as possible from neighbouring property and by arranging that other buildings, such as warehouses, will be placed between the plant and the community. Too much cannot be expected of this since desirable attenuations may need appreciable distances greater than available on some sites. Sound will go round or 'flank' buildings or barriers and their efficiency is somewhat limited. The height of the noise source and of the potential sufferer are both factors in considering this form of control. The lower the source and the receiver, the more the attenuation. In high rise flats, for example, above, say, the sixth floor there may be little difference in noise level at the front or the back of the building and reflection from other buildings may reduce the expected attenuation.

The type of support structure provided for equipment may be important particularly for the lower frequencies in the 125–500 Hz bands. Suitable isolation of the equipment may reduce this. Heavy supports are needed for some equipment, e.g. gas turbines, and care should be taken that these supports do not become large transmitters of noise.

Standby equipment should be considered carefully since, as a plant expands operations, something which was expected to operate only very occasionally, e.g. an emergency generator, may come into continuous operation. Otherwise intermittent noise sources may be located with more freedom although unless extremely infrequent, they can be a great source of annoyance. Examples are relief valves, audible alarms and public address systems.

In considering layout to reduce noise other factors, such as safety, have to be taken into account. An acoustic barrier to prevent noise being transmitted to a nearby building cannot be brick-built if an explosion hazard exists which could cause a 'brick barrage'.

In many process plants the equipment is mounted in free air not just to save costs on buildings, but because it reduces the danger from toxic or flammable materials. This, of course, puts a restriction on possible noise control, although new methods are becoming available which overcome some of these problems.[4]

In assessing the noise produced by various parts of the process or plant equipment it will be necessary to decide whether any areas of the plant site will require special restrictions to avoid hazard to hearing. It would obviously be inconvenient to allow such areas to cover commonly used access roads, for example.

A point to bear in mind is that a site built in a sparsely inhabited area may, after ten years or more, be closely surrounded by housing (see section 3.10 (p. 24) and Table 3.1 (p. 25). It is prudent, therefore, to avoid use of any part of the perimeter of the site for other than offices, laboratories, etc. and to maintain informal contacts with local and regulatory authorities (see section 5.7.2).

Typical noise generators in plant include:

(a) *Moving machinery*: pumps, fans, compressors, mills, presses, stirrers, turbines, etc.
(b) *High velocity fluids*: in pipes or discharging to atmosphere, either continuously (e.g. vents and ejector exhausts) or intermittently (e.g. steam traps, relief valves).
(c) *Electrical equipment*: motors, transformers, high-voltage power lines, emergency diesel generators.
(d) *Traffic*: road or rail, particularly when starting, stopping, shunting, or being weighed.
(e) *People*: slamming doors, shouting, leaving plant doors and windows open, not using noise suppression equipment, car parks and bus stops.
(f) *Audible alarms*: e.g. mandatory high- and low-level alarms in boilers etc.
(g) *Drum handling*: particularly empty drums.

Table 13.1 gives typical noise levels both for plant items and levels in the community (ignoring frequency effects). A rough equation for preliminary design is given to estimate the effect of distance on noise levels.[4-7] However, generally it is better to reduce noise at source by insulation or better engineering design. In all cases of doubt, expert help should be sought, especially as equipment suppliers and users can be under legal obligation to minimize noise nuisance.

Table 13.1 Reduction of noise levels by siting

Level in dB	Subjective impression	Environmental example	Plant example	Suggested site boundary levels			
				Continuous		Intermittent	
				Day	Night	Day	Night
125			Steam vent				
120	Painful						
115		Jet plane at 200 m					
		Pop group					
110	Deafening		Natural draught furnace burners				
105		Car horn	Steam leaks				
100		Inside plane	Turbine/compressor				
95			Heaters				
90	Very loud	Busy street	Forced draught furnace				
			Air fin coolers				
85			Pumps and motors			Main road	
80		Workshop				Industrial	
70	Loud	Car		Main road		Offices	Main road
65		Loud radio		Industrial		Housing	Industrial
60		Speech		Offices	Main road		Offices
55			Control room	Housing	Industrial		Housing
50	Moderate	Office			Offices	Hospital	Hospital
45		Quiet house		Hospital	Housing		
35					Hospital		
30	Faint	Public library					
25		Whisper					
10	Barely audible						

Attenuation approximation
Subtraction in dB $\approx K + 20 \log_{10}$ (distance in m)

where $K \approx$ 8dB for open country or source height > 3 per cent of distance
13dB for plant areas and source height < 1 per cent of distance
18dB for screened areas and source height < 1 per cent of distance

REFERENCES

1. API RP 521 *Guide for Pressure Relief and Depressurizing Systems*. American Petroleum Institute 1980.
2. Anon. 'Incinerating toxics', *Processing*, 44, Aug. 1982.
3. Bland, W. F. and Davidson. R. L. (eds) *Petroleum Processing Handbook*, section 8–70, *et seq.* McGraw-Hill 1967.
4. Walters, J. K. and Wint, A. (eds) *Industrial Effluent Treatment*. (2 vols), Applied Science Publishers, London, 1981.
5. Dept of Employment *Code of Practice for Reducing the Exposure of Employed People to Noise*. HMSO, 1972.
6. OCMA *Procedural Specification of Limitation of Noise in Plant and Equipment for the Petroleum Industry*. Oil Companies Material Association, 1972.
7. Green, S. D. 'The engineering response to health hazards', *I. Chem. E. Sym. Ser.*, **58**, 137, 1980.

UTILITIES

14.1   BOILER HOUSE AND POWER STATION

The situation of the boiler house and power station is one of the earliest and most important matters to be decided in the overall site layout, particularly if the boilers are coal-fired. As the most economic method of coal delivery is by rail or ship, it is desirable that the coal-fired boiler plant should be sited so that rail or dock access for delivery of coal. and road access for dispatch of ash gives minimum interference with the rest of the site (see Ch. 11). It is advisable that ash truck routes within the site should be as short as possible, even if an additional perimeter gate has to be manned during a period in the day. Wind-blown ash will normally be found near the routes.

Dust emission from a solid fuel storage pile can be a serious nuisance, and the effect of prevailing winds should be considered. Freak winds may be created by large buildings near the fuel pile. This aspect of design may need investigation in the light of knowledge of the effect of atmospheric turbulence upon wind patterns near large buildings.

Emission of grit from a boiler stack will normally be subject to limits set by the appropriate regulatory authority. However, even if the dust-removal system is well designed, it is desirable to position the stack so that occasional poor operation does not cause difficulties either with other parts of the site, or with local communities. Thus a suitable position for boiler plant may be found on the perimeter of the site with the prevailing wind taking any residual dust emission and cooling tower emission away from the site, and from the local communities. On the other hand, the minimum piping cost usually occurs when the boiler is in a central position. In addition it is desirable to position generator, boiler plant, etc. a safe distance from hazardous plant and so that access to them is not through a process area. Clearly the siting of the boiler house will depend on the particular case.

Although fuel oil is not easily ignited, fuel storage for oil-fired installations should be placed at a distance from the plant so that there is no risk of ignition if there is a prolonged fire in the boiler house. The tanks should be located to give easy access for delivery (sections 9.4, p. 144) and 10.9 (p. 157) which is usually by rail (section 9.6, p. 145) or road (section 9.5, p. 144). The tank farm may take up considerable space as shown in Fig. 10.4 (p. 151). Depending on the quality of the fuel oil, the emission from the boiler stack may be obnoxious and need treatment (see section 13.3, p. 198).

Gas-fired boilers need a gas receipt station and can be noisy but overall give fewer environmental problems and take up less space than coal- or oil-fired boiler houses.

The internal layout of a boiler house and power plant (see Figs 14.1 and 14.2) is a specialist undertaking but preliminary layouts can be done by treating the boiler as a furnace (Ch. 20) and the turbine as a compressor (section 23.3, p. 342). Normally, boiler plant, water treatment plant and equipment for power generation are housed either in one building, or in closely spaced separate buildings since operating experience shows that the risk from this type of plant is acceptably small.

14.2   COOLING TOWERS

Cooling towers always produce large volumes of very wet air (see Figs 2.1 (p. 7) and 4.2 (p. 30) which can cause serious problems of fog, precipitation, freezing and corrosion in areas downwind of them. They also require water basins, and in the large-throughput natural draught design, need substantial foundations (see Fig. 14.3a). These features strongly influence tower location, in addition to considerations of cooling-water flow round the site.

Towers should be located to minimize the effects of wind drift on roads, rail, plant and the neighbourhood of the site. Smaller towers should be laid crosswise to the prevailing wind to minimize recycling of air from the discharge of one tower to the suction of the other. Emissions from plant vents, flares and boiler and other chimneys in the neighbourhood of the towers should be checked to prevent entraining corrosive vapours and sulphur-bearing fumes from fuel combustion. The placing of large, natural-draught towers needs special care to ensure that resonant forces are not generated by through-winds between towers, and should not be spaced closer than 1.5 times the tower base diameter between centres to prevent increasing the drag forces found at close spacings. The load-bearing capacity of the soil at several alternative locations should be checked to find the most economic location. Large towers may also present an aerial hazard and require authorization by aviation authorities.

Mechanical-draught towers are generally smaller and present fewer problems in construction, but they require power and may generate considerable fan noise. If the site is near a residential area, slow fans should be considered and buildings or other sound screens should be placed between the towers and the residential area. Crane access may be required around forced-draught towers for maintenance of frames and replacement of packing. Space for storage of packing and machinery may be provided near the towers. However, in some cases the packing may last for 15–20 years and so replacement provisions can be minimal. The use of mechanical-draught towers allows decentralization of cooling capacity (as in Fig. 14.3b) and the use of forced-draught air-coolers for cool process streams will reduce the cooling-water load. Thus it may mean that a decentralized arrangement is more economical than a large natural-draught installation, and the problems of tower location

Fig. 14.1 Boiler house layout (Courtesy: Babcock Woodall-Duckham)

Fig. 14.2 Turbine house layout (Courtesy: Babcock Woodall-Duckham)

(a)

Fig. 14.3 Cooling tower location (a) large central tower (Courtesy: Babcock Woodall-Duckham) (b) small local mechanical draught tower (Courtesy: Nottingham University)

(b)

are usually much less. Equipment may have to be provided for treating cooling water to reduce corrosion and fouling of heat exchanger surfaces and if this is only required on some plants, then decentralized cooling arrangements may be best.

If there is a chance (which should be small) that flammable materials are released into the cooling water and then into the atmosphere at the cooling tower then there must be no ignition sources near the tower.

14.3   WATER SUPPLIES

14.3.1   GENERAL USE

The supplies for each process area have to be sufficient to ensure that the normal requirements for demineralized, process and cooling waters are always met and that at the same time sufficient additional water is available both for firefighting[1,2] and for fire protection, such as cooling of equipment in the neighbourhood of a plant fire.

If water is taken from a river for any purposes the intake should be upstream of plant outfalls to avoid the intake of contamination or burning liquids. Note, however, some water authorities insist on the intake being below the discharge, to ensure that the discharge is clean. River water will need treatment before use as drinking, boiler feed or feedstock water (see Fig. 14.4). When sea water is used for cooling or firefighting, special precautions will be necessary to combat corrosion and the development of marine growth in water mains, especially at the intakes. In the case of tidal waters, allowance should be made for the extra lift required at low tide.

Fig. 14.4   Adsorption vessels and regeneration furnace for river water treatment (Courtesy: Humphreys & Glasgow)

Boreholes require heavy-duty pumps which will probably run intermittently, e.g. on level control, and can be noisy. It is usually best to locate these pumps near the biggest user of the water. Public main supplies present no particular layout problems.

To enhance reliability of supplies, ring main systems and intermediate storage reservoirs should be considered. Water distribution about the site is considered further along with other services in section 14.6 and within the plot in section 31.6 (p. 446).

14.3.2 FIREFIGHTING WATER SUPPLIES

The quantity of water required for firefighting and protection will depend upon the individual plant design and fire risk, but will seldom be less than a 150-minute supply with a pumping rate of 0.15 m^3/s.[1] Far larger quantities of water may be needed for fire emergencies requiring the use of continuous water sprays for several days. A convenient method of estimating the water discharge rate required from hose streams on to a hazard is to provide 0.01 m^3/min. per square metre of ground area covered by the hazard, plus a similar rate for the area of imaginary simple figures, e.g. vertical cylinder, cube, etc. defining the exterior boundary of each free-standing structure or tank in the hazard area. The duration of discharge should be related to the quantity of fuel present and its burning rate (see sections B.3.3 (p. 503) and B.4.2 (p. 513)) and to whether the fire may be extinguished, or just contained for a period of time. The anticipated rate of discharge should also include water provided as curtain sprays to reduce the dispersion of flammable vapours.

The rate at which water can be drawn from the public mains should be ascertained. Except for very small works, this will usually be insufficient for firefighting, and it will be necessary to provide independent water supplies that will be effective, even in the event of mains failure. These may consist of natural and artificial sources supplemented by large-scale static reservoirs (Fig. 14.5).

Natural water sources include rivers, streams, lakes, ponds and, of course, the sea in coastal areas. The strictures on river and sea water given at the beginning of the section also apply to firefighting water supply. Artificial water sources include wells, ponds, canals, dams and other forms of impounded water. Wells and natural streams may be used to supply artificial ponds even if the natural flow of the well or stream is too small for direct supply. Where practicable, indications of stored volume should be shown on emergency water supplies.

14.3.3 FIREFIGHTING WATER DISTRIBUTION

Firefighting mains are preferably buried and their ability to survive both natural and incident-caused shockwaves must be considered. The mains are normally laid as ring mains with hydrants placed adjacent to, but at a safe distance from, the plant they serve. The main network should be designed

(a)

Fig. 14.5 Firefighting water reservoir as landscaped feature (a) aerial view (Courtesy: Humphreys & Glasgow) (b) model (Courtesy: British Gas)

(b)

so that fractured pipework can be isolated by means of suitable valves without stopping the supply to other regions. Where there are alternative sources of supply, the feeder systems should be independent of each other and capable of being separately controlled. If both mains and non-mains supply are connected to a ring main or equipment it is important to ensure that water cannot enter public mains and contaminate the supply. This must be prevented by having break tanks between the mains supply and the points of use.

The layout of mains and hydrants and details of connectors should be reviewed with the local authority fire officers to obtain their advice and to ensure their staff are able to assist in firefighting. To this end, the larger and more complex sites may have their own firefighting service, manned by both full-time and part-time personnel.

Fire pumps may be either fixed or mobile. Mobile trailer pumps may be taken to any convenient point in the works, and are useful on a large site where water supplies are adequate. They are most often used to supply water directly to fire hoses than into fixed mains. Mobile trailer pumps should have a towing vehicle capable of carrying the crew and all auxiliary equipment. The available natural or artificial water sources within a works should have good access and hardstanding for major mobile pumps including the equipment of the local fire brigade.

A method of providing water for mobile pumping apparatus that may be considered is to link a low-pressure, high-volume main with a series of strategically sited water sumps. Each sump might have a capacity of about 14 m^3. Hardstanding arrangements should be made around each sump so that up to three appliances may be supplied (see section 9.3, p. 143).

Fixed pumps installed on pump slabs in the open, or in service pumphouses, should be near to the water source to minimize suction side pressure drops. Both stations and water sumps should in safe areas, in which non-flameproof motors and starters are permitted because the pumps are often of high horsepower. Power cables should follow safe routes. Ideally, pumps should be remotely controlled or self-starting. Standby pumps having a power supply source other than the main electrical supply are usually essential.

It may be appropriate to provide elevated tanks for emergency water supply, particularly for enclosed chemical plants protected by sprinkler systems. Tanks may be sited on the roofs of buildings housing chemical plant. The amounts of water stored should be sufficient to sustain water flow to sprinklers and hoses connected to the supply for a period of time that is sufficient to deploy plant firefighting equipment, and to provide for further supplies of water. This time should normally be more than 1 hr. It is usually necessary to provide for replenishing water in elevated tanks from pumps connected to low-level emergency supplies. The pumps may be activated by level control in the elevated tanks.

Firefighting water arrangements within plots are further discussed in section 17.7 (p. 255).

14.4   ELECTRICAL DISTRIBUTION

14.4.1   TRANSFORMER SUBSTATIONS

Normally the incoming public supply will be at 33 kV or more and a separate public utility substation will contain the main transformers and switchgear for the whole site. This is often put on the edge of the site. Electrical power will usually be distributed at high voltage (5.5 or 11 kV) with transformer and switchgear substations located to serve areas (or individual plants, if large power users).

All substations are high-energy sources and must be treated with respect. They should be located in areas of the lowest electrical hazard rating to allow use of standard equipment and to avoid damage during an emergency. Substations should always be located well above flood level and away from fume and cooling tower or steam trap drift. Oil-filled transformers and circuit-breakers contain appreciable volumes of oil which can catch fire. Although fires are very rare, appropriate firefighting equipment should be provided. Vapour-free ventilation is needed for enclosed substations.

The inside walls of substations should be kept dry, any rainwater spouts should be routed down outside walls, and substation floors should not be connected to site drains. The rainwater should be run-off to an apron around the outside of the building, and the apron connected to a drain. Buildings should be generously sized to facilitate maintenance.

Good crane and maintenance access is needed with a concrete standing above flood level for storage of electrical gear removed from the substation. Transformers often stand in the open and a good security enclosure is required around the transformer and substation area to prevent unauthorized access. Transformers can also emit a buzz, which is recognized as a noise nuisance, and may need screening or placing in a sound absorbing enclosure. Lightning protection may be needed.

14.4.2   TRANSMISSION LINES

Transmission lines should be run parallel to the road system and if there are adequate service reservations, it is often possible to run cables at ground level. Underground cables are generally safer than overhead or ground lines providing accurate records of their position are kept and used so that cables are not accidentally dug up or severed during construction work.

Underground cables should be run in sand-filled trenches covered by concrete tiles or a coloured concrete mix. These cables should, if possible, be run at the high point of paving leaving room for draw boxes. It is essential that cables, along with underground pipes, are put into position at the same time as foundation work is being undertaken.

Overhead lines, together with the poles and guide wires or props should not hinder maintenance equipment and should not be capable of being touched by cranes. They should be protected from damage by collision, fire or explosion and may need lightning protection. Unless fully insulated they must avoid flameproof and hazard areas. It may be desirable to make the poles strong enough to support floodlights.

Similar considerations apply to lighting standards which should be made flameproof for installation in hazardous areas.

14.5   OTHER UTILITIES

Boiler water treatment plants are usually situated within the boiler house and similarly process water plants may be put near the process. However, very large treatment plants may be contained in a separate building or area to provide better access for treatment chemicals and for locating water storage and effluent drainage systems.

Fig. 14.6 Water chilling equipment installed near plant (Courtesy: British Nuclear Fuels)

Service equipment such as air compressors, refrigeration, inert gas units, metering stations, condensate return pumps and similar equipment are usually small enough not to present problems in site layout. The equipment may be located in service buildings in or near the boiler house for operational convenience, but installations near to or inside manufacturing plants are also quite practical, particularly for refrigeration (see Fig. 14.6).

If air-conditioning plant is required for a particular building it should be installed near to, and treated as part of, that building.

14.6 UTILITY CENTRALIZATION AND DISTRIBUTION

A single large central utility plant has lower running and maintenance costs than a series of smaller plants. The equipment and building costs are also lower for a central plant, but there is the cost of the site distribution network and associated heat losses. A breakdown in a site service affects the whole site, unlike a local plant service whose effects are restricted. In choosing between a site or a decentralized system the following considerations apply:

(a) Economic and reliability studies.
(b) Layout of the distribution mains.
(c) Advantages of ring main systems.

A service required by only one plant should be located in or nearby that plant. Nearby, but off-plot, locations are preferable if there is the chance that a future plant may need that service as it is good practice to ensure that plant operation is not hampered by activities on neighbouring plants.

Site expansion plans should be considered when positioning central services plants. These services are often centrally placed so that site expansion can proceed in all directions. However, this may conflict with the policy of locating low-hazard plants, such as the utilities, on the site boundary. It may also mean that after extension the utilities will be undesirably surrounded by hazardous areas (see Ch. 8).

Service distribution from the central buildings should follow the site road system and run parallel to roadways. Service runs should not pass through plant areas. If this cannot be avoided, the services should be well protected against corrosion, fire or explosion damage within the plant area. Most piped services (steam, water, gas, condensate) with some medium- and low-voltage cables and some instrumentation lines are run on piperacks (see sections 31.4 (p. 436) and 31.5 (p. 439)). Long water mains may be buried to prevent freezing, instead of using lagging. Open pipe trenches may be used in places where there is no risk of flammable vapours collecting in them or of the material in the pipe freezing, e.g. steam mains. It is cheaper to have pipes at ground level on sleepers but this method should only be used where access by both vehicles and pedestrians is not hindered. At road crossings, therefore, pipes should either drop below ground or be raised sufficiently to provide headroom.

The distribution of services within a plot is discussed in section 31.6 (p. 446).

REFERENCES

1. BS 5908 'Code of Practice for Fire Precautions in Chemical Plant', British Standards Institution, London, 1980.
2. Balemans, A. W. M. *et al.* 'Check-list guidelines for safety design in process plants', in: Buschmann, C. H. (ed.), *Loss Prevention and Safety Promotion in the Process Industries*. Elsevier Scientific Publishing Co., Amsterdam, 1974.

CENTRAL SERVICES

15.1   ADMINISTRATION

Administration requires main offices for site operational management, accounts and personnel and for the communications network. Visitors are received at the administration offices so that the building should be designed to give a favourable impression of the company and the site operations (see Figs 2.1 (p. 7), 4.2 (p. 30) and 14.5 (p. 212)).

Only those production operators, local production supervisors and engineers with specific local technical responsibilities should normally have offices near the plant. Personnel with more general site responsibilities should be housed in an administration building sited in a non-hazard area near the main entrance. For major hazard sites it should be outside the main fence, although there may be an internal fence to give the necessary degree of local security. Personnel traffic between the administration building and the operating plants will usually be of a small but steady volume.

The building should be clear of drifting fumes, smoke or steam from plants and noise from road and rail operation. It should have adequate parking for staff and visitors preferably outside the site fence. A small waiting area for taxis and similar vehicles is useful near the main building entrance preferably within sight of the reception area in the building.

The building will contain substantial amounts of vital and confidential management information and some parts, at least, must be secure against blast, fire and firefighting damage, plant or off-site hazards, pickets and unauthorized entry or occupation. Apart from these threats, the building must conform to the national and local fire protection standards for offices.

15.2   AMENITIES

Canteens, the medical centre, employee shops and other general facilities are used by staff from all parts of the site. Thus they should be located together in a central area, so that the time spent by staff travelling to and from the area is minimal. Local enquiry and information offices for important employee matters such as wages, personnel or trade union affairs can also be situated in this area. However, it may be necessary to have separate canteen facilities for the administration building, if the latter personnel have to traverse a process area to reach the canteen.[1]

The amenities should obviously be in a safe area, upwind from drifting fumes and noise and should permit smoking, cooking, etc. to be unrestricted. Means of escape from amenity and other buildings, such as workshops and laboratories, should be on the side of the building furthest from the hazard and should lead away from the hazard.

The buildings and their immediate surroundings should be attractive to staff using the amenities. Essential activities, such as offloading food and the collecting of refuse bins and waste should be unobtrusive and not interfere with access to, nor detract from the enjoyment of, the amenities. One idea is to have the back-up firefighting water supplies as a landscaped feature (see Fig. 14.5, p. 212) perhaps near the amenities. The canteen should be usable as an overflow medical centre in the case of multiple casualties. Proper fire-protection facilities must be installed.

If possible, it should be arranged that the amenities are sited outside the main plant area, perhaps near to the main administration building. Access for food suppliers and waste-disposal wagons must be isolated from hazardous areas, and this may conveniently be arranged if the amenity area is separated from the main plant site with independent access. It is the practice to locate plants at distances from the boundary fence that depend upon the hazard rating of the plant, so safety requirements for personnel using amenities are always met if the amenity area is placed outside the boundary fence. If this is so it may not then be necessary to provide separate canteen facilities for the administration building. Screening of the administration building and amenity area from the main plant site by trees and landscaping is more economic if the areas are adjacent.

After hours sport and social facilities for the employees should preferably be off the site altogether or, at least, outside the plant fence area.

15.3 LABORATORIES

One of the main functions of the laboratory is to provide an analytical service both for routine and extraordinary samples of raw materials and products. The laboratory may comprise of analytical services and groups of staff carrying out pilot-plant or laboratory-scale investigations of the research and development type which have been sited to make use of available feedstocks or other site facilities. Thus the particular requirements for research and development work may influence the location of the laboratory, but otherwise the laboratory should be strategically placed to maintain the range of services with minimal transportation of personnel and supplies to all parts of the site.

The cost of transportation of routine samples from points of sampling to a central laboratory may be prohibitive. If such a problem arises, the siting of subsidiary laboratories in individual plants, possibly of the lock-up type, should be considered (see section 17.10, p. 261). In this way some samples may be analysed and recorded locally whilst other samples, particularly those that require expensive instrumental analysis, are transported to a central laboratory. The location of laboratories on plants may conflict with the concept of minimizing the number of personnel in the process areas. Thus a balance between hazard cost and convenience has to be struck in each case.

Central laboratories will contain flammable and potentially hazardous materials, and may house small-scale, but potentially hazardous experiments and chemical reactions.[2] The risk of fire must be studied and good firefighting access be provided at appropriate points. Liquid effluent may contain undesirable material or flammable solvents. The requirements of effluent handling should be studied, and adequate facilities be provided, if possible combined with the site facilities.

15.4   WORKSHOPS AND STORES

Services such as main workshops and garages require pedestrian and vehicle access to all parts of the site, including the remote corners. They should each be located after a traffic/travel study of the frequency and load and type of vehicle to be used. Access should be direct and avoid passing through plant areas.

Workshop and vehicle maintenance require external areas for equipment storage and cleaning or for large fabrications and repairs. Welding torches and naked flames will be used inside and around the workshop. The risk of drifting sparks must be studied and adequate distances between the workshop and adjacent plants must be maintained.

The central stores will often be located close to the central workshops, as like the latter they are normally in a position which reduces the distances pedestrians and vehicles travel to all parts of the site, including outside vehicles delivering materials. Security of the buildings can be vital if hazardous or valuable items are stocked. Good visibility to all store doors, windows and roofs will help security staff and management as well as discouraging intruders.

For both workshops and stores, off-road loading and unloading facilities should be designed and adequate parking space should be provided to avoid interference with moving traffic. Appropriate fire-protection measures should be installed.

It may be necessary to locate subsidiary workshops and stores at process plants to reduce delays in transport and repair. The local workshops may be of the lock-up type, equipped with the smallest range of machines and tools as the occasion demands. Lock-up stores containing supplies in frequent demand may also be housed near process plants.

15.5 EMERGENCY SERVICES AND CONTROL

Ambulance and fire stations should be positioned to give rapid vehicle access to all parts of the site without hazard to factory traffic. This requirement may be met by locating the stations outside the main fence, but close to the main site entrance, although security gates may be suitable provided the entrances can be opened for or by the services at all times.

Adequate space should be allowed around the station for crew-training exercises, hose-drying and other activities. The building must be in a safe area and be free from risk or damage by plant fire or explosion.

A disaster-control point equipped with telephone and ideally a radio tannoy communication system should be planned. This, like the fire station, medical centre, telephone exchange and emergency stores must be located away from hazardous sources. It should have quick access to the main entrance, which must also be in a non-hazardous area. This will enable external aid to be brought in and deployed under control of the site emergency controller. A rendezvous point for emergency vehicles near the main entrances should be considered. The layout of the main gateway should discourage the gathering of spectators. The layout of the site road system including emergency requirements is considered in section 9.5 (p. 144). The layout for fire-fighting water supplies is discussed in section 14.3.2 (p. 211).

It is desirable to have a number of additional subcontrol points about the site containing emergency equipment and connected to the main point by telephone and, ideally, by radio. Process control rooms can fulfil this function (see section 17.9.1, p. 257).

Safe areas (possibly with shelter from weather) which can easily be reached on foot, should be set aside as emergency assembly points for staff. Accounting for personnel after an incident is then both simpler and quicker. These locations should be equipped with emergency equipment and telephone/radio or within range of a tannoy system. The advantages of also having refuge rooms should be considered. These (and the main control point) should be capable of being easily sealed and have sufficient air supplies and heating/cooling and humidity-adsorbing capacity for the duration of a likely emergency.[3]

In planning for emergencies the effect of an incident spreading to or from a neighbouring site must be considered. In particular, safe areas should be outside the falling range of adjacent off-site buildings and structures.

Further details for dealing with emergencies are given by the Chemical Industries Association.[4]

REFERENCES

1. HSC *Second Report of the Advisory Committee on Major Hazards*. HMSO, 1979.
2. I.Chem.E. 'Safety in university chemical engineering departments', *Chem. Engr. Lond.* **357**, 439, 1980.

3. Duff, D. M. S. and Husband, P. in: Buschmann, C. H. (ed.), *Loss Prevention and Safety Promotion in the Process Industries*. Elsevier Scientific Publishing Co., Amsterdam, 1974.
4. CIA *Control of Major Emergencies*. Chemical Industries Association, London, 1980.

CONSTRUCTION AND LAYOUT

16.1   SITE CONSIDERATIONS

Construction needs should be taken into account during site layout in order to ensure that the various plants are located on suitable ground, e.g. heavy or high plants on good load-bearing ground, or plants needing substantial excavation or pits are not placed in high water table areas. For plants where heavy lifting gear is needed, the construction access must also be on good ground.

In laying out the site, the road network should take account of construction traffic and equipment (Figs 16.1, 16.5 and 31.11 (p. 446)). Bend and kerb radii, pipebridge and cable track heights should allow passage of cranes, scrapers, vessels on bogies, etc. Temporary roads may be needed and allowed for in the layout, but must avoid crossing buried cables or drains. These roads should preferably be capable of upgrading to permanent site roads. Widths of temporary and permanent gates should be checked against equipment to be used.

Fig. 16.1   Road needs for construction (Courtesy: British Petroleum)

Fig. 16.2   Construction storage compound (Courtesy: Burmah-Castrol (UK))

Security needs start when construction begins. A construction site contains much valuable material and equipment (see Fig. 16.2) and is also, because of the equipment, a dangerous place for intruders. The sizes of items of equipment are no safeguard against theft, as many former owners of earthmovers can testify. The construction site must be treated for security purposes as if it were a fully operational plant. Thus, a good security fence around the site is necessary with adequate gate control. The fence need not be at the site boundary unless all the site is under development since extra fencing is costly to build and supervise. The fence is intended to keep out intruders for their own safety and for the security of materials on site. A weighbridge is essential on a large site to check deliveries, to control outgoing vehicles and to help prevent theft. Facilities for car parking and public transport should be provided near the site but outside the fence. Approach roads to the site should not be narrow and thus easy to block by heavy vehicles and site traffic.

On large, remote or overseas sites, it may be necessary to consider living accommodation for the whole construction crew or for a small number of expatriate specialists. This should be near, but generally outside, the site.

Access around plants for construction equipment must be allowed within the site (see Fig. 16.3). Hardstanding for vehicles, construction equipment and for temporary storage of process equipment will be needed. Locked compounds and stores buildings for valuable construction equipment and commodities must be provided and separate, isolated locked compounds for gas and liquid fuels are essential. Site management and design offices, workshops and amenities will be required near the plants by main contractors and sub-contractors' construction crews (see Fig. 16.4). All these facilities are temporary and plans must be made for their rundown and use as plant space as construction tails off. It is better if these activities can be located in a separate compound outside the plant fence, but this ideal situation is rarely possible.

On a site where plants are spaced well apart, the interplant spaces can be used for construction needs. When plants are started up, these activities must be moved to safe areas. If space on site is limited the pace of construction will be reduced because men, equipment and materials cannot be deployed extensively and effectively. This consideration may affect choice of site. Similarly shortage of areas to lay out construction materials can present a difficult problem if a site inside a works is being redeveloped or expanded.

Some planning assumptions must be reflected in the layout. Progressive

Fig. 16.3 Construction access (Courtesy: Babcock Woodall-Duckham)

excavation and temporary spoil heap areas must match construction procedure. Individual plants must not be isolated by surrounding excavation. If any major equipment items will close-off access after installation, alternative access for subsequent work must be planned.

Commissioning requires offices, laboratories and, perhaps, materials storage. The programme should allow temporary use of permanent facilities for these purposes. On a new site, however, completion of these facilities may follow production plant completion and temporary facilities will be needed and must be allocated space in the layout. Redundant construction facilities can sometimes be used but some construction work inevitably proceeds dur-

Fig. 16.4   Construction site facilities (Courtesy: British Petroleum)

ing commissioning. Also safety considerations may mean that such facilities are in unsafe areas when the plant commissioning starts.

Where new plants are constructed on existing sites with plants in operation, means of safely accommodating the additional number of construction workers can be difficult. The amenity facilities should be provided at a safe distance from the work so that the construction workers are only exposed to any process risk whilst working. Emergency warnings and procedures and means of escape should be clearly recognized and understood and adequate training given.

A datum point for the site must be selected and used as a master reference for setting out all the plants, roads, levels and other site features. All site drawings must show the datum and all plant drawings must carry reference co-ordinates back to the site datum. This single decision will ensure that all design disciplines, contractors, planners, lawyers, etc. are working to a common, unambiguous reference for the positioning, dimensioning and setting-out and for legal purposes and will avoid many disputes during construction.

The method of selection of the datum point is discussed in section 7.2 (p. 69).

16.2   PLOT CONSIDERATIONS

The layout of a plot must allow for:

(a) Access for construction.
(b) Working conditions during construction.
(c) The sequence of construction and critical equipment delivery dates.

During the initial stages of layout these important construction considerations must be included amongst other factors in making decisions on layout. For example, a building may be preferred by process and operations needs, but it could hamper construction access while affording protection from the weather during construction. Similarly, a multi-level plant in an open structure may be more difficult to erect than a ground-level plant.

Access space is needed for a number of purposes:

(a) Storage area just off plot for equipment deliveries.
(b) Route to equipment location including room for the means of transport.
(c) Temporary lay-down area at the location.
(d) Room for construction tools and equipment, e.g. cranes.
(e) Room for the tradesman to work safely and comfortably.

The plot should be so designed that adequate access is available to lift large items of equipment or columns into place. In the case of a tall column, the vessel is raised to the vertical position at its location on the site and space must be available for it to be laid down on the ground adjacent to its foundation (Fig. 16.3). When such equipment is positioned close to boundary limits so that erection must take place from outside these limits, a careful check must be undertaken to ascertain whether space will be available at the time of the erection for positioning cranes or lifting tackles – and the load-bearing properties of the ground checked if not done as part of the site survey.

If a building or structure is needed (discussed in Ch. 18), lifting wells, removal panels and temporary lifting beams should be considered and equipment positioned to make use of these facilities (Figs 18.7–18.9, pp. 272–4). However, heavy items should be near main stanchions and the lifting facilities have to accommodate this.

Access space planned for operation, maintenance and catalyst loading, etc. should be utilized for manoeuvring and storage during construction but additional space may be needed for these purposes and for items such as scaffolding, rigging, etc. Some space may also be needed for dismantling construction equipment and removing it from within and around the plant. The standards laid down for headroom and gangways designed for operation should be reviewed so that construction equipment (e.g. fork-lift trucks) can use the spaces. If maintenance equipment such as cranes, lifting beams, davits, etc. are designed into the plant, their potential use during construction should be considered, but the devices must then be carefully tested before being used. A plant that has good construction access most probably will have good maintenance access (as in Fig. 18.7, p. 272). Therefore, it is essential to maintain specified spacing and headroom clearances during all phases of plant layout and piping design.

Consideration should also be given to long-delivery items of equipment which it is known may well arrive fairly late or out of schedule in the construction programme and therefore have to be fitted into place after most of the surrounding equipment has already been installed. However, equipment that has to be removed for maintenance should present no problem if it arrives late in construction, other than completion of interconnecting pipework, ducts, etc.

Consideration must be given to protection from the weather. If the building is to be enclosed, then the walls, etc. should be erected as early as possible. The permanent floors and stairways should likewise be installed promptly. Permanent lifting equipment should be available. Good layout planning will ensure that these facilities do not interfere with the later part of the construction and, in fact, can be used safely for this.

On extensions to existing plant (see south-west corner of Fig. 2.1 (p. 7) and east side of Fig. 4.2 (p. 30)) care should be taken to ensure that construction work interferes as little as possible with the running of the existing plants. Passage of long jib cranes, vehicles carrying large overhanging loads, erection of high plant or equipment and welding near hazardous installations all give cause for concern. Adequate permit-to-work systems combined with plant shut-downs are necessary but their need can be reduced by proper planning of the layout of the extension.

Construction workshops and amenities should be as near the plot as possible. This may mean having satellite facilities on large construction sites (see section 16.1).

Good working conditions are difficult to achieve on construction sites and this is one of the problems that has led to the use of modular construction described in section 16.3.

The sequence of construction is often:

(a) Clearance.
(b) Erection of construction facilities.
(c) Roads and drainage.
(d) Foundations.
(e) Structures.
(f) Equipment.
(g) Piping and electrics.
(h) Instruments.
(i) Lagging.
(j) Testing.
(k) Painting and tidying.
(l) Removal of construction facilities.
(m) Commissioning.

In order to decrease construction times the above activities will overlap. Layout models, drawings, etc. will be used in order to plan the construction programme. The objective is to avoid sharp peak-loading for any particular trade. The concept of few men working for most of the contract is the ideal. A tradesman, say a pipefitter, should progress across the plot following another trade, e.g. the equipment erection team.

The layout should be examined to see if such a programme is possible. This will involve splitting the plot into sections and deciding in which order sections will be started. Generally, the more complex parts or the sections near the plot centre are the first to be erected. To enhance this scheme, modifications and extensions to the layout may be indicated. For example, high or heavy equipment should be located near the construction accesses. A favourable balance in the cost of the extra structure, pipework, etc. against the cost of construction time must be found to justify such layout changes.

The commissioning requirements should also be planned. These will closely follow operational requirements (see sections 17.4.1 (p. 246) and 17.5 (p. 249)). However there will be the need for extra instrument, sample and drainage points, blanking off facilities etc. and convenient and safe access to these are needed.

16.3   MODULAR CONSTRUCTION

The purpose of modular construction is to have as much construction, inspection and erection work as possible done in the workshop and as little as possible on site.[1-3] The technique for doing this is to divide a plant into units or modules which are fabricated and assembled in specialist assembly yards remote from the construction site. They are then transported to the site where they are connected together to complete an installation (e.g. Fig. 16.5). The verb 'modularize' and associated noun 'modulization' are used to describe this process, particularly the choice of modules which is done at the plant layout stage.

Fig. 16.5   Typical modularized site (Courtesy: John Brown)

A module consists of the process item, lagged if required, plus the local structure, local pipework and electrics, instrumentation, etc. Piperacks, control rooms, laboratories, etc. can also be modulized.

The construction work on site consists of the major structure and foundation (Fig. 16.6) plus roads and underground work. The modules are lifted into position on to the major structure and connected up. For schemes with large modules, these will contain all the structural steelwork. Thus the module is placed directly on to the foundations.

Some of the instances when modular construction is used are:

(a)   On-site construction areas which are non-existent or very restricted and expensive, e.g. in offshore installations.

Fig. 16.6 Foundation for modules (Courtesy: British Petroleum)

(b) When on-site construction time is limited, e.g. in remote areas. The climate is frequently not conducive to long working hours in the open and may even be hostile enough to make construction work impossible except for short periods (polar regions, jungle, desert, etc.). Other factors giving rise to low productivity are, for example, religious customs and short days in winter.

(c) Local construction labour does not exist and has to be trained or imported, e.g. in developing countries. Expatriates require high rates of pay and accommodation in specially built camps supported by an expensive commissariat service.

(d) Difficult labour relations, e.g. in developed countries. These contribute to low overall productivity and hence protracted and unpredictable construction times. Reduction in the number of site activities and workers has been found to improve industrial relations.

The advantages of modular construction include:

(a) It is found that modular construction can reduce construction costs (typically by 10 per cent or more[3] by increasing reliability of target times and by decreasing the construction time (e.g. by 3 months). The relatively high productivity of the fabrication yards compensates the time lost in shipping.

(b) A more reliable and better-engineered plant is also obtained thereby providing a plant which runs more efficiently. The yard organization is permanent with perfected working and management procedures, whereas the construction organization on site is more-or-less renewed for each contract and thereby more liable to have flaws. Yards and workshops are normally located in industrial areas where there is an availability of skilled labour. The workforce lives locally so there are no accommodation problems and travelling and living costs are accounted for in the wage rates. Much of the work is carried out under cover in reasonable conditions. It is thus less subject to weather, season, minor industrial disputes and is more easily and effectively planned and programmed. Wage rates are not subject to the usual uplift for site working.

(c) Erection of steelwork and installation of equipment, piping, instrumentation and major cable runs can be carried out off site while civil works are taking place on site.

(d) There are also fewer delays on sites with existing plant, due to the need to obtain permits to work, provide protection for the existing plant and to suspend work during leaks of flammable material, blowdown, obnoxious operations, etc.

(e) The size of a construction camp is reduced.

(f) Extensive pre-commissioning work can be done off-site, thus reducing the site commissioning time.

(g) It is found that foundations for modules are simpler and can be standardized. Plot space may be saved.

(h) The effects of industrial action can be minimized by careful selection of workshops and by spreading the work to more than one pre-assembly site. Industrial unrest particularly near completion is a problem on a construction site and the whole site is affected.

The principal disadvantage of modular construction is an increase in design material and transportation costs due to:

(a) Earlier and a greater quantity of detail design required.

(b) Additional supporting steel framework, piping fittings and increase in the number of special design items.

(c) The increase in organization and planning needed in selecting and co-ordinating the work of the various yards.

(d) The preparation, shipping, off-loading and location of modules on site.

These four points are amplified below.

16.3.1   EARLIER DESIGN

The adoption of modulization enables the work on equipment, etc. to commence at an earlier date than would be possible in conventional construction since module fabrication can go on simultaneously with the civil work on site. However, to enable the construction of several modules to proceed in parallel, possibly at different locations, the design and mechanical detailing have to be completed and frozen earlier than usual, resulting in an earlier design peakload. The inspection organization is principally employed off site rather than on site.

The programme benefits can only be realized provided the detailed engineering and procurement activities are achieved to plan, which in turn requires a more sophisticated project organization than that normally associated with conventional construction.

16.3.2   ADDITIONAL SUPPORTING STEEL FRAMEWORK

Additional steel goes into the module frames to make them rigid, transportable units, if the permanent local structure does not have sufficient strength. The module can, in transit, suffer temperature changes and will assume different positions to its eventual permanent position (e.g. a distillation column will be shipped horizontally but installed vertically) so stress calculations on structures, equipment and pipework must be made for all foreseeable positions. The dynamic effects of sea transport must also be considered; apart from strength considerations, rotating parts of equipment must be braced to prevent bearing damage. The lagging and paint finishes have to be made vibration and abrasion proof. Any temporary packaging has to be easily removable after the module has been fixed into position.

16.3.3   INCREASE IN ORGANIZATION AND PLANNING

Points to consider in the selection of the fabrication yards are:

(a) Position relative to site.
(b) Waterside facilities or road/rail access.
(c) Adequate space for the module and for stores.
(d) Skill and flexibility of, and relationship between, labour and management.

16.3.4   PLACING MODULES ON SITE

The cost of transporters, tugs and barges or roll on/roll off (RoRo) ships specially designed for the purpose of carrying these heavy loads and expensive to charter is a major factor to be considered. In some cases there is the cost of taking the land transportation vehicles by sea to remote sites for moving modules from the sea transportation to site. There is the cost of possible strengthening and widening of jetties to support these heavy loads to be allowed for.

Large modules are moved over foundations on their transportation and accurately positioned. Jacking legs are attached to the modules, and the weight taken on the jacks and the modules lowered hydraulically into position under controlled conditions. Small modules are lifted into position onto pre-erected structures from their transport by cranes or perhaps helicopters. There is an alternative to land transportation on site. This is to float the module into a flooded pit which is then drained.

Modules are normally set with a minimum distance of 1 m between adjacent modules, piperacks or equipment. The metre distance allows for movement during transport and for taking up fabrication and setting down tolerances by using pipespools.

It is possible to pre-erect the whole plant offsite and then split into modules. This is costly and not particularly satisfactory, as the modules may alter shape slightly during travel. Certainly, if an item of equipment spreads over two or more modules it should be pre-erected in the works. However, where pipework or ducting provides the connections between modules, the modules should be connected for the first time on site.

To save on transport costs it is possible to dismantle the components of a module and make a 'kit' to be reassembled on site. However, the transport saving is probably outweighed by the cost of dismantling and reassembling.

It is essential in layout to allow for flexibility in the sequence of module installation. Although a planned delivery sequence will be established by a site installation programme, for a variety of reasons the actual module delivery may not follow the planned sequence, and in this event site work should not be held up because of non-delivery of a particular module. It is also essential to maintain access for module movement on site. This can influence the routing of underground services.

Modular construction of process plant can only be justified after careful evaluation of all the costs involved, including allowance for the risk of loss or damage during transportation.

16.4 MODULE DESIGN AND LAYOUT

An important evaluation phase must take place early in the design sequence to determine which equipment best comprises each module and equally, which equipment may not be modularized. The physical relationship of non-modularized equipment to modularized equipment needs to be settled at the same time. This evaluation is based on a review of the process flow diagrams to identify compatible groups of equipment for accommodation into individual modules whilst generally adhering to the natural process sequence. Careful consideration of the consequences to safety, operability, maintenance and cost is required before removing an item of equipment from its natural location in the process sequence in order to utilize free space within a module. In particular, for a well-tried process being modularized for the first time, it is probably best to retain the original layout because of the operational experience, etc. inherently contained therein, rather than trying to improve the modularization.

All types of plant can be modularized to some degree and first the block and then the piping model are usually used as aids to decide on the modules (Fig. 16.7). Plants which contain items of equipment of a similar size are easier to lay out in modular form and result in the most economical use of structural steelwork (see section 18.2, p. 267). Plants with tall columns or long vessels are more difficult to modularize. Such vessels can be pre-assembled with their internal and external attachments and transported to site horizontally as single units.

Heavy machinery such as large compressors are particularly difficult to incorporate into modules owing to the weight of steel required to support them.

Whilst elevated piperacks can be economic in pre-assembly, ground-level pipeways normally requiring little or no steel support would need extensive

Table 16.1 Typical sizes of modules (Courtesy: *Processing*[3])

Type of pre-assembly	Structural features	Typical max. weight gross	Typical dimensions (m)	Transport mode	Storage features	Typical application
Vendor Packaged Units (VPU) or packaged units						
(a) Pre-dressed	Free-standing equipment item with associated items, e.g. columns with appurtenances, instruments, etc.	Determined by transport mode		Often by road but could be sea		Free-standing equipment items Part of main plant
(b) Containers	Standard cage type suitable for lifting	25–30t	12 × 2.5 × 2.5	Standard container modes	Stackable 5–6 high	Small water treatment and service units
(c) Skid mounted	Open or cage types complex arrangements of small plants or sections, ground erection or stackable	60–70t	14 × 4 × 3.4	Road or barge	Stackable up to 4-high	See water desalination units, small gas treatment units, sulphur recovery plants. Large compressors.
On-shore modules or **preassembled units (PAU)**	(a) Suitable for ground level and lifted installation	300t	25 × 15 × 10	Usually sea. Road transport needs careful definition	Limited	Oil, gas and chemical plants. Power stations.
	(b) Suitable for ground level installation using self-propelled or towed trucks. Support structure may be buried.	2000t	40 × 25 × 25			
Barge mounted	Ship designed 'bedded' at site	variable (36 000t –260 000t)	184 × 44 × 13.8	Sea	Suitable depth berthing	Oil, gas and chemical processing

(a)

Fig. 16.7 Models of modularized plant (a) development of module (b) a complete module (Both courtesy: Petrocarbon Developments) (c) assembled modules (Courtesy: John Brown)

(b)

(c)

steel framing to make each section transportable and the cost would usually outweigh the saving in site welding costs.

Most storage tanks are too large and flexible to be moved any distance and are therefore best erected on site. Short movements of tanks on air cushions have been achieved but the idea does not really have a practical application to pre-assembly considering the normal transport distances involved.

A great deal of site work can be saved by maximizing the instrumentation located in modules, but early specification of control valves and other instrumentation is needed. Module control rooms are now an accepted form of construction especially where distributive control systems are adopted.

Electrical aspects are mainly concerned with lighting and small power supplies on process modules. The location of cable routes and junction boxes can be important in plant layout considerations in modules.

The suitability of equipment for incorporation into modules depends on their elevation, size and weight and how these affect the height of the centre of gravity of the module. Should the equipment result in a module which is unstable during sea or land transportation, in spite of ballasting and careful structural design, then it is unsuitable for modularization.

Actual module size may be less than the maximum allowed. The ideal for modularization is to obtain the minimum number of modules consistent with the minimum amount of framework and the minimum number of connections on site. Some sections of the plant will form obvious modules, smaller than the maximum size.

235

The maximum sizes and weights of modules are largely determined by the transportation considerations from the fabrication yard to the foundations on site, being smallest for air transport, standard container size for road and rail and very large for transport mainly by water (see Table 16.1 and Fig. 16.8). Size may also be restricted by the desirability of being able to get at all the equipment on each module from the outside.

As a general guide, the larger the module the greater the reduction in site work, i.e. in providing structures and in connecting up. Also larger modules give a better ratio of equipment, permanent structure and piping weights to the weight of additional framework needed for transportation. On the other hand the larger the module the more planning has to be done for the transfer.

For public transport (rail or road) modules are confined to the standard container size 12 m × 3 m × 3 m. Larger modules can sometimes be carried on public roads by arrangement with the authorities. Normally though, large modules can only be carried by water and where there is private access between fabrication yard and the ship and between ship and site. For these marine-conveyed modules, dimensions are:

(a) *Width*

Roll on/roll off ships allow unit widths up to 18 m. Barges have widths up to 30 m which is usually larger than can be used. Road transporters can take module widths up to 20 m on unrestricted roads. At least 1 m has to be added to the width of a water borne module to allow for the site bogey transporters (see Fig. 16.9). Normal site roads are only 7.5–10 m wide.

(b) *Length*

There is unlikely to be a constraint on length of modules transported by sea. On land 40 m is the normal maximum length. This is determined by the bend radii of the roads.

Fig. 16.8　Transportation of modules (a) by road (Courtesy: APV Hall International) (b) roll on/roll off ship (c) by towing barge with pre-assembled piperacks (Both courtesy: John Brown)

(a)

(b)

(c)

Fig. 16.9 Site transportation of modules by trailer (Courtesy: British Petroleum)

(c) *Height*

The height is determined by the height of the centre of gravity. This affects the stability of the ship or land transporter. Marine insurers may insist on certain design criteria. On land, gradients up to 1 in 15 can be negotiated but road camber and side tilts must be kept as low as possible.

(d) *Weight*

The weight is not a limiting factor unless the module is to be lifted by crane on site as ships can accommodate any likely module weight and the weight limitation with land transportation is 1,500–1,800 t. However, the greater the weight, the heavier the supporting steelwork has to be and the more likely the module will change shape during voyage.

16.5   PACKAGED PLANTS

A number of plants may be bought as complete packages mounted on skids. Typical of these are service plants such as water treatment, inert gas, refrigeration (see Fig. 16.10), boiler and cryogenic[4] plants. Process plants for producing acids or for various kinds of recovery, are also sold as packages. Specialized electrochemical plant such as halogen-producing equipment[5] or reducing plant is provided as a package (see Fig. 16.11).

The user has to provide space and connections for the package. The layout within such packages, incorporates the experience of the vendor and commonly is more compact than normal plant (see Fig. 16.10b). Although the layout offered may not be the optimum for a particular situation it is usually best to accept it as non-standard layouts cost extra.

Fig. 16.10 Example of package plants (a) a refrigeration package in transit (b) a refrigeration package after installation (Both courtesy: APV Hall International)

(a)

(b)

Fig. 16.11 Example of specialized plant (fluorine cells) (Courtesy: British Nuclear Fuels)

REFERENCES

1. I. Mech. E. *Modular Design and Construction of On-shore Process Plant*. Inst. of Mech. Engrs, 1980.
2. Anon. 'Packaged plant conference and exhibition supplement', *Process Engineering*, June 1981.
3. Anon. 'Package plant', *Processing*, 27, Jan., 1982.
4. BCC *Cryogenics Safety Manual: a guide to good practice*. British Cryogenics Council, (2 edn.), Mechanical Engineering Publications, 1982.
5. Hine, F. 'The optimum number of cells in a amalgam process chlorine plant', *Electrochemical Technology*, **117**, 139, 1970.

DETAILS OF PLOT LAYOUT

17.1 PROCESS REQUIREMENTS

Plant layout should first be considered from a process point of view since the efficiency of the plant operation is of the greatest importance. The equipment, certainly initially, should be laid out using the flowsheets and generally in the order of process flow requirements.

The process design should have been subjected to hazard and operability analysis. In particular the advantages of gravity flow will have been reconciled with the danger of having hazardous materials at elevation on structures. Also the dangers to personnel of man-handling chemicals, etc. would have been considered.

Process considerations may suggest that some items be elevated to provide gravity flow of materials, to accommodate pump suction requirements, to provide drawing head around a thermosyphon reboiler, or to allow the use of barometric legs. When, after consideration of economics and process factors, gravity flow is advantageous, a multi-storey or, at least, a tall, single storey structure is provided to support the equipment at its correct level. It should be noted that hills and mounds can be used to provide a safer means of gravity flow than a structure.

Other process considerations include limitations of pressure or temperature drop in transfer lines to and from heat exchangers, furnaces, reactors, and columns. Isolation of hazardous areas was discussed in Chapter 8.

Space has to be provided for handling facilities. Typical needs are solid chemicals storage near a process vessel in bags or drums, tote bins for filter cleaning and manual solids transport between items of process equipment.

If liquid is near its boiling point, a static head is required upstream of an orifice meter (and, possibly, a control valve) to avoid flashing in the line. For the same reason pumps should be put close to, and a sufficient distance below their point of suction, subject to having a valve between pump and supply for emergency purposes. Orifice meters each need to be placed in a straight length of pipe. Where a pipeline operates at extreme high or low temperatures, allowance must be made for the line to be flexible enough to expand or contract without damage both to the pipes and the equipment connected by the pipes. (Pipe stressing and instrument layout are discussed in Ch. 31.)

Process requirements tend to be forgotten during the later layout stages and the process designers should frequently review the layout during development.

17.2 CHOICE OF PLANT STRUCTURE

Economies can be made in layout by reducing interconnecting and supporting equipment, i.e. steelwork, concrete, piping and electric cables. The plant should be laid out as compactly as possible, consistent with maintenance and safety clearances and operational access.

To achieve savings the guiding principle of detailed layout design[1,2] is to eliminate, combine and minimize structures.

Equipment located on the ground saves the use of structural material. The cost of positioning equipment in structures of more than two floors is inflated by the need to provide access platforms and stairways, for both operation and maintenance, whilst the compensations from saving ground space and enclosure costs become less important. Above 30 m the cost of operator safety becomes increasingly heavy. Likewise, removal of heat exchanger tube bundles or other internals from process vessels may influence the positioning of equipment above ground level, and this can increase the cost because extra supporting structure and extended process and utilities pipework become necessary. On the other hand, spreading equipment out in the ground can require extensive pipework.

It is then necessary to decide whether to have enclosed buildings. These can either be at grade around a particular item or group of items or can enclose a complete plant.

Enclosed buildings are to be avoided if possible for the following reasons:

(a) Extra cost both of the cover itself (walls, roof, windows) and of the need to strengthen the structure and foundations to take the weight of cover and counteract windloads induced by the cover.

(b) Ventilation may be required to disperse toxic and flammable fumes.

(c) Operator, construction, maintenance and emergency access may be impeded.

(d) Fires may spread quickly and firefighting is impeded.

Valid reasons for having an enclosed plant include:

(a) Continuous or very frequent operator or maintenance attention is required.

(b) The process requires a sterile, stable or special environment.

(c) Process materials cannot be exposed to wind, rain or strong sunlight.

(d) The process is secret.

(e) The structure is very high and operator safety and morale are better in an enclosure.

(f) Plants in buildings are sometimes taxed less than open plants.

A compromise between the above conflicts may be decided. For example, Figs 4.1 (p. 29) and 23.3 (p. 343) show cases where processing is partially

Fig. 17.1 Partial cover (Courtesy: British Gas)

enclosed. Alternatively, simple protection may be used such as roof (Fig. 17.1) and enclosure only on the windward side.

The particular problems of housed plants are discussed further in Chapter 18.

17.3   ECONOMIC SAVINGS

The need to keep plant buildings and structures to the minimum has already been stressed.

Many more savings, modest individually, but large in total, can be made by applying some thought to the layout.

17.3.1   PUMPS AND ELEVATION

Moderate elevation is needed for transferring a saturated liquid with a pump needing a high net-positive suction head (NPSH) (Fig. 23.2, p. 339). An alternative pump that works with a lower NPSH is usually more expensive. Occasionally, a cost comparison between equipment support and cost of pumps can point to an economical solution. For example, steam jet vacuum pumps which must be elevated to provide barometric legs for gravity flow from the interstage condenser can be supported from columns or other tall equipment.

The selection of a vertical pump (see section 23.2.5, p. 341) can bring process drums close to grade, eliminating the possible need for high footings, access platforms and ladders, while improving operation and maintenance access. On the other hand, pumps related to surface condensers can increase the floor height for a turbine/compressor set. A reboiler with pump elevates the reboiler and, in turn, the related distillation column.

17.3.2   MINIMIZING CONNECTIONS

Equipment should be located to minimize extensive pipe runs as these increase costs by reason of the additional pipe used, the extra insulation needed, greater heat losses, more devices to accommodate expansion and by greater pumping costs. The cost of alloy piping can lead to perhaps the most compact and unusual arrangements for process equipment. Some examples are: nozzle-to-nozzle location between vessel inlet and exchanger outlet; head-to-head connection between exchangers; and stacked exchangers and drums.

To minimize the length of incoming power lines and to optimize power distribution, the motor control centre should be as near as possible to the incoming power line and close to the major electric-power users. Also, in non-hazardous areas large electrical junction boxes can be located so that optimum cable runs can be designed. However, electrical area classification rules must be observed (see Chs. 6 and 8) and normally power distribution is done from a centralized switchhouse. If possible, major cooling water users should be located toward the cooling water towers or supply point.

17.3.3   SAVING SPACE AND STRUCTURES

Consideration should be given to optimizing the use of structures by using them to support more than one item of equipment and ensuring that accessways and platforms have more than one function.

Pairing condensers between towers enables a common structure (and access) for supporting the condensers and reflux drums to be used. Single water feed and return lines can be designed for grouped condensing equipment.

Most reboilers are at grade and sometimes condensers are also at ground level (see Ch. 21). Space may be saved if these two items are located adjacent to each other. Another common arrangement is to have condensers and reboilers on either side of distillation columns (Fig. 21.7, p. 322).

Often, particularly in refineries, surge drums, storage drums, coolers, and heaters are placed between groups of distillation column groupings in a process flow sequence. Reactors are usually grouped as near to furnaces and/or compressors as hazard considerations permit. Suction and knockout drums, intercoolers and aftercoolers are located adjacent to compressors.

Space can be saved by locating small equipment, such as pumps, under pipebridges or racks, providing the area is not needed for access. However, pumps are prone to seal leaks and pipework to joint and flange failure. So pumps carrying hazardous fluids should not be put over or near sensitive equipment. Pumps placed outside the piperack increase the distances between the piperack and process equipment resulting in additional piping and supports to connect the pumps and equipment.

A one-sided arrangement is usually more expensive because only one side of the piperack is used. However, if only a narrow area is available or future expansion is planned this arrangement may be the optimum solution.

Small vessels, e.g. air-cooled fin exchangers and similar items of equipment may be located over the main pipeways. This is provided that the increase in expense of additional steel sections, with possible extra access platforms and heavier foundations to the bridge or rack, can justify the saving in ground space, and that there is not an increase in construction difficulties or delays (Figs 22.4 (p. 329) and 22.5 (p. 330)).

Larger exchangers may be elevated but unless the process demands it this should only be carried out if economies in pipework and floor space balance the cost of elevation.

17.3.4   MATERIALS FOR THE STRUCTURE

Although not strictly a layout problem, the materials for the structure are often chosen in association with the layout design.

The use of concrete versus steel as a support material should be based on the location of the site, availability of materials locally, importation problems, and the need to clad steel for fire protection.

17.3.5   FOUNDATIONS

Foundations are usually expensive. Often, equipment regroupings (in co-operation with a structural designer) can result in a more economical foundation design. The cost of additional piping is usually much smaller than the saving in foundation costs.

There is a cost balance[3] for columns between a high tower skirt with a small plinth or a high concrete plinth with a short skirt. The former can affect shell-wall thickness, windload moments, and access to the skirt interior. A slender tower might require a more expensive long, flared skirt; a tall plinth can make possible a short and straight skirt. A tall tower skirt is more prone to collapse in a pool fire unless insulated.

The size of foundations depends on the bearing properties of the soil, as well as on the actual loads to be carried by the foundations. Compact equipment spacing may allow combined footings. Large, separate foundations might restrict minimum spacing for equipment. Support, though, for reciprocating compressors must be independent of building-steel and pipe supports. Compressors should be as near the ground as possible. If variations in load-bearing properties exist over the plot area, heavy or rotating equipment should be located over the best load-bearing surface. Grade elevations should be at a level where a minimum of earthmoving will be required.

Occasionally, equipment spacing and location can be influenced by large-diameter sewer and cooling-water lines. Footings should not be positioned over buried lines. Since it is so difficult and expensive to gain access to anything buried under a plant or building it is desirable to run any important buried lines outside the plant area.

17.4   SAFETY DETAILS

The philosophy of designing for hazard containment is discussed in Chapter 8. In particular, adequate distances between plants are needed to minimize escalation of incidents.

Here some detailed aspects on safety are outlined.

17.4.1   OPERATOR PROTECTION

Consideration should be given in the equipment layout to provide sufficient clearance around critical and mechanically dangerous or high-temperature items to ensure the safety of operating and maintenance personnel. It is sometimes necessary to insulate pipes for operators' protection as well as for energy conservation. Machine guards may be adequate for operators but not for maintenance fitters. (Where guards have to be moved for maintenance, a 'permit' system has to be used.) The layout of lifts, hoists, etc. should be made with operator and maintenance safety in mind.

Glass equipment, such as sight glasses and rotameters can break and should be fitted with transparent guards. Other glass items, such as vessels, should be guarded or bound with tape to avoid shattering, or situated so that if they break they do so without danger to operators. However, where possible, glass equipment should be avoided.

Lighting must be arranged to give adequate illumination throughout and extra lights installed near equipment where physical and chemical hazards exist and where instruments are read (Fig. 17.2). Congested and ill-lit plants lead to frequent tripping and bumping accidents.

Access requirements are given in sections 17.5–17.7. Valves, pipe flanges, etc. should not be situated over access ways because of the danger of leaks falling on personnel.

17.4.2   SPILLAGE CONTAINMENT

Solid floors or impervious ground should be used where liquids are liable to leak. The equipment should be laid out so that it is on the tops of a series of humps and so the surface slopes towards the appropriate effluent collection points situated in the valleys; a slope 1 in 80 is the usual minimum, but 1 in 40 is advisable for corrosive liquors to minimize run-off time. The maximum fall in any one direction should not exceed 150 mm. If this fall does not give adequate slopes on large areas, a valley drainage system should be used. The valleys should be in the spaces between items of equipment so that spillages will drain to a predetermined point (as in Figs 17.3, 16.11 (p. 240) and 23.1

Fig. 17.2 Example of plant illumination (Courtesy: British Nuclear Fuels)

Fig. 17.3 Drainage under vessels (Courtesy: The Boots Company)

(p. 337)) and, should it catch fire, damage will be minimized especially if the spill is also smothered by an appropriate spray. If sprays are not provided the size of any possible pool of flammable liquid should be estimated and then the heat received by surrounding equipment from a subsequent fire calculated (see section 8.4.2, p. 113). This will indicate the likely effects and damage, e.g. tank rupture, liquid boiling and/or overflowing through vents, liquid flash point exceeded, chemical reaction started, etc. Vulnerable items should be moved or insulated. All plant and structures likely to be exposed to fire, especially up to a height of 6 m, should be insulated.

Liquid puddles lingering in the valleys should be avoided and the drains should not carry the spillage into other vulnerable areas nor allow fire to spread to the effluent system. Traps can be fitted to prevent this.

The valleys should not coincide with the access ways. It may be necessary to use kerbs to prevent leaks, particularly of corrosive or toxic liquid from reaching the access ways. If fluids from two different leaking tanks can react together in an undesirable way the kerbs should be placed to prevent inter-mixing. Equipment containing acids, alkalis or toxic materials is usually grouped within a paved and curbed area that is drained to a treatment pit. Site effluent systems are discussed in section 13.2 (p. 195).

Storage of liquids in tanks and other vessels on roofs should be confined to small amounts of substances that are harmless if spilled or which can be effectively contained.

It should not be assumed that open structures are sufficiently ventilated. Any possible leak must be dispersed safely (i.e. be below the danger limit before reaching the nearest ignition source) by a 2 m s^{-1} wind.

Slippery floors, whether from process liquids or rainwater, ice, snow, etc. must be avoided. Equipment should be provided for cleaning, e.g. washing-down hoses, hot water, good drainage. Slip-resistant floor coverings should be considered.

17.4.3 VIBRATION

Rotating and reciprocating machinery can set up vibrations which can loosen connections and cause leaks or, even worse, shatter pipes, equipment and structures. Adequate foundations and damping devices are needed. The higher the forces, the nearer the ground the equipment should be.

17.4.4 HEATING STRESSES

Plant including pipework expands on heating and so stress calculations should be made (section 31.3, p. 429). These may decide what expansion loops, flex-ible couplings (or bellows) or expansion joints are needed to prevent leakage or breaks. Loops take up more space but are the most reliable. Bellows can burst and expansion joints can jam due to misalignment, dirt between the moving surfaces or uneven heating-up of the two moving surfaces. Both should be avoided if possible.

17.4.5   ACCIDENTAL IMPACT

Pipes can be broken under impact from trucks, cranes, etc. and so adequate clearances and barriers must be provided for these vehicles.

Glass equipment is particularly vulnerable to falling tools, lengths of piping being carried, etc. If it must be used, it should be suitably situated and/or protected.

17.5   OPERATIONAL REQUIREMENTS

Thought should be given to the location of equipment requiring frequent attendance by operating personnel and the relative position of the control room to obtain the shortest and most direct routes for operators when on routine operation. However, the control room should be so sited to avoid the need for pressurization, reinforcing, etc. This conflict is discussed further in section 17.9.1.

In grouping and lining-up similar equipment, it should be ensured that operation remains easy and has not been sacrificed to visual appearance or construction convenience.

The layout engineer should try to make the plant easier and more convenient for the operators to undertake their routine tasks, and attention to apparently trivial detail can often help considerably. More conscientious and reliable operation can normally be expected when the operators' life has been made easier. On a very high plant, an elevator or lift should be considered to ensure operators are fit for immediate work on reaching the top level.

Valves and control handles should be placed so that they are easily accessible (e.g. from platforms) and indicators placed at a height so that they can be easily read. Emergency shut-off valves and controls should be protected from hazards, including those which they are designed to cut-off. These points are discussed further in section 31.8.2 (p. 455).

Generally batch or semi-batch processes, e.g. batch reactors, batch dryers, plate and frame presses and packaging lines, need more operator attention, and therefore more consideration has to be paid to the ergonomics of the layout. Even in continuous plants it helps to locate starter buttons so that the operator can see or hear the pump, mixer, etc. start or stop when he operates the switch. Where materials are added to or removed from vessels manually, attention should be paid to the height of discharge points and loading chutes, the area of material heaps, etc.

On small batch processes whole items of equipment may be rearranged, removed and replaced by other items, either for easier cleaning, or maintenance or for making a different product. Sufficient room for this should be provided and adequate lifting equipment installed.

Access, and general safety requirements, should conform to company standards which, in turn, should comply with approved industrial standards and statutory requirements designed to give easy access and escape under emergency conditions. These, it must be remembered, differ markedly from ideal design conditions.

17.5.1 HORIZONTAL ACCESS

Clear routes for the operators should be provided avoiding blind spots, sharp corners, kerbs and other awkward level changes. Operating and workspaces should have two separate non-intersecting access routes for escape (section 17.7.2 p. 256). Safe routes must be planned for fork-lift trucks. General access ways should be:

(a) 750 mm wide to allow passage for a single person.
(b) 1.2 m wide for two persons.
(c) 2 m wide for up to six people.
(d) Greater than 2 m for trucks or a large number of people.

Railings should be 850 mm high. Where breathing apparatus may be needed in emergencies, access ways should allow for the extra bulk and their effect on operators' balance. Route capacities should be determined by the maximum load which is often during maintenance (Fig. 17.4).

17.5.2 VERTICAL ACCESS

When plant items require to be supported at elevated levels (defined as 3.5 m above grade for vessels, 2 m above grade for valves, sample points, sight-glasses, instruments or 2 m above another platform) to suit the process lay-out, platforms must be provided for operation and maintenance. These are reached by stairways, ladders or lifts. The number of intermediate floor levels should be kept to a minimum. This is helped by an intelligent arrangement of manholes, operating valves, instruments, maintenance and inspection points. In multi-floor plants, vertical penetration of floor spaces by random pipes or cables, should be avoided. Co-ordinated penetrations for groups of these items should be planned to avoid clutter and limit cascading from leaks above.

Floors are generally not less than 3 m apart. Minimum headroom under pipes, cableracks, etc. should be not less than 2.25 m. This can be reduced to 2.1 m vertically over stairways (Fig. 17.5). In establishing the distance between floors it is important to allow for the removal of agitator shafts and other vessel internals. Pipes, steel trimmers, cables, etc. are trip hazards and should not run horizontally away from walls above floors at levels below 600 mm.

For main vertical access, stairways are preferred to ladders, which are used for emergency escape routes on outside equipment (but not on buildings) and to isolated positions where attention is required infrequently (Fig. 17.6).

The recommended angle for steel stairways is 35–40° and an overall width of 1 m should be allowed for layout purposes. Railings should be 850 mm high and headroom 2.1 m (Fig. 17.5). The maximum height of single flights without intermediate landing is 4.5 m.

Ladders must not be attached to the same supports as those that carry hot pipes. Expansion/contraction forces can be transmitted which could distort the ladder or pipes. Ladders should be so positioned that the user faces a structure and does not look into space. Ladders may be vertical (Fig. 17.6) or inclined at an angle up to 15°. They should be supported 225 mm out

Fig. 17.4 Typical access for maintenance

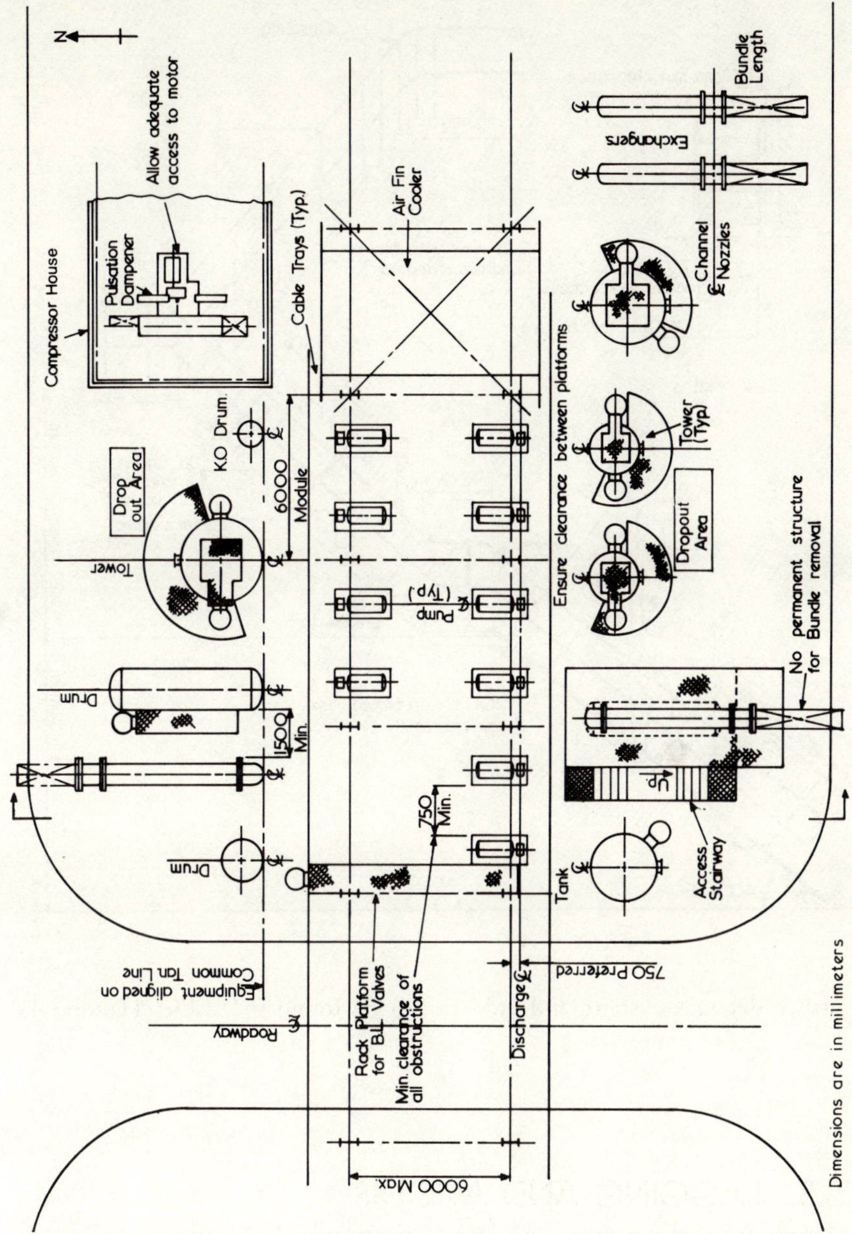

from the wall with a plan of 1 m square on the climbing side allowed in the layout. They should be arranged for side exit, though step-through ladders may be used for runs from grade up to 7.5 m or for elevated runs of 3 m. Maximum ladder height without landing is 7.5 m. Back cages or safety hoops must be provided for ladders over 2.5 m high (Fig. 21.5, p. 319).

Intermediate steps are needed for elevation changes in excess of 400 mm.

Fig. 17.5 Stair layout

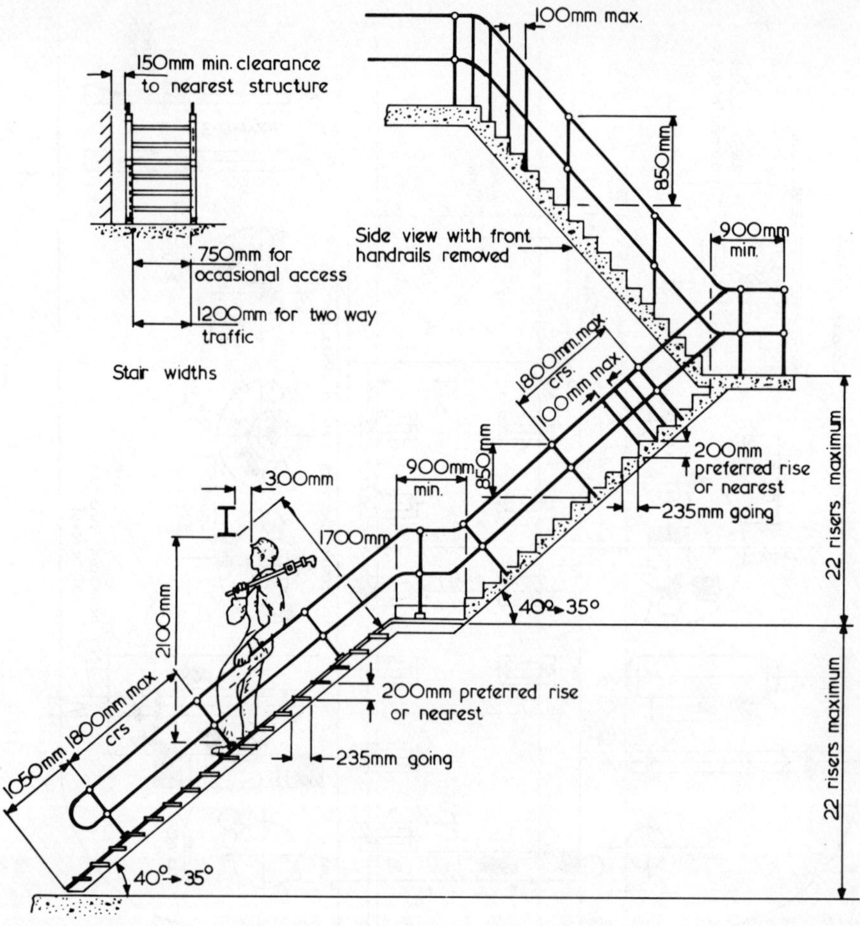

Further details on stairs and ladders can be found in EEUA Handbook No. 7[4].

17.5.3   LAGGING AND ACCESS

It is important that insulation requirements are considered during the layout of plant. Insulation decreases the available free space for access, affects the sizing of tank supports, valve spindles, etc. and alters the siting of pipework, instruments or electrical equipment. Space is also needed for installing the lagging itself. In particular, the thickness of insulation of very high-temperature or low-temperature piping may be considerable, thus effectively increasing the outside diameters of pipes or sizes of fittings.

Fig. 17.6 Fixed ladder layout

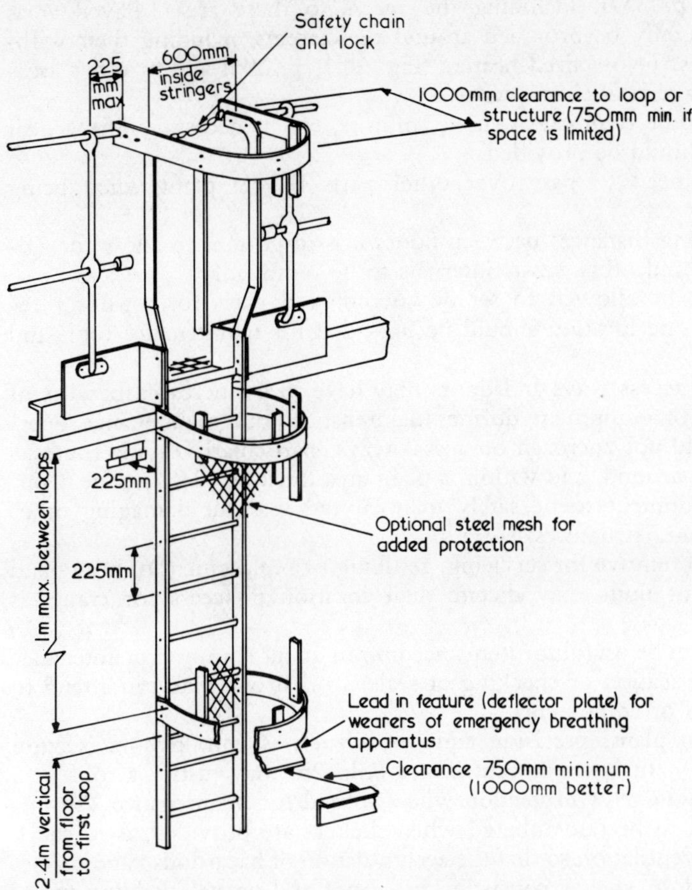

17.6   MAINTENANCE REQUIREMENTS

Safe and effective routine and preventive maintenance procedures should be planned[5] during the design stage as they will affect layout since spacing distances and space allocation around most equipment items assume some degree of maintenance (Fig. 17.4). Other general points concerning layout are:

(a)   The plant must be made safe for maintenance by such measures as:
   (i) adequate valving for equipment isolation;
  (ii) vents/drains to suitable collection points;
 (iii) adequate wash-out facilities using either water or solvents according to solubility;
 (iv) adequate purging with air or nitrogen.
(b)   Consideration must be given to withdrawal areas, e.g. shell and tube exchanger bundles (Fig. 22.2 (p. 327), 22.3 (p. 328) and 22.6 (p. 332)),

tower internals (Fig. 21.3, p. 317), pumps and compressor parts (Fig. 23.1, p. 337), including the access to these areas. Paved areas should generally be provided around these items including their withdrawal areas, below fired heaters (Fig. 20.1, p. 306) and in other areas where spillage is likely to occur.

(c) Where periodic removal of shafts, motors, etc. is necessary lifting gear provision should be provided.

(d) Parts may need to pass over other parts of the plant when being removed.

(e) In establishing distances between floors it is important to allow for agitator shafts and other vessel internals to be removed.

(f) Space must be allowed to set down removed parts for repair or removal and safe location should be provided for maintenance tools and materials.

(g) The size of access ways and doors may have to be increased because of movement of equipment during maintenance, but maintenance operations should not encroach on access ways, particularly escape routes.

(h) Road access around, and within, a plant area must be adequate for maintenance equipment to be safely manoeuvred without damaging other equipment and structures, etc.

(i) The need to remove for servicing, retubing or replacement of heavy and integral plant units may dictate their location if access for cranes is needed.

(j) Grouping can be useful for items needing frequent common maintenance (such as lubrication or checking of seals) so that one man can attend to many pieces of equipment.

(k) On an open plant, precision equipment such as compressors (section 23.3, p. 342), turbines or centrifuges (Ch. 25) may justify a roof or a separate building for protection when opened for maintenance. An important part to be remembered when shelters are provided, is to check for reliable ventilation so that the accumulation of hazardous vapours etc. is prevented. A shelter consisting of a roof and partial cladding meets the requirements of protecting personnel and equipment from inclement weather without creating the need for special ventilation.

(l) Piping and structural steel should not interfere with maintenance clearances but should not be weakened in order to favour maintenance work.

(m) In removing equipment parts, pipe dismantling should be minimized. Where this is necessary break-flanges should be provided. Short spool pieces should be fitted between valves and items such as pumps and heat exchanger channels to allow items to be withdrawn without removing valves or springing pipes. Spool pieces can also be used for physical disconnection before safe entry into equipment.

(n) Control valve positions must allow for removal of internals and stems.

(o) Valves and instruments should be laid out for ease of operation and maintenance (Ch. 31).

(p) Manholes should face gangways and provision made for lifting a man out of the equipment. They should be between 0.5 m and 1 m above the platform with a minimum clear access space of 1 m × 2 m around manhole.

(q) Distillation columns (Ch. 21) may be lined up, close to each other, in order to provide common manhole platforms, which also enable valves and instruments to be serviced. In this case, condensers can be arranged on the opposite side of the piperack, with all pumps under the piperack.

(r)   With packed columns (Ch. 21) there should be access for at least one man and a supply of packing material around packing nozzles and nearby space for new materials and removed packing. Similarly, allowance should be made for catalyst removal from reactors (section 19.2, p. 289). Temporary lifting gear for drums or skips of packing, etc. may be needed.

(s)   When removing equipment, e.g. centrifuges or filter baskets (Chs. 24 and 25), space is required between the trolley beam and equipment flange for a hoist, hook, sling and lifting beam, in addition to overall clearances.

(t)   For some heat exchangers (Ch. 22) space may be needed for cleaning tubes and rodding-out *in situ*. On grouped exchangers, a gantry is often useful for channel removal, bundle pull-out, etc. Space is also needed for chemical cleaning pumps and tanks. Water- or air-cleaning facilities may be needed (section 31.6.8, p. 449).

(u)   Drying facilities are needed if the process materials are water–sensitive.

17.7   FIRE FIGHTING AND ESCAPE

17.7.1   FIRE FIGHTING

Good access for firefighting means plots restricted to 100 m × 200 m with approaches preferably on all four sides, with 15 m between plots and/or buildings.

An independent system of mains water supply should be used for fire services and should be available from at least two sources (section 14.3, p. 210). Firefighting water ring main systems should be arranged to ensure that hydrants cannot be isolated except at the entrance to the plot area. Firefighting water is supplied to hydrants or sprinkler systems or both. The hydrants should be spaced not more than 60 m apart along the ring main system around the plot, usually near the plot boundary. They must be positioned so that jets can reach any fire, without having to enter a plant area that may become highly dangerous during an emergency. It is not normal practice to run fire mains for hoses into open process structures (though it is for buildings (Ch. 18)), since access to valves within the plant in an emergency can become impossible.

Attention should be given to the location of fire buckets and extinguishers at strategic and conspicuous points. Some extinguishers should be placed on escape or access routes when a man has the option of either continuing to safety or deciding from the safe location to go back to fight the fire. The units must not be placed near the hazard where they could be damaged at the outset. The layout of the above plus other firefighting systems such as sprinklers and foam equipment should be made or approved by experts.

Main switchgear should be readily accessible in the case of fire. Emergency buttons should be arranged (say, in the control room or in an escape route) so that the operator does not risk his life to stop pumps, etc. It is essential that local by-laws and regulations should be checked with the authorities prior to seeking planning permission.

17.7.2   ESCAPE ROUTES

Except in areas of small fire risk, a minimum of two escape routes is required from any workspace. The two routes should be true alternatives, and to allow safe escape no workplace should be too far from an exit. The appropriate distance has to be judged for each case according to the height and degree of hazard of the plant and is often in the range 12–45 m. The length of a dead-end should not exceed 8 m. Escape routes across open-mesh areas should consist of solid flooring. Open-mesh flooring should not be used over areas of high fire-risk. Escape routes should be signposted and those across flat roofs should be guarded with handrails.

Escape stairways should be in straight flights, preferably on the outsides of structures and on the least hazardous sides. Fixed ladders can be used for escape from structures for ten persons or less.

Escape routes should be adequately illuminated (with emergency lighting, if need be) and route-marked for use in smoke or poor lighting. They should be at least 750 mm wide, preferably 1.2 m, to allow a rate of passage of 40 persons per minute on the horizontal and 20 persons per minute down stairways. The maximum height between landings should be 4.5 m (9 m on columns) and enclosed stairwells may be justified in hazardous plants. In laying out the escape routes, the escape times of all personnel should be estimated including those likely to be wearing breathing apparatus. Particular allowance has to be made for personnel in elevated situations such as cranes and distillation columns, etc. (Fig. 21.5, p. 319). Bridges between columns should be considered (section 21.5, p. 323).

Emergency assembly points should be designated for each building or plot. They should be at least 100 m from any hazard, but this distance should be checked by the methods indicated in Chapter 8.

17.7.3   CONSULTATION

There are numerous regulations (see BS 5908[6]) concerning safety and fire-fighting, and the safety officer should be consulted in all stages of design. It is also necessary to consult, as early as possible, the appropriate regulatory and fire authorities and insurers. At an early stage in the preparation of the emergency plans for a new plant, it should be decided how much action may be safely taken by employees in firefighting and what should be left for the emergency services.[7]

17.8   APPEARANCE

As indicated in section 4.8 (p. 33) layout can be used to enhance appearance and lead to better operation. Some notes on this follow.

Pipeways

Outside buildings, pipeways are run at common elevations. Changes of direction and elevation in pipeways will ordinarily be made with 90° bends. Where a minimum difference in elevation is necessary, two 45° bends may be used. The elevation is taken as being at the bottom of the pipe (or underside of shoe for insulated lines). This standardization should help to achieve also a common elevation for off-takes from pipeways (section 31.5, p. 439). Inside buildings, piping may run in vertical banks and flat turns may be used.

Towers

Towers (Ch. 21) and large vertical vessels should be arranged in rows with a common centre-line if of similar size, but if the diameters vary they should be lined up with a common face. Due note must be taken of building lines. Manholes or adjacent towers should be at similar elevations and orientation to present uniformity of platforms and of appearance generally.

Piping should be located radially about the column on the pipeway side and, when possible, manholes and platforms on the access side. Valves and flanges should not be located inside vessel skirts. Valves should be located flange-to-flange on elevated nozzles, the downward piping turned straight down after the valve, and run as close as possible to the column.

Exchangers and pumps

As far as possible the centre-lines of exchanger channel nozzles should be lined up, as should the centre-lines of pump discharge nozzles (Chs. 22 and 23). Piping around pumps, exchangers and similar ground-level items of equipment should be run wherever possible at set elevations for north to south piping, and another elevation for east to west piping.

Sequences of plant

Handed arrangements should be avoided. This principle should be followed where there are similar equipment sequences within the process stream, i.e. fractionating column with its overhead condensers, reflux drum, pumps and reboiler, etc. These sequences should be repeated in an identical way. This gives a better appearance as well as saving on design costs and on the costs of spares. It also prevents the dangerous fitting of wrong spares.

17.9   CONTROL ROOM

17.9.1   SITING

The siting of a control room is determined by the needs of normal operation and by the requirement for protection during emergencies.

The nearer the control room is to the plant, the more often operators are likely to be on the plant. Consequently, they are more likely to notice the start of a malfunction, e.g. through change of noise level and pitch, dripping and hissing. Sight of plant through a control room window also helps to serve this purpose. Thus, dangerous situations can be avoided. At distances greater than 35 m and particularly over 100 m, operators will find reasons for not going into plant, particularly in inclement weather. Also, the nearer the control room is to the plant the shorter the instrument cables, etc.

On the other hand, the nearer to the plant the greater the danger to, and therefore cost of protection of, the control room and the personnel in an emergency.

The danger to the control room can be reduced by:

(a) Improving the safety of the process by the various means described in section 8.6 (p. 123).

(b) Protecting the control room against blast (see section B.7.3 (p. 525) and CIA[8], HSC[9]).

(c) Protecting the control room against fire (see section B.5 (p. 516) and BS 476[10]).

(d) Intelligent layout so that tall structures do not fall on the control room, releases of solids and liquids flow away from the room and jets of liquids or gases cannot impinge on the room, particularly on the windows. No heavy equipment should be located on the control room roof.

(e) Having the ability to isolate the control room and provide, if desirable, an emergency air supply for the duration of the emergency. Back-up personnel breathing sets should also be available in case windows or doors are destroyed.

(f) Avoiding the need to have windows by installing modern instrument and control systems which can give a very adequate understanding of the state of the process.

(g) Having the standard fire protection measures against fires starting within the control room from faulty electrics, etc.

The problem therefore reduces to one of placing the control room as close as possible to the plant consistent with economic protection against explosion and fire.

The policies for protection are, in order of priority:

(a) The personnel should be protected from unacceptable levels of risk.

(b) The control room equipment should remain functional, particularly emergency shut-down devices.

Thus, within reason, any plant which is not itself damaged can still be controlled from the room. The consequence of this policy is that the control room structure can be distorted but must not collapse or holes appear. It does not matter if the building has to be replaced afterwards. The room should provide refuge for about 30 minutes while rescue operations can be mounted from the unaffected parts of the site. In that time it can act as the forward control point and provide an emergency assembly point for key plant personnel only. (Others should be able to assemble in a safer position as described in section 17.7.2.)

The building should be so sited so that a minimum of vibration is transmitted to the structure of the room.

The intakes for air for ventilation should be in a safe area. The room

Fig. 17.7 Example of local control station (Courtesy: The Boots Company)

should operate under a slight positive pressure in order not to draw-in toxic, flammable or obnoxious vapours and gases.

Direct access should be provided from the control room to the plant and if possible to rooms provided for computers, dataloggers and associated equipment for instrumentation. The access should be arranged so that the control room will not become a thoroughfare. In a safe zone cabling may be reduced by putting a switchroom under the control room. Also in housed and safe plants, local control stations can be employed as shown in Figs. 17.7, 19.9c (p. 298), 25.1 (p. 355) and 25.2 (p. 355).

Provision for the possible extension of the control room should be considered. An adjacent area of plot should be reserved and structural features such as internal cable trenches arranged so as to connect to the extension.

17.9.2   PANEL ARRANGEMENT

The instruments needed for the area of control exercised by any one process operator should be grouped together.[11] A clear demarcation should be made on the panel between process units. Instruments should be arranged so that the variables requiring the most critical attention are in a position of optimum visibility and manual accessibility. Related variables should be displayed adjacent to each other. As a secondary criterion, instruments should occupy relative positions according to the sequence and situation of process items.

Windows should give the operators any necessary view of the process area. They should face north if possible. This avoids direct sunlight on instruments

Fig. 17.8 Control panel and consol layout (a) petrochemical plant (Courtesy: ·BP Chemicals) (b) pharmaceutical plant (Courtesy: The Boots Company)

(a)

(b)

and annunciators, an important factor since direct light shining on to a panel can result in lights not being visible on the panel face. However, as already indicated, for hazardous plants, windows should be avoided and good instrumentation systems used instead. The additional wall space available gives economies in panel layout.

In general (Fig. 17.8), the height from the floor to the top of the highest point of recording instruments should be no greater than 2.1 m and the top of the lowest instrument not less than 1.2 m. Most multi-point recorders

Fig. 17.9 Dimensions of comfortable
seating

have to be viewed from above and so should be in the middle of the panel.
Panel space between 2.1 m and 2.4 m may be used for the clock, pressure
indicators, visual alarm systems and similar instruments. For free-standing
or U-type panels, the space behind the panel must give an unobstructed passage of 1 m. In front of the panel, 3 m should be allocated, leading to an
overall width of control room of 16 m allowing for panel width. The length
is determined by the number of instruments. Figure 17.9 shows dimensions
most suitable for providing comfortable seating for the operator.

17.10   OTHER PERSONNEL BUILDINGS

Plant accommodation other than the control room may be needed for ancillary rooms such as locker rooms, mess room, toilet facilities, offices for supervisors and clerical personnel, test room (laboratory tests), instrument
servicing room, electric relay and switchrooms plus possibly local workshops
(section 15.4, p. 219).

Whether some or all of these facilities are required depends on the main
site facilities and the distance to them, on the facilities in adjacent plants and
on the needs of the plant itself.

For operational purposes, these ancillary activities are often combined with
the control room. However, in hazardous plants, these activities should be
a safe distance from the plant. In this way the number of people exposed to
the hazard is kept to a minimum.

Safe distances above which no special building features are needed are fixed
from the fire and explosion considerations discussed in Chapter 8 and sections
B.5 (p. 516) and B.7 (p. 525). Amenity buildings placed within these distances because of the necessities of operation must conform to the same stan-

Fig. 17.10   Control building layout in a safe area

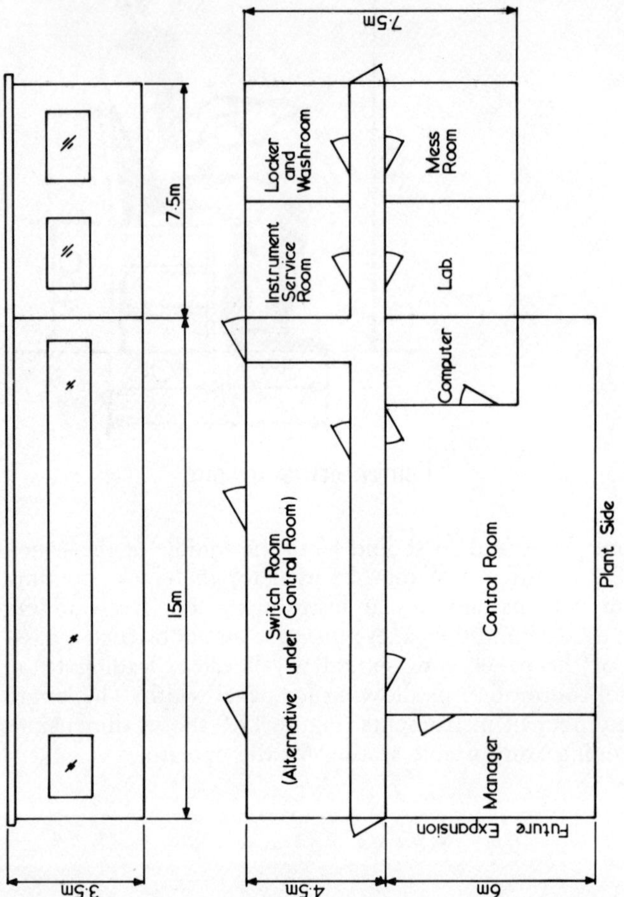

dards of building as control rooms (section 17.9). This means incurring extra construction costs which must be justified by increased operation efficiency. Application of the standards usually indicates that buildings cannot be higher than two storeys, and preferably no higher than one.

A typical arrangement for building in a safe area is shown in Fig. 17.10. When there is much coming and going between control room and plant the former should be at ground level. Alternatively, in a multi-storey building it may be sensible to put the control room near the centre of gravity of activities (Fig. 17.11). There is the advantage of a good view of the plant if the control room is above the ground floor but this diminishes accessibility and windows are a source of danger even in relatively safe areas.

Transformers can be a fire risk and should be placed away from the switch-room in the control building. The control building itself will generate its own dangers and all normal building standards for fire protection, evacuation, etc. must be observed. However, evacuation routes must be away from the plant. Access to the site road system must be planned and, if allowed, parking areas allocated. The possibilities of future expansion must be considered. For the last reason, these buildings are often planned as a series of modules.

Fig. 17.11 Example of internal control room (Courtesy: Babcock Minerals Engineering)

REFERENCES

1. Kern, R. 'How to manage plant design to obtain minimum cost', *Chem. Engng*, 23 May, 130, 1977.
2. Kern, R. 'Controlling the Cost Factors in Plant Design', *Chem. Engng*, 14 Aug., 141, 1978.
3. Kern, R. 'How to arrange the plot plan for process plants', *Chem. Engng*, 8 May, 191, 1978.
4. EEUA Handbook No. 7, *Factory Steel Stairways, Ladders and Handrails.* Engineering Equipment Users Association, 1962.
5. Thomas, B. E. A. (ed.) *Preparation of Plant for Maintenance.* I.Chem.E., London, 1980.
6. BS 5908 'Code of practice for fire precautions in chemical plant', British Standards Institution, London, 1980.
7. Balemans, A. W. M. *et al.* 'Check-list guidelines for safe design of process plants', in: Buschmann, C. H. (ed.) *Loss Prevention and Safety Promotion in the Process Industries.* Elsevier Scientific Publishing Co., Amsterdam, 1974.
8. CIA *An Approach to the Categorization of Process Plant Hazards and Control Building Design.* Chemical Industries Association, 1979.

9. HSC, *Advisory Committee on Major Hazards, Second Report.* Health and Safety Commission, HMSO, London, 1980.
10. BS 476, 'Fire tests on building materials and structures', Parts 2–8, British Standards Institution, London, 1970–81.
11. Edwards, E. and Lees, F. P. *Man and Computer in Process Control.* I.Chem.E., London, 1973.

LAYOUT WITHIN BUILDINGS

18.1 APPROACH TO LAYOUT

Generally speaking, space is more restricted inside a building than in an open plant. The plant should be laid out first and then the building added around it. This means that the penalty for an uneconomical and badly planned layout is high building costs, so a greater amount of planning is required to obtain an economical layout satisfying the requirements of operation, maintenance and safety.

Particular problems encountered in enclosed, compared with open, plant include:[1]

(a) Plants are usually put in buildings because there is a large amount of manual operation, e.g. batch reactors and dryers, plate and frame filter presses, spinning machines, filling and packaging. The ergonomics of these operations have to be satisfied.

(b) Fire can spread more easily in a building and it can be more difficult to escape and there are often larger numbers of people to evacuate (because of manual operation). Buildings, however, can, at some cost, be made safer with respect to fire and evacuation than open plants.

(c) Ventilation is needed to remove toxic and flammable fumes and (probably) heating and (possibly) air-conditioning is required. The ducting can take up significant amount of space.

(d) Building erection must be planned carefully to enable large items, e.g. columns and large tanks, to be installed at appropriate times. This will call for early design and ordering decisions on the plant item involved.

(e) Piping and cables can occupy considerable space and can interfere in particular with removal of equipment for maintenance. Generally, maintenance access is more difficult to plan in buildings than for open plants.

(f) The lighting, both natural and artificial, has to be considered in greater detail.

The normal order in which many layouts in building can be developed is:

(a) Establish the process functions and the services, environment and people required to support them.

(b) Find the best equipment positions. This involves minimizing within the constraints of the plot size available, the building height, the number of floors, the building area and the distances chemicals and services have to be piped.

(c) Determine the operating arrangements. The area around and access to items needing manual operation is planned. Operator routes to work stations and escape routes in emergencies are decided.

(d) Assign the areas for routing piping and ducting in conjunction with the maintenance arrangements for accessing and removing equipment.

(e) Determine the lighting requirements.

(f) Subject the layout to a hazard assessment.

(g) Reiterate the above steps to reconcile conflicting requirements.

(h) Complete the detailed piping, ducting and lighting arrangements.

The best aid to initial layout in buildings is the block model (either physical (Fig. 18.1) or computer), especially one where floor heights can be adjusted. Cutouts or computers (Fig. 18.2) can be used to produce plans of preliminary arrangements on each floor. Subsequently, detailed piping models are produced.

Fig. 18.1 Block model of housed plant (Courtesy: Gunn[2])

Fig. 18.2 Computer produced plan at a floor level (Courtesy: Gunn[2])

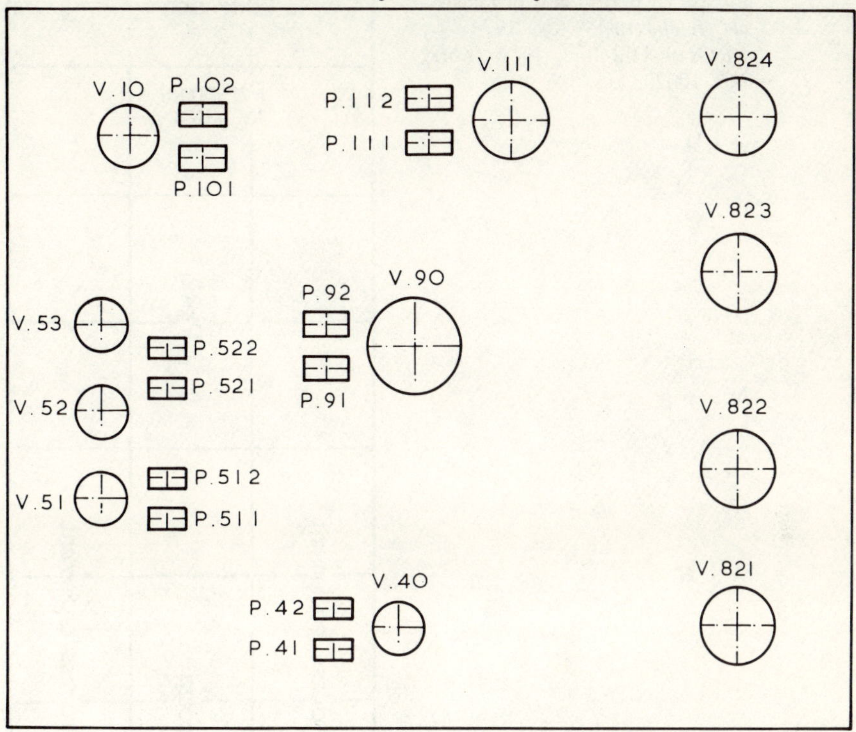

18.2   OPTIMUM EQUIPMENT ARRANGEMENTS

The first layout is based on the process flowsheet. The usual practice, then, is the use of skilled intuition with the aid of a block model in order to find an economical arrangement. Less frequently employed are the more formal techniques of correlation charts, travel charts or sequencing which apply to layout in buildings (section A.3, p. 478). Computer programs such as CRAFT, ALDEP and CORLAP are available for layout optimization of factories and further details are also given in section A.4 (p. 482). They suffer from the constraint that the building size is fixed. This means reiteration with different building sizes if building cost is also to be optimized. Gunn, with his student Al-Asadi,[2] have developed computer methods of optimizing the layout of process plant which can be applied in buildings and their work is summarized in section A.1.2 (p. 463).

If the plant is similar throughout the building the modular[3] concept of design is useful (Fig. 18.3). The plant is split-up into a series of parts in which piping cables, etc. and lifting equipment are put in standard positions. This aids and standardizes design philosophy throughout the plant including deciding a standard structural steel grid which best suits the individual sections

Fig. 18.3 Modular arrangement for a building (Adapted from Kern[3] by special permission of *Chemical Engineering*; © 1978 by McGraw-Hill Inc., New York, NY 10020, USA)

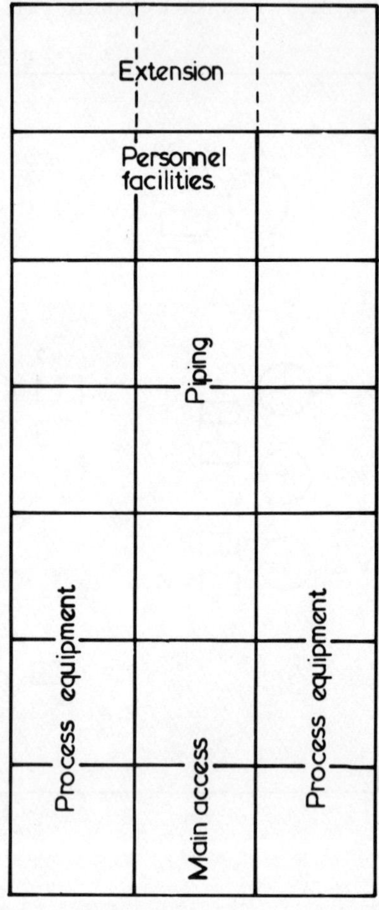

of the plant. Other advantages are economies in construction work and in future expansion.

A useful guide is that the equipment on its own should occupy no more than 5 per cent of the floor space. Exceeding the value often leads to a congested plant when pipework, ducting, cabling, structural supports, etc. are added.

18.3 OPERATIONAL AND EMERGENCY ARRANGEMENTS

Intelligent use of the block model is the main tool in deciding operational and emergency arrangements. Employment of modelscopes to obtain the operators' view can be rewarding (Fig. 18.4).

The requirements of good operational access, working conditions and access and local storage areas for chemicals, filter cloths, packaging materials,

Fig. 18.4 Use of modelscopes (Courtesy: Gunn[2])

etc. are discussed in the chapters on particular items, e.g. reactors (section 19.2), dryers (Ch. 27), filters (Ch. 24), packaging (Ch. 30). Often the result of such studies is that equipment is set through floors to give better access (see Figs 18.1, 18.10, 18.11, 19.3 (p. 287), 19.5 (p. 289), 19.7 (p. 290) and 19.8 (p. 292)).

The building should be easily cleaned and dust, dirt and spillage not encouraged to accumulate. For biochemical processes that are carried out in sterile conditions, concrete and ceramics are used for wall and floor treatments since they can be shaped to give non-dust-collecting and easily cleaned surfaces.

The provision of firefighting facilities and emergency escape routes is outlined in section 17.7 (p. 255). Extra points for enclosed buildings concern escape routes and firemains.

Escape routes should be clearly marked and only hinged doors should be used. They must open in the direction of escape and they should not open so that they block passage ways. Escape routes down staircases should be on the outsides of buildings. If inside they should be enclosed to give protection to escaping personnel. In areas of restricted access, e.g. cells for very toxic materials, the security arrangements should not interfere with emergency escape provisions.

In high-risk plant of any height and other plant over 6 m, wet riser mains are used and the outlets on each floor should be accessible. Hosereels should be sighted in prominent and accessible positions at convenient heights on each floor level on exit routes, preferably in corridors and, if possible, permanently attached to a pipeline. The length of a hose on a hosereel should not normally exceed 36 m and no part of a floor should be more than 6 m from the free end of a hose when fully extended.

In addition, in plants of over 18 m high it is desirable to fit dry riser mains. The inlets on the ground floor and the outlets on the other levels must be readily accessible.

18.4   PIPING AND CABLING

Piping that carries liquids under gravity should usually go, as far as is feasible and pipe stressing allows (section 31.3, p. 429), directly from one vessel to the next; the actual route is chosen to preserve access and, in general, to maintain a good fall in the direction of flow. The latter is especially true if any solids are present in the fluid, e.g. see Fig. 25.4 (p. 358) for shallow bends. Chutes for solids always go directly and as near vertical as possible (Fig. 19.8, p. 292), although the use of line vibrators or fluidizing air pads on ducting can alleviate shallower falls.

Service piping and piping carrying materials that are pumped can take more indirect routes in order not to clutter-up the space (Fig. 18.5). Nor-

Fig. 18.5   Piping and cable ways in building (Adapted from Kern[3] by special permission of *Chemical Engineering*; © 1978 by McGraw-Hill Inc., New York, NY 10020 USA)

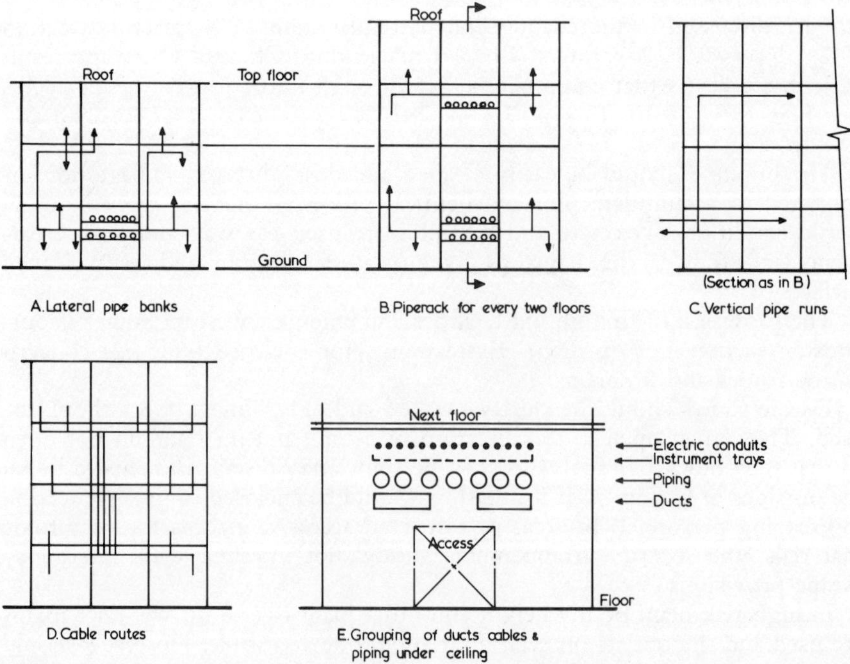

Fig. 18.6 Double piperack with takeoffs (Courtesy: Nottingham University)

mally, piping, electrical and instrument cables should go up the walls or support-columns and along, underneath floor beams (Figs 16.11 (p. 240) and 18.6). There is the possibility of using only alternate floors to take piping. In this case the piping goes up to the floor above and down to the one below. With piping under every floor, piping can either go up from every floor or come down from every ceiling; which system to use will depend on particular plants. Generally, though, piping coming out from a wall directly to a vessel, wastes space. Another facet to consider is whether to have single racks at each level running along the centre or double racks near the wall.

Pipe runs and cable trays should be easily accessible. They should not run over equipment but to the side of access ways so that they can be easily maintained but without blocking access ways.

The size of piperacks depends on the number and size of pipes and their load should be considered in the design of the beams or structures supporting them. The depths of instrument and cable runs is between 0.3 and 0.45 m. Special consideration should be given for space requirements for the thickest cabling going to the control rooms and switch room. However, heavy cabling needs minimum maintenance and access so can be routed with greater flexibility.

18.5 DUCTING AND HEADROOM

Ductwork for heating and ventilating systems should preferably be of a flat long cross-section[3] with the bottom elevation level maintained throughout the distribution system. At the same time, a suitable clearance of approximately 300 mm between the top of the ductwork and the underside of the

271

floor beams should be considered. This will permit cables and piping to pass over the top and prevent interference. In some cases it may be convenient to put large pipes under the ducting.[3] Duct sizes vary throughout the distribution system and will be largest adjacent to the air-conditioning room where the blowers and associated equipment are located.

For very large ventilation systems a more economic cross-section than a flat one may be used and this will lead to layout treatments specific to the installations.

Spacing between floors is usually 3–4 m with the first floor level being set high at 6 m because the heavier and therefore large items of equipment are usually put there. The minimum clearance should be 3 m, including ducting and pipework. However, this may have to be increased for maintenance of large items.

18.6   MAINTENANCE

Maintenance must be carefully planned. Items can either be repaired *in situ* or in the workshops.

If possible, the assumption should be made that every item will have to go to the workshops or be replaced during the life of the plant. The routes

Fig. 18.7   Equipment removal in buildings (Adapted from Kern[4] by special permission of *Chemical Engineering*; © 1978 by McGraw-Hill Inc., New York, NY 10020, USA)

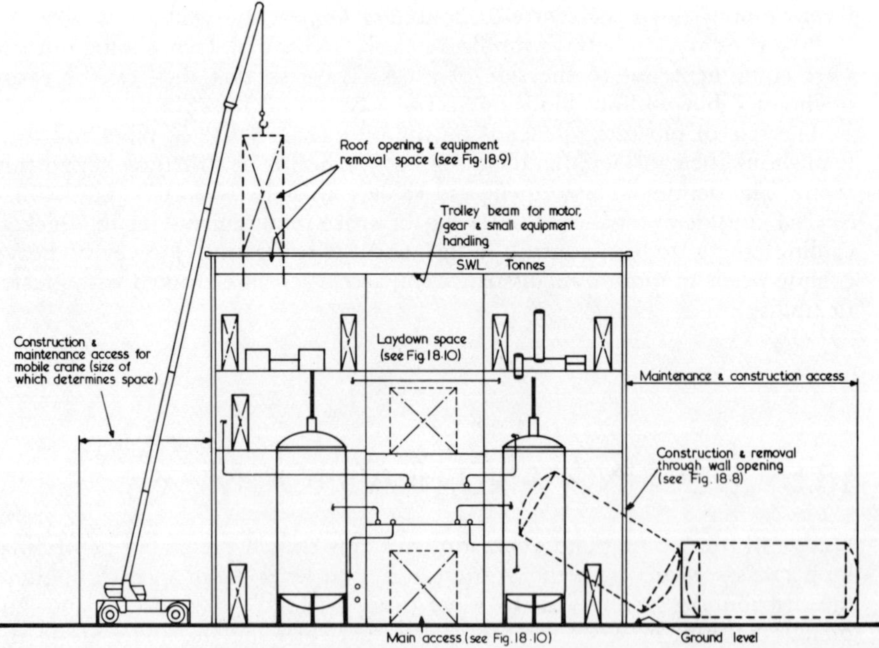

Fig. 18.8 Access for large equipment through wall (Courtesy: APV Hall International)

used for construction may not necessarily be available so for each item the maintenance route out of the building must be planned, clearing other equipment. This may be out of the roof or down to ground floor both via wells or out of the side (Figs. 18.7–18.10) but appropriate lifting equipment must be chosen. This may mean installing lifting beams or allowing access room for cranes, 'A'-frames, mobile platforms, fork-lift trucks, trolleys, etc.

273

Fig. 18.9 Access for large equipment through roof (Courtesy: APV Hall International)

inside the building. Outside the building there must be space for putting items on to lorries, etc. and positioning large cranes.

If it is decided that some major maintenance work will be done within the building then lay-down space for dismantling parts and tools is needed and appropriate lifting methods should be planned, such as lifting (trolley) beams as illustrated in (Figs 18.7 and 18.10). Consideration should be given to seeing whether the operating space needed for batch reactors, etc. is sufficient and will be available for maintenance.

18.7 PLANNING PIPES, DUCTS AND EQUIPMENT REMOVAL

Routes for pipes, ducts and equipment can conflict and so should be planned together. The ideal is to allocate specific space regions to piping and ducting, leaving the maximum (and most usable space) for access. Aspects to consider include:

(a) Wells to remove equipment can conflict with horizontal piping, etc.
(b) Side access to remove equipment can clash with vertical piperuns.
(c) Horizontal runs of ducting and piping can run across the building or

Fig. 18.10 Smaller equipment removal in buildings (a) upper floor (b) lower floor (Both courtesy: The Boots Company)

(a)

(b)

along its length. They can run in the centre or near the wall. Similarly, lifting beams for servicing several items can run across or along the building. Clearly, there can be conflict.

(d) Any arrangements decided must not violate the access needs of operation and emergencies. Handrails, etc. are needed for the protection of wells and wall openings while being used. They also are needed when a maintenance removal crosses an operational access route. However, these should be kept to the minimum.

18.8   SAFETY IN BUILDINGS

Ideally, plants having toxic, flammable or dust risks should not be installed in enclosed buildings except where the enclosure may in itself limit releases to the environment and/or aid in tackling releases from the plant. If an enclosed building is chosen in such circumstances the following actions should be taken:

(a) There should be no ledges for dusts to collect and spillages of liquids should drain quickly and safely from the building.

(b) Ventilation should reduce concentrations to below the appropriate flammable or toxic limits (Fig. 18.11). Air intakes should be positioned remote from adjoining process plants to avoid the risk of drawing-in toxic

Fig. 18.11   Local ventilation (a) from and over vessel (Courtesy: The Boots Company) (b) in UF_6 diffusion plant (Courtesy: British Nuclear Fuels)

(a)

(b)

or hazardous fumes. Similarly, exhaust ducts should be so placed to prevent dispersion of fumes into other plants. There should be no short-circuiting between inlet and exhaust. The exhaust air may require treatment by washing, carbon bed adsorption, filtering, etc. and inlet air may need heating and conditioning.

(c) Toxic areas may have restricted access including installing connection 'barrier rooms' with showers where clothing has to be changed.

(d) Electrical hazard area classification is more demanding inside buildings because leaks are not readily dispersed as in the open air (see sections 8.7 (p. 127) and B.8 (p. 528)). There will therefore be more extensive Zone 1 areas but these can be controlled if:

 (i) there are no admissions of hazardous materials into the building;

(ii) where system containment has to be open for maintenance or operational purposes, extraction ventilation local to the emission point should be installed (Fig. 18.11).

277

The separation between different electrical classification zones could well affect the plant layout and switchroom siting by the need to have solid walls and floor barriers, air-locks and sealed doors (see BS 5345[5]).

(e) Explosion relief methods may be necessary[6,7]. They include hinged panels, louvres and weak sections of roofs or walls. The direction of relief should be away from plants and people. Section 26.4 (p. 365) discusses relief for dusts.

(f) The possibility of collapse of high structures on to adjacent plant and buildings must be considered.

(g) Fire may spread in a building by lift shafts, corridors, conveyors, ducting, box girders and these should be kept to a minimum. It may be necessary[8] to provide fire stops, sprinkler curtains, fire resistant self-closing doors, heat-acting shutters, etc. Wells for lifting equipment should be fitted with solid covers when not in use. Floors in enclosed structures should be solid fire-resistant and not usually open mesh (though this is satisfactory and common in boiler plants).

(h) Fire and smoke vents can be used to channel the products of fire to a safe position in the open.

(i) Inventories, especially if elevated, of toxic, flammable and corrosive materials must be kept to the minimum.

In addition, all the safety aspects of open plants such as spillage containment also apply to enclosed ones (see section 17.4, p. 246). The need for a proper safety audit of the layout is very necessary.

18.9   ILLUMINATION AND APPEARANCE

Natural illumination may be obtained by use of patent glazing, windows or translucent sheets (see Figs 18.10a, and 11.7a (p. 172)) in the side walls or the roof. Rooflights may also be individual skylights. North light roof construction gives a good uniform natural light without the disadvantage of direct sunlight glare. South roof lights can be shaded to diffuse light and can be advantageous when solar heat gain is needed. Artificial lighting must be arranged to give adequate illumination throughout, and extra lighting points are needed near equipment with physical or chemical hazards and where instruments are read. Emergency lighting is needed on escape routes. In plants working 24 hours a day artificial lighting location is more important than the use of natural illumination.

Flat roofs can be used to store (harmless only) liquids in tanks and other vessels on or above the roof. The roof can also be a convenient place to mount ejectors, fans, air-cooled heat exchangers, vent scrubbers, etc. providing it is designed properly to stop leaks. However, pitched roofs enable snow and water to clear quicker and make the installation of roof lighting and ventilation easier.

The layout can be altered to optimize the lighting power consumption although in many cases other power requirements are much larger. Similarly, the layout can be changed to give a pleasing appearance to the building. Local planning laws may require special architectural treatments or impose limits on heights of buildings, stacks, etc. These factors must not overrule process safety, maintenance and other considerations. However, early consultation with lighting engineers, architects and planning authorities is advised. It raises operators' morale, and therefore productivity, to work in a well-lit, good-looking, clean plant.

REFERENCES

1. Balemans, A. W. M. *et al.* 'Check-list guidelines for safe design of process plants', in: Buschmann, C. H. (ed.) *Loss Prevention and Safety Promotion in the Process Industries.* Elsevier Scientific Publishing Co., Amsterdam, 1974.
2. Al-Asadi, H. *Computer Aided Layout of Chemical Plant.* Ph.D. thesis, University College, Swansea, 1980.
3. Kern, R. 'Arranging the housed chemical process plant', *Chem. Engng*, 17 July, 123, 1978.
4. Kern, R. 'Specifications are the key to successful plant design', *Chem. Engng*, 4, July 123, 1978.
5. BS 5345, Part 2, 'Classification of hazardous areas' 'Code of practice for the selection, installation and maintenance of electrical apparatus for use in potentially explosive atmospheres (other than mining applications or explosives processing and manufacture)', British Standards Institution, 1983.
6. Maidstone, R. J. *et al. Structural Damage in Buildings Caused by Gaseous Explosions and Other Accidental Loadings 1971–1977.* Building Research Establishment HMSO, 1978.
7. Baker, W. E. *et al. Explosion Hazards and Evaluation*, Elsevier Scientific Publishing Co., Amsterdam, 1983.
8. BS 5908, 'Code of practice for fire precautions in chemical plant', British Standards Institution, 1980.

PART III

DETAILED LAYOUT OF EQUIPMENT AND PIPEWORK

PLANT VESSELS

19.1   PROCESS VESSELS

The sizes of vessels and drums are mainly determined at the stage of process design when the principal features of internal and external heat exchangers, mixers or agitators are also specified. Nozzle connections are specified in the process piping diagram. Nozzle dimensions are fixed when sizing pipelines for process flows, meeting standards for relief valve fittings and matching instrument connections.

In the first stage of vessel layout, platform levels and then details of vessel elevations are set from process requirements (net positive suction head (NPSH), gravity feed, barometric legs, etc.) and from considerations of access for safe and convenient operation and maintenance. Methods of supporting vessels and operating platforms are detailed. Access for lifting equipment or overhead hoists and trolley beams is arranged for removal of motors, mixers and internal heat exchangers from the process vessels. A platform should always be provided for the removal of such heavy items of equipment and for access to manholes.

The detailed mechanical design of the vessel may be carried out once methods of support have been determined and nozzle sizes and positions have been fixed. However, economic pipe layout and access to valves and nozzles also depend upon the position and orientation of vessel nozzles; which therefore may be repositioned to give a better layout. The evolution of vessel design from process design specifications is illustrated in Fig. 19.1 and further discussed by Kern.[1]

In addition to the relative placing of process vessels, the economy of vessel layout depends upon the supporting structure.[1] Heavy vessels and other loads should be placed near vertical supports. Beam spans should be kept short by putting heavy loads near one another. This sometimes allows the number of vertical supports to be reduced. The possibilities of saving on support needs are illustrated in Fig. 19.2. Further savings are possible if vessel structures can be combined with ones supporting ancillary equipment such as pumps and heat exchangers, or with piperacks (Fig. 19.3).

Drums when horizontal may be supported on saddles. A single saddle is sufficient when the drum is less than 2 m long, but otherwise two saddles may be used, positioned about one-fifth of the drum length from each end of the drum (Fig. 19.4).

Fig. 19.1 Stages in vessel layout

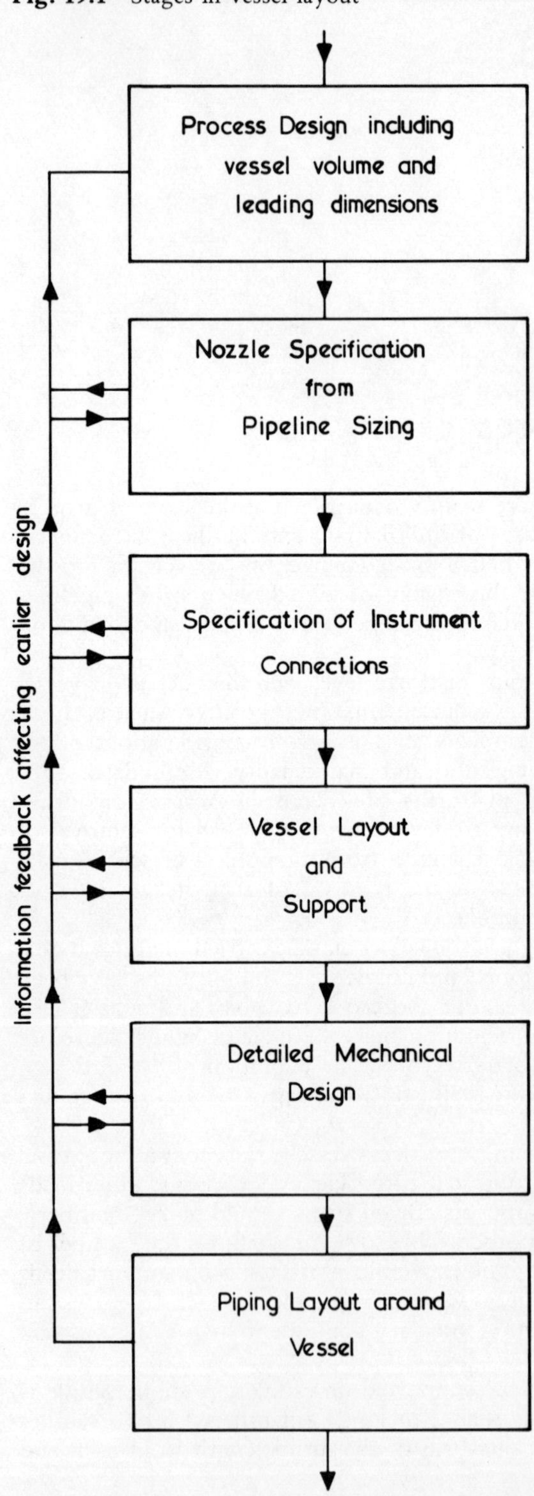

Fig. 19.2 Optimizing structures for drums (Kern[1])

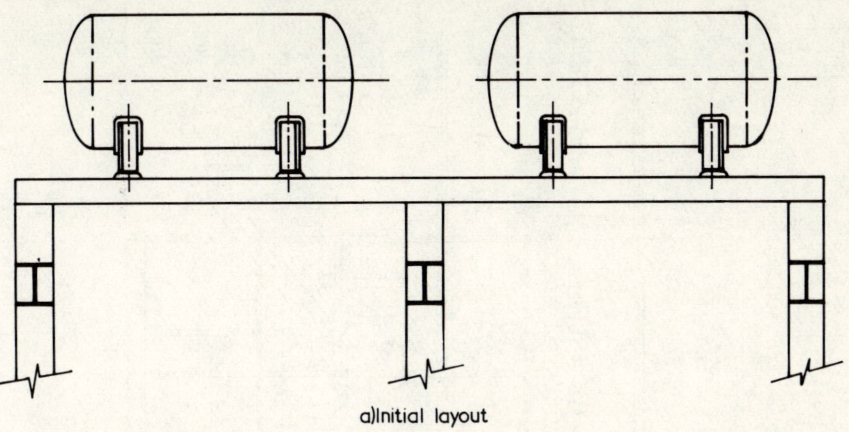

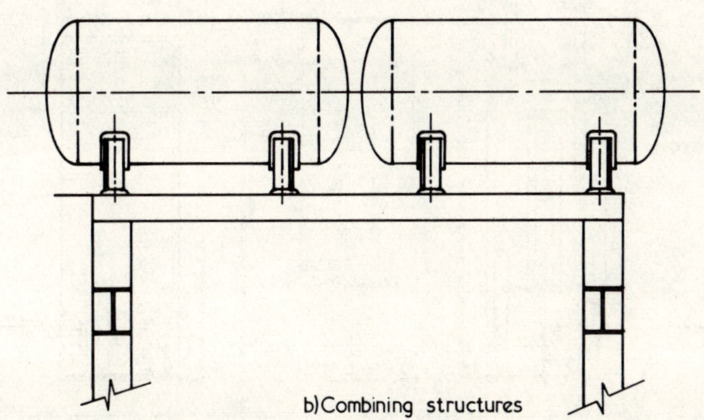

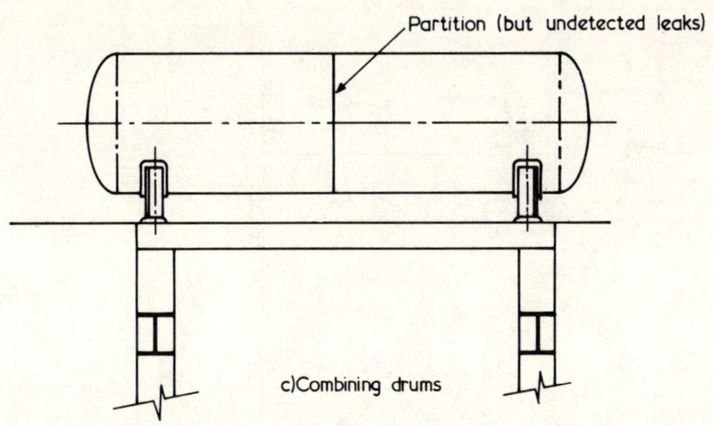

Fig. 19.3 Typical batch reactor layout (Courtesy: APV Hall International)

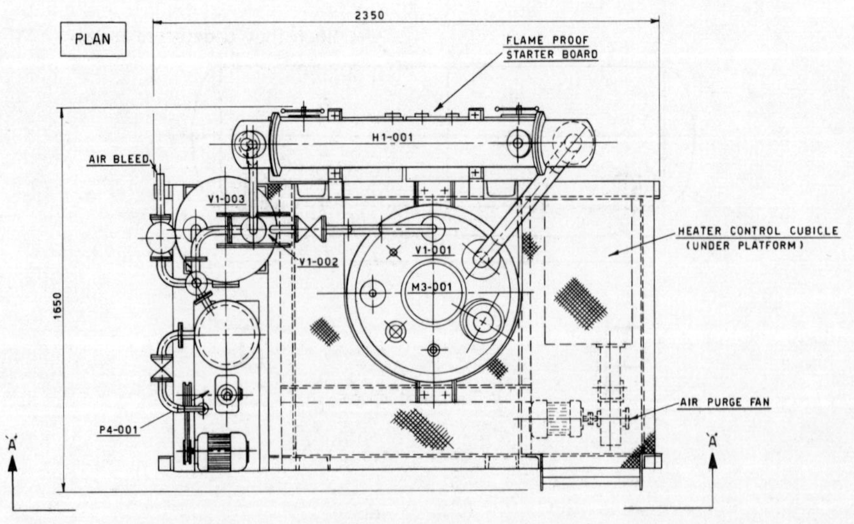

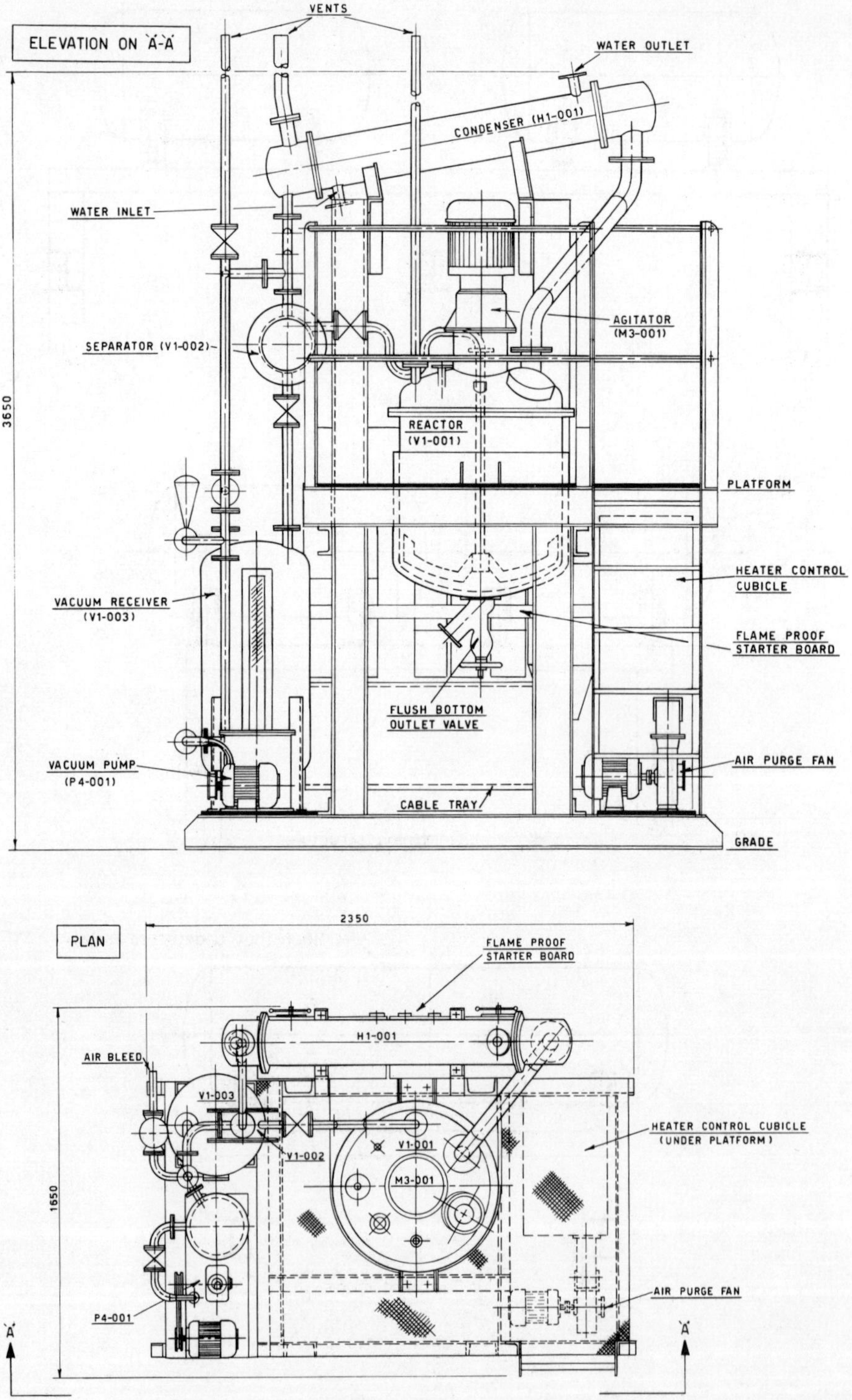

Fig. 19.4 Nozzle and manhole arrangements on drums (a) standard arrangement (b) typical example (Courtesy: The Boots Company)

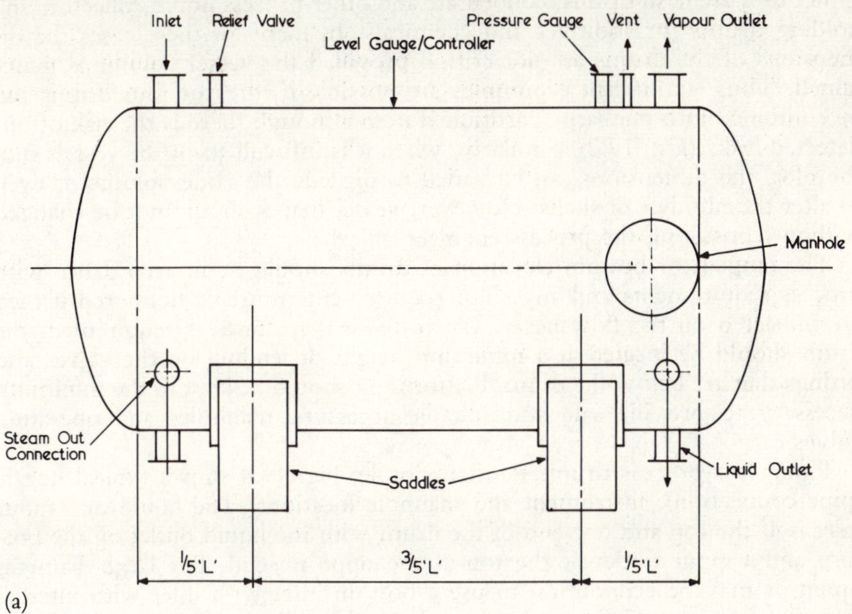

(a)

(b)

Vessels are often used to provide surge volumes, for liquid–vapour separations on distillation columns, or separating mixtures of immiscible liquids. Other uses are flash drums, condensate and other process liquid collectors and holding drums for additives and chemicals. In many of these cases the dimensions of the drums are not critical provided the vessel volume is maintained. Thus, significant economies are possible if three or four drums can be combined into one large partitioned item although there is the risk of undetected leaks (Fig. 19.2). Similarly, when it is difficult to fit the vessels into the plot, the dimensions can be varied to provide the same volume or even to alter the number of shells. However, vessel shapes should not be changed without consulting the process engineer.

The tangent or bottom elevation of drums should be in accordance with process requirements and any slope requirements must be honoured if they are indicated on the flowsheets. Where there is no process requirement the drum should be located at a minimum height depending on the valves and fittings that are below the drum. Platforming should be kept to the minimum necessary to provide safe and suitable access to manholes and operating valves.

Piping for process drums is often simple; Fig. 19.4 shows typical nozzle (pipe connection), instrument and manhole locations. The liquid or vapour inlet is at the top and one end of the drum with the liquid outlet on the bottom and vapour outlet at the top at the opposite end. For large-diameter piping it may be economical to use a bottom inlet with inlet with internal standpipe if pipe and fittings are saved. Vent nozzles are located on the top of the drum or on the manhole cover if a top manhole is used, with drum drain at the opposite end. Steam-out points may be placed with advantage at the opposite end of the tank from the vent line which should be open when steam-purging. An additional vent at the bottom may be useful when steaming-out as steam has a lower molecular weight and therefore is less dense than most compounds. Drain and vent lines may be located centrally or at the ends if the drum is horizontal, and if desired, the drain valve may be placed at the low point of the outlet piping. Vessels should always have small slopes towards the drain points.

The relief valve should be placed at a point on the top of the drum where the access platform can also provide access to other valves connected to the top of the drum.

When inlet and outlet valves are placed at the ends of the drum, the least agitated liquid region will be at the centre of the drum which is therefore the best location for gauge glasses and level-controllers. However, it is good practice to put level-gauges at the drain-point end and reduce turbulence by use of a stilling tube. A level-gauge should not connect with outlet pipe as this can subtract the velocity head from the reading. The pressure connection should be placed in the vapour space at the top of the drum, so that the face of the pressure gauge is visible from the ground or platform. The temperature connection is usually close to the bottom outlet, pointing towards the access aisle or platform.

Manholes can be positioned at the top, at the side or at one end of the vessel. A platform with good guard railings is necessary for access if the manhole is more than 3.5 m above grade. Some form of safe access assistance, temporary if need be, is required for horizontal manholes whose bottoms are more than 1 m above the ground or floor.

19.2   REACTORS

Liquid or gas–liquid reactors are often vertical and equipped with a vertically mounted agitator (Fig. 19.3). Heat transfer may be by an external jacket, internal coils, or an external heat exchanger. Side-mounted or bottom-mounted agitators are sometimes used and space must be allowed for maintenance and complete withdrawal of the mounting and agitator (Fig. 19.5).

Fig. 19.5   Typical layout for vertical mixer (Courtesy: APV Hall International)

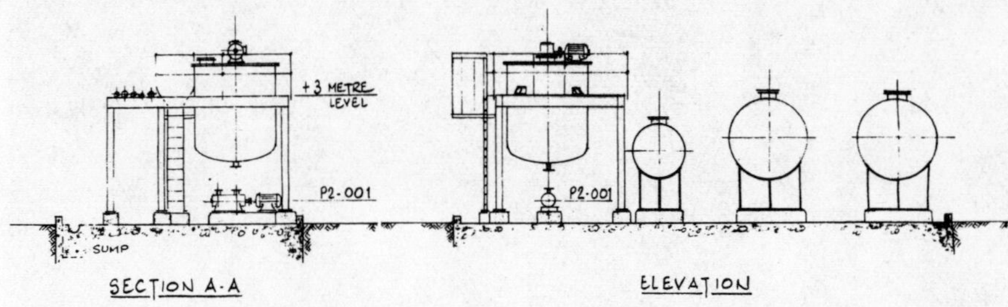

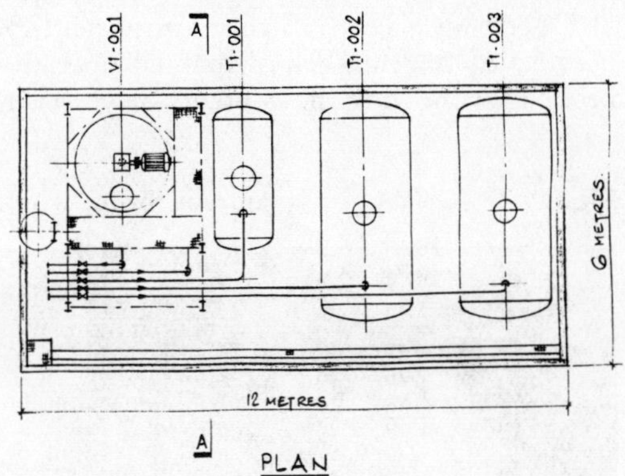

Reactors with large motors, gear boxes and agitators can cause problems by the transmission of vibrations to surrounding steelwork. Such problems are minimized by mounting the vessel on a foundation separate from the building steelwork. No building or floor steel may be attached to the reactor. The same principle applies to reactors mounted out-of-doors.

19.2.1   BATCH REACTORS

Batch reactions are often carried out as liquid phase or gas–liquid reactions. These may be at elevated temperatures and pressures in autoclaves (Fig. 19.6). Some autoclaves are thick-walled vessels often supported by lugs located on the vessel walls and placed on steel or concrete supports at low elevations

Fig. 19.6 Autoclave arrangements (Courtesy: British Nuclear Fuels)

because of the weight of the vessel. Vibration and thermal pipe stress problems have to be considered because of the weight of the vessel.

Operating access is a more important factor in the layout of batch plants because of the number of manual operations (Fig. 19.7). Platforms may be needed for viewing the contents or vessels through sight-glasses, for cleaning

Fig. 19.7 Reactor access (a) front (b) rear (Both courtesy: The Boots Company)

(a)

(b)

and for addition of materials by the operators. Hoists should be provided for lifting chemicals from the floor level and there should be space on the platform for their temporary storage. Often additions are made by chute from the next floor (Fig. 19.8). Lighting, steam and water for cleaning should be conveniently placed, and space for waste bins should be allowed at the side of the platform. The floor should be graded and drained (Fig. 17.3, p. 247); extra ventilation may have to be provided during cleaning. For some batch

291

Fig. 19.8 Chutes for solids addition (Courtesy: Humphreys & Glasgow)

processes the whole reactor may be removed and replaced by another if the saving in the time of on–line cleaning or maintenance is justified by increased production.

Batch production is often scheduled for runs of different products and it may be necessary to rearrange vessels between different production runs. Flexibility in replacing vessels may be provided by floor openings and hoists, by supporting the vessels on split rings and by the use of flexible connections and hoses where possible and safe. Hose manifolds and vessels should be placed so that hoses foul neither walkways nor lifting shafts. Each reactor may be permanently connected to its own heat exchanger, condenser and centrifuge or filter to form a reactor unit and flexible connections may be made between units. Layout has to be considered within each unit and between units. Ventilation ducts should be so arranged that flowbacks of dangerous fumes into vessels is prevented.

19.2.2   FIXED-BED REACTORS

This type of reactor is loaded with packing or catalyst in bulk between supports inside the reactor vessel, or in baskets if the temperature is not critical. The reactors may be constructed in several stages with heating or cooling between stages. When temperature control is more critical the catalyst may be packed within tubes mounted in a shell with heating or cooling fluid circulating between the shell and tubes or heating may be radiant as in furnaces (see Ch. 20). The elevation of the reactors will be determined by the method adopted to remove the spent catalyst or clogged packing.

If the catalyst (or packing) is to be removed from the base the reactors must be elevated sufficiently to allow removal by mechanical transport, belt conveyors or hand trucks. Catalysts may also be removed by fluid conveying. Space should be allowed for the introduction of mechanical drills and other equipment if coking, sintering or other hard agglomeration of the catalyst is likely to be a problem.

Loading of small quantities of catalyst may be carried out by a davit or winch, but larger quantities require a monorail and hoist, a skip hoist or a pneumatic conveyor depending upon the quantities. An access platform is required for charging the catalyst and for removal of internals so that the whole of the top of the reactor can be serviced. Removal of catalyst or internals from side manholes is carried out manually so that each manhole requires an operating platform of at least 2 m^2. At least 4 m^2 must be left at the base of each reactor for the transport and temporary holding of fresh and spent catalyst. Granular catalyst can be removed from some tubular reactors by large vacuum cleaners which avoids both elevating the equipment and providing large openings at the bottom.

Several reactors, whether in series or parallel, should be arranged in line with common support structures and an overhead monorail to serve all the reactors. The same overhead structure may be used for loading catalyst and for removing internals.

19.2.3 GAS-FLUIDIZED BED REACTORS

The layout of fluidized bed reactors is dominated by the requirements of handling solid catalyst or reactant under gravity flow conditions, and piping for the considerable quantities of gas or air (or possibly liquid) used in the reactor.

Solid reactants are fed from overhead hoppers with chutes usually directed into the top of the fluidized bed. Solids are removed through a standpipe or side exit controlled by a valve actuated to maintain pre-set upper and lower solid levels within the bed. The exit line is usually diametrically opposite the feed entry. Chutes should be as near vertical as possible, of generous dimensions, and provided with access points for rods to remove blockages in feed or product lines. The top of the reactor should have an access platform and overhead lifting equipment for the removal of reactor internals such as heat exchanger coils, candles or gas–baffles. Sufficient overhead clearance must be maintained for the withdrawal of internals.

The elevation of the reactor is fixed by the requirements for removing solids from the reactor through the discharge line to product hoppers, or conveyors. There should be provision for clearing solids that have fallen through the distributor plate into the base of the reactor, and from the gas discharge lines and heat exchangers.

When the solid phase is the catalyst which does not need regeneration *in situ*, there must be provision of catalyst at the start, make-up during the course of production, and removal of catalyst at the end of a production run. Equipment for the handling of solid catalyst is often similar to that described for non-catalytic fluid bed reactors. If regeneration of catalyst is required as in a fluid bed catalytic cracker, the fluidized bed may be combined with a gas transport loop. In one arrangement, catalyst is transported into the fluidized bed section and then passes by gravity through a regeneration section before entering the transport loops for recycling. With such transfer-line reactors there may be considerable problems of abrasive wear in transport lines particularly at elbows and bends.

19.2.4   REACTOR SAFETY

Reactors using elevated temperatures and pressures are one of the most potentially dangerous parts of a plant. It must be process design philosophy that the size of reactors should be as small as possible and that the reactors should be segregated. Apart from the usual hazards associated with flammable and toxic materials, the additional hazards of reactors are due to exothermic reactions. The control of these reactions is often sensitive and depends on the sophistication of the control system to be able to cope with variations in feed, cooling medium temperatures and the amount of coking. The principles described in Chapter 8 should be applied to the layout of these potentially hazardous reactors.

19.2.5   AGING

Products can be left for periods ranging from hours to years in order that reactions are completed naturally (aging or maturing). The maturing is done in storage areas (see Chs. 10–12) which, however, often have to have special temperature and humidity control and security, inspection and sample facilities.

Fig. 19.9   Layout of evaporators (a) rising film (b) falling film (c) forced circulation (d) vessel type (e) plate evaporator (All courtesy: APV Co.)

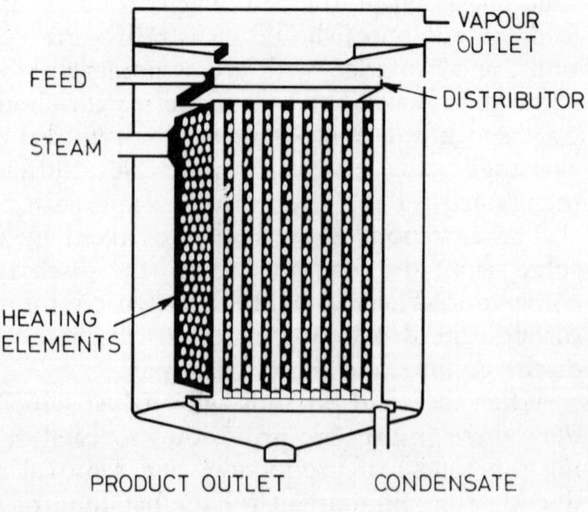

(a)(i)

(a)(ii)

19.3 MIXERS

Mixers are divided into those for processing solids or pastes and those for handling liquids. Equipment used for mixing liquids into solids falls within the solids mixer category, while that used for mixing solids into liquids comes within the category of liquid mixers.

19.3.1 SOLIDS MIXERS

The first group of the solids mixers contains those having a rotor within a stationary container, such as the ribbon, the single and double rotor mixers and the planetary type. Materials are fed in at the top of the mixer or at one end and the product is removed either from the middle or the opposite end at the bottom of the mixer. The conveyor feeding the materials can be laid out above the mixer from any plan angle with a chute discharge placed at the correct slope to allow free flow of material. The layout of the discharge side is made in a similar fashion.

Mixers with heating or cooling, must have the necessary piping connections, valves and instrumentation. In all cases space must be allowed for opening the equipment in order to clean the agitator and casing, and to remove the agitator completely for maintenance.

Solids may also be mixed by a pneumatic mixer which can have a subsidiary rotor. Space is needed for the cyclone above the mixer.

The second group contains those having a flat horizontal pan with vertical mullers on the surface of the pan. Feeding is from the top, discharge from the bottom, but since those mixers are almost always without a cover, there is no problem of access to the inside of the mixing pan. A monorail or hoist should be provided to enable the whole muller turret to be removed for maintenance. As there is no cover there must be adequate safety guardrails, etc. to protect the operators.

The third group of solids mixers comprises those with conically shaped rotating tumblers. Feeding is through covered holes in the top of the drum (while it is stationary) and access platforms need to be provided in the case of large mixers. Apart from occasional cleaning of the inside of the drum there are no internals to be maintained. However, guardrails, etc. must keep the operators away from the tumblers whilst in motion.

Paste mixers can be of vertical, horizontal or angular types. Some have hinged agitators for drum removal and agitator cleaning. Kneader- and

Fig. 19.9(b)

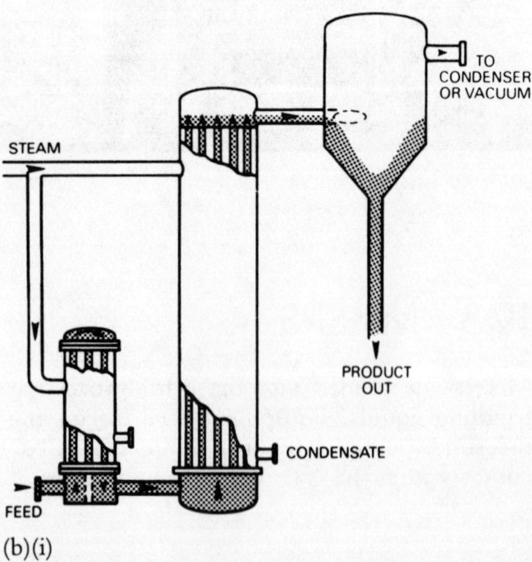

(b)(i)

(b)(ii)

Banbury-type mixers are heavily built horizontal machines. They should be sited on firm foundations, preferably at ground level. They can be fed by hand or by conveyor and for the former, local storage is needed. If they tip for emptying, the travel should not be fouled by pipeways, etc. nor should it extend into gangways. Space is needed for removing heavy internals for maintenance.

Fig. 19.9(c)

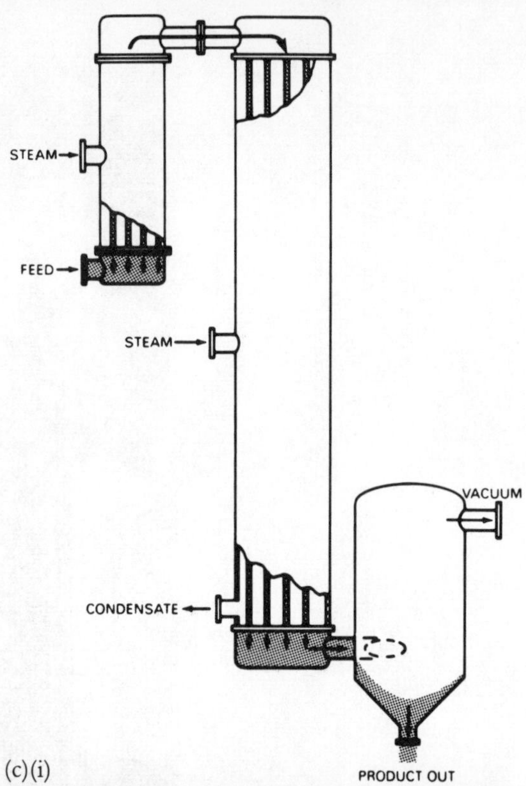

(c)(i)

(c)(ii)

19.3.2 LIQUID MIXERS

Liquids mixers are tanks having vertical, angular and sometimes horizontal agitators. The most common type consists of a vertical tank and a top-mounted vertical impeller unit.

Access to manholes and to the agitator drive mechanism should be provided on the large tanks, as well as the space to remove the whole agitator system. Such space can be saved by using a split shaft with a coupling connection. However, couplings are a source of dynamic imbalance, unless they are properly designed and installed.

Groups of mixing tanks can be laid out in a straight line, in pairs, or staggered. The last arrangement is useful for pairs of connected mixer–settlers, since the tanks can be physically close together, and have the short connections. Gravity flow mixer–settler arrangements have sloped launders or pipes connecting each pair, and it is necessary to provide sufficient head for gravity flow. Other types of mixer–settlers include the 'box' arrangement of horizontal rectangular boxes with mixers in alternate compartments, which mix and pump at the same time, thus eliminating the need for differences in elevation. As they tend to be rather long, sufficient space must be provided for erection. Otherwise maintenance and access is as described above.

Mixer-type reactors have already been described in section 19.2.

19.4 EVAPORATORS

The minimum height of an evaporator is set by the NPSH requirement of the product pump. Because of cost and inconvenience and the danger of vapour collecting it is not good practice to put the pump in a pit to obtain the required NPSH although it can be done. A pump should not be placed directly under the evaporator as it is sometimes necessary to lower the heating elements (calandria) (Fig. 19.9a). Barometric legs should be at least 10.5 m from the vessel base to the level in the barometric sump. This is usually situated on the ground floor. Horizontal and sloping sections should be avoided in barometric legs.

Sight glasses, instruments and sample points should preferably be reached by access platforms. These should also be provided, for cleaning purposes, at manholes, allowing 4 m^2 of free platform per form per manhole opening. Platforms may be needed for bundle-cleaning and repair together with hoisting equipment. Room must be left for the use of mechanical tube cleaners (when required) and for the removal and replacement of tubes. For calandria coaxial with the shell (Fig. 19.9a) this may mean having a removable panel in the roof above the evaporator. Space may be needed for additional pipework and valves so that one evaporator and/or external heater can be blanked-off for repairs and cleaning while the others remain on-stream. For chemical cleaning, room must be left, usually on the ground floor, for the cleaning liquor tanks and pumps. It may be necessary to provide additional ventilation during cleaning due to toxic fumes, etc.

For multiple-evaporators it is desirable to place the individual effects as close as possible to minimize vapour lines and share access and maintenance facilities. However, space must be allowed for insulation and its maintenance. Vapour liquid separators should be accommodated in the layout without increasing the distance between effects. It is better to stagger the effects at 60° centres in parallel rows or have a simple 'U'. The 60° desired layout allows one pan to be cut out as the vapour pipes can be easily rearranged. The structure and access platforms should be common for all the effects (see Figs 19.9b and 19.10).

Where spillages are liable to occur the floor should be bunded and drainings channelled to the effluent pit for sampling and necessary treatment.

As vapour piping is always of large diameter, it is necessary that layout of the evaporators is not finalized until the detailed pipe layout has been executed. In this detail future plant expansion should be considered, i.e. whether an extra effect will be added, or whether extra heat-transfer surface will be added to each evaporator either internally or externally with circulation pumps.

Fig. 19.9(d)

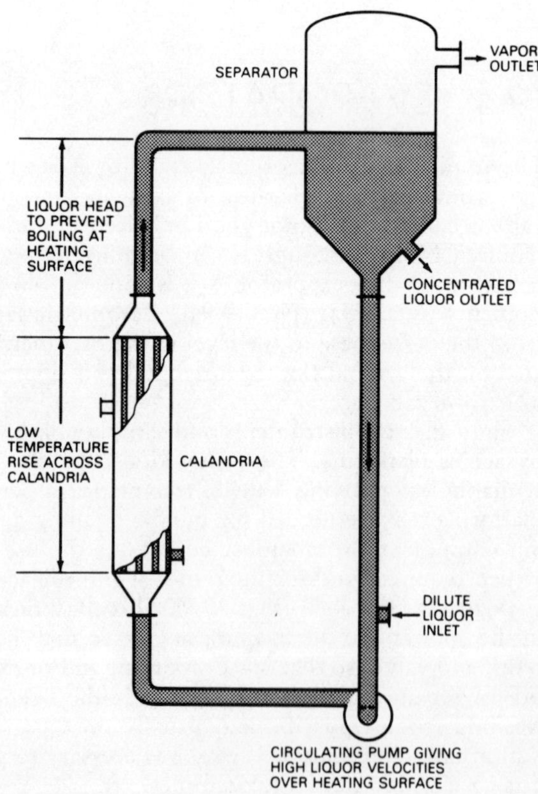

(d)(i)

(d)(ii)

19.5 CRYSTALLIZERS

Some crystallizers are similar to evaporators (Fig. 7.15b, p. 91) in constru-
tion and in their ancillary equipment. Otherwise, there are the agitated batch
crystallizers which can be treated as mixing vessels, the double-pipe crystallizers which can be laid out as heat exchangers, and the trough-type continuous crystallizer which is comparable to a ribbon mixer.

Layout requirements for crystallizers are similar to those for evaporators but there are often additional agitators. For slurry piping care must be taken to use long-radius elbows (Fig. 25.4, p. 358) and to provide plenty of cleanout facilities. Such pipes should be sloped for ease of drainage.

While it is desirable to place the crystallizers as close as possible to one another, the use of horizontal heaters or coolers precludes this, as they are placed between the crystallizers; on the other hand, vertical exchangers need not as they can be located above the crystallizers.

A large number of crystallizers can be arranged either all in one row or in a double bank with the separators and heat exchangers located between the two rows. This leaves the outside of the double bank completely clear, and allows one central platform for easy operation and access. This results in a neat piping arrangement with very short connections.

19.6 THICKENERS

Thickeners are large rectangular or circular vertical tanks with sludge rakes at the bottom of the tank rotating slowly about a vertical shaft centrally mounted in the tank. The tanks have a shallow cone-shaped base, and the blades of the rakes are inclined to direct the thickened sludge to the centre of the tank from which it is removed, usually by a screw conveyor. Sludge can also be removed by suction through pipes supported on the overhead rotating gantry. Slurry is fed to the feed well in the centre of the tank, and

Fig. 19.9(e)

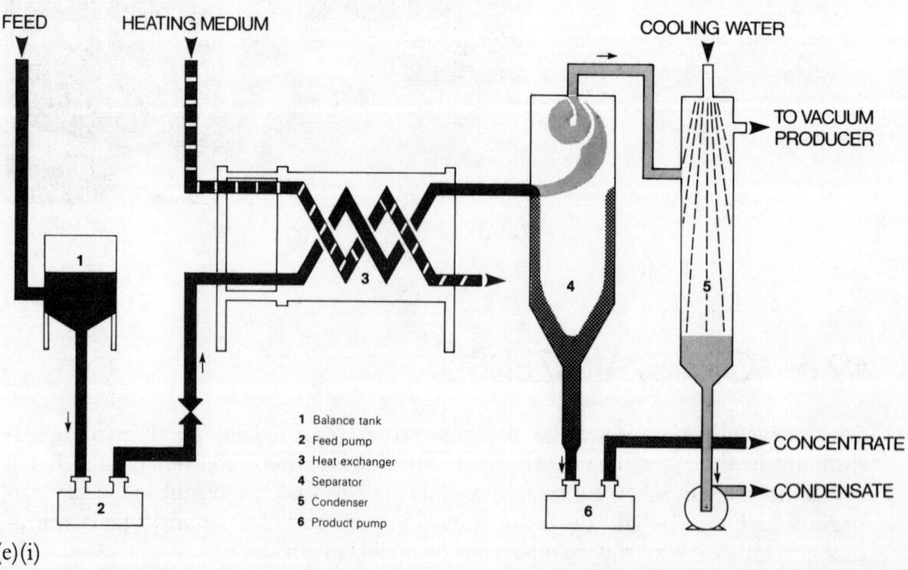

(e)(i)

(e)(ii)

supernatant liquor overflows over a peripheral launder and outlet pipe. It is important that the launder should be horizontal, otherwise the effective settling volume of the tank is reduced. Due to their large diameter, thickeners are often placed out of doors and away from the main process in order not to take up valuable process space (Fig 19.11).

A gangway constructed over the top of the tank supports the motor, gearing and cables, and provides access for maintenance. Maintenance of the sludge rake is usually carried out *in situ*, after emptying the tank.

Continuous counter-current decantation is frequently carried out with thickeners. In this case differences in elevation between the first and subse-

Fig. 19.10 Layout of multi-effect evaporators (Courtesy: APV Co.)

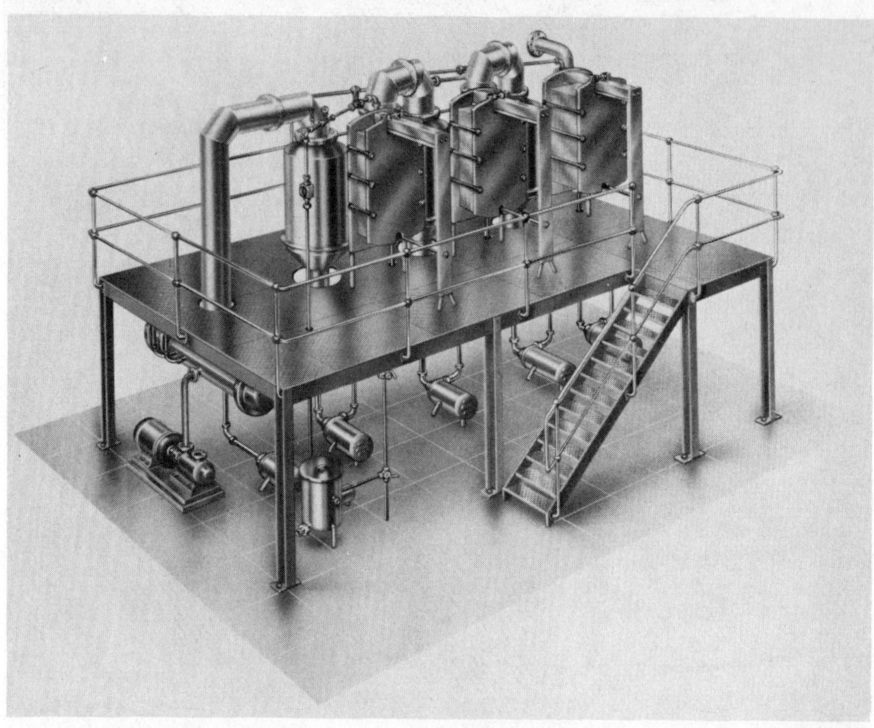

Fig. 19.11 Thickener showing maintenance gangway (Courtesy: Babcock Minerals Engineering)

quent thickeners are made in the layout to allow gravity flow from one to the next. Use should be made of natural variations in ground level to reduce the cost of elevating tanks.

Thickeners are also used as primary concentration devices for centrifuges, filters and cyclones. The thickened sludge is pumped from the thickener to the secondary separating device, and there is no need for the thickener to be close to the filter, centrifuge or hydro-cyclone (see Chs. 24 and 25).

REFERENCE

1. Kern, R. 'Arrangements of process and storage vessels', *Chem. Engng*, 7 Nov., 93, 1977.

FURNACES AND FIRED EQUIPMENT

20.1   FURNACE LOCATION

The first consideration in the layout of fired equipment is safety and a thorough study should be made of local and specific codes and standards. The location of such equipment may depend on the raw material flows from the previous stage or storage, and the product flow to the next stage in the process. Furnace transfer lines should be short and consideration should be given to common stack policy. Other factors affecting locations are the disposal of liquid and gaseous effluents relative to the other plants and the neighbourhood, and the proximity of services. Furnaces are placed in upwind locations so that flammable gases or vapours are less likely to be blown toward the furnaces and be ignited.

Furnaces in chemical plants usually occupy large areas of the plant site, and are located at the outskirts of the battery limits (Fig. 20.1). Process equipment (such as reactors, primary fractionators and distillation columns (Fig. 20.2)) connected to furnace outlets is located as close as possible to the furnaces so that transfer lines are short and simple; subject to the requirements that:

(a) Sources of ignition should be separate from possible flammable release.
(b) Hot lines may need special routing to achieve flexibility.

Fig. 20.1   Furnace location in relation to other units

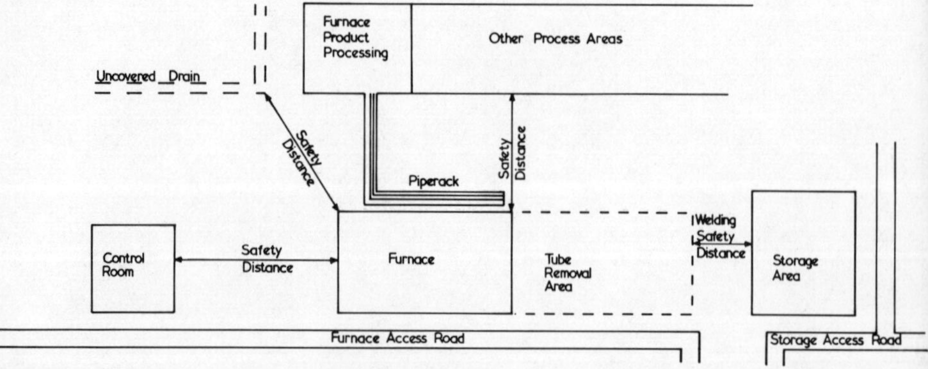

Fig. 20.2 Furnace with associated distillation columns (Courtesy: Humphreys & Glasgow)

The general rule is that fired equipment should be located at least 30 m away from other equipment which could be a source of spillage or leakage of flammable material. Detailed hazard assessment may indicate a greater separation. Underground drain-points and manhole covers should be sealed within the furnace vicinity, i.e. 30 m measured horizontally from furnace walls. No pits or trenches should be permitted to extend under furnaces or any fired equipment, and should be avoided in furnace areas if they may possibly contain flammables. The floor should slope so that spills drain away from the furnace. Consideration should be given to the provision of fire-resistant floors and partitions in buildings and the safe location of workplaces.

Firefighting space, and safety distances, are required between furnaces and process units, buildings or storage areas. Electrical control apparatus and cabling, associated with the furnace should, as far as possible, be placed and protected so that they survive a furnace fire.[1]

20.2   FURNACE GROUPING

Several process furnaces are usually grouped together. Furnaces should be spaced about two widths apart (centre to centre). They should be arranged with the centre-line of the stacks on a common line wherever possible and the stacks should be located at the end or side which is not used for access. Economics or air pollution control (through the regulatory authorities) will dictate whether single or common stacks are required for groups of heaters. From a layout point of view single stacks are preferred as they do not box-in the furnaces with breaching pieces to the common stack, with the consequent problem of access by cranes for tube removal. Single stacks also allow easy isolation for operation and maintenance. However, a common stack (Fig. 20.3) saves capital.

The flue gases can pass through the furnace and stack under natural draught. In this case, the stack must be tall, as this aids the dispersion by giving more lift and less interference. It can be shorter if an induced draught fan is placed (usually at grade but can be on the top of the furnace structure) between the outlet of the convection section and the inlet to the stack. However, stack heights also depend on operator safety and environmental protection, as discussed in section 13.3 (p. 198). For example, a sulphur-free fuel would not need a very high stack but a noxious flue gas would.

20.3   FURNACE AND HEATER DESIGN

Sizes may vary from small radiant helical coil units of 1.8 m diameter to large crude oil radiant/convection heaters of up to 20 m tube length, and methanol or ammonia reformers of 50 m^2 plan area.

Furnaces and fired heaters basically consist of tubes, burners, refractory-lined firebox and a stack. The firebox and the tubes within it is termed a 'radiant section'. In large units it is desirable to operate at high efficiency to reduce the fuel cost and this is achieved by adding a tubular convection section downstream to the radiant section to recover heat from the flue gas. In small units high efficiency is less important and a convection section unjustified.

Burners provide the means of heating and can be 'up firing', i.e. housed in the floor of the heater. Alternatively, oil- and gas-fired boilers and ethylene crackers are usually side-fired and reformers commonly down-fired. Liquid fuels may require an atomizing medium of either air or steam. Gas-burners do not require an atomizing medium, though there will be a gas pilot flame. The burner flame can be viewed through a sight glass or through peep doors. The latter are preferable, although these may involve expensive provision of platforms where multi-burners are used. It is often necessary to remove a burner in order to change or clean tips and atomizers, which requires sufficient clearance around the burner. In many cases the only means of entry into a furnace will be by removal of a burner. Piping should be arranged so as to make access as easy as possible.

Fig. 20.3 A series of furnace heaters and column reboilers discharging to a common effluent stack (Courtesy: ICI Petrochemicals and Plastics Division)

Furnaces can be classified into three types: helical, cylindrical and box furnaces. Helical coils are used mostly but not exclusively for small duties like start-up conditions and are usually vertical and cylindrical in construction with helically coiled tubing. The inlet to the coil is usually at the top and the outlet at the bottom. The furnace is supported on legs with the burners in the floor. Peepdoors are not often found in the shell since the helical coil construction prevents a clear view into the heater. Access is by means of removing the burners.

Cylindrical furnaces (Fig. 20.3) comprise a large vertical shell housing vertical tubes around the circumference. There is usually a cluster of burners in the floor and access through one of the burner holes. Peepdoors are feasible with platforms positioned around the shell. The convection section houses horizontally supported tubes each with an inlet at the top of the convection section, the outlet being at the top or bottom of the radiant section. Tubes are removed through a hole in the arch and then lifted by a davit mounted on the stack.

Box furnaces (Fig. 20.2) and heaters have horizontal or vertical tube layouts for both the radiant and convection sections, with burners firing either from the top floor or side walls. The horizontal type of heater requires extra plot plan area for tube removal. The advantage of this type is that a bridge wall may be incorporated if required to give control to different coils in the radiant section, there being burners either side of the bridge wall.

Construction should be substantial and of non-combustible materials. Structural steelwork should be located or protected by lagging so as not to collapse in the event of a serious fire. Inspection and access doors should be secured so that they are able to withstand the same pressures as the equipment and not become dangerous missiles in an explosion.

20.4   ACCESS TO FURNACES

Furnaces are usually elevated so that headroom is available under the steel of the furnace floor and under the under-floor piping for the floor burners. In the arrangement of furnaces, care should be taken that there is ample room at the firing front for the operation and maintenance of the burners, and for a burner control panel, if required. In the case of bottom floor fired furnaces this would mean adequate headroom (2.5 m) underneath the furnace. In the case of wall-fired furnaces a platform width of at least 1 m with escape routes at each end is required. With top-fired furnaces adequate exit routes from each end of the furnace are necessary, one of which must be a stairway. Staircases and ladders should connect platforms, and also serve as escape routes. Slide poles from elevated platforms are sometimes provided for emergency escape. Peepholes and observation doors should only be provided where absolutely necessary. Access platforms must be provided to these and other points of operation and observation that are 4 m or more above grade.

Access is needed for relining, tube cleaning and burner removal and other repairs. The roof clearance has to be consistent with crane and lift requirements and the support structures should not interfere with maintenance, mobile cranes, etc. Platforms are needed for maintenance of soot blowers.

Ventilation should be provided in the working area, particularly where sulphur-bearing fuels are used and high temperatures may be experienced.

20.5   PIPING FOR FURNACES

Piping is required for process fluids, steam generation (if used) and fuel supply. It is also needed for supplying steam for snuffing fires, blowing down process piping in an emergency and for cleaning and decoking. Air heaters and associated ducting can take up space. Usually only process piping is shown in flow diagrams, but typical diagrams for both process and service piping for furnaces are given by Kern[2] and an outline in Fig. 20.4.

Piping and valving between reactors and furnaces are expensive. The size and location of connections are determined by:

(a) Furnace and reactor designs.
(b) Pipe used for expansion compensation.
(c) Space and configuration limitations.

Thus economic piping layout depends on the designer's ingenuity. For reactor/furnace arrangements the central piperack provides utilities to both items. The hottest process lines must be flexible to stay within allowable thermal stresses and must have a very simple layout. Large-diameter lines are occasionally water-jacketed, and internally or externally insulated. This is a useful method of reducing the expansion and, hence, the problems of flexibility. The tendency of furnace transfer lines to coke can be kept to a minimum by short lines. In some petrochemical units quench oil is injected at the beginning of the transfer line to minimize coke formation. In transfer lines,

Fig. 20.4 Equipment and piping associated with furnaces

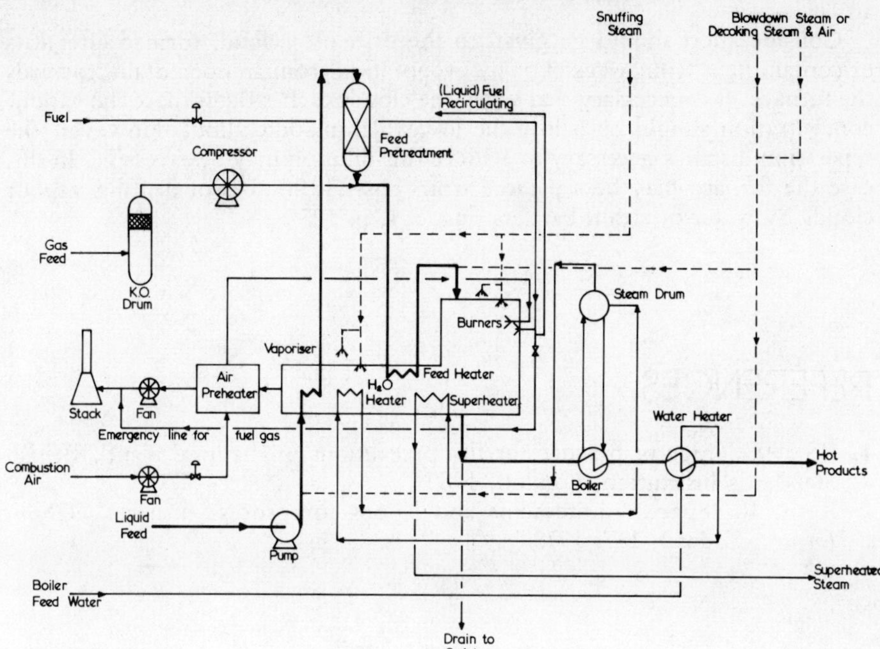

where coking is expected, flanged fittings and elbows are specified to facilitate cleaning. Control valves and valve manifolds are usually lined-up under the furnace piperack with convenient access provided. Air and fuel piping layouts are discussed in sections 31.6.2 (p. 447) and 31.6.4 (p. 448) respectively.

The valves for snuffing and blowdown steam must be at least 15 m from the furnace, well marked and accessible since both are used in emergencies. The main feed valves for the process fluid and the fuel supply should be similarly situated in a safe position.

20.6 CALCULATION OF SAFE DISTANCES FOR FURNACES

The general rule that furnaces should be 30 m away from other equipment should be checked for particularly severe cases, using the methods of Chapter 8 and Appendix B, and if necessary, spacings should be increased.

For calculating the effects of a furnace fire, it may be assumed that the fire burns from the roof upwards. It is necessary to estimate the thermal radiation levels at nearby vessels, buildings, and points of human access and take appropriate action (see Ch. 8). It is also necessary to estimate whether open-mesh drain covers are heated to the ignition temperature of possible liquids in the drain. Another probable fire scenario is that liquid fuel will spill on to

311

the ground and burn. Thus the spills should run away from plant to safe areas.

Consideration should be given to the drift of a cloud, formed after loss of containment from a vessel or by evaporation from an open drain, towards the furnace. It is necessary that when the cloud reaches the furnace the vapour concentration should be below the lower flammability limit. However, the separation distance necessary to achieve this dilution may be excessive. In this case the furnace may be separated from possible sources of drifting vapour clouds by water or steam barriers (Fig. 8.1, p. 125).

REFERENCES

1. BS 5908 'Code of practice for fire precautions in chemical plant', British Standards Institution, London, 1980.
2. Kern, R. 'Space requirements and layout for process furnaces', *Chem. Engng.*, 27 Feb., 117, 1978.

COLUMNS AND TOWERS

21.1   ARRANGEMENTS

A column arrangement should receive early investigation since the layout of other equipment usually depends upon it.

Columns are often much higher than most other items of equipment and are usually delivered to the site complete. If this is not possible, two or more sections have to be welded together on site and an area near the final location which does not hinder other work must be available. There must be adequate access for the column to enter the site and for unloading and erection (see Fig. 16.3, p. 224). Thus, columns are best located if possible at one side of a process area.

Lifting points should be provided on all but the smallest columns, situated above the centre of gravity of the column or section.

An economical arrangement is often provided by placing columns in line. Horizontally mounted reboilers and manholes (man ways) face out to the back with the space in front of the columns left relatively clear to provide access to reboilers and to provide space for the loading and unloading of column internals.

Groups of towers can be lined-up on their centre-lines if they have the same diameter (Fig. 21.1) or otherwise tangentially on the rack sides of the towers.

The distance between columns is governed by:

(a) Platforms, whether common or separate and whether they overlap.
(b) Foundation sizes and whether a common foundation can be used.
(c) Wind interaction.
(d) Insulation thickness, especially with very high or low operating temperatures.
(e) Access requirements.
(f) Dropout areas for receiving removed internals.

The minimum column spacing is about 3 m and 1 m space minimum is needed between column foundations and adjacent plinths. Plinths are not necessarily the same size as the underlying foundations.

The dropout area at grade should be located to give easy access. Columns should be designed to be self-supporting if possible. They are mounted on

Fig. 21.1 Typical distillation column layout (Courtesy: BP Chemicals)

a skirt on spread or piled foundations. However, for columns less than 0.75 m in diameter it is seldom possible to support annular platforms and vertical ladders or to provide enough resistance to wind loading. Therefore, small-diameter columns have to be supported in a structure which provides access platforms (Fig. 21.2). The column may be mounted on a skirt (see section 17.3.5, p. 245) or supported entirely in the structure if convenient, perhaps elevated to allow gravity flow to a subsequent BP unit. Partial support from existing towers or structures should be considered when plant is being modified. In the case of very tall towers, guy ropes may be required as well as aerial navigation warning lights.

The minimum elevations of the columns are normally governed by pump net-positive suction head (NPSH) requirements and the thermosyphon re-

Fig. 21.2 Layout of supported columns (Courtesy: APV Hall International)

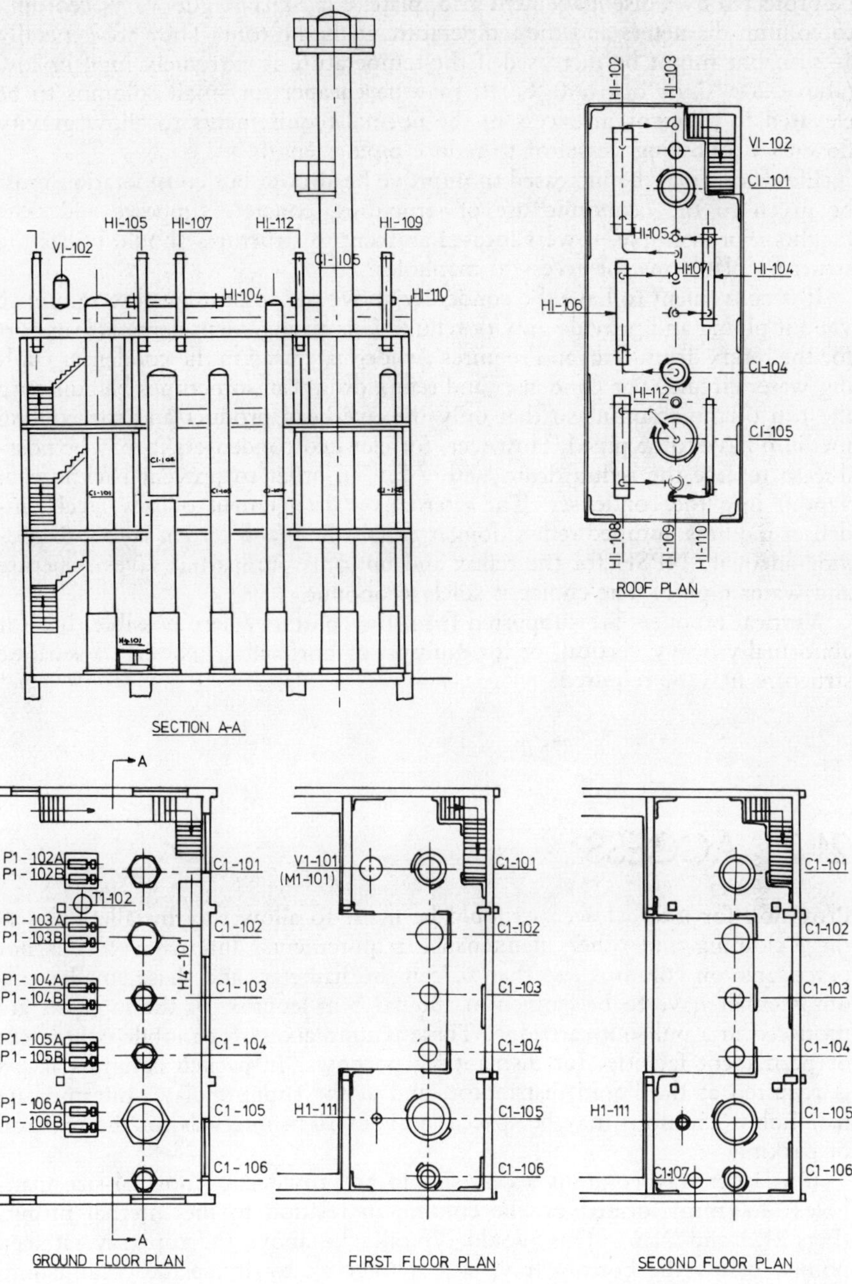

lationship to the reboilers with respect to drawoff and return nozzles (Fig. 23.2, p. 339). There also must be sufficient height (500 mm) between the reboiler return nozzle and the first plate in order to disengage the liquid from the vapour. Similarly, there must be sufficient height for disengagement between trays and between the top tray and the vapour outlet although internal access requirements may determine these heights. The vapour pipe may be protected by a disengagement grid, plate, etc. Skirt heights vary according to column diameters and the temperature at the bottom. They are generally 1–2 m, but might be increased if the temperature is extremely high or low (above 200 °C or below 0 °C). It may be cheaper for small columns to be elevated to a height in excess of the normal requirements to allow gravity flow to a following vessel or to reduce piping length.

Elevations may be increased to improve headroom but consideration must be given to the economic use of structures, concrete supports and skirt heights. For example, towers located adjacent to structures should utilize the structure platforms for access to manholes, etc.

It is convenient to have the condenser above the column head to minimize vapour piping and provide gravity reflux. This arrangement demands support for the reflux drum, etc. and requires an adequate head in the condenser cooling water circuit. The condenser and reflux drum can sometimes be built into the top of the column, so that only the overhead product and the cooling medium have to be piped. However, for elevated condensers it may be possible to replace the reflux drum with a lute in order to prevent backflow of vapour into the condenser. The alternative arrangement of low-level condenser requires pumped reflux, longer vapour lines and care in layout to provide adequate NPSH for the reflux and bottoms pumps but saves structure and water piping. The choice is solely economic.

Vertical reboilers are supported from the column where possible, but for abnormally heavy vertical, or for multiple or horizontal, reboilers a separate structure may be required.

21.2 ACCESS

Provision for internal access should be made to allow tray installation, setting, cleaning and other maintenance requirements. Internal access is not practicable on columns less than 0.75 m in diameter, and these smaller columns either have to be flanged in about 1.5 m lengths, or the internals are mounted in a pull-out cartridge. These require access for mobile equipment or permanent facilities for dismantling purposes. In packed columns access is required at the liquid distributor, and at the support plate. Intermediate handholes (250 mm) may be spaced at about 10 m intervals for the removal of packing.

On larger tray columns access should be provided by normal-size manholes (460 mm) located on the column in relation to the internal fittings (Figs 21.3 and 21.6). This would typically be above the top tray, at feed points, below the bottom tray, and at intervals to divide the column into sections containing not more than ten trays. It is common to leave a removable section in each tray to form a vertical accessway. This is designed so that a man can work downwards when removing segments. The minimum tray spacing for access is 460 mm, but this may be increased at manhole locations. The centreline of the manhole should be 0.75–1.2 m above the access platform elevation and there should be at least 2.2 m clear headroom over a plat-

Fig. 21.3 Equipment arrangement and access

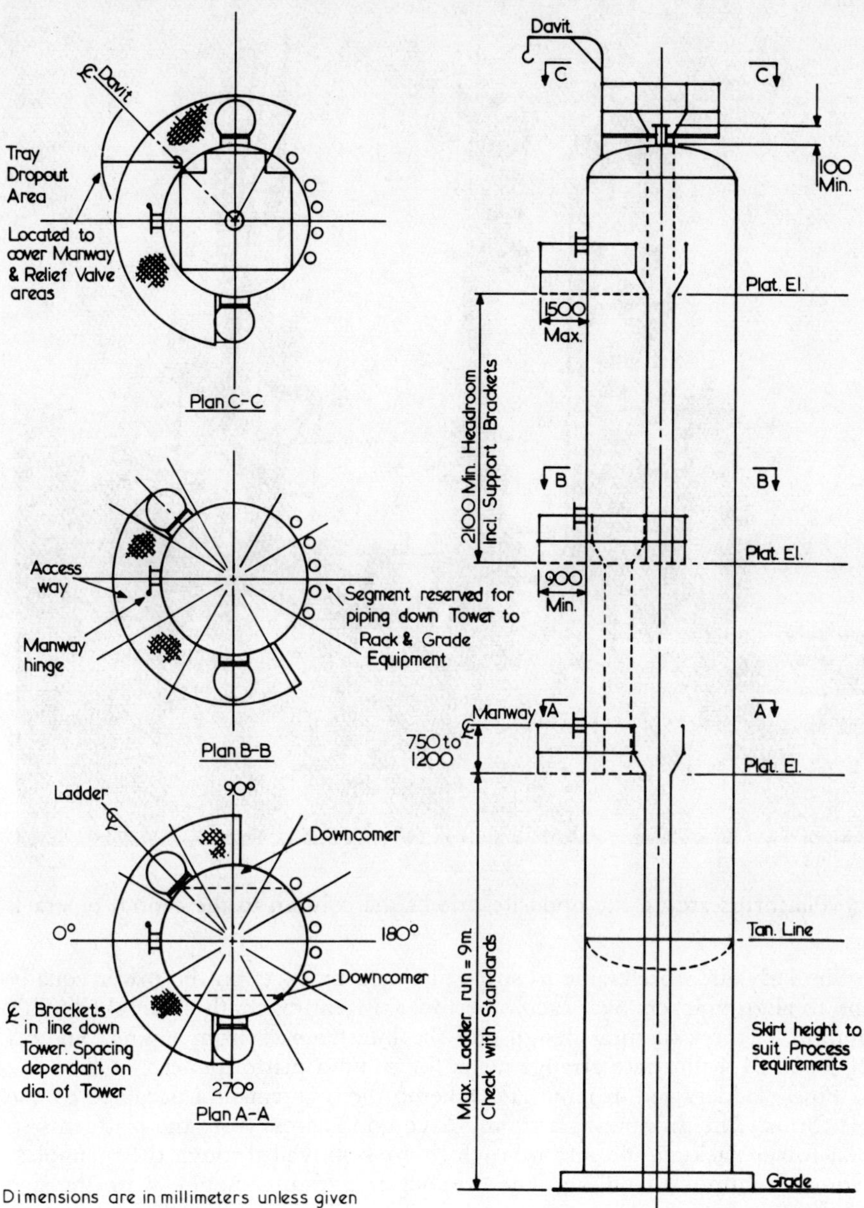

form. Instruments, valves and other fittings should not protrude from the column at elevations and positions likely to cause a hazard.

It is usual to provide an annular platform at each manhole. Unless there are contrary reasons, the platforms and manholes should all be orientated in the same direction, preferably so that:

(a) Manholes give access to be contact area of the tray.
(b) Platforms face the main accessway.

Fig. 21.4 Example of access to columns (Courtesy: Humphreys & Glasgow)

(c) Platforms are on the opposite side of the column to the central piperack or structure.

Similarly, it is preferable to space platform brackets on the tower equally and to align brackets over each other over the entire length of the shell. This minimizes the structural design and the interferences from piping. Figures 21.3 and 21.4 illustrate arrangements for column platforms, etc.

Pipes, ladders and dropout area take up the segments not occupied by the platforms. The dropout area should have good access from the platforms so that tower packing and internals which are removed through the manholes, require minimum handling. The dropout area ideally should be on the side of the column away from the piperack. Where mobile handling equipment cannot be used, davits should be provided for lifting and holding manhole covers, column heads, or similar large, heavy items intended to be removed during maintenance. Adequate clearance must be left for the swing of the manhold cover on its davit and entrance to the manhole must not be obstructed by piping. Thus, pipes should not be passed through the platform, but occupy the segmented gaps, nearest to the piperack. Where a tower is a packed column a permanent runway beam is normally installed with adequate space at grade for handling the packing ($\sim$40 m^2).

Fig. 21.5 Safety ladder with safety bar

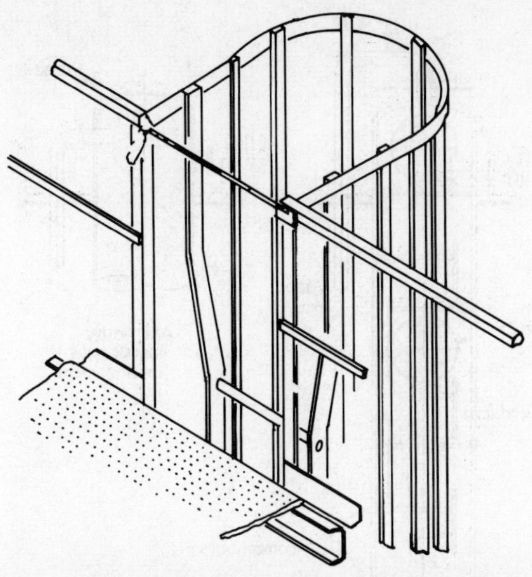

Two further segments of the tower are occupied by the ladders interconnecting the tower platforms and are ideally 180° apart. This further restricts the areas available for piping. The maximum allowable straight runs of vertical ladders between platforms vary according to company standards but will generally be between 7.5 and 9 m, dependent upon the acceptability of intermediate access to platforms from the same ladder run. Safety hoops should always be fitted to ladders more than 2.5 m high, and where a ladder starts from an elevated platform, back protection should be provided right down to the platform handrail. An arrangement of a safety ladder with self-closing safety bar is shown in Fig. 21.5.

Access is required outside the column to flanges, relief or other valves and instruments and should be provided either by the manhole platforms or intermediate platforms (if there is enough headroom). Instruments can sometimes be located in adjacent piping accessible from the main structure and this may simplify access arrangements. On free-standing columns, access for major maintenance to insulation or painting will usually require the erection of temporary scaffolding. Room for this must be allowed at the base of the column and it is useful to fit cleats in the shell to facilitate scaffold erection.

21.3   PIPING AND NOZZLES

The location of fittings within the tower determines or constrains the location of tower nozzles, instruments, piping and steelwork. The design of column internals is a somewhat specialized area[1] so the following discussion is only an outline.

The depth of liquid in the tower bottom is mainly controlled by the re-

Fig. 21.6 Relation between internals, manholes and nozzles

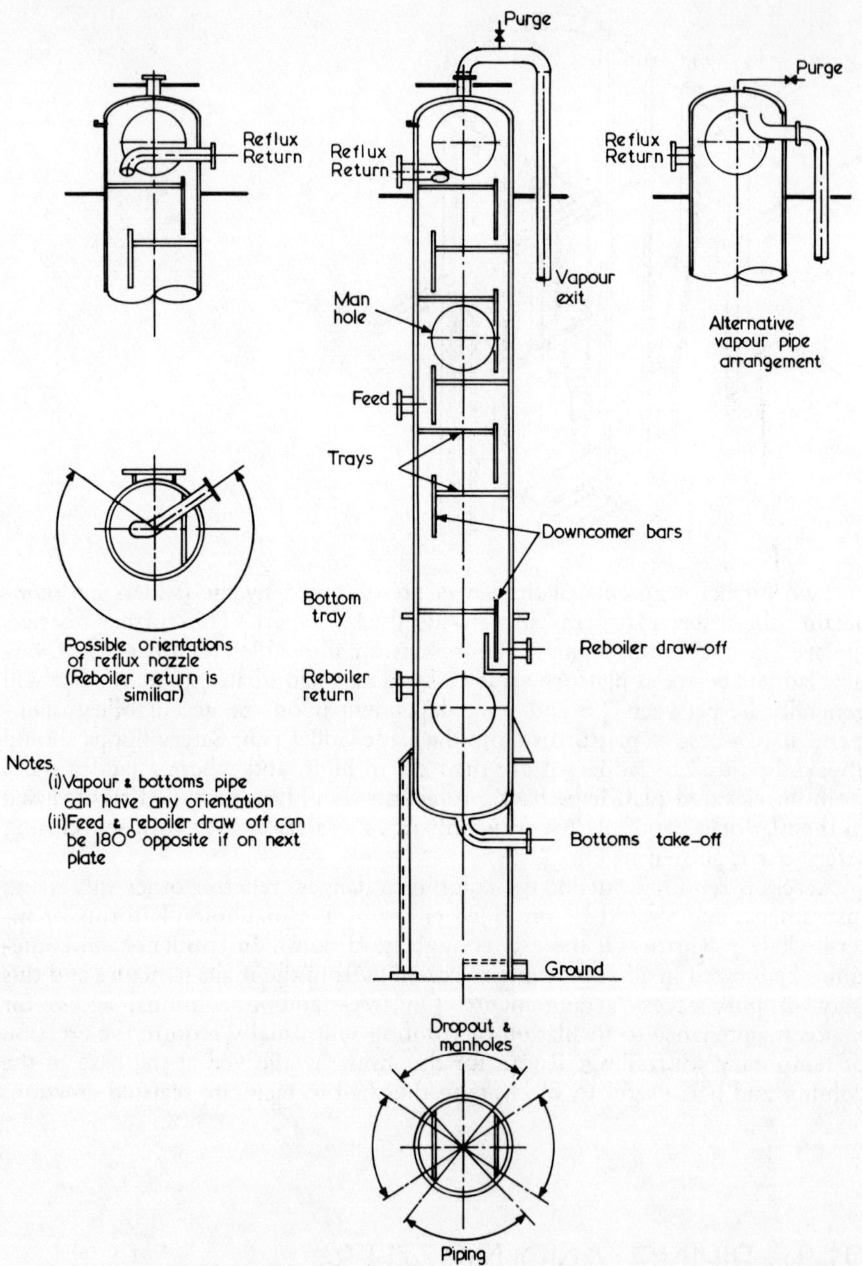

quired holding time for the bottom product. Figure 21.6 illustrates a reboiler drawoff connection to the downcomer of a single flow tray. The return connection from the thermosyphon reboilers is also shown. These lines should be as simple and as direct as possible, consistent with the necessary requirements for thermal flexibility.

Two simple reflux pipe arrangements are shown in Fig. 21.6; if tray orientation has been fixed by other factors this nozzle may be reorientated to give a better arrangement provided the horizontal leg clears internals.

For reflux and feed connections illustrated in Fig. 21.6, an enlarged feed tray spacing should be provided for the introduction of the distributors.

The vapour outlet nozzle may be arranged overhead, or an internal vapour takeoff may be chosen as shown in Fig. 21.6. The latter arrangement eliminates the need for a small platform above the head of the column.

Internal piping for vapour outlets, reflux inlets, inlet distributors and tray draw-offs must clear obstructions such as bubble caps or weirs, and be capable of passing through manholes for assembly in the tower. The manhole openings must not be obstructed by internal fittings.

The main features of the piping layout outside the tower are outlined in Fig. 21.7. For economical and simple support, pipes should rise or fall close to the nozzle and should run parallel and close to the tower allowing for lagging and leaving room to get at flange bolts. The horizontal height of the piping is governed by the main piperack elevations being 0.6–1 m above or below it, to allow for rack takeoff. Pipes that run directly to equipment at grade often have the same elevation as the rack.

21.4   LAYOUT PROCEDURE

The process department provides the basic design information for towers. This includes the length and diameter of the tower, the number and spacing of trays, the number, size and relative locations of nozzles and instrument connections and the positions of condenser, reflux drum and reboiler. With this process information, the plant designer must orientate the nozzles (which will depend on the internal design) as well as decide platform sizes and elevations and determine handling facilities for tower internals, manhole covers, relief valves, etc. Platform sizing enables the structural engineer to locate and select types of platform support clips for welding to the vessel shell.

The detailed layout of a tower can then commence. Firstly, the specification of the internals, handling facilities and process requirements described in the above paragraph require special study and then probable reference back to help resolve layout problems. The specifications should then be summarized.[1]

Next, the designer should draw out to scale the elevation of the tower and associated exchangers, etc. showing all nozzles, manholes, and instrument connections. Using this elevation the location of platform levels can be determined and lengths of ladder runs ascertained. When the elevation locations are determined, a plan at each platform level should be prepared. The plan views in Figs 21.3 and 21.6 show the manner of allocating segments of the tower, to platforms, ladders, dropout area and piping. The piping runs can then be designed. Finally, the layout engineer should consult with the construction department and modify the layout to accommodate erection needs.

Fig. 21.7 Piping layout around column (a) elevations (b) plan view

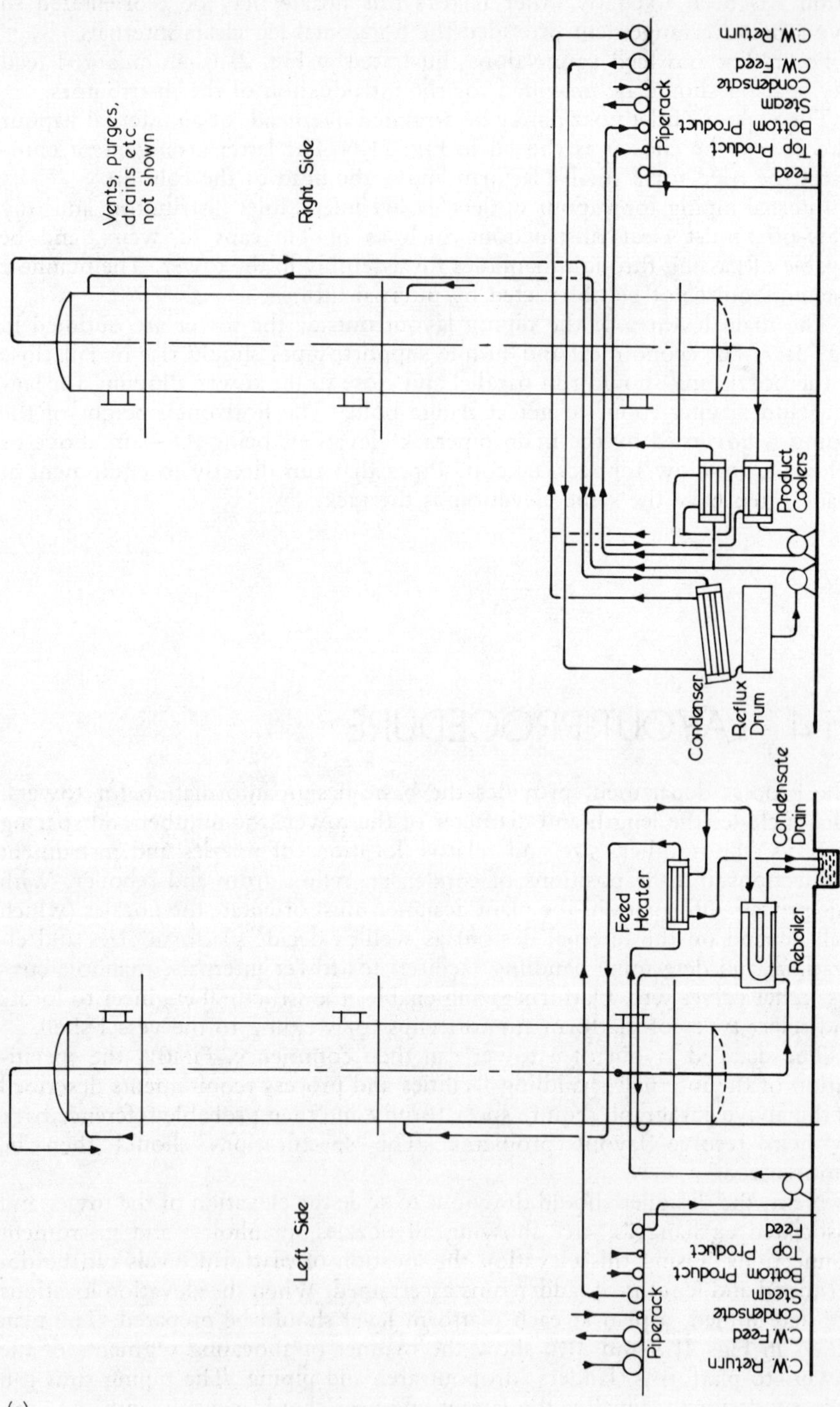

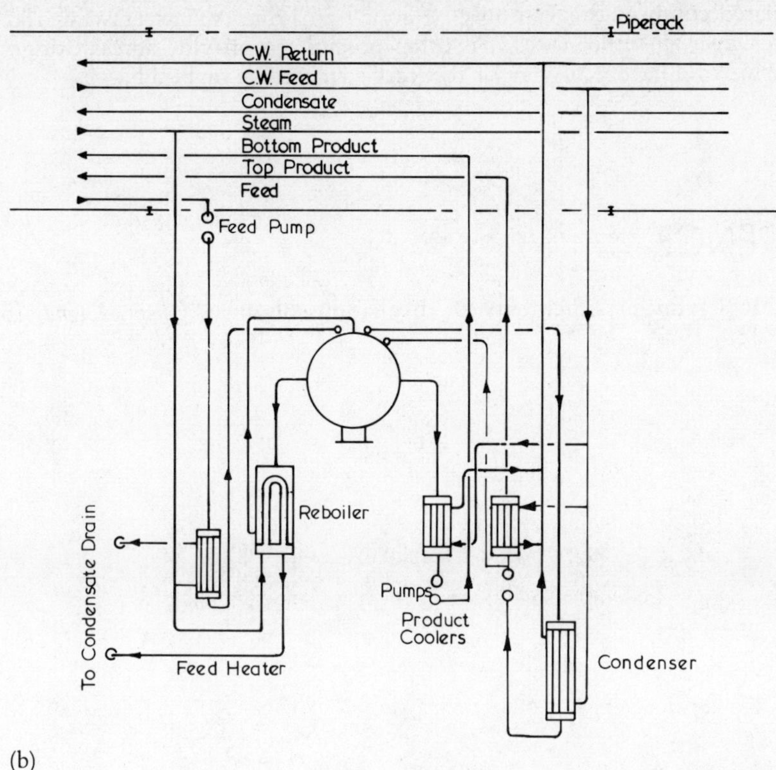

21.5 COLUMN SAFETY

A column or group of columns must be sufficiently separated from other plant, particularly buildings containing personnel. A fire starting in the column area must not spread to other areas and columns must not collapse onto other plant. Consequently feed and product tanks should not be near the columns.

It is uneconomic to protect columns within a group from each other by means of spacing. Instead, protection is achieved by good process and engineering design such as instrumentation, shutdown procedures and use of drench, steam curtains, etc.

Even though some distillation columns work under elevated pressures and temperatures, they are only considered moderately dangerous if there are low liquid inventories and no chemical reaction. For continuous columns, the lowest possible hold up is used to prevent degradation. However, with batch distillation there is a large amount of liquid in the boiler and this can mean more safety considerations. The methods discussed in Chapter 8 for hazard assessment can be employed.

Platforms must have two means of escape (see section 17.7.2, p. 256). With isolated columns this can only be done by having two accessways 180° apart. However, in other cases it is often possible to provide access bridges to platforms of adjacent towers or to nearby structures or buildings.

REFERENCE

1. Kern, R 'Layout arrangements for distillation columns', *Chem. Engng*, 15 Aug., 153, 1977.

HEAT EXCHANGERS

22.1   TYPES OF EXCHANGER

There are a number of types of exchanger including shell and tube, aircooled (fin fan), plate, coil and jacket. Vessels with coils and jackets should be treated as process vessels (see Ch. 19) and plate exchangers (such as illustrated in Fig. 22.1) have similar characteristics to plate and frame filter presses (see section 24.2). In this chapter we consider shell and tube exchangers and aircooled exchangers.

Fig. 22.1   Example of plate exchanger (Courtesy: APV Co.)

In addition to the different mechanical types, the process duties have some bearing on layout. Examples are:

(a) Exchange of sensible heat between two process streams.
(b) Cooling with air or water.
(c) Chilling with refrigerant.
(d) Heating with steam or hot water or other thermal fluids.
(e) Condensing by water or a cold process stream.
(f) Vaporizing by steam (e.g. column reboiler) or a hot process stream.
(g) Generating steam from a hot process or waste gas stream.

Fixed tube exchangers are employed when the temperature differences between shell side and tube side fluid are small, depending on the shell and tube materials. An expansion joint in the shell extends the range but where there is a significant temperature difference, floating head or 'U'-type exchangers are employed. Kettle-type reboilers are used for evaporation when pressure drop is limiting and when no depression of the boiling point by imposed static head is permissible. Otherwise, vertical tubular reboilers can be employed for evaporation.

Typical and standard internal tube and baffle arrangements are given in Kern, D. Q.[1], Kern, R.[2], TEMA[3] and Perry and Chilton.[4]

22.2 LOCATION

Exchangers should be placed close to the major equipment with which they are associated on the flowsheet. Reboilers are located next to their respective towers, and condensers are placed over reflux drums. Exchangers between two distant pieces of process equipment should be placed at optimal points in relationship to pipe tracks.

Most exchangers should be located at grade where possible with elevations kept to a minimum. So most exchangers are on a base about 1 m above ground level. Piping generally determines this elevation by setting the lowest possible pipe elevation of the largest-diameter pipe from a bottom nozzle. This includes allowances for drain connections. There are, however, exceptions, e.g. some exchangers have a condensate or holding pot after an outlet. Such cases should be arranged so that the top of the pot is at least in line with the bottom of the exchanger to avoid flooding the tubes and adversely affecting the exchanger duty. In some other cases, flooding of tubes is necessary (e.g. a total condenser) and the relative level of the control pot to the exchanger is important and will determine the position of the exchanger. The positioning of condensers and reboilers relative to columns is discussed in section 21.1 (p. 313).

Elevation of exchangers may be necessary because of a net-positive suction head (NPSH) requirement of a following centrifugal pump. Where there is a large number of exchangers, they are often put together in one or more groups (Fig. 7.16b, p. 98). By this means savings are often possible on service pipework, pipebridges, structural work and in the provision of lifting and other maintenance facilities, but may entail additional process piping and ac-

Fig. 22.2 Access around heat exchangers (Kern[2])

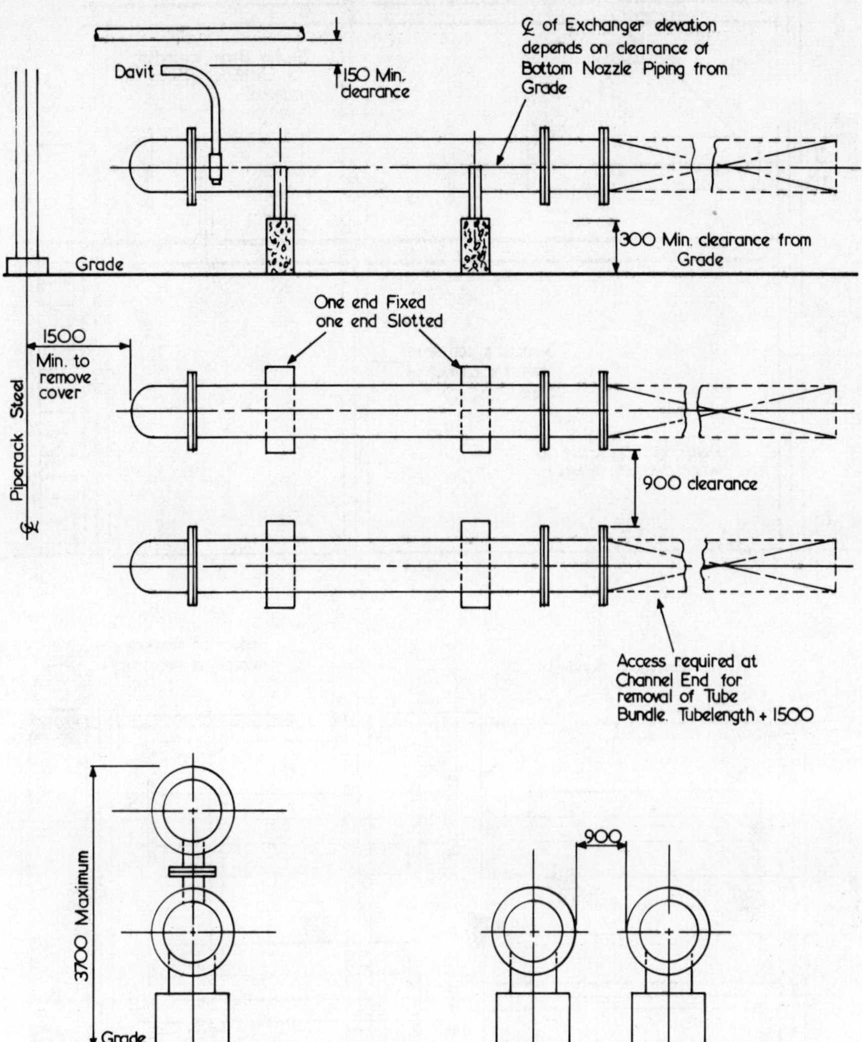

cess steelwork. An economic balance has to be struck and the location should result in a layout which is convenient and comfortable to operate and maintain (Figs 22.2 and 22.3). The exchangers need not be in the same service but the structure (steel or concrete) should be limited such that the maximum height to the top of the shell is 3.7 m and ideally no more than two-high so that mobile equipment can handle the tubes. Fixed tube exchangers should be placed above floating head exchangers, if possible, to ease maintenance.

Groups of exchangers should generally be located by the alignment of the channel nozzles in a vertical plane in order to present a neat appearance and make pipe detailing easier. When the tube lengths in a group are the same, the centrelines of the support feet should also line up. Each shell has two

Fig. 22.3 Handling of heat exchanger equipment

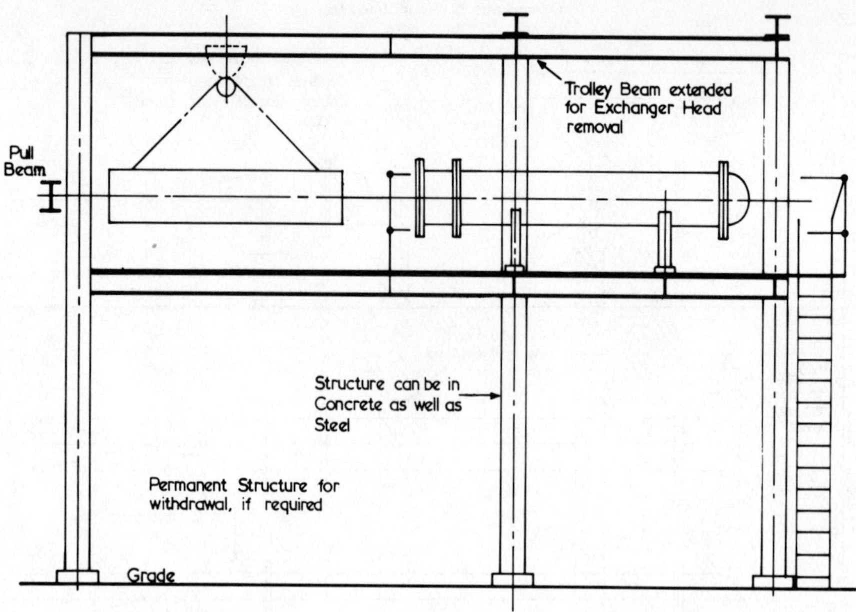

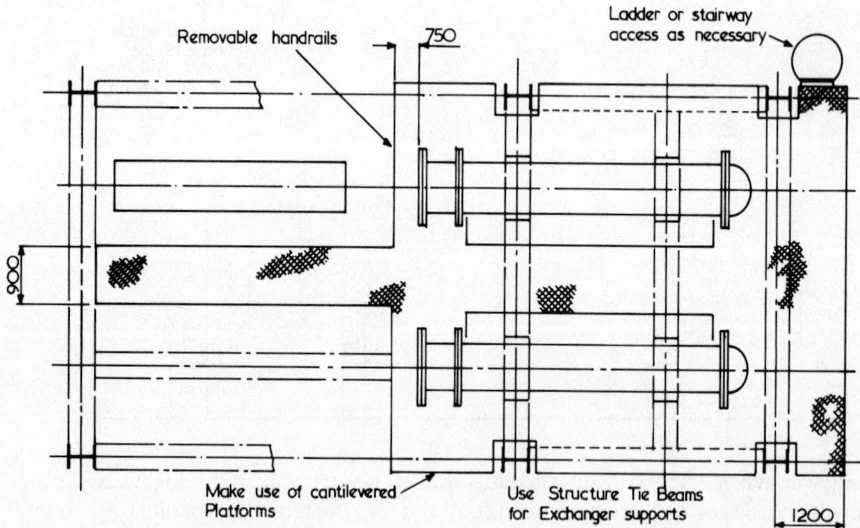

support feet, one with elongated holes to permit thermal expansion. These are normally located on the foot farthest from the channel end but the final location depends on the plant layout and the stress analysis of the associated piping.

Matrix-type aluminium heat exchangers are usually light enough for each to be regarded as a pipe fitting with little or no support other than from the pipework itself. Room should be available to replace the exchanger. When process and ground space conditions allow, aircooled exchangers can be lo-

Fig. 22.4 Forced draught aircooled exchangers (Courtesy: APV Spiro-Gills)

(a)

(b)

cated at grade. Generally, however, they are put adjacent to the plant they serve and can be conveniently mounted on top of structures or above pipe-racks (Figs 22.4 and 22.5). In these cases the supporting legs of the cooler should coincide with stanchions of the supporting structure where possible so that loads can be directly transmitted. This might affect the module spac-

Fig. 22.5 Induced draught aircooled exchangers (Courtesy: APV Spiro-Gills)

ing of steel columns. Consideration must be given, though, to mobile maintenance and access for tube removal (see section 22.3). The locating of aircooled exchangers over compressors, electrical switchgear, control rooms, hot pumps or any other equipment giving-off heat should be avoided. Similarly, the fans associated with the coolers should be selected so as not to cause difficulties with noise and vibration which could adversely affect control rooms and other delicate instrumentation.

22.3 ACCESS

Figure 22.2 shows arrangements of heat exchangers and the space required for access.

Horizontal clearance of at least 900 mm should be left between exchangers (flange to flange, see Fig. 22.2) or exchanger flanges and piping. Where space is limited, clearance may be reduced between alternate exchangers but in no case should the clearance over insulation between channel flanges be less than 0.6 m. For example, at the channel end of a typical 'U'-tube or floating head exchanger, about 0.6 m should be allowed for removing the channel cover. Thus a floating head exchanger with 5 m tubes requires an installation length of approximately 12–13 m to allow for tube withdrawal. The tubes may be cleaned and maintained *in situ* or removed to the workshop. Vertical exchangers should be set to allow lifting of the tube bundle. Similar considerations apply in the case of fixed tube plate exchangers as adequate space must be left for the insertion and withdrawal of individual tubes or cleaning rods.

It is thus important at the initial layout stage to identify the clearance and working space required all round an exchanger shell. These spaces should be kept clear of any piping and its components. The channel ends of exchangers should face their local access road for bundle removal and the shell cover should face the piperack. However, pulled-out bundles should not extend over main access roads. Where possible, mobile equipment should be used for the handling of tube bundles and covers at grade and expensive built-in handling facilities, such as lifting beams, kept to a minimum. Nevertheless, convenient anchored points may be provided in front of the exchanger for connection of rope and pulley blocks to aid bundle removal. Maintenance operations may be helped if small sections of the pipe connections to the exchanger can be removed to give access for removing channels and tube bundles.

Kern[2] describes a number of devices to service a row of single or stacked exchangers. These include a travelling gantry, tubular legs and hydraulic-lift trucks. Sufficient space is needed in front of the row for this equipment (Fig. 22.3).

When *in situ* tube cleaning is planned, good washdown and drainage facilities are needed.

For aircooled exchangers, platform arrangements must suit maintenance access requirements. Consideration must be given to tube bundle removal, tube rodding out at header boxes, motor and fan access.

22.4 PIPING ARRANGEMENTS

Figure 22.6 illustrates the piping arrangements around an exchanger. Heat exchanger design should be related to the piping layout, especially for multi-bundle systems and stack arrangements. Therefore there may be a requirement for reiteration between design and layout to achieve the optimum solution as the layout is often improved by changes in fluid direction, and nozzle orientation (Fig. 22.7). To ascertain whether a change is possible, a thorough knowledge of the operation of the heat exchanger is necessary. For example, in the case of a single shell, single tube pass heat exchanger the direction of both fluids (but not just one) may be reversed without affecting the thermal performance of the heat exchanger. However, for vaporizers, condensers and 2, 4, 6 tube pass, single-shell pass floating head heat exchangers, the direction of flow of either fluid may be reversed without affecting thermal performance. Multiple-pass arrangements on shell or tube side are employed to match the available fluid flow rates with the desired shell or tube-side velocity and yet retain compact dimensions for the heat exchanger. Cooled streams usually flow downwards and heated streams flow upwards; this arrangement is mandatory when there is a change of phase, desirable when the streams are liquid,

Fig. 22.6 Piping arrangements for exchangers (Adapted from Kern[1] by special permission of *Chemical Engineering*; © 1978 by McGraw-Hill Inc., New York, NY 10020, USA)

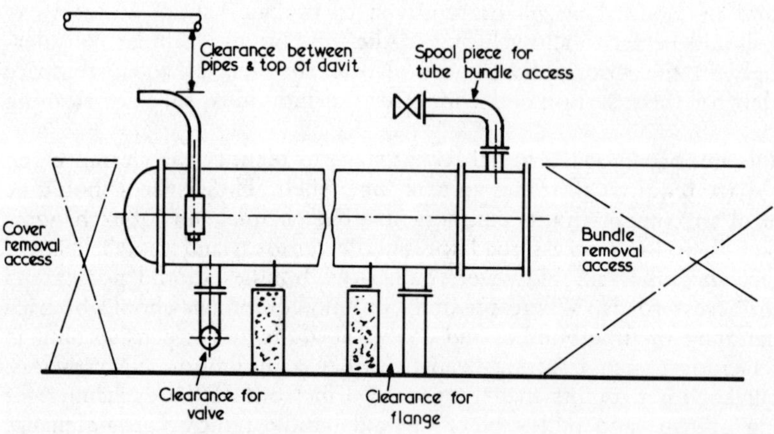

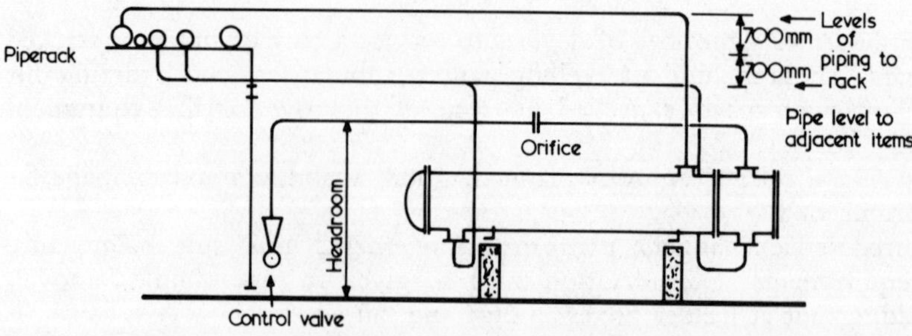

Fig. 22.7 Space saving layout for exchangers (Kern[2])

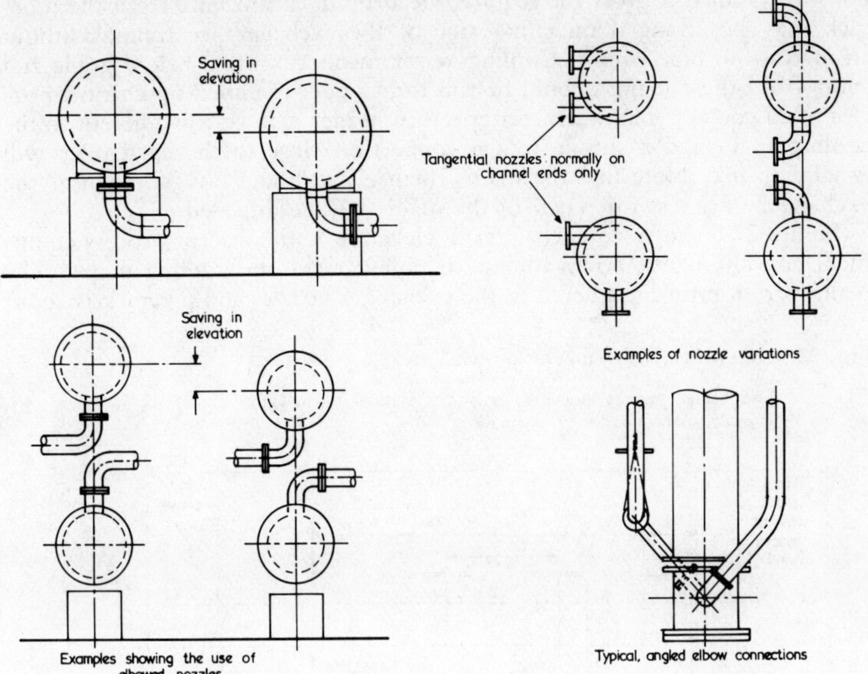

but not important when the streams are gas or vapour and there is no super-cooling or superheating. Thus cooling water enters at the bottom inlet of exchangers and leaves at the top, and vapour enters the top of condensers and condensate leaves at the bottom. Condensation usually takes place in the shell because dirt deposition and heat transfer is governed by the water velocity which is higher if put through the tube side.

Tube-side nozzles may be freely rotated with respect to shell-side nozzles when there is only one shell-side pass, but the possible rotation is constrained when there are two or more shell-side passes. Shell-side nozzles may be re-located longitudinally on the shell if the operation of the heat exchanger is not affected. Elevations and stacking heights may be reduced by changing straight nozzles to elbowed or tangential nozzles. However, pipe configurations must be designed to allow for the correct flow conditions at nozzles and with a flexible and well-supported system to meet manufacturers' allowable stresses on the exchanger. It should be noted that tailor-making of nozzle positions to suit piping can militate against standardization of design which is desirable for interchangeability and low stock of spares.

If piping is arranged on one elevation only between the exchanger and the piperack, the pipeline to the top shell-side nozzle may be located over the centreline of the exchanger. Pipelines turning right into the rack should then be run to the right of the exchanger centre-line and pipelines turning left should be run to the left of the centre-line. Pipelines from bottom connections of exchangers should be turned up on the appropriate side of the centre-line. However, piping should not foul exchanger removal either in the horizontal direction for bundle-pulling nor in the vertical for whole exchanger removal.

Pipelines connecting exchangers with adjacent process equipment can run point to point if it gives the required headroom. Steam lines from the pipe-rack may be arranged on either side of the exchanger centreline without an increase in pipe length. Cooling water mains run often below grade and spurs from these mains should be run right under the lined-up channel nozzles of all coolers. The warm water return header may be run adjacent to the cooling main and a simple return connection close to the exchanger will usually suffice. Note that the mains themselves should not be too near the exchanger or access for repair of the mains will be impeded.

Figure 22.6 shows an exchanger in elevation with adjacent process equipment and alternative arrangements to piping racks and cooling mains. The main elevation for lines between the exchanger nozzles and piperack is about

Fig. 22.8   Layout of exchangers: dimensions

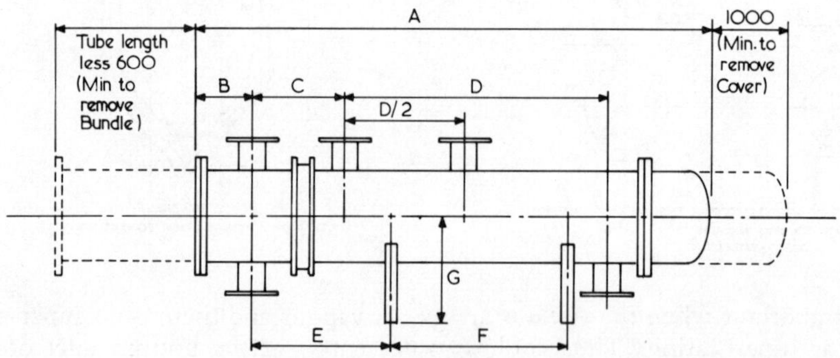

Nom. Shell dia.	B	C	G	4800 Tubes				6100 Tubes			
				A	D	E	F	A	D	E	F
305	200	380	355	5335	4295	1525	2440	6555	5510	1680	3660
355	230	380	380	5385	4295			6605	5510		
380	230	430	420	5410	4265			6630	5485		
405	255	510	430	5485	4165			6705	5385		
455	280	560	485	5560	4115			6780	5335		
510	280	560	510	5615	4115			6835	5335		
560	305	560	535	5665	4115			6885	5335		
610	305	585	560	5690	4040			6910	5260		
660	355	660	560	5790	3990			7010	5210		
710	380	685	585	5840	3990			7060	5210		
760	380	685	620	5865	3960			7085	5185		
815	380	685	650	5865	3935			7085	5155		
865	405	710	685	5920	3910			7135	5130		
916	405	710	710	5945	3910			7165	5130		
965	430	785	725	6020	3835			7240	5055		
1015	455	785	750	6070	3835			7290	5055		
1065	455	815	785	6095	3835			7315	5055		
1102	485	815	815	6120	3835			7340	5055		
1170	485	840	840	6145	3810			7365	5030		
1220	485	840	865	6200	3810			7420	5030		

Dimensions are in millimeters

0.6–1.0 m lower than the rack elevation. This elevation may be used for connecting to equipment below the rack and for discharge lines of pumps. Fig. 22.7 gives some space-saving nozzle orientation and Fig. 22.8 shows typical layout dimensions.

Orifice flanges in exchanger piping are usually in horizontal piperuns just above headroom level, but they may be placed at a lower level for convenience in attaching manometers and inserting orifice plates. Locally mounted pressure and temperature indicators on equipment or process lines, sight glasses, and level controllers should be visible from access aisles, and valves should be accessible from the aisles.

REFERENCES

1. Kern, D. Q. *Process Heat Transfer*. McGraw-Hill, New York, 1950.
2. Kern, R. 'How to find the optimum layout for heat exchangers', *Chem. Engng*, 12 Sept., 169, 1977.
3. TEMA *Standards of Tubular Exchanger Manufacturers Association*. Tubular Exchanger Manufacturers Association, New York.
4. Perry, R. H. and Chilton, C. H. (eds) *Chemical Engineer's Handbook*. (5th edn), (Ch. 11), McGraw-Hill, 1973.

FLUID TRANSFER EQUIPMENT

23.1   TYPES OF EQUIPMENT

Pumps, fans and compressors fall into three basic types: momentum, centrifugal and positive displacement. A detailed description of the various forms would be out of place in a book on plant layout but the best type for the process duty should have been chosen before detailed layout of plot commences.

Steam ejectors (momentum type) present no particular layout except the suction lines and barometric legs which are discussed in section 24.3 (p. 350).

Most centrifugal units are driven by electric motors. An alternative power unit is the turbine, quick-starting with easily controlled variable speeds. It is particularly important that all piping at turbines be designed to allow for sufficient support and flexibility to prevent overstressing of turbine branches (section 31.3, p. 429).

There are two main types of positive displacement unit, namely, reciprocating and rotary. Both types must have pressure-relieving arrangements either built-in to the machine or piped between the suction and discharge lines.

Compressors are often driven by steam turbines or by turbines driven by gases being depressurized. The above stricture on overstressing the turbines also applies to the branches of the compressors as well (section 31.3, p. 429).

23.2   PUMPS FOR LIQUIDS

The various requirements for the layout of pumps are summarized in Fig. 23.1 and also Fig. 17.4 (p. 251).

23.2.1   LOCATION AND PIPING

A common location of pumps in chemical and petrochemical plants is under

Fig. 23.1 Pump layout requirements

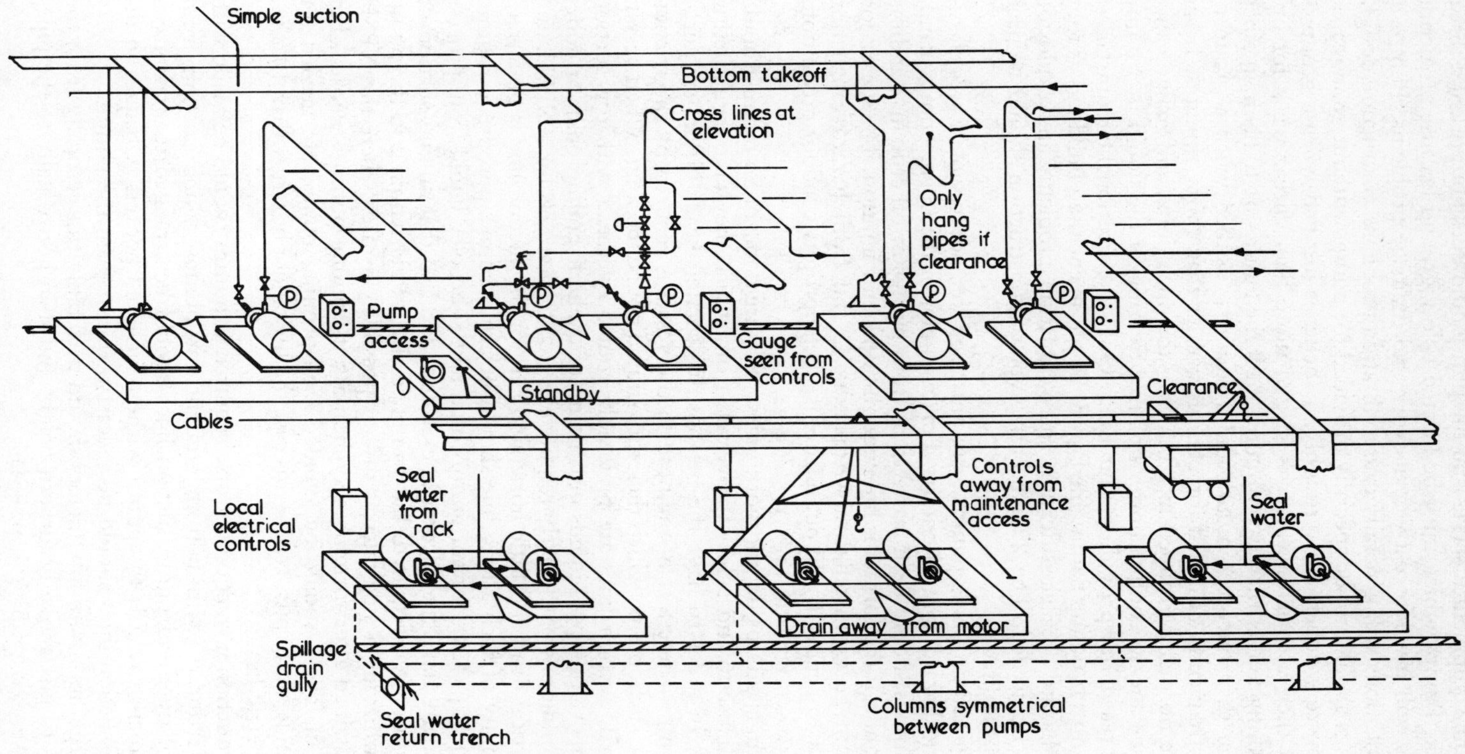

the piperacks at grade but above flood level providing the pump is not too far from the equipment it serves and the pump fluid does not present a threat to the rack. Pumps which must be located below ground because of suction conditions will involve costly civil work and drainage problems. Pumps at elevated locations will generally cause vibration problems in structure design.

Buildings should not be specially provided for pumps without good reason such as severe weather. Very roughly, a shelter is justified when a group of pumps or their associated valves is visited more than once a week for operations taking longer than two minutes, or if the pumps are to be maintained more often than once per month and a room is justified when a group of pumps is frequently attended for periods greater than thirty minutes. Large pumps are normally grouped in a building for operational and maintenance convenience and to confine noise. Possible freezing of the liquid in a pump is another reason for providing a building. Only in exceptional circumstances should pumps handling flammable liquids be put in enclosed buildings as they will give hazard area classification and ventilation problems and explosion relief will be needed. The 'Dutch barn' type of construction should be used if cover is required.

In siting and grouping pumps, care must be taken not to violate area classification and hazard control criteria. For example, in the open, pumps handling hot liquids should be at least 7.5 m from pumps or other items handling volatile liquids if the former is hot enough to provide ignition. The separation distance can be estimated from Appendix C and from hazard calculation (Ch. 8 and Appendix B). In pump rooms, the appropriate indoor precautions should be taken (section 18.8, p. 276).

Centrifugal pumps do not work unless they are flooded. So pumps should be placed close to and below the vessels from which they take their suction (Fig. 23.2) in order to have the net-positive suction head (NPSH) required by the pump. The effect of the head in the vessel is reduced by the friction head in the suction pipe which therefore should not be too long. When vessels are elevated, suction lines are preferably routed overhead with top suction connections to pumps. Liquids near their atmospheric boiling point or under vacuum conditions particularly require elevated feed vessels for a NPSH, and a long vertical drop and little horizontal run for the suction line is preferred. The flexibility of such a long line should be achieved by appropriate routing (section 31.3, p. 429). Indoor pumps, that take subcooled liquid from process vessels usually have the suction line at floor level so that an end-suction or side-suction inlet is required, but to obtain the NPSH required in the pump may mean having a side inlet as low as possible. In cases where the NPSH is only obtained after forming a syphon a self-priming pump is needed, but this kind of layout should be avoided. Self-priming pumps are needed for lifting liquids out of pits, etc. In this case the lift cannot be larger than 10.5 m, the barometric head.

Any reduction in suction line size required at pumps should be made with eccentric reducers with bottom straight for pumps taking suction from below. All overhead pump suction lines should be arranged to drain from the equipment towards the pump without inverted pockets. Changes in direction of suction lines should be at least 600 mm from the pumps.

Discharge lines with flowmeters should preferably run vertically from the pump top to just above headroom height and then horizontally to the piperack (but see section 31.7.4, p. 451).

Fig. 23.2 Suction head requirements of pumps and thermosyphons

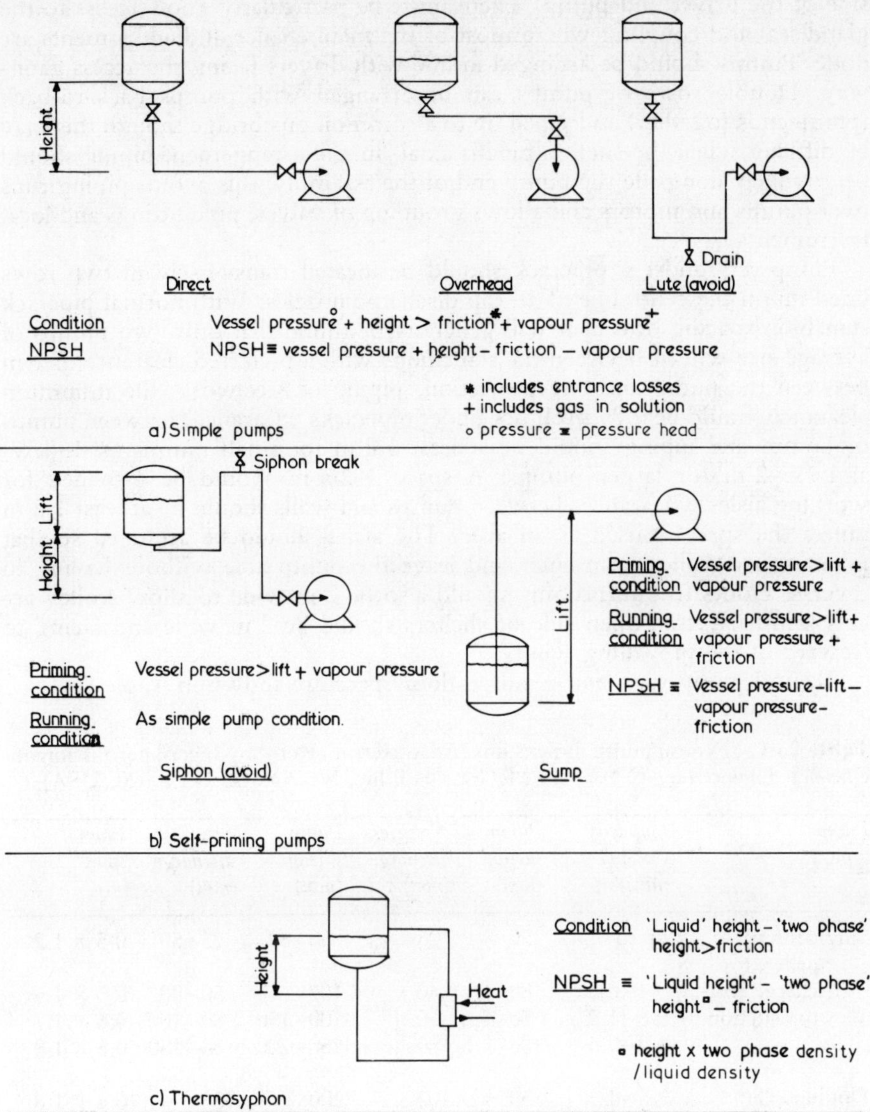

Provision should be made to isolate the pump from the feed vessel when it leaks or otherwise malfunctions so that it can be replaced without draining the vessel. There must also be space allowance for the lubricating oil piping, seal vent and drain piping and the flow measurement and control apparatus when these items are fitted.

Groups of pumps should have their discharges and discharge valves lined up and, if possible, their suction valves and seal water pipes also lined up (Fig. 23.1). Space may be saved by mounting two pumps, motors and starters on the same plinth if maintenance and operation of the pumps are not impaired (Fig. 23.1).

As far as possible, clearances and piping should provide free access to one side of the driver and pump. There must be particularly good access to the gland/seal and coupling where most of the maintenance and adjustments are done. Pumps should be arranged in line with drivers facing the access gangway. Double rows of pumps can be arranged with pumps back-to-back (pump ends together) and piped up to a common pipebridge though this may be difficult when the suction pipe is axial. In any arrangement piping should be arranged alongside the pump end of the assembly; this avoids piping runs over pumps and motors and allows grouping of valves, pipe fittings and local instruments.

Pump sets under a piperack should be located transversely in two rows lined up on the centreline of all top discharge nozzles. With normal piperack stanchion spacing of 6 m it will generally be found that only two pumps of average size will fit between the stanchions with a preferred clearance of 1 m between the pump and any projection, piping or steelwork; the minimum clearance should be 0.75 m. If not under piperacks, clearances between pumps or pumps and piping should be at least 1.2 m for small pumps (< 18 kW) and 1.5–2 m for larger pumps. A space 2–2.5 m should be provided for working aisles. Clearances between pumps and walls should be at least 1.2 m unless the space is used as an aisle. The aisles should be arranged so that maintenance trolleys can enter and leave the pump area without having to reverse. Doors to pumprooms should also be 2 m wide to allow trolley access. Similarly, the open side of shelters should be 2 m wide and facing to leeward of the prevailing wind.

Typical pump sizes and required floor space are shown in Table 23.1.

Table 23.1 Typical pump dimensions (Adapted from Kern[1] by special permission of *Chemical Engineering*, © 1978 by McGraw-Hill Inc, New York, NY10020, USA)

Pump type	Capacity (m^3 per min.)	Pump suction (mm)	Nozzles discharge (mm)	Pump suction (mm)	Sizes discharge (mm)	Floor space (m)
Single inlet impeller, top	up to 0.4	50	25	50–80	25–50	0.5 × 1.2
discharge end	0.4–0.8	80	40	100	50–80	0.5 × 1.5
or top suction	0.8–1.2	80	50	100–150	80–100	0.6 × 1.7
	1.2–2.5	100	80	150–200	80–150	0.6 × 1.8
Double inlet	2.5–4	150	100	200	150	0.6 × 1.8
impeller	4–6	200	150	250–300	150–200	0.8 × 2.0

23.2.2 LIFTING ACCESS

Means of lifting should be provided for pumps or motors weighing more than 25 kg. This can be either a fixed lifting beam or portable 'A' frame. Lifting beams should run from the pump to the centre of the nearest aisle, and their use should not be hindered by pipework. It should be possible also to place and use 'A' frames without interference from pipework.

Overhead clearance should be a minimum of 2.5 m and should be increased to 3.5 m if 'A' frames are used. For small pumps, a minimum clearance distance of 1.5 times the maximum height of pump should be allowed for removal of parts.

23.2.3 FLOOR AND DRAINAGE

Pumps should stand on plinths raised at least 150 mm above the floor. No part of the pump or motor should overhang its plinth. Plinths supporting several pumps should be graded (1 in 120) to allow spillages to drain from the pump and away from the motor into drain points or gullies. The floor should be provided with a fall (1 in 80) so that washings and spillages drain into the nearest gully (section 17.4.2, p. 246). These gullies should be connected to the correct factory effluent system. Pumps, which are liable to leak corrosive or flammable liquids should not be placed under or near vulnerable equipment such as piperacks.

23.2.4 ELECTRICS

Normal practice is for starter switchgear to be in a non-classified area with only stop/start buttons in the field. These should be adjacent to the pump (Fig. 23.1) and their pressure gauges and each may incorporate an ammeter. In addition, there should be provision for remote stopping of pumps, usually at the starter switchgear and, for emergencies, in the control room. Start/stop buttons should be heights 1.2 m to 1.8 m above floor level. They and the cables should be in the same relative position to each pump motor.

Pumps other than those driven by belts must be mounted with motor shaft and pump shaft carefully aligned.

23.2.5 NOTES ON PARTICULAR PUMPS

The simple single-stage pump has one impeller with single-end suction. A wide capacity range is available from a few interchangeable impellers, and rotating parts can be removed without disturbing piping, casing or motor. This type of pump may be used when the suction line is near-grade, the liquid is subcooled, or the available NPSH is low. Alternative suction arrangements may be provided with the inlet set at the top of the pump for connection to an elevated suction vessel.

Multi-stage pumps have two or more stages connected in series and provide a high discharge pressure. The casing may be longer and the suction and discharge nozzles are usually vertical. Pumps with horizontally split casings should have access from both sides for convenience in maintenance. Vertically split casings require removal space in front of the pump for shaft and impeller during *in situ* maintenance.

Large capacity water pumps usually have horizontally split casings with a double inlet impeller. Inlet and outlet are horizontal and suction piping should be as simple and as short as possible. Considerable space is required around the pump for maintenance because of the large-diameter piping, fittings and valves.

Most pumps are horizontal although close-coupled pumps are compact and economical and may be mounted in any position including overhead. Large in-line pumps should be located at ground or floor level and the supports should accommodate vibrations and out-of-balance forces as for compressors.

Most centrifugal pumps are designed to be stable back to shutoff. However, overheating and flow instability causing vibration and shaft metal fatigue can occur at low flows particularly with high-speed pumps and those with high power inputs. In these cases a return line to the suction vessel is necessary.

Vertical shaft pumps occupy small floor areas but require access space for removal and vertical space for lifting motor and impeller. They may each require a non-return valve at the inlet to maintain flooding, and a common maintenance operation is to lift the pump to clear the non-return valve.

There are several types of vertical pumps. The submerged pump has a single radial impeller with a long vertical shaft. Deep-well pumps have impellers with radial or mixed flow discharge. A large number of impellers may be mounted in series so that the pumps may develop high head. A long shaft may be avoided by using a submersible motor. The dry-well pump is the type of self-priming pump which is mounted above the fluid and so can be located next to, instead of below, a suction vessel. Provision should be made for mobile crane access or pulley-block lift. A common vertical pump is the in-line pump which is small and supported from pipework, thus saving foundations. The motor has to be removed to service the pump.

23.3   COMPRESSORS

23.3.1   GENERAL

Compressors are normally located inside a permanent shelter or building (compressor house) for weather protection. The building can be fully sheeted to grade if handling non-hazardous materials such as air. For compressors handling flammable materials, weather protection is usually of the 'Dutch barn' type where ventilation is assured by having significant openings ($\sim$ 2.2 m) at grade together with roof ventilators (see Figs. 23.3 and C.3 (p. 572)). Except for lighter-than-air gases, trenches, pits and similar gas-traps should be avoided within gas compressor houses. Fire protection measures may be needed and if relief valves are used they should discharge safely (section 13.3, p. 198) outside the compressor house.

There should be a clear space of about one-half the width of a compressor (subject to a 1.5 m minimum) between compressors, between rows of compressors and at the end of each row subject to any special maintenance considerations (Fig. 23.4). Built-in maintenance equipment, such as travelling gantries with overhead cranes and separate dropout areas, should be included in the buildings. The clearance above the compressor should be at least 3 m more than the longest internal part to be removed.

Fig. 23.3 Compressor building with open sides (Courtesy: ICI Petrochemicals and Plastics Division)

Fencing, lighting or remote monitoring may be required for the security of isolated compressors.

23.3.2 RECIPROCATING COMPRESSORS

Reciprocating compressors create a considerable amount of vibration due to unbalanced forces, pulsation, etc. For this reason they should be located as close as possible to grade. Consequently, the floor level of the building is near the top of the compressor foundation. The building and compressor foundations should be separate to avoid transmitting vibrations to the building structure.

Much attention should be paid to the piping which interconnects pulsation dampeners, knockout pots, intercoolers and aftercoolers. Knockout pots and intercoolers should be located as close as possible to the pulsation dampeners which in turn should be located on or below the compressor nozzles. The pulsation dampeners are used to eliminate pulsation in suction and discharge piping and to separate the source of vibration from the piping system. Knockout pots and intercoolers can either be in the building or more usually, because of lack of room in the building, lined up outside and supported from grade. However, if the gas being compressed is wet and there is a chance of freezing, these pots and coolers should not be put outside.

Piping should be run at grade for ease of supporting with the minimum of changes in direction. It should not be supported from any building steel-work since this will transmit compressor vibrations. The grade supports should be spaced unevenly to reduce harmonic motion in the piping. Piping

Fig. 23.4 Compressor layout (a) Courtesy: APV Hall International (b) Courtesy: Humphreys & Glasgow

(a)

(b)

which is routed simply and with short runs is less prone to vibration but at the same time it must be designed to provide flexibility to prevent over-stressing of compressor nozzles due to expansion of the exchangers and pipes, etc.

23.3.3   CENTRIFUGAL COMPRESSORS

The vibration in centrifugal compressors and piping must be minimal, otherwise the bearings will wear and other components will fail by fatigue. The layout and piping considerations are similar to those for centrifugal pumps except that often: (a) pipe sizes and components are very much larger and therefore more critical; (b) temperatures can be higher, and (c) the permitted force and moments on the compressor casing are lower.

A centrifugal compressor has one or more rotating assembly, handles large volumes, and can be electric motor, steam or gas driven. A large-size multi-stage machine usually has a horizontally split casing. Those with top-nozzle connections can be located at grade whereas those with bottom connections are elevated. In the former case the machine cover can only be lifted after the piping has been disconnected which is less convenient than inserting spades before maintenance work. A barrel compressor has only end-closures and space must be allowed for the horizontal withdrawal of the rotor.

Compressors inside a building usually have foundations independent of those of the building. The machine is generally mounted on a concrete table supported on concrete columns. Further access can be provided by canti-levered platforms. If there is one machine in the building, maintenance is gen-erally by a trolley beam mounted over the centreline of the compressor extending into the dropout area. If the building houses several machines then a hand-operated travelling crane is normally supplied. The height has to be carefully estimated so that machine covers and rotors can be lifted over ad-jacent equipment.

In a similar way to reciprocating machines, knockout pots and interstage exchangers can be placed at grade outside the compressor house with auxili-ary equipment consisting of lubricating, seal and control oil systems occupy-ing large areas of space adjacent to the machine. The piping should be designed to accommodate flexibility requirements. Clearance must be pro-vided for internals withdrawal, operation and maintenance.

23.4   FANS

The inlet and outlet ducting local to a large fan (say 700 m^3/min.) must be planned in detail because of its large size, together with the space required for bends and the clearance around valves and filters (Fig. 23.5 and section 18.5, p. 271). Tight bends or fittings on ducting near the fans should be avoided because of their harmful effect on air flow. Fans may be housed in a separate building to isolate any noise problem and provide access under

Fig. 23.5 Layout of fans (a) single inlet fan (b) double inlet fan (Both courtesy: Sturtevant Engineering Products)

(a)

(b)

cover for operation and maintenance. Adequate space and headroom must be allowed for removal of impellers, shafts, motors, etc., particularly for dirty or corrosive duties. Generous foundation masses may be needed to attenuate any vibrations and the fans should be located away from any external source of vibration.

REFERENCE

1. Kern, R. 'How to get the best process-plant layouts for pumps and compressors', *Chem. Engng*, 5 Dec, 131, 1977.

FILTERS

For the more complicated types of filter the layout may with advantage be discussed with the manufacturer.

24.1   LINE FILTERS AND STRAINERS

These are filters designed to remove small amounts of solids. There should be access to take the element out for cleaning. Filters which have to be serviced more frequently than once a week must be accessible from the operating level (Fig. 28.11, p. 400) or platform. At lower service frequencies a fixed ladder may be sufficient, although strictly speaking, it is bad practice to work from ladders. For all but the smallest elements a hoist must be provided if ladders are the only means of access.

24.2   BATCH FILTERS

Batch or semi-continuous filters are usually operated under pressure. Plate and frame, leaf and bed filters are usually operated in batch mode, and are often installed in a building.

Plate and frame filters (as illustrated in Fig. 24.1) may remove considerable quantities of solids which build up on the filter cloth or other filter medium as a cake which is periodically removed, possibly after washing. Such filters may be piped in parallel with some filtering, some washing and others being de-caked in rotation. Access to the changeover valves, pressure gauges, sight glasses and sample points is needed.

Plate and frame and bed filters may also be used to remove small quantities of solids. In the case of bed filters the deposited solids may be removed by back-washing of the filter.

Fig. 24.1 Layout of a plate and frame filter press (Courtesy: Babcock Minerals Engineering)

In filters that have to be opened for cake removal, layout is extremely important for ease of operation, but less so where cake is removed by air-blowing or other mechanical method, when maintenance considerations may predominate. A space of at least one filter's width should be left around the filter and if trolleys are used for bringing up new cloths, removal of the cake or transporting the filter plates, the free space should be at least 1.75 m on one side; in the case of leaf filters, cranes or hoists should be arranged to allow discharge of the cake on to a conveyor or into a vessel or chute. Filters are sometimes staggered thus saving space and effort in handling trolleys because right-angled turns are avoided.

To aid discharge the filter may be elevated above floor level but generally it is placed on a floor discharging to the level below. Chutes must be of generous dimensions and as near vertical as possible.

Special ventilation arrangements such as spray nozzles, vapour hoods and extractor fans may be needed if the liquids are toxic or flammable, but clear access should be maintained for removing filter cake. The floor underneath plate and frame filters, and traversed by leaf filters should be channelled, drained and graded to restrict liquid overflow. There should be provision for washing-down the floor.

For larger filters having internals such as plates, overhead lifting beams may be needed. A room may be needed for storing cloths or filter media and for cutting filter cloths. Provision may be needed for vessels and piping for handling filter aid, etc.

24.3 CONTINUOUS FILTERS

Typical of this class is the rotary vacuum drum filter shown in Figs 24.2 and 24.3. Other types include the rotary vacuum disc filter. If the cake and/or filtrate are not adversely affected by the weather it may be placed in the open, with or without a shelter.

It is usually easier to transport and store slurries rather than filter cake and so liquid–solid separation equipment should normally be located near the point at which solids are finally discharged (i.e. dryer or packaging). It is better to have one pump delivering to a feed vessel and a separate filter pump

Fig. 24.2 Access to a rotary vacuum filters (Courtesy: Babcock Minerals Engineering)

Fig. 24.3 Layout of rotary vacuum filter

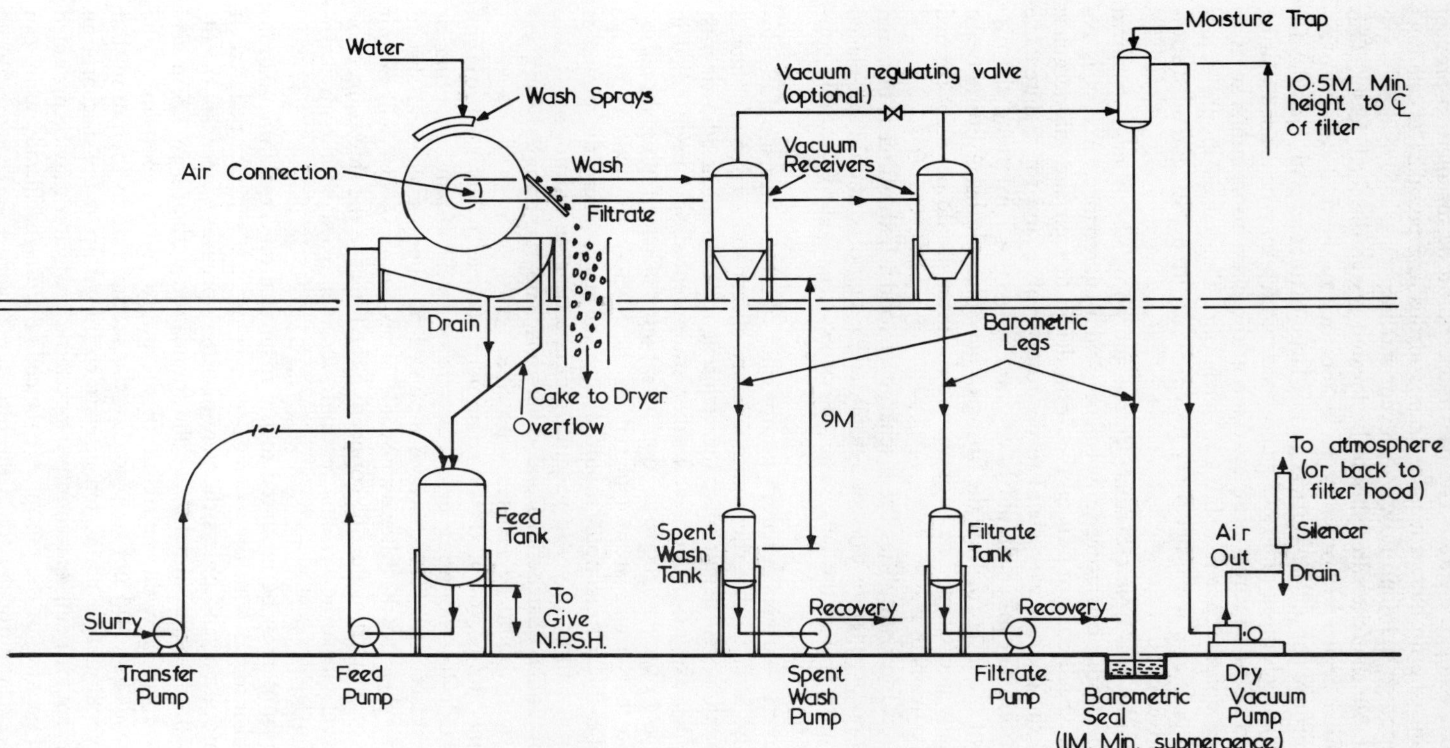

rather than utilizing a single pump for transporting the slurry over a long distance and feeding the filter. The feed lines should be installed to prevent settlement and should have adequate drain and flushing points. Recycle lines may be needed to maintain flow to prevent settling.

It is desirable that the filter should be located above the feed vessel. This allows the filter drains and overflow to be returned to the feed tank by gravity. The overflow should be vertical and of large diameter, with facilities for rodding out and cleaning of any tees and bends which may be used. Isolating valves should be fitted direct on the trough at the drain connections so as to give the least volume where solids can settle out. The valves should be accessible for operation, but if necessary they may be fitted with extended spindles or chain wheels.

There should be ample clearance around and above the filter to perform routine inspections, operating adjustments and maintenance. Access is required to the speed control unit on the drum drive, to weir and submergence controls, to the control valves and to the rotary valve, and for maintenance and lubrication. The filter should have good access (> 1 m) along the discharge side of the filter to allow the fitting of new filter cloths. A lifting beam is not normally necessary. The floor around the filter should be easily cleaned and properly drained.

An elevated filter simplifies the discharge of solids. If the filter is elevated above the floor, concrete or wood foundation supports should be placed under the entire area of the tank base, and access arranged to the drain connections on the tank.

The receivers, blowers, vacuum pumps, and filtrate pumps should be located near the filter so that the piping is short, direct and as free of joints as possible, but allows sufficient space for maintenance and routine adjustments. Valves, piping, expansion loops and other equipment should be installed in such a way as to prevent the formation of vapour pockets. Piping should be supported so that strains are not placed on the filter or its auxiliary equipment, and flexible connections should always be used where the piping is connected to the filter valve. In vacuum piping a riser should not be used to pass over or under obstructions and diversions, and right-angle bends, especially vertical, should be avoided if possible.

In locating pumps, or auxiliary equipment such as receivers, provision should be made for mounting on a concrete or steel base and grouting or bolting into place. Receivers may be mounted on support legs and should be located as close to the filter as possible at an elevation that will permit drainage by gravity from the line connecting the filters to the side connection on the receiver. In some systems, a single receiver is used for both wash-liquid and filtrate.

Filtrate pumps may be mounted to the manufacturers' recommendations on a pedestal integral with the vacuum receiver or on a separate pedestal. If separately mounted, and unless the manufacturer specifies otherwise, the filtrate pump should be mounted on a solid foundation with at least 1.5 m vertical clearance between the centreline of the pump and the bottom of the receiver. A valve should not be placed in the suction line to the pump, but a check valve should be connected directly to the pump at the discharge in order to prevent air being drawn into the system. Seal water may be provided to the gland packing of the pump. A balance line from the filtrate pump can be connected to the upper part of the filtrate receiver to prevent the filtrate pump losing prime due to surges in the volumetric flow of filtrate. If there is at least 10.5 m vertical clearance below the bottom of the receiver a barometric leg may be used instead of a pump for discharging filtrate from the receiver.

All controls, valves and switches connected with the operation of the filter should be located conveniently so that they can be reached by the operator without leaving the filter station. Wherever possible it is recommended that a control panel be provided near the filter. However, when pumps, blowers and other accessories are located on a floor below or away from the filter auxiliary switches must be placed nearby to allow for local adjustments and maintenance and to prevent accidental starting.

Moisture traps in vacuum systems must be installed wherever the vacuum pump is of the dry piston type to avoid pulling filtrate or fine solids into the pump. To ensure this (see Fig. 24.3) the uppermost side connection on the moisture trap that is directly connected to the pump should be at least 10.5 m above the filter. The lower side-connection on the moisture trap is connected directly to the vacuum receiver. The bottom drain connection on the moisture trap is connected to the tail pipe which should extend as a barometric leg downwards approximately 10.5 m to a seal box where it must be submerged to a depth not less than 1 m below the overflow weir in the seal box. There should be provision for examining the box at frequent intervals to inspect for sediment. The minimum heights of moisture trap above the filter and seal leg should be maintained even if the trap has to be placed some distance from the filter because of restrictions on equipment height nearby.

When a rotary-type wet pump is used, the trap is only necessary when the solutions are corrosive or when some constituents have a tendency to precipitate and form deposits in the pump.

A silencer should be placed in the discharge line of the vacuum pump to reduce noise when the pump is indoors or in populated areas. Water seal pumps require separators in the discharge line, as well.

In the filtering of heated solutions where excessive vapours are produced, a condenser may be required. It may be installed instead of the moisture trap (wet vacuum pump only), or between the receiver and moisture trap provided that the minimum vertical clearances of 10.5 m between the top connection of the moisture trap to the filter and the bottom of the condenser and the overflow of the seal tank, are maintained.

CENTRIFUGES

25.1 TYPES OF CENTRIFUGE

The two principal types of industrial centrifuge are the sedimenting (clarification) centrifuge and the filtration centrifuge.

In a typical sedimenting centrifuge (illustrated in Fig. 25.1) feed is introduced by an axial tube into the centre of the machine, is accelerated up to the speed of the bowl so that the solids sediment collects against the bowl wall and are ploughed out by the scroll conveyor driven at a reduced speed. The liquid residence time is controlled by adjustable weirs at the liquid discharge. Rinsing of the solid may be arranged by adding rinse liquor near the solids discharge. Particle size classification may also be arranged.

The performance may be enhanced by adding a stack of cones or discs within the bowl shell. The distance a particle must travel before settling at a surface is reduced, and therefore smaller particles may be removed. This type of centrifuge may also be applied to the separation of immiscible liquids.

Filtration centrifuges are of several types. The variable speed automatic batch centrifuge is usually a vertically mounted basket which may be either bottom or top drive. The operating cycle of the machine is controlled manually or by timers which govern the load, spin, wash and unload lines of the cycle. The machine is particularly suitable for the separation of slow-draining particulate solids because of the flexibility of the operating cycle.

The constant speed batch automatic or peeler centrifuge, also operates on a timed cycle, and is primarily used to process materials having a medium to fast drain rate. The axis of the basket is horizontal or inclined.

The conical scroll discharge centrifuge, has a vertical inverted screen and is mainly used for the dewatering of coarse crystals or fibrous solids. The angle of the screen is chosen to be substantially greater than the angle of repose of the solid and the rate of advancement of the separated solids across the screen is controlled by the screw conveyor which rotates at a different speed to the bowl.

In the pusher discharge continuous centrifuge (Fig. 25.2) slurry is deposited onto a horizontal cylindrical screen. The liquor drains through the screen and the solids are moved over it by either a reciprocating pusher or the reciprocating primary dewatering cone.

Fig. 25.1 Layout of a plough type centrifuge (Courtesy: ICI Petrochemicals and Plastics Division)

Fig. 25.2 Layout of pusher centrifuges (Courtesy: ICI Petrochemicals and Plastics Division)

25.2 FOUNDATIONS AND LOCATION

Different types of centrifuge have different requirements of foundation and layout. Vibration and its effect upon surrounding plant can be severe so that both the foundations of centrifuges and layout should always be discussed with the manufacturer. Adjacent plant should be considered as centrifuges should be installed away from external sources of vibration. Centrifuges can damage each other, in particular the hardness of a standby centrifuge's bearings can be affected.

Liquid separators and sedimentary centrifuges may be mounted on a foundation mass which is of sufficient size and rigidity to reduce the amplitude of vibration to acceptable levels. Alternatively the equipment may be mounted on vibration isolators allowing a less rigid and lighter supporting structure, but to suppress vibration on startup and shutdown, dampers are also used. Other recommended methods are to embed the baseplate in concrete, or to connect the baseplate to the concrete foundation mass by foundation bolts. The bolts may incorporate an elastic element to suppress vibration (Fig. 25.3) but to be effective the natural frequency of the foundation should be higher than the operating frequency of the machine. When several centrifuges are installed it is economical to construct a joint vibration foundation. Lining-up a group of centrifuges also allows a more orderly arrangement of pipes and services. Flexible connections must be made to the centrifuge to cope with the amplitude of vibration. Centrifuges elevated to aid discharge of solids must be carefully installed to avoid vibration problems, etc.

For filtration centrifuges, particularly the pusher type, the out-of-balance displacements are likely to be considerably greater than those in sedimenting centrifuges because of the uneven distribution of cake. The centrifuge should be mounted on an inertia block of sufficient mass so that even in the event of the loss of a 30° segment of cake the amplitude of vibration is still reduced to an acceptable level. Anti-vibration mounts may be used beneath the inertia block to reduce the amplitude of vibration transmitted to the building. For large centrifuges the inertia block may have a mass of several tonnes. For pusher-type centrifuges an additional consideration is the possible coincidence of the frequency of the reciprocating action with the resonant frequency of the support structure.

Centrifuges should be installed away from sources of spillage, e.g. not under piping. Spillages can corrode safety interlocks, motors, control wiring, etc. Spillages and leakages from the centrifuge, tanks and piping should be contained in a curbed area provided with a fall to a gully connected to the plant effluent system.

25.3 CENTRIFUGE FEED SYSTEMS

Ideally, the feed to the centrifuge should be uniform in concentration and of constant flowrate. The slurry velocity should be sufficient to prevent settling

Fig. 25.3 Centrifuge foundations (a) vibration isolators (b) vibration dampers (c) bolts with rubberized isolators

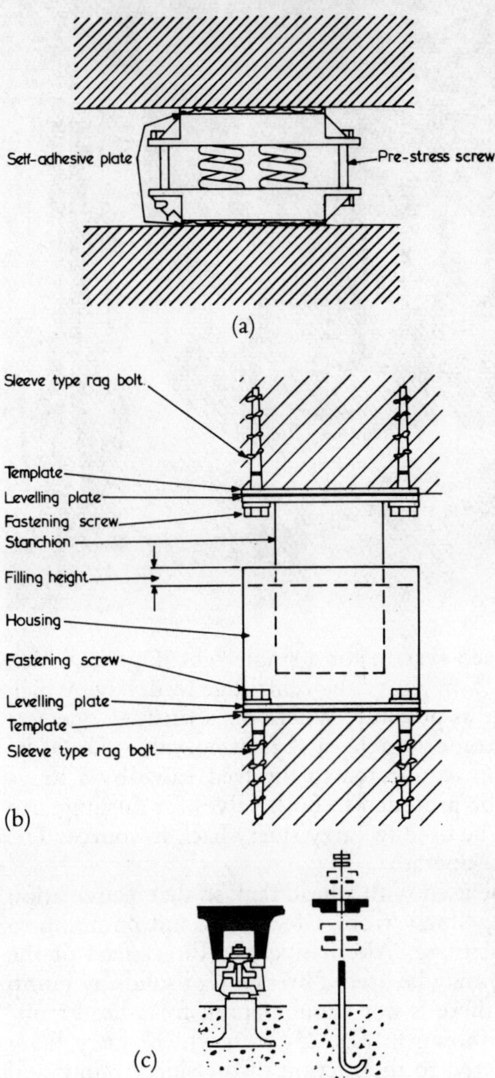

(1.5 m/s is often sufficient) and all lines should be as short and direct as possible with large-radius bends at least eight to ten times the pipe bore as illustrated in Fig. 25.4. In addition, flushing points for water or steam should be fitted at points where blockages might occur. Any fittings or branches should be placed in vertical pipe runs and angled so that after drainage no residual liquor or crystals can be trapped.

Gravity feed systems are recommended when the settling rate of solids is low enough that lines are not likely to block, or when it is desirable that solids should not be exposed to the action of pumps. The cone bottom feed tank should be mounted at least 2.5 m above the centrifuge to have sufficient head to ensure unimpeded flow of slurry.

Fig. 25.4 Large-radius bends in slurry feed line (Courtesy: The Boots Company)

Figure 25.5a shows a gravity feed system for a single centrifuge in which the feed tank is mounted at least 2.5 m above the centrifuge feed valve which should be as close to the centrifuge as possible. Where the centrifuge operates in cycle, and the feed valve is an on/off valve, a regulating valve should be placed directly below the feed tank connected to the feed valve by a long-radius bend. Connections should be provided to both valves for flushing. An overflow from the head tank may be used to carry slurry back to source. The feed valve is usually automatically operated.

Circulation feed systems may be used with a head tank so that a circulation loop is set up by a pump drawing slurry from a feed tank and pumping to a head tank 2.5 m above the centrifuge. Alternatively, a direct feed of the centrifuge from the pump delivery may be used if break-up of solids by pump impeller is not important, and if there is not enough room for a head tank.

A recirculation feed system is shown in Fig. 25.5b in which slurry from a crystallizer or feed tank is delivered to the suction of the slurry pump and recirculated around the feed loop. A regulating valve below the feed tank controls the pump delivery, and a second valve in the return line placed after the centrifuges controls the feed to the centrifuges.

Centrifuges may be used as second-stage thickeners in which the underflow from a (hydro) cyclone or primary thickener (section 19.6, p. 302), is fed to the centrifuge, while the overflow bypasses the centrifuge. Figure 25.5c shows a cyclone connected to a centrifuge in such a manner. The diagram also illustrates the use of a diversion valve when short-term interruptions of slurry supply to the centrifuge are necessary. Slurry recirculation is maintained to avoid problems due to settling in the lines.

If the slurry feed to the cyclone contains oversize particles or foreign matter that may damage working parts in the centrifuge the centrifuge may be

Fig. 25.5 Centrifuge feed systems (a) Gravity feed system (b) closed recirculation feed system (c) system with cyclone

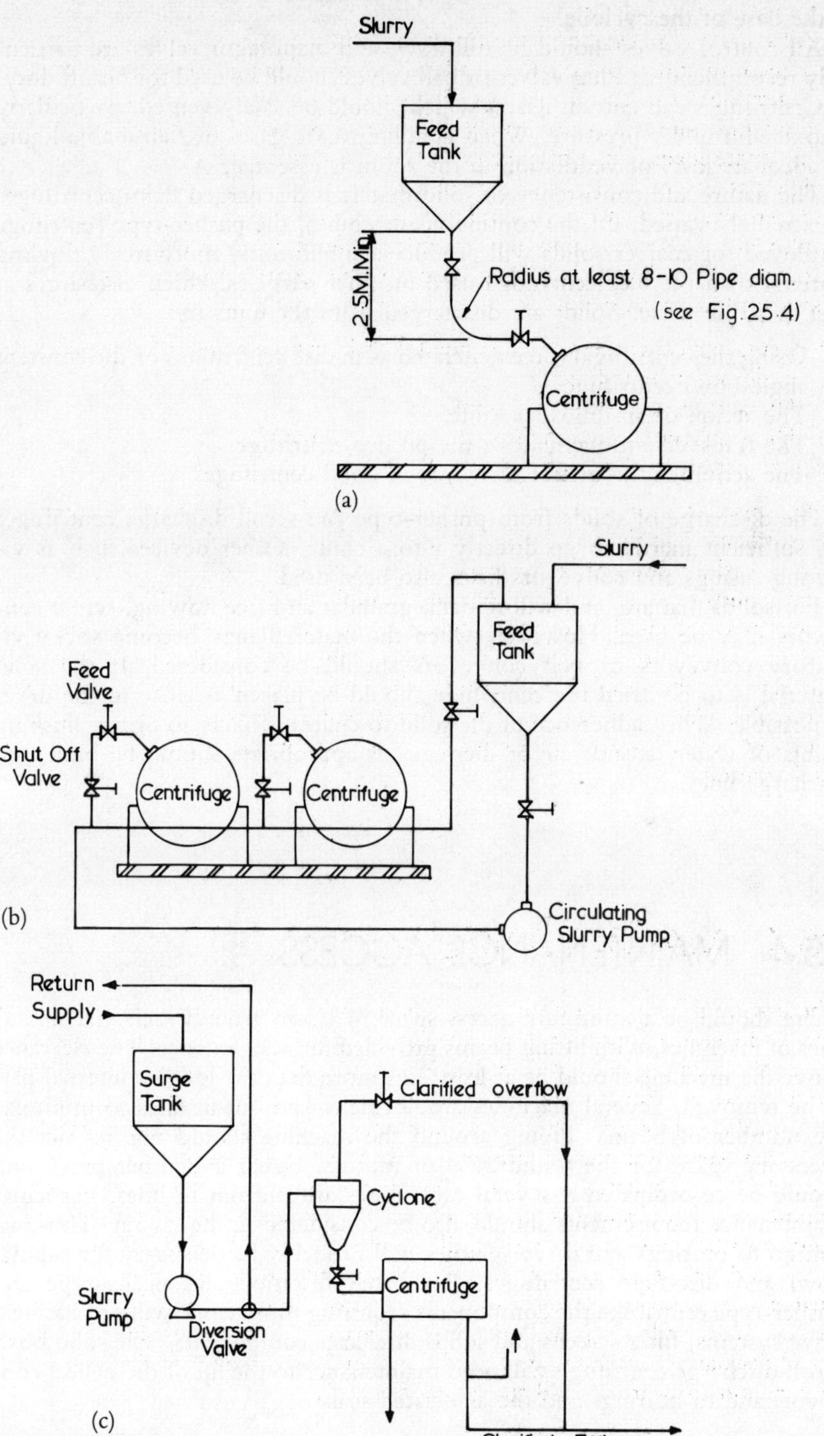

installed to take feed from the overflow of a cyclone; oversize particles are removed in the cyclone underflow, or periodically emptied from a grit box at the base of the cyclone.

All control valves should be full flow, and diaphragm valves are particularly recommended. Plug valves or ball valves should be used for on/off duty.

Centrifuges can entrain gases, which should be safely vented, particularly if toxic and under pressure. When handling toxic gases or flammable liquid an adequate level of ventilation in the room is essential.

The nature and consistency of solid materials discharged from centrifuges is extremely varied. Of the continuous machines, the pusher-type centrifuge employed for coarser solids will produce a significantly more freely flowing material than the disc centrifuge used on finer particles which discharges at best as a thin paste. Solids are discharged from the units by:

(a) Using the centrifugal force generated as in disc centrifuges or the constant angle bowl centrifuge.
(b) The action of an unloader knife.
(c) The transverse movement of the pusher centrifuge.
(d) The action of the helical conveyor in bowl centrifuges.

The discharge of solids from pusher-type and scroll discharge centrifuges has sufficient inertia to go directly into a chute. Other devices such as vibrating casings and conveyors have also been used.

For solids that are, and will remain, granular and free flowing, screw conveyors may be used. However, when the material may become sticky vibratory conveyors or belt conveyors should be considered. If the solid material is to be dried the centrifuge should be placed as close to the dryer as possible. When adherence of the solid to chutes is likely to occur, flushing points of water, steam, air or inert gas as appropriate should be built into discharge lines.

25.4 MAINTENANCE ACCESS

There should be a minimum access space of 1.5 m around each machine or lines of machines, with lifting beams provided for maintenance. The clearance above the machine should be at least 2 m more than the longest internal part to be removed. Several machines are normally laid out in lines to minimize the number of beams. Piping around the machine should not restrict the necessary space for the withdrawal of motors, bowls and other parts, and should be co-ordinated if several centrifuges are laid out in lines. Particular maintenance requirements should also be considered in the layout. Thus, attention to bearings and drive spindles will probably be necessary for tubular bowl and disc-type centrifuges. For batch filtration and peeler-type and pusher-type centrifuge the components requiring most work will be bearings, drive systems, filter screens and solids discharge components. The solid bowl scroll discharge centrifuge will need maintenance to the lip of the helical conveyor, and to bearings and the associated seals.

25.5 CENTRIFUGE SAFETY

Centrifuges rotate at a very high speeds and faults of balance, assembly, bearing failure or corrosion must be guarded against by adequate inspection procedures. An operating procedure, incorporating mechanical interlocks where possible, is necessary so that the centrifuge cannot be opened until the bowl has stopped rotating.

When processing volatile and flammable liquids, nitrogen or inert gas purging should be used to prevent the formation of explosive mixtures in operation or during a part of the cycle in batch operations. Static electricity from the machine or from the operator can be a considerable hazard so that electrical earthing must be effective with flammable liquids.

Hazards of toxicity are specially severe in batch machines with manual handling and it is essential to provide for operator safety by adequate gas venting, and fresh air masks for operators handling vapour-phase toxins. The *User Guide for the Safe Operation of Centrifuges*[1] gives comprehensive guidance on providing for safety in layout as well as operation and should be consulted for further details of the recommended safety procedures.

REFERENCE

1. I. Chem. E. *User Guide for the Safe Operation of Centrifuges with particular reference to hazardous atmospheres*. Institution of Chemical Engineers, 1975.

SOLIDS HANDLING PLANT

26.1 PROCESS LAYOUT

Solids handling plant contains equipment both for processing, and for transportation between processes. Layout must take into account the total operation of all these activities from feedstock supply to product dispatch (Fig. 26.1).

Process flow and process equipment specifications are a significant factor in determining layout, and the feasibility of practical layout should be considered when these are being prepared. A balance has to be struck between the level of capital investment; quality of plant; and labour, energy, and distribution costs. Work study principles should be applied, particularly when considering manning and distribution requirements.

Continuous operation of process plant, giving maximum utilization, makes the best use of capital invested. However, continuous operation of associated activities such as delivery, storage and dispatch are not always practicable or economic (Fig. 26.2). Consequently, the size of each activity varies according to their respective periods of operation and in addition, there has to be appropriate buffer storage.

26.2 ENVIRONMENTAL CONSIDERATIONS

Solids handling plants are liable to create dust and noise, but implementation of health and safety legislation demands that these are kept within defined limits. Heavy machinery, such as crushers and ball mills, should be separately housed or located in clearly defined areas to reduce the effect of noise, and other equipment such as fans and blowers should be lagged or baffled where isolation is not possible.

Dust emission can be minimized by using totally enclosed and vented equipment, but where this is not possible dust-collecting plant should be installed. Buildings should be designed so that they are easily cleaned and have no dust-collecting ledges, etc.

Fig. 26.1 Solids handling plant location

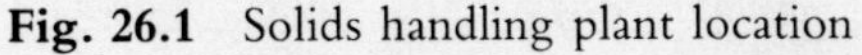

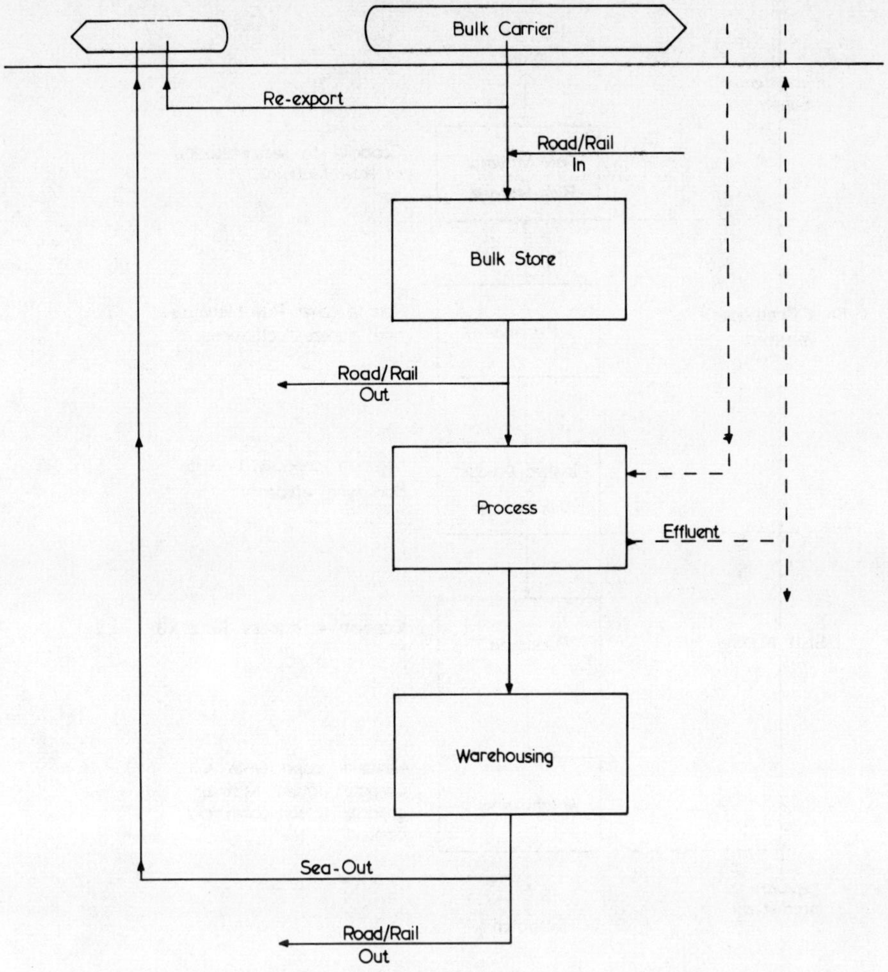

Open-belt conveyors can be dusty at their feed and discharge points. This can be minimized at the feed point by sloping the feed chute bottom to suit the material repose angle, and by the use of soft rubber sealing skirts and flaps (Fig. 26.3). Dust at the discharge point can be contained by limiting the free fall of discharging material, by fitting dust skirts and flaps around the hood, and by using multi-point or single-unit dust collectors.

Totally enclosed equipment, such as bucket elevators, can become slightly pressurized and require venting, either through air-relief filter panels or multi-point dust collectors. Dust-collecting ducting should be inclined for self-clearing in the event of particle separation in the ducts, with rodding ports and cleaning doors which are accessible from operating platforms. Surface heating may be necessary if dust is hygroscopic. Because of the dust it is important that provision must be made for the hygiene requirements of the operating and maintenance personnel, i.e. changing rooms, showers, etc. Storage and servicing facilities are needed for personal protection equipment, e.g. face masks, ear muffs, etc. if they have to be used.

Fig. 26.2 Operation of solids handling plant

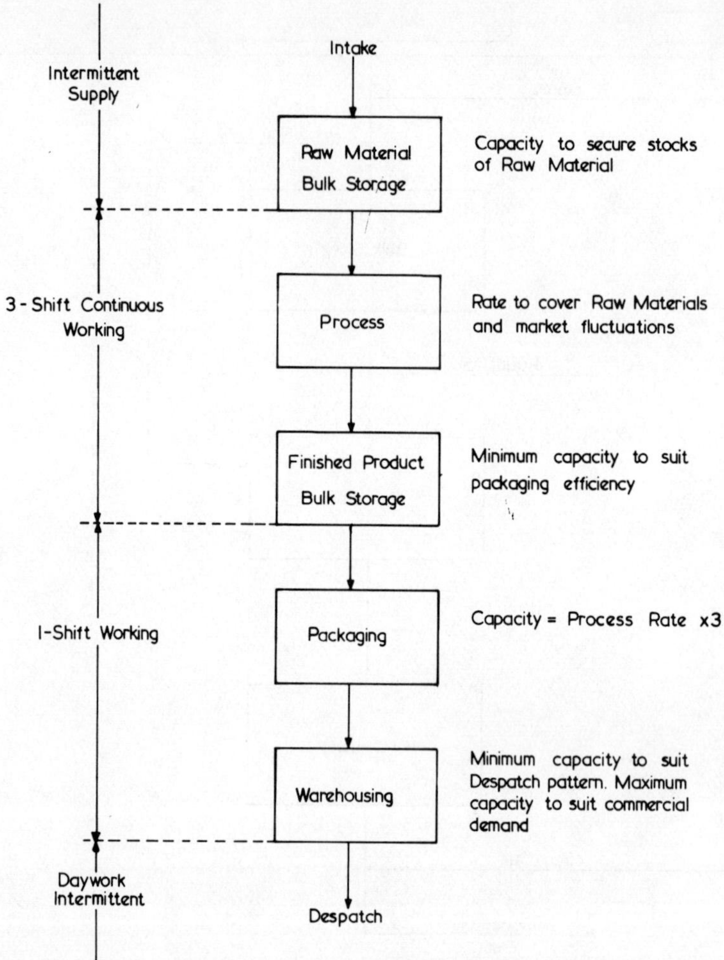

26.3 BUILDINGS AND ACCESS

The layout of buildings for solids handling plants requiring housing is influenced by a number of factors.

When permanency in the life of the plant is expected, intermediate floors and plant supports are constructed as an integral part of the building (Figs. 26.4 and 26.5) but where frequent layout changes are necessary to facilitate switching processes, free-standing plant and structures within the building shell permit easy alteration or replacement.

Access doors should be provided, which allow for replacement of the largest plant item, with separate personnel doors at strategic points. Access is also

Fig. 26.3 Dust precautions for belt conveyor transfer chutes

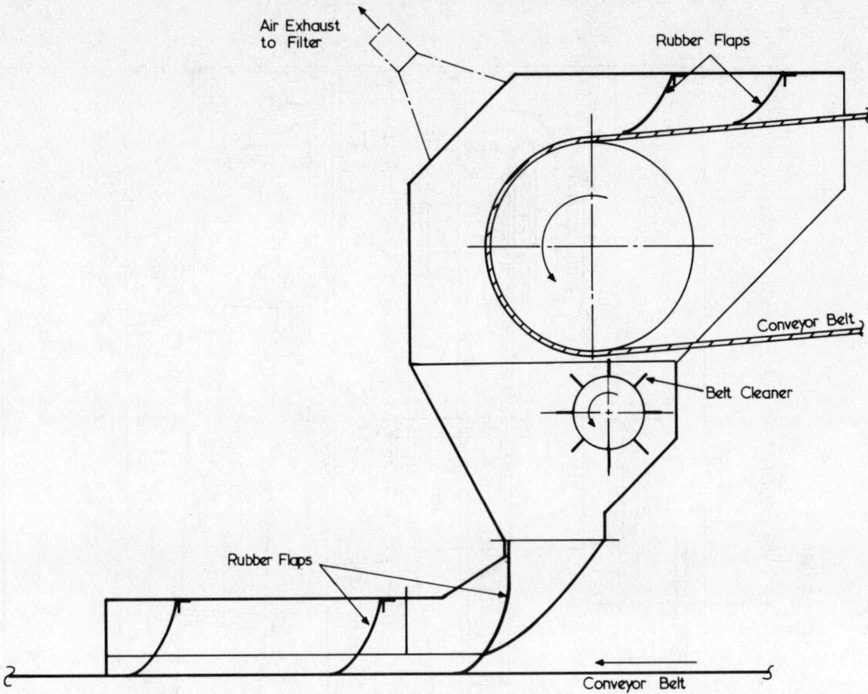

required to the roof, which should have suitable catwalks for inspection of vents, process exhaust stacks, and gutters.

Materials of construction will also be influenced by process conditions and should allow for wet or dry operations. Reinforced concrete floors and walls have a long life, but are difficult to modify, whereas steel must be protected against corrosion and possibly fire, but is amenable to change. The choice of cladding will depend on internal and external environment, and door and window frames should be protected against corrosion.

Lifting beams should be provided, with lifting wells passing all floors to facilitate plant construction, maintenance, and operation. If space is not available for internal lifting, an external cantilevered hoist should be provided, with doors to permit entry at all floor levels.

26.4 EXPLOSION PROTECTION AND PREVENTION

Where there is a dust explosion risk the equipment is isolated from the other items and fitted with explosion protection.[1] By isolation is meant either separation by distance or by walls, plus some device in the ducting to prevent explosion transmission.

Fig. 26.4 Layout of housed solids handling plant (a) elevation (b) section A–A (c) section B-B (Courtesy: Norsk Hydro Fertilizers)

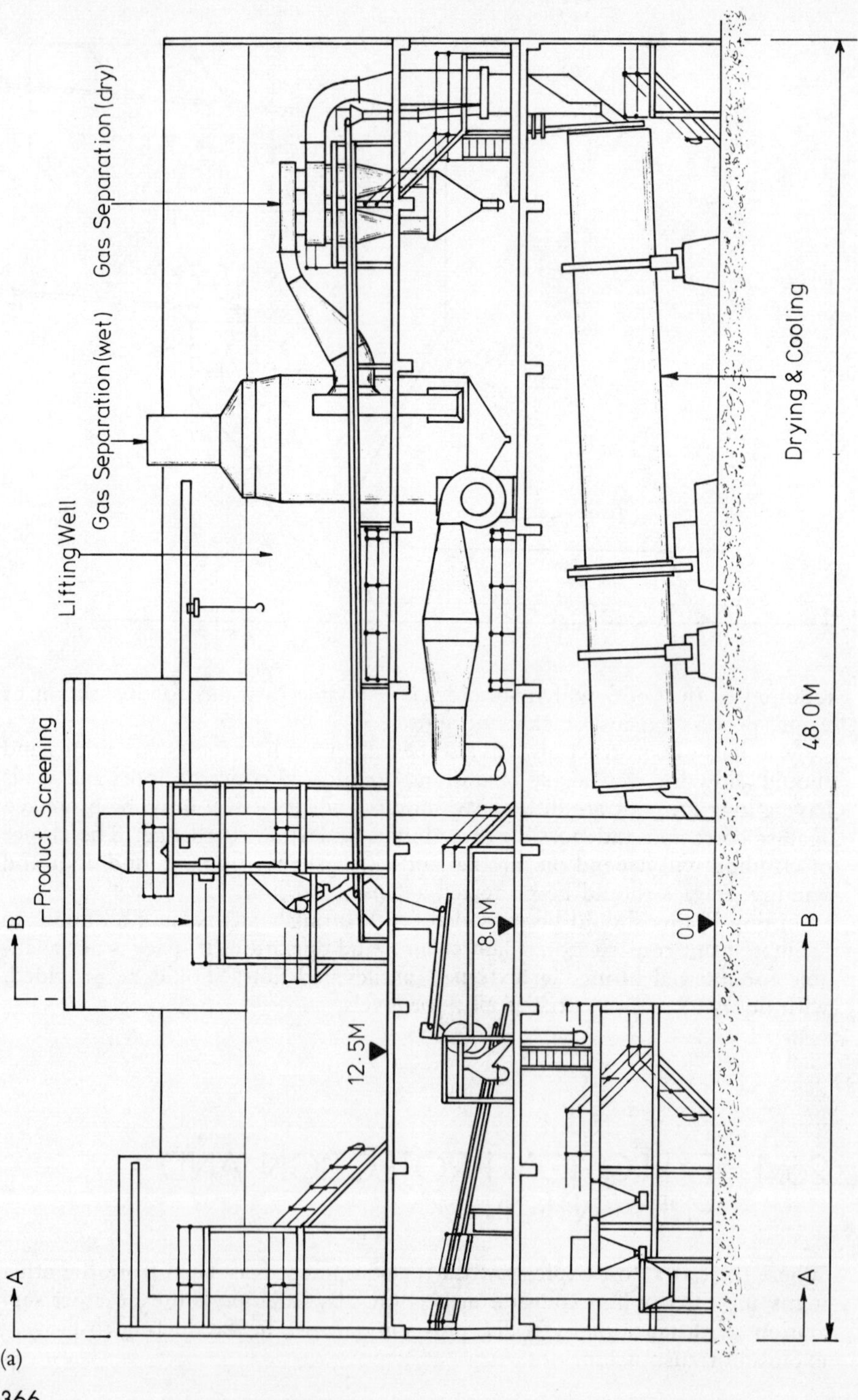

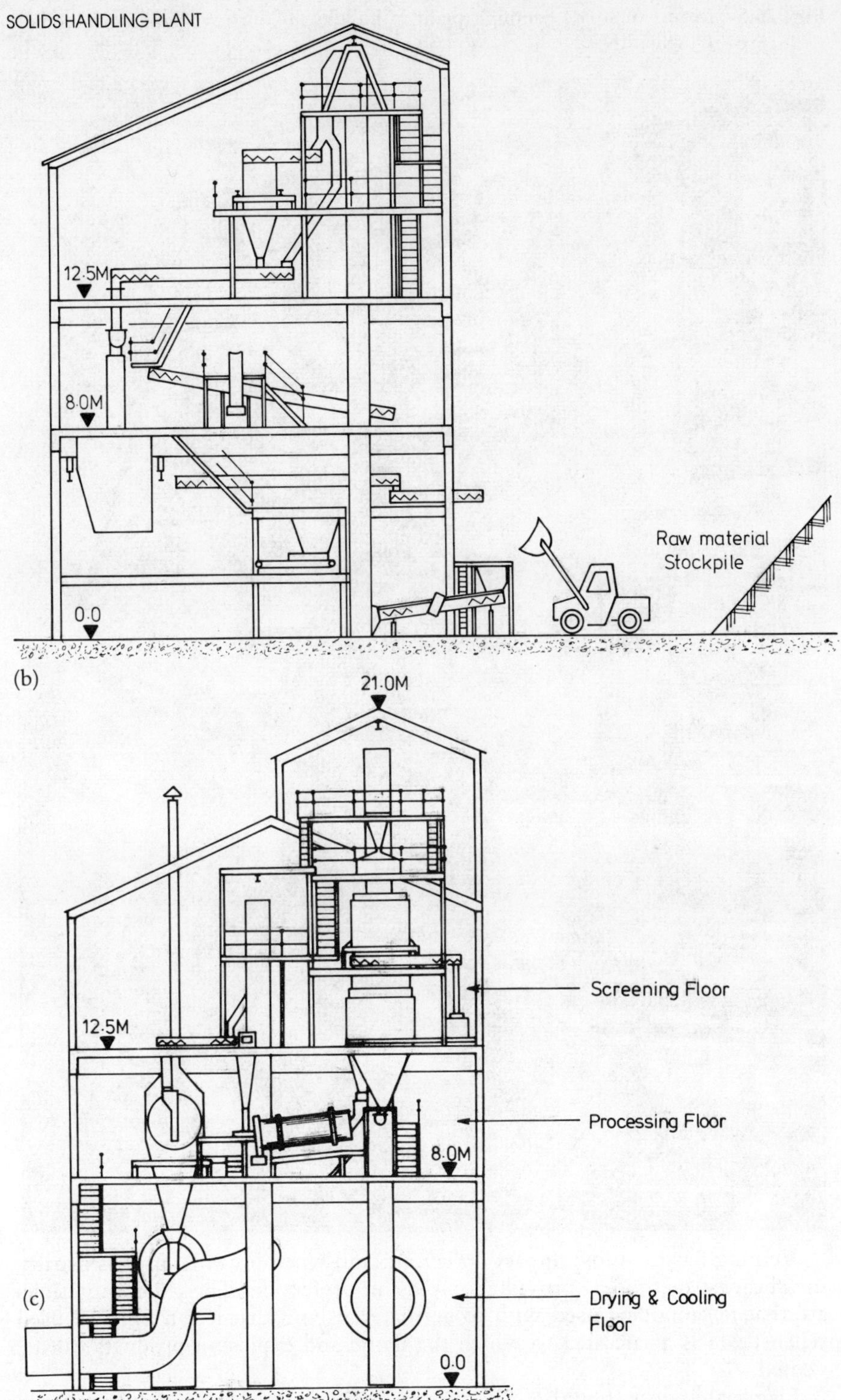
SOLIDS HANDLING PLANT
12.5M
8.0M
0.0
Raw material
Stockpile
(b)
21.0M
12.5M
Screening Floor
8.0M
Processing Floor
0.0
Drying & Cooling
Floor
(c)

Fig. 26.5 Model of solids handling plant in building (Courtesy: Babcock Woodall-Duckham)

Venting has the most impact on layout, and when it can be used is usually the cheapest method of providing explosion protection. The chief limitations are that it cannot be used with toxic materials and that it can only be used when there is a safe area to which the flame and explosion products can be vented.

Flammable dust should not be vented to an area inside a building because of the danger of secondary fires and explosions initiated by burning dust. This leads to many dryers and dust-collection facilities being sited outside buildings. Vents should never discharge to an area where flammable materials may be stored.

Ducting may be used to carry flame and explosion products from a vent to a safe area, e.g. from a vessel inside a building to a safe area outside the building, but the peak pressure in the vessel during a vented explosion is then increased. The increase in the peak pressure may be kept below 10 kN m^{-2} by the following combination of measures:

(a) Restrict the length of the duct to not more than 3 m.
(b) Make the cross-sectional area of the duct at least equal to the area of the vent.
(c) Have not more than one bend in the duct, and keep this bend very shallow.
(d) If a hat is placed over the end of the duct for weather protection, raise it at least one duct diameter above the top of the duct and make it strong enough and secure enough to withstand the pressure wave.

The increase in the peak pressure is proportional to the square of the duct length. Hence, if the duct length is 6 m, the increase in the peak pressure may be up to 40 kN m^{-2}. Ducts longer than 6 m should have additional vents in the duct walls. These figures quoted for pressure and duct length are only typical as they depend on the nature and concentration of the dust and the flow properties of the system.

There are alternative methods of explosion protection but these do not concern layout appreciably. Full details are contained in I. Chem. E.[1] but briefly these methods are as follows.

(a) *Suppression* is a system where the incipient explosion is detected by means of a pressure transducer thus triggering a discharge of a suitable inerting agent.
(b) *Containment* means achieving protection by making the plant strong enough to contain a dust or vapour explosion without damage.
(c) *Inert blanketing* relies on the provision of a non-flammable atmosphere so that combustion cannot occur.
(d) *Avoidance of dust* is achieved by having low air velocities.
(e) *Elimination* of all ignition sources is difficult to achieve completely.

Whatever method of explosion protection is used, the spread of an explosion to or from equipment must be prevented as far as possible. Rotary valves or some designs of double-acting flap valves are effective chokes especially if they are stopped automatically. They will, however, not stop the spread of fire.

REFERENCES

1. I. Chem. E. *User Guide to Fire and Explosion Hazards in the Drying of Particulate Materials.* The Institution of Chemical Engineers, 1978.
2. *Pow Tech '83. I. Chem. E. Sym. Ser.* 69, 1983.

DRYERS

27.1 CLASSIFICATION

For the purposes of layout design, dryers may be considered in two classes. In the first there is low airflow. Heating is usually by indirect heat transfer into the material being dried. Airflow is only needed to carry away evolved vapour. Typical examples are:

(a) Screw conveyor (trough).
(b) Paddle.
(c) Drum or roller.
(d) Some freeze and vacuum dryers.
(e) Some types of tray dryer.

These dryers tend to be compact standard units, mounted essentially on one floor level.

The second has medium to high airflow. Most commonly, these use heated air as the drying medium and the air-heating systems and the exhaust-gas cleaning and handling systems are often as large as the dryer itself. These dryers may be subdivided into categories according to shape and the method of handling the material being dried.

(a) Material enters, passes through the dryer, and is removed, in a continuous stream.

    (i) *Horizontal*, e.g.   rotary
                             band
                             vibrated fluid bed
                             some plug flow fluid beds;

    (ii) *Vessel-shaped*, e.g.   spray
                             turbo-tray
                             mixed fluid bed
                             spouted bed;

    (iii) *Vertical*, e.g.   pneumatic conveyor
                             some spray dryers.

(b) Material is batch-loaded and discharged, often manually, e.g.
                             small fluid bed
                             some types of tray dryer.

27.2 LOCATION

The location of a dryer is often determined by environmental and safety considerations, in addition to the requirement to be convenient to associated plant.

Depending to some extent on local climate, a dryer handling a particulate that is flammable, or presents a dust explosion hazard, is preferably mounted outdoors along with its associated cyclones, etc. If a dryer with explosion relief vents is inside a building, the vents need to be ducted outside the building, and this increases the explosion relief designer's problems (see section 26.4, p. 365 and I. Chem. E.[1]).

If buildings are used their size and layout is determined by the access required for erection and maintenance in addition to the feed and product flow arrangements. The building needs to be carefully designed to be as free as possible from ledges and places where dust might accumulate, in order to minimize the risk of secondary explosion.

Plant which requires frequent access for operation, cleaning or maintenance is more convenient indoors. Dryers of the batch-loaded type should be indoors in order to prevent the exposure of dried product, and the operators, to the elements. A large rotary dryer (see Fig. 27.2, p. 376) (e.g. kilns up to 230 m long × 7.5 m diameter for cement, lime or other calcination process) is often weather-proofed and placed in the open to save building costs with only the filter bags and instrumentation housed in a small building.

On the gas exit stream of dryers, with direct contact between the gases and solids, there are cyclones or some form of dust removal (section 28.3, p. 395).

On smaller installations the air heater can be placed outside the building. The heaters on larger installations are often integral with the dryer. Steam heaters can be considered as exchangers (see Ch. 22) whilst oil- or gas-fired heaters must be treated as furnaces (see Ch. 20) particularly with regard to safety.

Sufficient elevation is needed in vessel and rotary dryers to enable the product to fall by gravity to the product conveyor.

For environmental noise reasons, some or all of the drying plant may need to be acoustically enclosed. This applies especially to fans, and to some grinding mills incorporated into drying systems.

27.3 PROCESS FLOW

27.3.1 PRODUCT FLOW

Dryers are rarely installed in isolation. The plant supplying the dryer feedstock, and the subsequent handling of the dried product, influence the dryer layout. Most dryers operate with a solid feed material which is often wet and sticky, and therefore difficult to handle. Hence, the dryer feed point location is often determined by upstream plant. Similarly, the dried product may go to silos, be packaged, be cooled, or go for further processing and these locations must be considered in drying plant layout.

27.3.2 AIR FLOW

Except for small dryers, the drying air is usually drawn from outside the building (see Fig. 23.5, p. 346) as otherwise too great a load is placed on the building's heating and ventilation system. The air inlet must be protected from rain, prevailing winds, and high dust loads. In some areas attention must be paid to freezing fog conditions. Air inlets must not be located where they can draw moist air from a nearby cooling tower, or unclean air exhausted from another process (or from the dryer itself).

For the air exhaust, pollution laws do not allow the discharge of air with a high dust loading. However, exhausts can contain some dust, if only under occasional failure conditions, as well as odorous or otherwise objectionable gases. If the dewpoint of the exhaust is high, and especially when a scrubber is used to clean the exhaust, a steam plume can be present, and this may precipitate water droplets locally. Exhausts can also emit noise. The dryer exhaust stack location and height must take these factors into account, in conjunction with consideration of prevailing winds and the local environment.

Heat recovery systems for dryers are becoming more economically desirable. Where transfer of heat from the dryer exhaust to the inlet air is by air/air heat exchange, or heat pipe unit, correct siting of the inlet and exhaust systems can save money and energy. These comments also apply to dryers using partial recirculation of exhaust gases, and fully closed cycle dryers, including units using inert gases.

To meet the requirements of the pollution laws some dryers incinerate the exhaust gases. For economy, such dryers usually employ partial recycle, and incinerate the bleed-off gases. The hot incinerator exhaust is then used to heat the make-up air. If, in addition to the incinerator, there is also a main air heater, it will be found operationally convenient to site the two as close together as possible.

27.4 DRYER SUPPORT

Nearly all dryers operate at temperatures different from ambient, and allowance must therefore be made for thermal expansion. Large high-temperature dryers may have thermal expansions of several centimetres. Careful choice of support points, anchor points, and positioning of components can minimize relative movement between components, and hence reduce the problems involved. Where several platform levels are involved, the drying plant must be free to expand without imposing loads on the platform structure. Stairway and ladders anchored to the drying plant must have one end free to move.

27.5 ACCESS

Access to parts of a drying plant is required for:

(a) Initial construction of the dryer.
(b) Operating.
(c) Cleaning.
(d) Inspection and maintenance *in situ*.
(e) Removal of parts for off-plant maintenance.

Where explosion vents are used,[1] accessways required during plant operation must not allow an operator to pass within the blast area. On the other hand, access to explosion doors must be provided so that they can be inspected and reset, and also to check the operation of micro-switches frequently fitted to explosion doors for interlocking purposes.

All access should be made convenient as items difficult to reach will in practice be ignored leading to operational and maintenance difficulties.

Most drying plant needs cleaning from time to time, especially near the feed point where the material may be sticky. Access doors must be located so that all dryer parts are easily reachable and sufficient platform space (1.5–2.5 m) must be provided to allow easy and safe use of cleaning equipment.

Consideration must be given to the possible removal of some dryer components for major overhaul or replacement, e.g. ring gear and tyres on rotary dryers, atomization equipment for spray dryers, rolls for drum dryers, fan impellers, fluidizing grids, large motors, etc. In some cases it is worth while installing permanent lifting beams for this purpose. The dryer manufacturer should be consulted for advice on frequency, weights and clearances required. At least 2 m extra should be added to the length of the longest internal part to achieve adequate clearance above and around the dryer.

Sufficient clearance and access should be provided to allow for the construction of the drying plant, and for the application of thermal and acoustic insulation. A minimum clearance of 1 m from any outer surface should be allowed wherever possible.

27.6 SPECIFIC DRYER TYPES

27.6.1 LOW-AIRFLOW AND VACUUM DRYERS

These are generally compact, and fixed in a standard design and size. The only real variable is location.

A cylindrical trough-dryer can be made strong enough to withstand the full pressure of an unrelieved explosion or exothermic decomposition. It is less easy to do this with a flat roof dryer. Particular attention must be paid to feed and discharge ports so that they withstand the pressure.

For structurally weak dryers running at atmospheric pressure, explosion vents will have to be along the full length of the dryer roof to ensure an unimpeded explosion path unless there is a large free space between the roof and the top of the agitator blades. They must discharge to a safe area.[1]

For batch vacuum dryers, provided there is no risk of exothermic decomposition with massive gas evolution, safety during drying may be based on containment of any possible explosion. In order to withstand the vacuum plus the pressure in the heating jacket, these dryers are designed to recognized pressure vessel codes (Fig. 27.1). If the normal absolute working pressure during drying is not more than one-tenth of the vessel design pressure, the vessel should usually be able to withstand any dust or vapour explosion which may occur at the working pressure. Particular attention should be paid to discharge ports and valves to ensure that they can withstand any possible explosion pressure.

Vacuum dryers not designed to contain any explosion, such as shelf dryers, must have vents installed (see section 27.6.8, p. 383, on tray dryers).

Fig. 27.1   Examples of vacuum dryers (a) Vacuum pan dryer (b) vacuum cone dryer (Both courtesy: APV Mitchell Dryers)

(a)

(b)

At the end of the drying cycle the vacuum should be broken with nitrogen or other inert gas. Pressure relief should be provided to guard against over-pressurizing the dryer at this stage. A bleed purge of nitrogen should be continued during the discharge. On pan dryers (Fig. 27.1a) and shelf or tray dryers (Fig. 27.7) the pressure relief may be on the dryer body, but on a rotating cone dryer (Fig. 27.1b) the only practicable location is on the inert-gas line.

When there is a risk of exothermic decomposition with massive gas evolution in batch vacuum dryers, protection can be achieved by venting. Suitable vent locations would be the back of a shelf dryer or the top of a pan dryer. There is no suitable location on a rotating cone dryer. Vents must relieve to a safe area.

27.6.2 ROTARY DRYERS

Rotary dryers (Fig. 27.2) are often large and heavy, with high static and dy-

375

Fig. 27.2 Rotary dryer layout (a) schematic (b) an example (Courtesy: APV Mitchell Dryers)

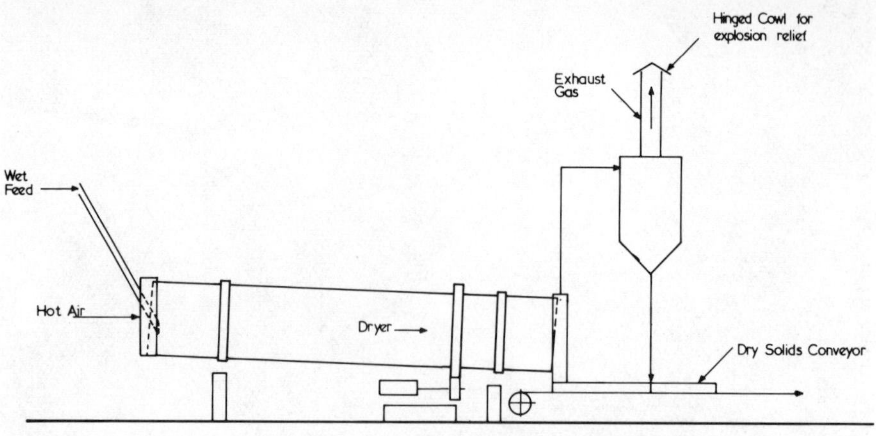

(a)

(b)

namic foundation loads. Therefore, they are usually mounted at ground level, with just sufficient elevation for dry product handling. Those units with internal chains or external hammers can create noise problems.

Access is required for installation and maintenance, including removal and replacement of drum sections. Access should also be provided to inspect and clean the feed chute. It is an advantage to have a platform with inspection ports and externally mounted spot lamp at the discharge hood for examining the drum internals. Access is also necessary for maintenance of drum hammers where these are installed to thwart internal build-up.

Provision should be made for the collection and recycling of spillage which may occur between fixed and rotating parts at the feed end.

To avoid the use of pits for discharge elevators or conveyors, lifters at the discharge end can be shaped to achieve a high-level drum discharge, provided interference with gas flow is taken into account.

Guarding is necessary for supporting trunnions and drives, but with safe access for maintenance and greasing. Guarding from the rotating drum should also be provided. This is usually done by the manufacturer, but also the layout engineer should arrange his design so that operators are kept away from the moving drum.

Figure 27.2 illustrates a rotary dryer with co-current airflow but counter-current operation is also used. The only practicable places where relief vents may be installed[1] are on the air inlet and discharge hoods at the ends of the drum. If the drum can withstand a pressure rise of the order of 250 kN m^{-2} and the vents open fully at a pressure rise of roughly 20 kN m^{-2}, then a vent at each end of area equal to the cross-sectional area of the drum will be adequate for protection against most dust explosions. Venting cannot be used if the design of the internal flights or baffles is such that it impedes free access of explosion products to the vents from any part of the dryer. Access doors and vents should be kept separate.

27.6.3   BAND DRYERS

These are generally all contained on one floor level (Fig. 27.3). Because of the mechanical nature, good maintenance and cleaning access is essential. The considerations in Chapter 29 on conveyors apply with allowance for insulation and thermal expansion.

Since there should not normally be a dust cloud in the oven and this type of dryer is not recommended for evaporating flammable solvents, explosion protection is only required when the oven is heated directly by burners along its length.[1] In that case it should have vents at least equal in area to the vertical cross-section of the oven and spaces with not more than 6 m between vents. If the oven is divided into zones by vertical baffles, there should be at least one vent per zone, again with vents being not more than 6 m apart.

Since unburnt fuel may accumulate below the band, it is preferable to locate the vents on the side of the dryer, protecting the space both above and below the band. Ducting should be provided outside the vents to divert combustion products upwards away from any area where personnel may be present.

These ovens have large rectangular cross-sections and are usually fairly

Fig. 27.3 Layout of band dryer (Courtesy: APV Mitchell Dryers)

weak, but distortion can be minimized by providing vents which open fully at a very low internal pressure rise,[1] say 3.5 kN m^{-2}.

If a fuel explosion occurs in a band dryer it will generate a dust cloud and if the dust is flammable, burning dust particles can be ejected from the vents and from the slots at the ends where the band enters and leaves the oven. In this case, therefore, the vents must discharge to outside the building, the ends of the band must be totally enclosed and the end enclosures provided with their own explosion protection. In view of these complications direct firing with burners along the oven should not be used on band dryers handling solids which can form a flammable dust cloud.

If a flammable dust cloud can occur above the discharge chute, the discharge end of the band should be totally enclosed by a dust hood and the hood provided with explosion protection. Either venting or suppression is suitable and should activate at the lowest practicable pressure rise, say 3.5 kN m^{-2}. As some distortion of the hood is still likely because of its weakness, it and the discharge chute should be isolated from the rest of the dryer, e.g. by a baffle, so that the explosion does not disturb the material on the band.

27.6.4 CONTINUOUS FLUID BED DRYERS

If relatively shallow hot air plenums are used, the dryer itself can usually be contained on one floor level, with perhaps the exhaust gas cleaning equipment elsewhere. For units with a conical hot air plenum, the principles relating to spray dryers apply.

Properly balanced vibrating dryers do not need heavy foundations, and they may be mounted anywhere to suit flowsheet and operational purposes. Unbalanced units may need heavy ground foundations.

Venting[1] in the roof is the most usual method of protection in fluid bed dryers handling water-wetted flammable dust. Plug flow fluid bed dryers should have vents distributed along the roof to give direct relief to all parts of the chamber.

Feeders and discharge valves should be of a type which acts as a seal in the event of an explosion in the drying chamber.

27.6.5   SPRAY DRYERS

The considerations in Chapter 19 on vessels apply and the feeding of slurries to spray dryers follows the practice indicated in Chapter 24 on filters and Chapter 25 on centrifuges. In sizing the building, at least 2 m clearance over and around the dryer should be left plus any additional room for the withdrawal of internals.

The drying chamber size and shape is generally fixed by process parameters, but the hot air and exhaust systems afford much scope for imaginative layout design depending on the particular situation. Those dryers having each a conical bottom drying chamber (Figs 27.4 and 27.5), are generally supported from a level just below the point where the cone meets the cylindrical section, and this is often also a convenient level for access to cham-

Fig. 27.4   Layout of a spray dryer (schematic) (Courtesy: Anhydro)

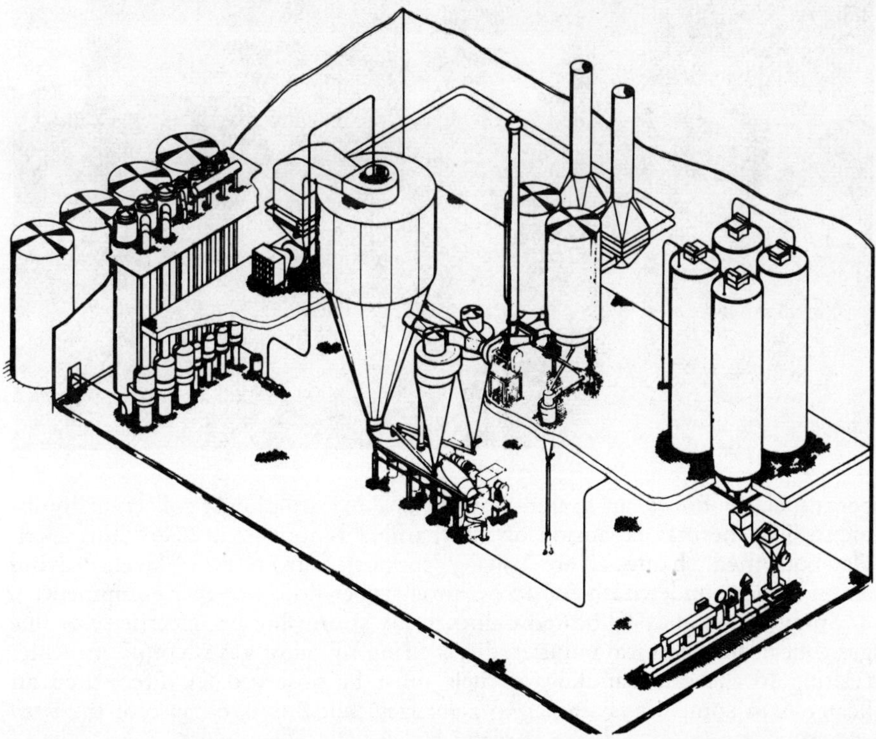

Fig. 27.5 An example of a spray dryer layout (Courtesy: Anhydro)

ber inspection doors, air heater controls, and the top of dust collection equipment. The thermal expansion of the chamber is above and below this level. Flat-bottomed chambers are usually supported from floor level, leaving enough space underneath for access to chamber-floor sweeper equipment.

Spray dryers may be heated indirectly by steam, hot oil, electricity or flue gas, but in the chemical industry direct firing by oil or gas is common. Rules relating to the safe handling of fuels must be observed. A direct-fired air heater is in some ways similar to a furnace, but differs because of the very high percentage of excess air usually available, and the relatively low temperatures usually encountered. Many spray dryer heaters are completely without refractory material, and are very light in consequence. More freedom is therefore possible with their support and layout.

All spray dryers require washing down from time to time. Good access to the bottom of the drying chamber is required for cleanout and drainage. A good chemical drain, sometimes in a small area, is advisable near this point. If the dryer has chamber solids discharge, access to the airlock at the bottom is required.

For small plants of low air-pressure drop, fans are small, and can often be mounted at high level, even on top of the drying chamber, to minimize duct length. But for large plants, especially if there is a high overall pressure drop, fans of several hundred kilowatts may be required, and foundation and acoustic considerations may cause the fan to be mounted at ground level.

Venting[1] is the most common method of providing safety in spray dryers handling water-wetted flammable materials. In the drying chamber, the roof is usually the most convenient place for installing a vent, although on some dryers the vents take the form of hinged doors on the side of the chamber. If the latter are used, they must be dust-tight as well as opening fully at the prescribed pressure rise, which means that considerable care is required in the design of suitable door catches. The weight per unit area of these doors should not exceed 40 kg m^{-2}. Explosion doors should fit flush with the inside of the wall so that dust cannot accumulate on ledges.

The roof is generally the weakest part of the drying chamber. Even with special bracing, its design strength is usually less than 40 kN m^{-2} internal pressure rise. A design strength of 20 kN m^{-2} internal pressure rise is typical.

The required vent area may be calculated by the methods outlined by the Institution of Chemical Engineers.[1]

It may be possible to give very small spray dryers sufficient strength to contain an explosion without damage. The exhaust air ducting and the cyclone should be at least as strong as the drying chamber.

27.6.6 PNEUMATIC OR FLASH DRYERS

The dryer itself is essentially a length of duct, offering great layout versatility. The solids feed system, however, is often complex, involving such items as backmixers, feeders, conveyors, airlocks, dispersers, screens, etc. The relative positions of these items is restrained by solids flow considerations.

The drying duct is often vertical, with the hot air inlet and solids feed point near the bottom. To avoid excessive total height, drying ducts are often folded or curved (Fig. 27.6) but generally it is wise to allow sufficient vertical height for the material to dry to a point where its surface is no longer sticky before entering the first bend, otherwise product build-up with subsequent fire risk can occur. Horizontal ducts should be avoided, to minimize the risk of solids settling out of the airstream. If classified recycle is used, the classifier should be positioned to give a good return path from it to the bottom of the drying duct.

Generally, the feed preparation equipment is mounted indoors, but the drying duct itself, as well as the product collection system, can well be out-

Fig. 27.6 Pneumatic conveyor dryer with cyclone (Courtesy: APV Mitchell Dryers)

side the building, particularly convenient if explosion relief vents are used.

Thermal expansion of the drying duct can be large, but can often be taken up by correctly designed bends in the duct.

Good access is required to points where product build-up can occur, and also to explosion vents, so they can be inspected and, if necessary, reset.

Venting[1] is the most usual method of protection with water-wetted flammable dust. It is not practicable to install vents along the length of the vertical tube, but protection can be achieved by a combination of a strong tube, vents at the top and bottom of the vertical lift, and a vent on the tube at the inlet to the cyclone. Each vent should have an area equal to the cross-sectional area of the tube. Since the vents at the top and bottom of the tube have to withstand a significant airflow pressure without leakage during normal operation, the pressure rise at which they are designed to open fully may have to be as high as 20 kN m^{-2}. If the vents open fully at this pressure, the vertical tube should be made sufficiently strong to withstand an internal pressure rise of the order of 250 kN m^{-2}. If the cyclone strength is in the region 35–70 kN m^{-2} then the vent on the cyclone inlet should open fully at a pressure rise of not more than 10 kN m^{-2}. Where the tube shape changes to a square cross-section at the top of the vertical section to accommodate the vent, the square tube should be strengthened to withstand at least the same pressure as the cyclone. In addition to the above vents, the cyclone should also have a vent in its roof and other dust recovery units should also have appropriate venting. If the exhaust gas is recycled, the recycle line should have a vent of area equal to its cross-section area every 6 m.

Some very large-diameter vertical tubes in pneumatic conveying dryers may not be able to withstand an internal pressure rise of 250 kN m^{-2} and so should not be used for drying flammable dusts. It may be possible to use them for dusts of mild explosibility if restrictions are placed on their length/diameter ratio, but expert advice should be obtained.

When venting is chosen for protection, it is essential that a feeder is selected which acts as a seal preventing an explosion in the tube from spreading back into the feed hopper.

27.6.7   BATCH FLUID BED DRYERS

Small standard batch-fluid bed dryers are commonly used in the pharmaceutical and fine chemical industry. The drying chamber, fluidizing grid and hot air plenum are mounted together and wheeled into operating position after charging with wet feed. Layout is mainly determined by handling requirements.

To provide protection[1] against a dust explosion a single vent on the side of the chamber, between the product container and the filter socks, will be satisfactory. The vent should be on the dusty side of the filter so that a dust explosion does not have to burst through the filter to the vent. Furthermore, a filter sock torn from its support could partially block the vent. However, if venting is used for protection against a vapour explosion there must be an additional vent on the clean side of the filter.

It may be feasible for a small batch-fluid bed dryer with a circular cross-section to be made strong enough to contain an explosion. Particular attention will have to be paid to inspection doors and to the method of clamping the product container in position. It will probably not be feasible to have windows in such a dryer.

27.6.8   TRAY AND TUNNEL DRYERS

In using tray dryers (Figs 27.7 and 27.8) a number of operations can be identified (Fig. 27.9). Tray drying is a production line operation like packaging and filling (Ch. 30).

If operation is manual (Fig. 27.7) then layout design is principally an ergonomic problem. For example, trays should be filled and emptied on tables and the trolleys suitably designed so that trays can easily be handled. If the

Fig. 27.7 Forced convection tray dryers (Courtesy: APV Mitchell Dryers)

(a)

(b)

operation is automated, then the type of dryer and the method of transporting the solids determines the layout, and the equipment manufacturer should be consulted. As an example, Fig. 27.8 shows a tunnel type of tray dryer. For both systems there should be areas for cleaning and storage of trays and trol-

Fig. 27.8   Layout of tunnel dryer (a) general view (b) trolley for drying bricks (Both courtesy: Babcock Woodall-Duckham)

(a)

(b)

leys. This prevents the working areas being cluttred with spare trays, etc.

The drying area may have to be ventilated and de-traying booths used if the powders are toxic, etc. Cross-contamination of products may be avoided by partitioning-off the drying area.

Venting is feasible and is normally the preferred method of explosion protection. Because of the weak construction of box ovens, the vent area needs

385

Fig. 27.9 Operations in tray drying

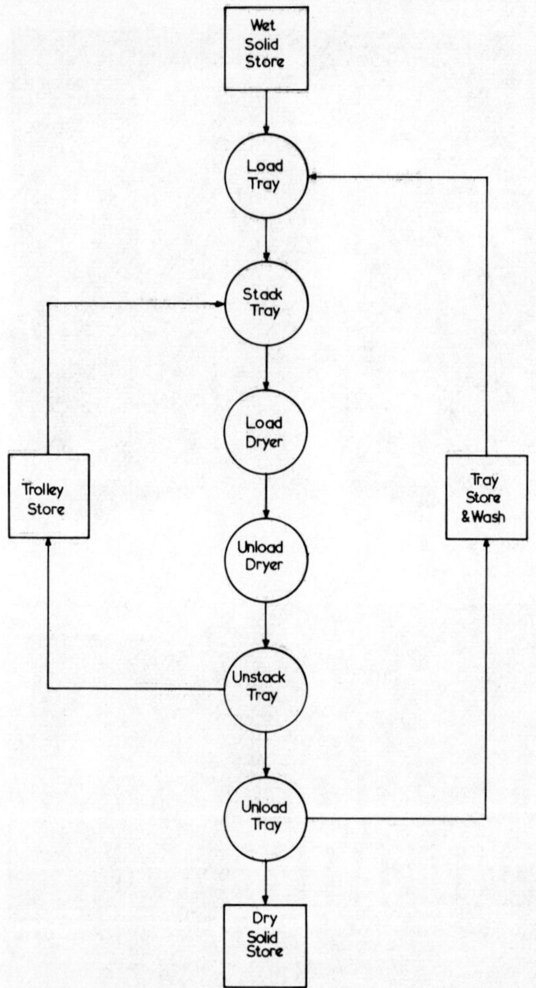

to be large and should be in the back of the oven, stretching the full height of that part of the oven occupied by trays, and with an area at least half the area of the back of the oven. The heater compartment should have direct access to the vent. The preferred vent construction is a lightweight panel of insulating material inserted into a channel frame. The vent should open fully at a pressure rise not exceeding 3.5 kN m^{-2} if serious damage to the oven is to be avoided, although with the design strength of some ovens being only about 7 kN m^{-2} some distortion is still possible.[1]

When the oven contains a flammable solid, the vent must discharge to outside the building. This means locating the oven close to an external wall and providing a hole in the wall for a short duct from the vent. The duct should have an area at least equal to the area of the vent. The space to which it discharges must not contain any obstructions and should preferably be fenced-off so that personnel are unlikely to enter it.

27.6.9   AFTER COOLERS

In large driers it is often necessary to cool the product when natural heat losses are insufficient. Pneumatic, rotary or vibratory conveyor coolers may be used. Similarly, considerations as for dryers apply to the layout of aftercoolers.

REFERENCE

1. I. Chem. E. *User Guide to Fire and Explosion Hazards in the Drying of Particulate Materials*. The Institution of Chemical Engineers, 1978.

SOLID REDUCTION AND SEPARATION EQUIPMENT

28.1 GRINDERS AND CRUSHERS

A wide range of equipment is available, and selection is governed by the physical properties of the material, size distribution of feed, rate of feed, and required size reduction.[1] Manufacturers' recommendations should be followed for selection of type, size and capacity. The types of equipment may be classified as follows:

(a) Coarse comminution – product range down to 50-mesh,

    (i) gyratory or cone crushers
    (ii) jaw crushers
    (iii) roll crushers
    (iv) hammer mills

(b) product range down to 350-mesh from approx. 50-mesh feed,

    (i) ball mills
    (ii) rod mills
    (iii) attrition pulverizers.

(c) Superfine comminution – product below 350-mesh,

    (i) vibration mills
    (ii) fluid energy mills.

Several crushing stages may be necessary to obtain the desired material reduction, and these are usually arranged in line, to simplify the feeding and delivery systems. The feed should be either in the direction of rotation of the mills, or vertical and at a rate controlled by conveyor or mechanical feeder. As a precaution to avoid damage to the machine, it may be advisable to install a magnet in the feed stream to extract tramp iron. Figure 28.2a illustrates an electromagnet suspended over the conveyor. The magnet is moved to one side along the runway beam to discharge the tramp iron to a hopper or trolley. Figures 28.1 and 28.2b show a magnetic head pulley which automatically discharges the iron to a chute.

Coarse crushers run at relatively slow speeds, but due to their size and weight, ground floor installation is preferred (Fig. 28.1). For the finer types of crusher, floor design should allow for excessive stresses which can occur through build-up on rotors or through blockages. Thus the machines and their drives are usually housed in a building, adjacent to an access road and supported at grade on concrete foundations. If the machines have to be el-

Fig. 28.1 Typical layout in a hammer-mill house (a) side view (b) front view

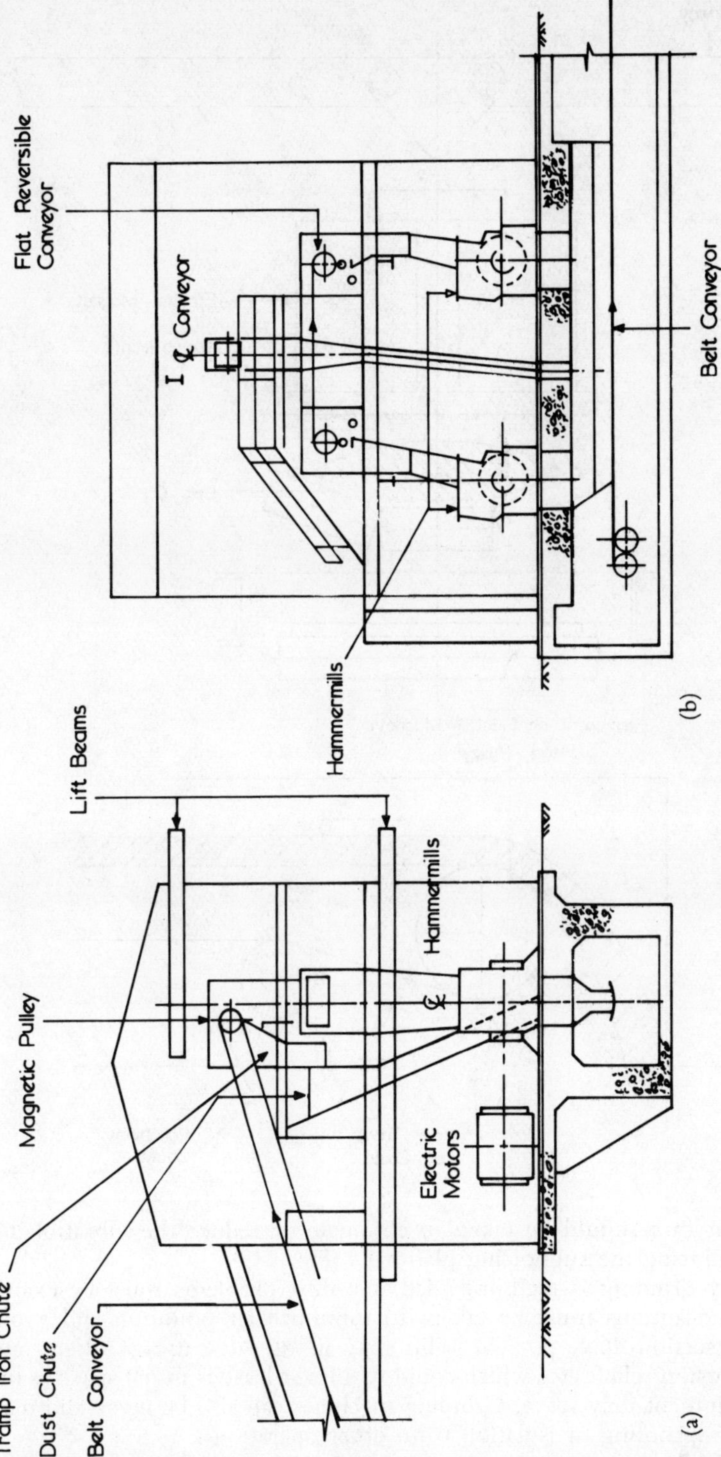

Fig. 28.2 Layout of tramp iron magnets (a) suspended magnet (b) magnetic pulley

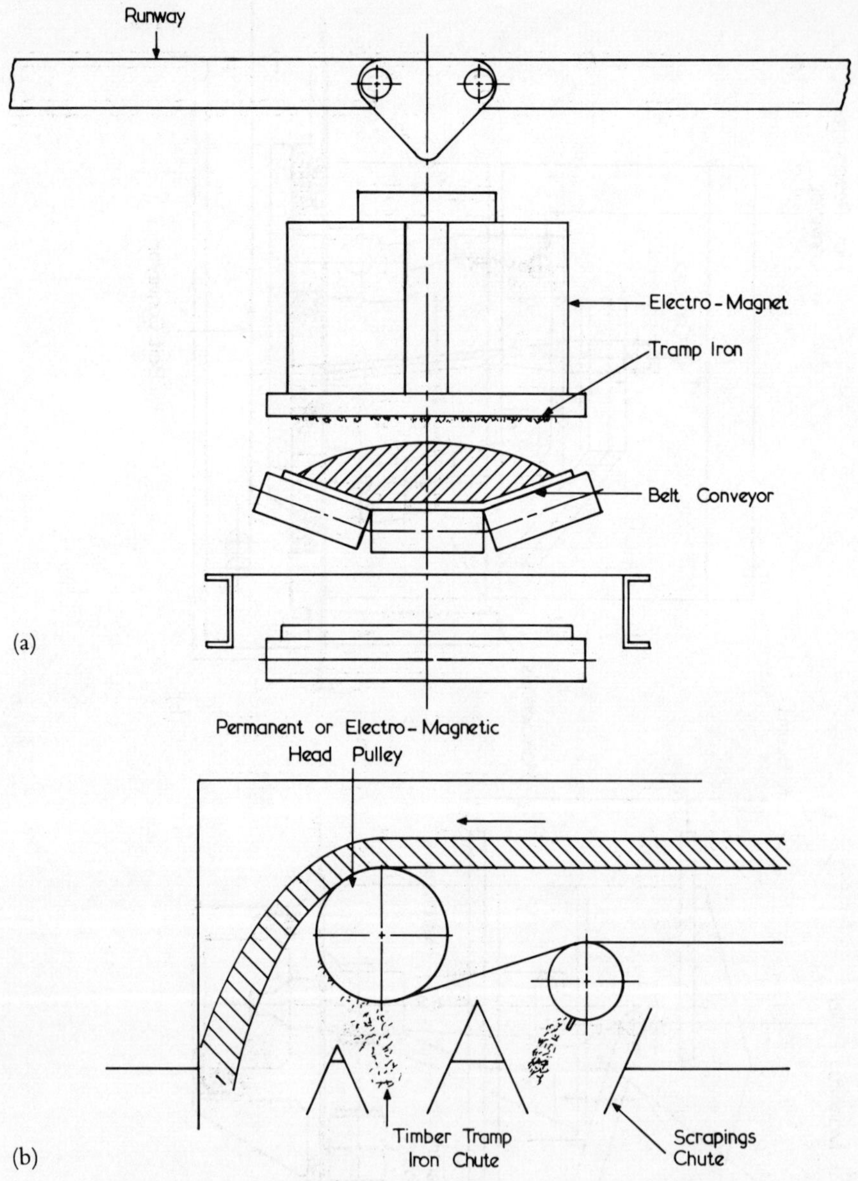

evated, great care should be taken to eliminate or reduce the vibration by adequately bracing the supporting platforms (Fig. 28.3).

When dry grinding is recommended, the dust problems must be examined, and precautions must be taken, to minimize air pollution and avoid explosions (section 26.4, p. 365). This may involve the use of filters, cyclones, explosion relief, etc. which could require at least as much space as the milling equipment they serve. Grinding machines can also be noisy and may require soundproofing or isolation from other operations.

Fig. 28.3 Ball mill installation (Courtesy: ND Engineering)

Closed-circuit air-swept systems are used for some fine-grinding oper-ations. These employ classifiers as in Fig. 28.4 or cyclones, or bag filters to remove products (see section 28.3).

Space of at least 1.5 m around each crusher is required for maintenance, cleaning, and replacement of wearing parts such as grids, rotors, grinding discs, rods, cones, rolls and jaw plates. Lifting beams will facilitate handling of heavy items. The clearance above the item should be at least 2 m more than the longest internal part to be removed. Provision should be made for storing and replacement of rod and ball mill charges, adjacent to the mills and for access to breaker plate adjustment screws, tramp iron boxes, inspec-tion doors, etc. The access around a crusher or mill can be taken as approxi-mately the width of the machine.

28.2 SCREENS

Screening can be generally divided into two classifications, i.e. sizing and scalping. Sizing is the separation of particles into a number of fractions, while scalping is the removal of oversize material too big for downstream plant operations to handle, or of unwanted fine material from the product.

28.2.1 VIBRATING SCREENS

Vibrating screens are invariably used for size control in association with crushing and grinding of dry material. Sizes vary considerably and it is im-

Fig. 28.4   Classification recycle stream

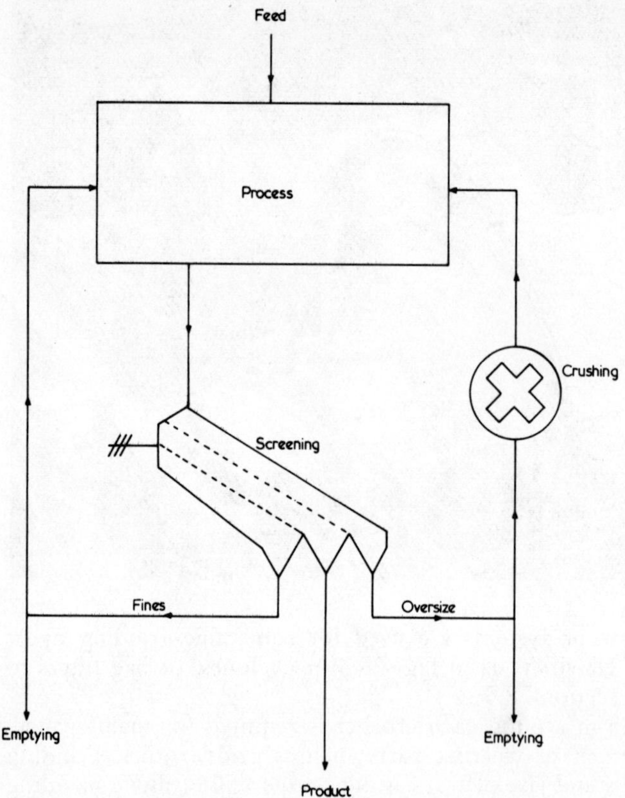

portant that each installation is carefully laid out, ensuring that minimum angles of feed and delivery chutes are used (Fig. 28.5).

Screens fall into three main categories:

(a) One system necessitates vibrating the whole machine, which can be horizontal or declined in the flow direction (down to 25°). Care must be taken to avoid transmission of vibration to supporting structures, and feed and discharge chutes should have flexible connections.

(b) Another system provides vibration to the screen mesh framework, which can also be horizontal or declined. Attention is again required to avoid transmission of vibrations. Chute connections are normally fixed to the static casing.

(c) The third type imparts vibration to the mesh only, through a series of electromagnetic vibrators, and the machine can be tuned by adjustment of the mesh angle and vibration amplitude. Chute connections are normally fixed to the static casing.

Screens may be supported either by suspension, from the floor below, or a combination of both. Totally enclosed screens should be used when necessary to minimize a possible dust nuisance. The feed and discharge chute design should allow space for maximum surge conditions and for the excessive movement of machines, which are supported on springs, which occurs on starting up and stopping.

For maximum efficiency of vibrating screens, material should be evenly spread over the whole mesh width by a mechanical distributor (Fig. 28.5b). It is usually desirable to place a screen above the equipment or silos into which it feeds. A screen takes feed from one stream of material and splits it into a number of streams. By elevating the screen, one stream only is to be elevated whilst the fractions fall through chutes or into hoppers (Fig. 28.5c).

Space must be allowed for changing screen meshes (Fig. 28.5b) and lifting beams are used to facilitate this operation on larger installations. For speed of operation it is normal to have the replacement screen mesh already mounted on a spare frame. Access is necessary to maintain the vibrator mechanisms, and for mesh cleaning. The surrounding area should be easily washed down.

In laying out screening and associated equipment, it is imperative to lay out the hoppers and chutes first, since these will dictate the relative elevation of the equipment. Furthermore, since belt conveyors are generally associated with screens and grinding equipment, the maximum elevation of the equipment will determine the distance between the various sections, and the overall size of the plant. For each additional metre in elevation, a belt conveyor needs to be increased by approximately 3 m in length, so it is important to determine elevations in this kind of layout.

28.2.2   GRIZZLIES

Grizzlies are used primarily for scalping or making a high-capacity primary separation. Two types are used, the fixed-bar type and the live-roll type. There are no problems involved in the layout of grizzlies, the size being dependent on the throughput, and the incline to suit the material. For layout purposes fixed-bar screens should be inclined at 45° and live-roll at 20°.

Access is required for cleaning and design should take into account the impact of dumped loads often associated with this type of equipment (Fig. 28.6).

28.2.3   ROTARY SCREENS

This type has mesh fitted around a rotating cylindrical framework, which is mounted on a central spindle or on trunnions, and is angled to induce material flow (Fig. 28.7).

Mesh apertures can be varied along the length of the cylinder to obtain the required number of cuts, with oversize discharge from the open tail end.

Facilities and space should be provided for mesh changing. Safety enclosures are necessary for protection against the rotating cylinder, and against dust emission. Dust collection should be provided for dusty materials and drainage provided for dewatering processes.

Weir plates should be fitted at the feed end of the drum to avoid back spill,

Fig. 28.5 Vibrating screen installations (a) parallel installation (Courtesy: Babcock Minerals Engineering) (b) access requirements (c) sizing examples

(a)

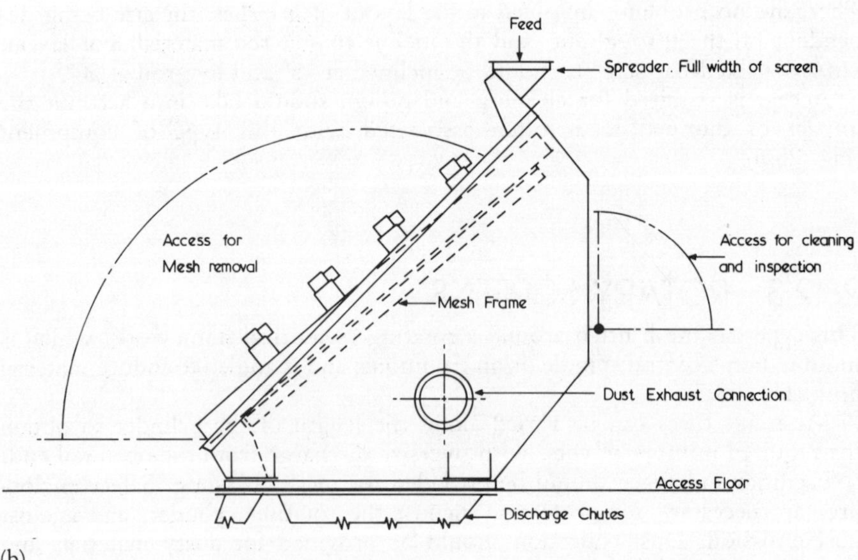

(b)

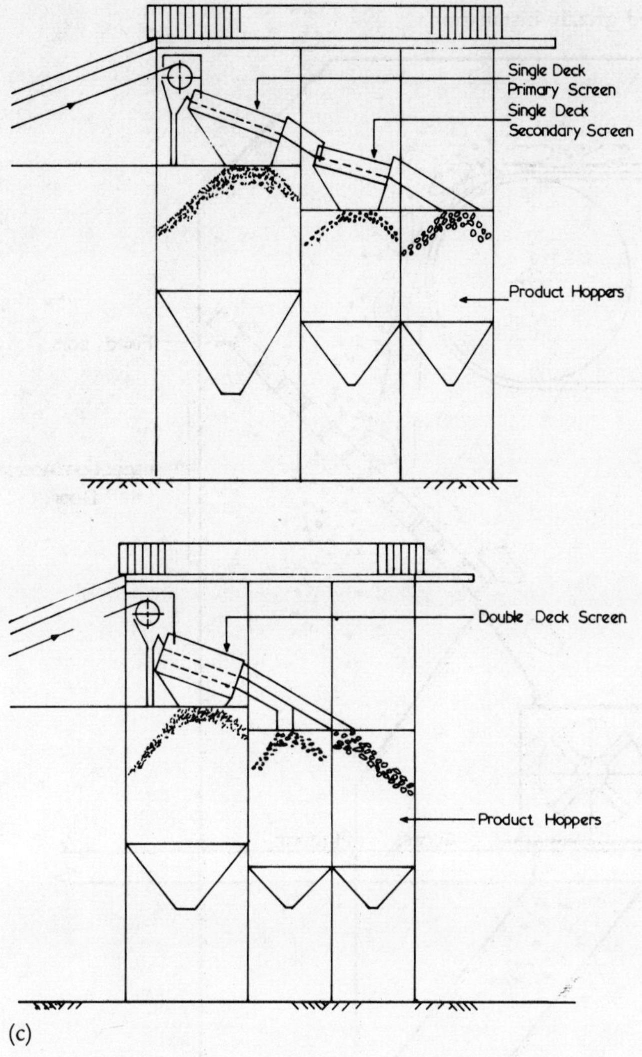

and discharge chutes extended to the drum surface, to avoid the contamination of separated fractions.

28.3   GAS/SOLID SEPARATORS

Removal of suspended solids from process and ventilation air is undertaken in three principal ways:

(a) Inertia separation.
(b) Mechanical separation.
(c) Scrubbing.

Fig. 28.6 Typical fixed grizzly installation

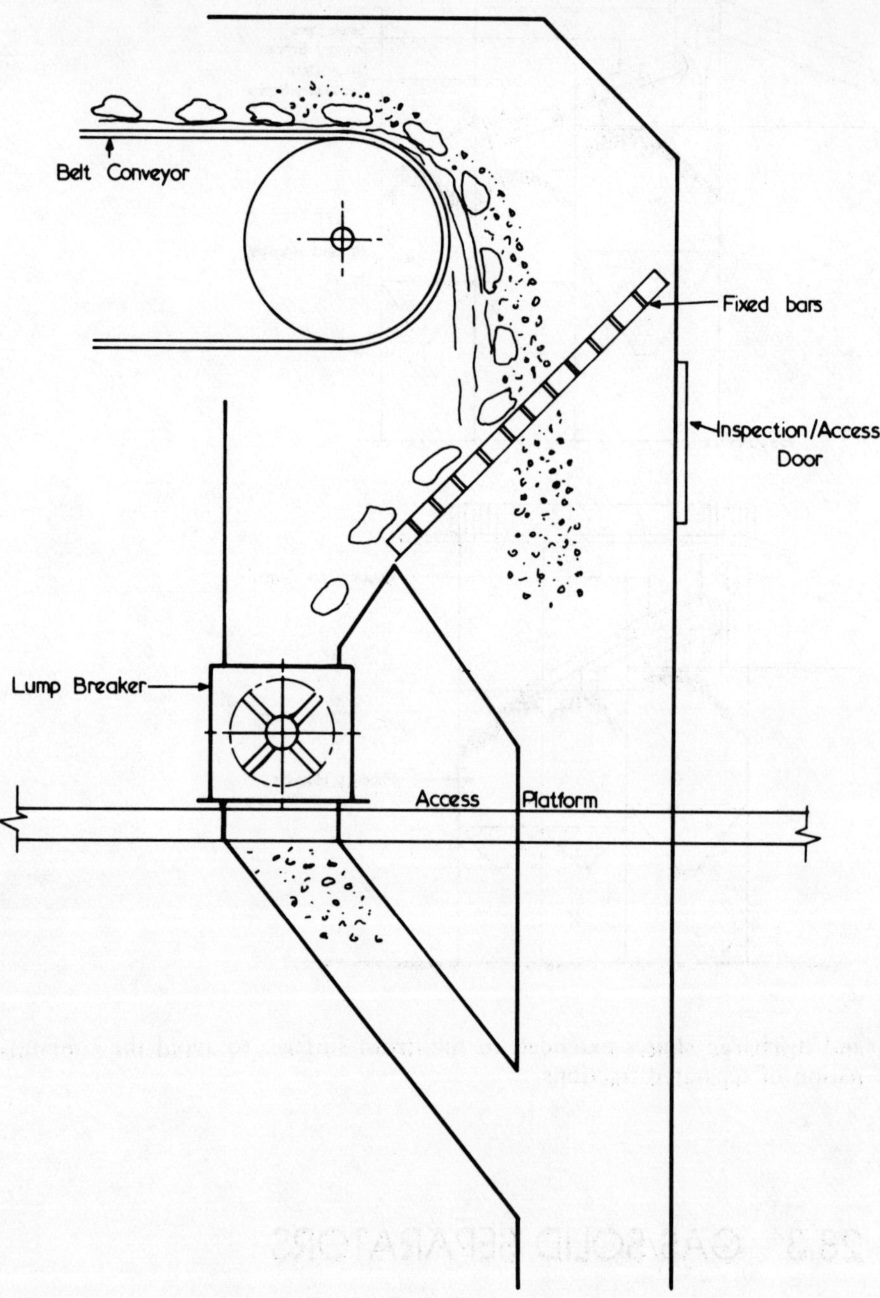

Layouts should take into account the duct runs to and from the separators, and recycling or disposal of collected solids. If there is an explosion hazard, separators should be isolated from the main process plant. Explosion venting and other protection measures should be taken as described in section 26.4.

Fig. 28.7   Rotary screen installation

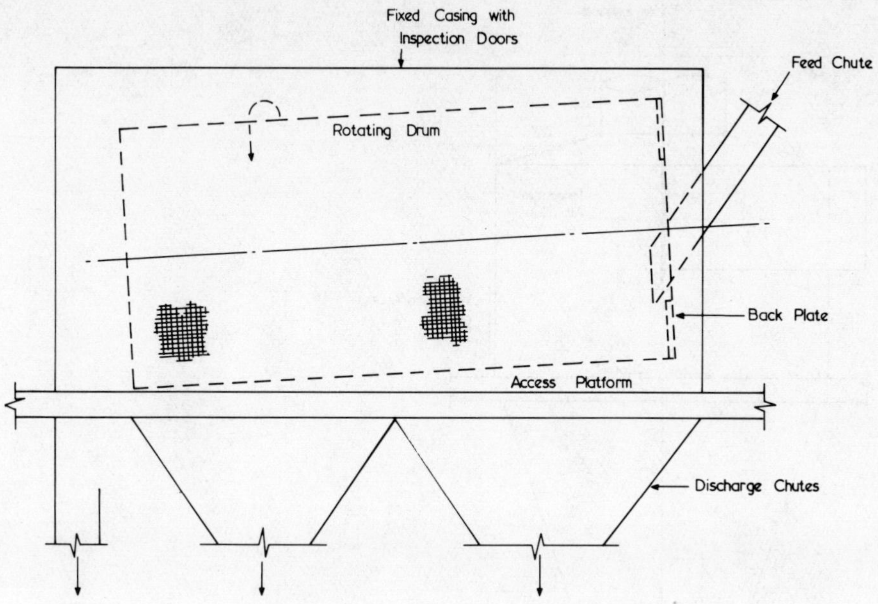

28.3.1   INERTIA SEPARATION

Cyclones are the most common type of inertia separator. These are relatively light and maintenance-free and can be located in any position convenient for duct runs and recycling of separated solids (see Figs 26.4 (p. 366), 26.5 (p. 368), 27.2 (p. 376) and 27.6 (p. 382)). Air seals should be provided at the solids discharge, which require access for maintenance, and cleaning (Fig. 28.8). A similar but simpler device is the gravity settling chamber for larger heavy particles.

There are mechanically driven centrifugal separators which are each shaped like a fan (section 23.4, p. 345) but with an additional dust outlet.

28.3.2   MECHANICAL SEPARATION

Mechanical separators are usually box-like units which contain internals such as filter bags (Fig. 28.9), filter pads, filter beds of granules, impingement obstacles or electrodes for electrostatic precipitation (Fig. 28.10). The cabinet or box can be designed for inside or outside use, and lagged if necessary. Its top should be load-bearing for maintenance and access.

Cabinets can have free-standing supports or be built-on to plant structures, and access is required for inspection, changing and maintenance of the internal elements and for cleaning. For electrostatic precipitators the access should be restricted because of the 30–40 kV electricity supply.

If the dust is hygroscopic, internal heaters may be required to keep the

Fig. 28.8 Cyclone installation

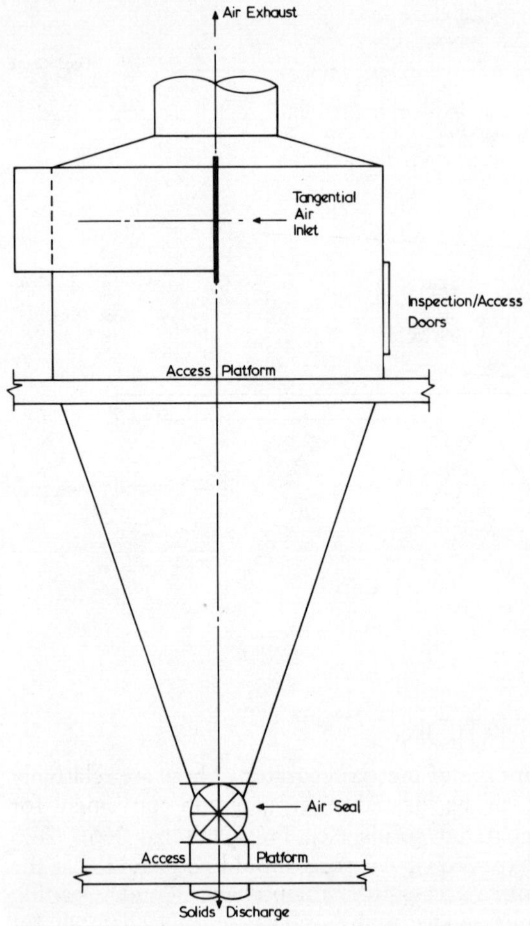

Fig. 28.9 Bag filter installations (a) small type (Courtesy: The Boots Company) (b) large pulse-jet type (Courtesy: Peabody Holmes)

(a)

(b)

elements dry when not in use. During operation, heating can be provided through the air-cleaning system (if installed) using air which should be filtered and oil-free.

Dust-hoppers should have inspection and cleaning doors, with bin vibrators for difficult materials and outlet air-seals. The size should be compatible with the method of removal of dust and frequency of emptying. If dust is recycled into a continuous process, space should be allowed in the hopper for separated solids to collect if the process plant is shut down but fan and filter allowed to run.

Fig. 28.10   Electrostatic precipitator for power station (Courtesy: Peabody Holmes)

Fig. 28.11   In-line gas filters (Courtesy: British Gas)

When the dust load is light the filter may be of the in-line type as in Fig. 28.11.

28.3.3   SCRUBBING

These take various forms such as the spray column and venturi. They should be laid out as plant vessels (Ch. 19). Scrubbers frequently produce a liquid and/or solid effluent and the choice of location should take this into account (see Ch. 13).

REFERENCE

1. *Pow Tech '83. I. Chem. E. Sym. Ser.* 69, 1983.

CONVEYORS

The choice of equipment for conveying loose bulk solids is determined by the type of material to be transported, rate of handling, and distance. Environmental considerations such as dust and noise must also be taken into account together with operating and maintenance costs.[1]

29.1 BELT CONVEYORS

These are the most common type of conveyor, being versatile, relatively cheap, and easy to maintain (Figs. 29.1–29.4).

Belting is available with rubber or plastic facings, and joints can be made by bonding or by use of belt fasteners. It is delivered on the reel, and space must be provided for installation and replacement. Belt widths range from 0.3 → 1.8 m. Proprietary cleaners are available for brushing or scraping the belt, and space must be provided in the discharge chutes for collecting cleanings.

Allowance must be made in the layout for tensioning (taking up) the belt (Fig. 29.1a). This is usually done with a tension screw at the tail end of the conveyor for belts up to 60 m long. For longer conveyors an automatic gravity weight method is used. This may be positioned at any convenient point

Fig. 29.1 Belt conveyor arrangements (a) typical troughed belt conveyor (b) typical chute detail (c) typical access arrangement to head end (Courtesy: Babcock–Moxey)

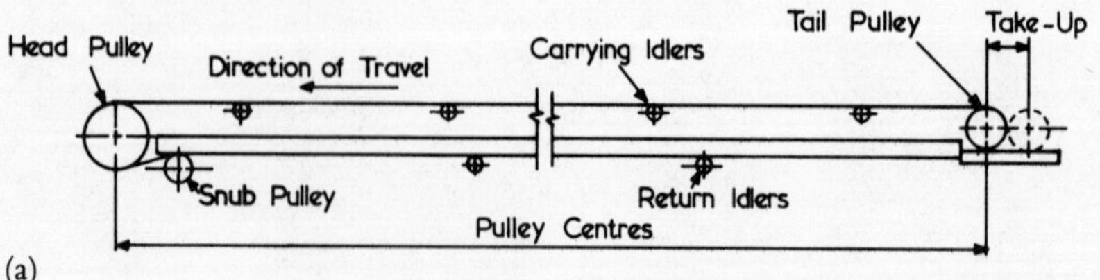

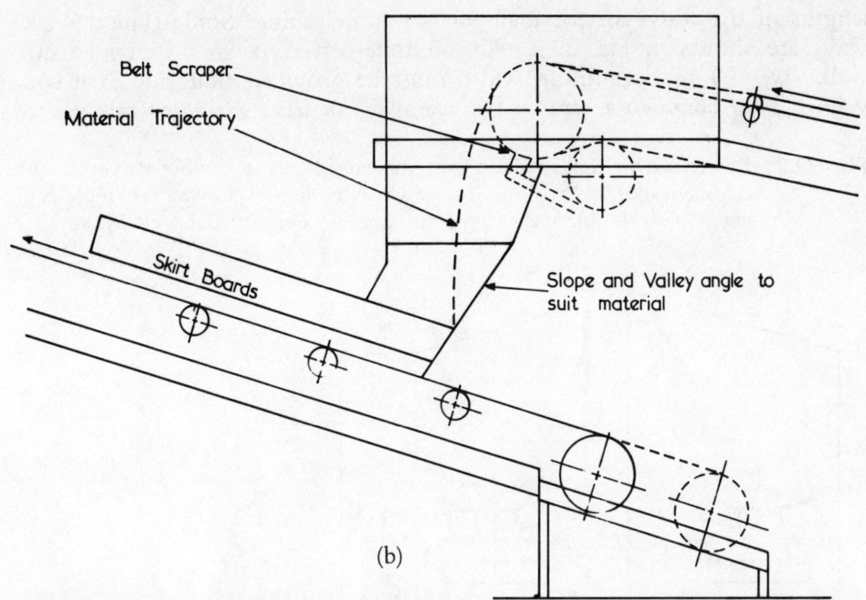

on the return strand of the conveyor (when the take up is vertical), or at the tail end (when the take up is horizontal like the screw). The amount of tensioning required can be taken as 2 per cent of the distance between the pulley centres.

Access is required around the head (Fig. 29.1c) and tail ends and along the lengths of the conveyor for maintenance and cleaning. Some typical walkways are shown in Fig. 29.2. For multiple-belt conveyors a maintenance walkway, 600 mm minimum width must be provided alongside each conveyor. Two conveyors arranged in parallel can have a single walkway of

Fig. 29.2 Belt conveyor walkway and gantry arrangements (a) double conveyor with central walkway (b) single conveyor with single walkway (c) single conveyor with double walkway (Photographs courtesy: Babcock-Moxey)

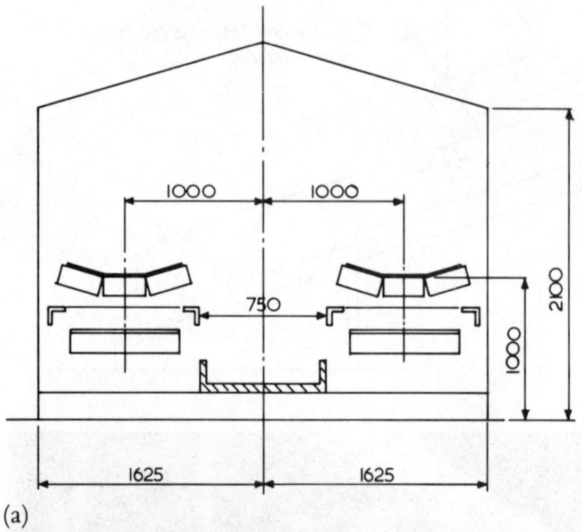

Fig. 29.2(b)

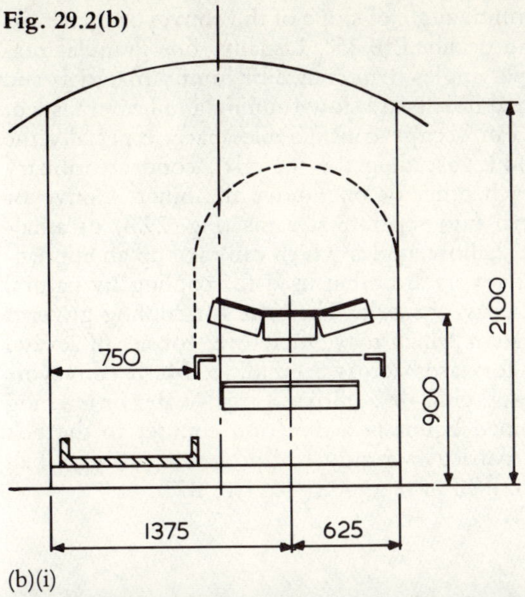

(b)(i)

(b)(ii)

750 mm between each of them. The walkway level should be approximately 750 mm below the belt line. When conveyors link process buildings or store at an elevated level (Fig. 29.3) supporting gantries are provided. These may be enclosed or open dependent upon the materials being conveyed (Fig. 29.2). Belt conveyors have a certain minimum distance at which they can be located

405

from the units due to the maximum angle of slope of the conveyor, generally inclined to angles up to 20° and declined to 15°. Usually, fine granular materials can be conveyed at steeper angles than course or lumpy materials and run-back of material is minimized if belts are slow running and heavily laden.

Multiple conveyor systems can occupy considerable space, especially the special conveyors which, without cascading the material, connect ordinary conveyors at right angles to each other or one above the other. Conveyor sections which split the material into separate streams (Fig. 29.4) or amalgamate streams usually require shallow angles which can take up an appreciable amount of room. Belt conveyors are often used for cooling by natural convection or by blowing air across the belt. The time for cooling governs the length of the conveying system which may, therefore, consist of several sections going backwards and forwards across a building. Mesh conveyors are used for processing such as spraying-on additives, wash-water or leaching liquors. Respraying of the drained liquor in a direction counter to the belt motion has been employed. A particular example of processing using a belt conveyor is the band dryer described in section 27.6.3. (p. 377).

29.2   PNEUMATIC CONVEYORS

These are suitable for powders or granular materials which are within a reasonably close size-range, and will not suffer from attrition or moisture pick-up.[2]

Two systems are used: air slide conveyors and pressure systems.

Fig. 29.2(c)

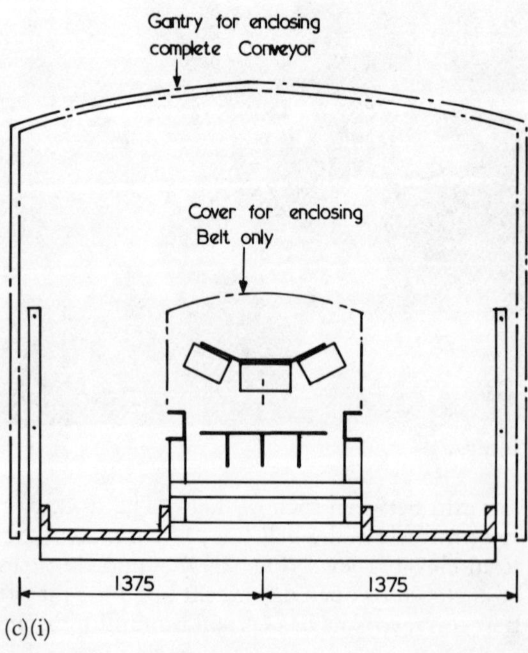

(c)(i)

(c)(ii)

29.2.1 AIR SLIDE CONVEYORS

Air slide conveyors are totally enclosed and can be declined at angles to suit specific design factors. The angle of declination is commonly about 5°. Material is fluidized by passing low-pressure air through a porous medium forming the bottom of the trough and flow is induced by sloping the trough. Directions can be changed and feed and discharges arranged anywhere along the length (Fig. 29.5).

A clean, dry, air supply will help to preserve the porous medium and fil-

Fig. 29.3 Conveyor layout between buildings (Courtesy: Humphreys & Glasgow)

tration should be provided for air leaving the conveying chamber.

Access is required to the blower unit and material diverters for maintenance and cleaning purposes.

29.2.2 PRESSURE SYSTEMS

Pressure systems are used to carry material in the airstream along pipelines (Fig. 29.6). These are totally enclosed, and single or multi-point delivery systems can be provided. Pipework can be routed at any angle, but long sweep bends are necessary to avoid blockage and wear. The system should be purged with clean air after use to avoid settlement. The air supply should be clean and dry, and conveying air must be separated from the material at the discharge point (Fig. 29.6a). Access requirements are similar to those for air slide conveyors.

A development of pressurized conveying is the dense phase system in which the solids flow along pipes in plugs. The material is fed via pressurized hoppers (Fig. 29.6b), often fluidized, which are filled in batches at atmosphere and discharged continuously under pressure. Thus, usually, there are dual hoppers, one filling, one discharging.

The layout of the pipeline is very important in avoiding fusing or breaking-up the plugs; in particular, the pressure drop must not be excessive and

Fig. 29.4 Travelling throw-off arrangements (a) example (Courtesy: Babcock-Moxey) (b) schematic

(a)

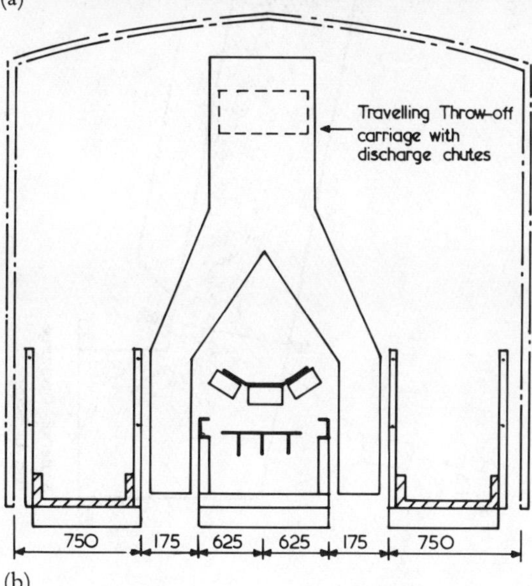

(b)

sharp bends must be avoided. There may be extra air-injection points along the line to maintain the plugs.

Cyclones, etc. are not required for separation as the plugs just drop out of the line into storage hoppers, bags, etc. For further details see Stadler,[3] Brown,[4] and Anon.[5]

Both air slide conveyors and pressurized systems can be used with inert gases instead of air to avoid the hazards of combustible powders and dust explosions.

409

Fig. 29.5 Typical air slide conveyor

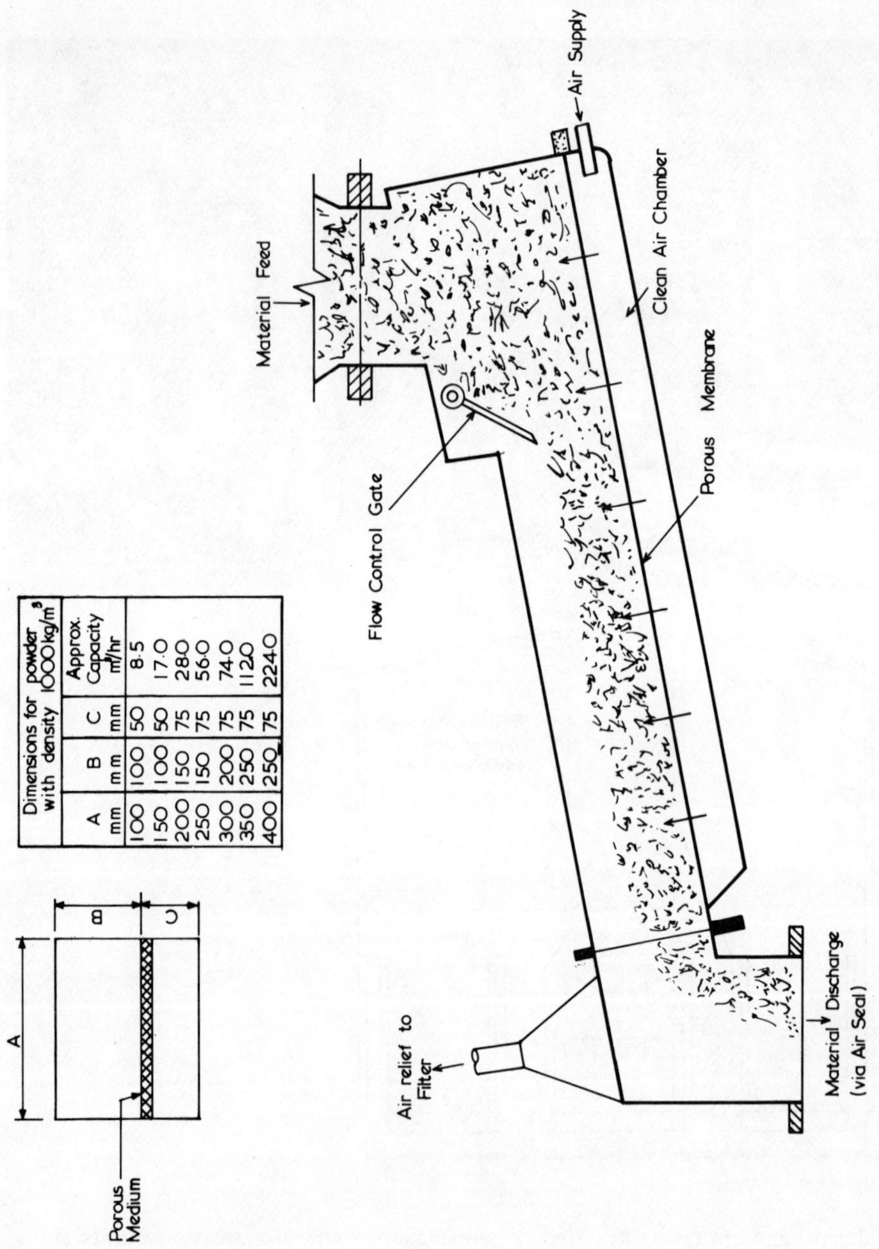

Dimensions for powder with density 1000 kg/m³			Approx. Capacity m³/hr
A mm	B mm	C mm	
100	100	50	8.5
150	100	50	17.0
200	150	75	28.0
250	150	75	56.0
300	200	75	74.0
350	250	75	112.0
400	250	75	224.0

410

Fig. 29.6 Pressure conveying systems (a) conventional system (b) dense phase conveying with pressurized hopper

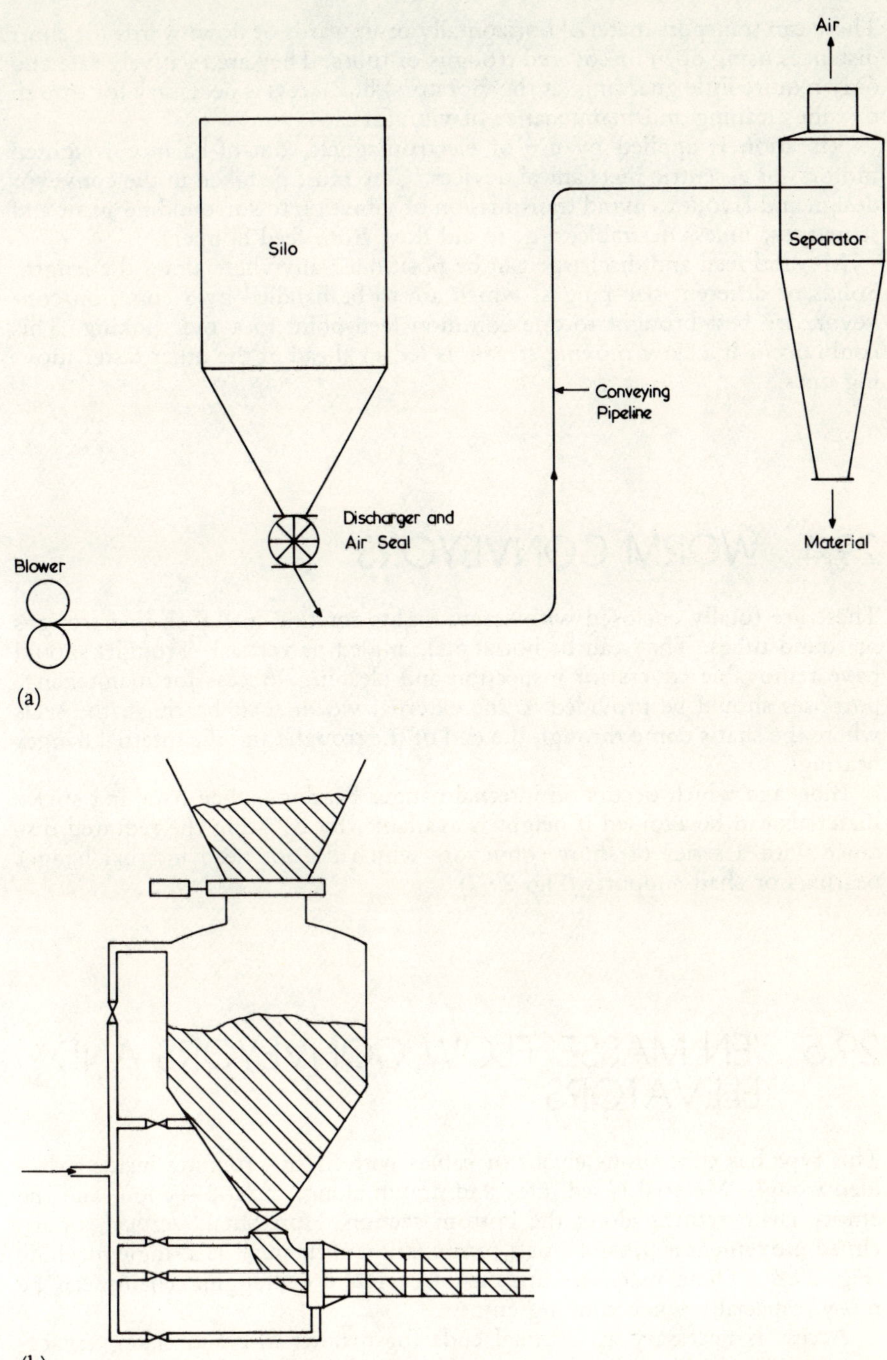

29.3 VIBRATORY CONVEYORS

These can transport material horizontally or upwards or downwards for short distances using open or covered troughs or tubes. They are relatively safe and only require little guarding (at the vibrators) but access is necessary for trough or tube cleaning and maintenance of vibrators.

Vibration is applied by use of electromagnets, out-of-balance weighted motors, or eccentric mechanical devices. Care must be taken in the conveyor design and layout to avoid transmission of vibration to surrounding plant and structures, unless desirable, e.g. to aid flow from feed hoppers.

Material feed and discharge can be positioned anywhere along the length. Solids of different size-ranges, which are to be handled by a common conveyor, are best brought to one common feed-point to avoid choking. This could occur if a slow moving stream is fed-in ahead of the other faster moving ones.

29.4 WORM CONVEYORS

These are totally enclosed with screw flights rotating in 'U'-shaped troughs or round tubes. They can be horizontal, angled or vertical. Troughs should have removable covers for inspection and cleaning. Access for maintenance purposes should be provided to the external worm shaft bearings, the seals where the shafts come through the end of the troughs and the internal hanger bearings.

Blockage which occurs on internal hanger bearings when handling sticky materials can be avoided if height is available, by covering the required distance with a series of short conveyors which do not need internal hanger bearings or shaft supports (Fig. 29.7).

29.5 'EN MASSE' FLOW CONVEYORS AND ELEVATORS

This type has continuous chains or cables with flights, running inside a divided trough. Material is fed into, and drawn along, the top section, and the empty chain returns along the bottom section. Horizontal, vertical, or inclined movement is possible, or a combination of them all as a single machine (Fig. 29.8). These machines are dust free but the dragging chains can be noisy, especially when running empty.

Access is necessary at terminal ends for maintenance and chain replacement, and at bends for replacement of wear plates if these are installed.

Fig. 29.7 Worm conveyor layouts

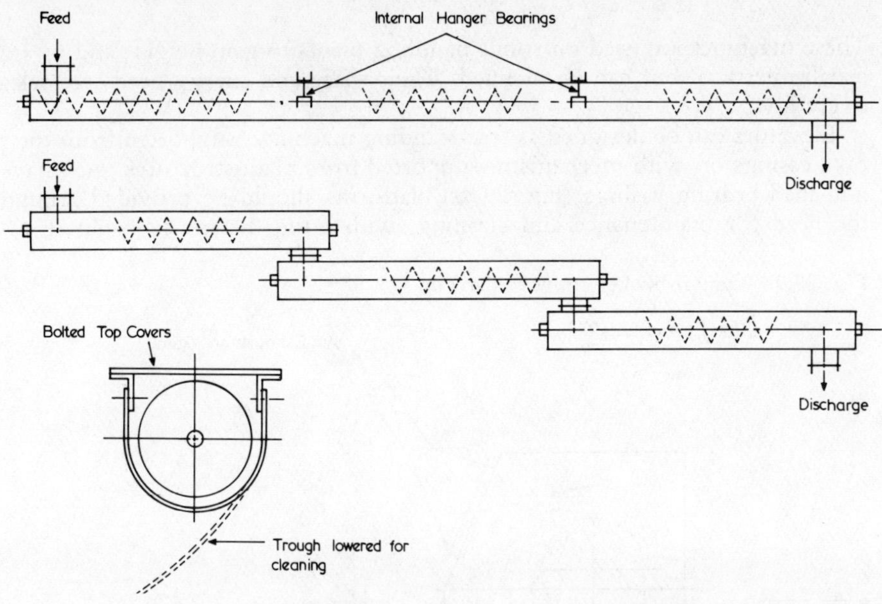

Fig. 29.8 'En Masse' chain conveyor/elevator (Courtesy: Redler Conveyors Ltd)

29.6 BUCKET ELEVATORS

These machines are used on solids handling plants to gain height, and so are usually vertical, but can be inclined. Their space and energy needs are relatively low.

Elevators can be designed as free-standing machines, supported from their own casings or, with mechanisms supported from plant structures, may have non-load-bearing casings (Fig. 29.9). Platforms should be provided around the head for maintenance and cleaning, with lifting beams to facilitate re-

Fig. 29.9 Typical bucket conveyor layout

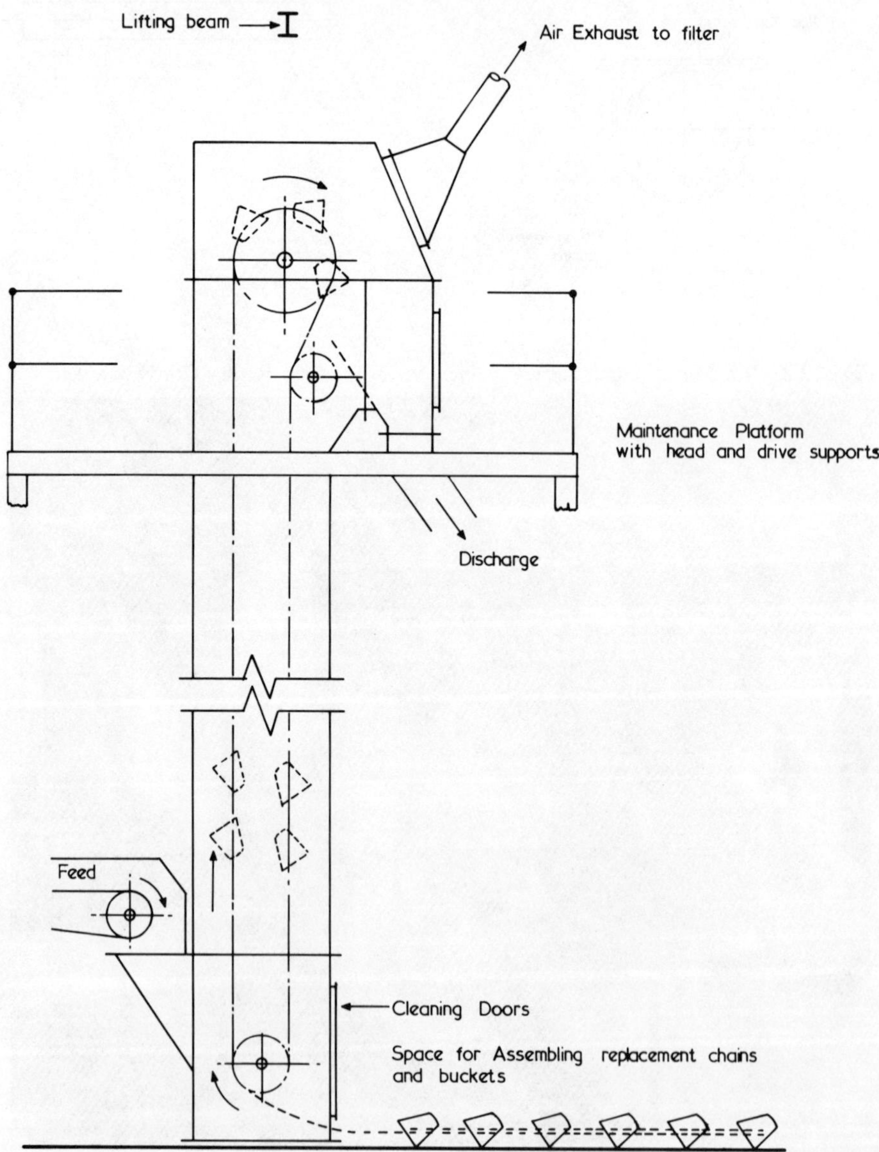

chaining. Space should be allowed on the floor for assembling replacement chains and buckets.

Economic layout of associated plant could require an elevator boot to be located in a pit, but this is a false economy if consequent lack of attention results in low utilization of the equipment, due to accumulation of spillage in the pit.

Material fed to the centre of buckets will prolong belt or chain life particularly if double strands of chain are used.

29.7   OTHER CONVEYORS

Skip hoists are suitable for lifting measured and unmeasured batches. The lifting well should be guarded and access provided for winch maintenance.

Aerial ropeways can be used in transporting difficult materials over long distances such as in the metallurgical industries.

REFERENCES

1. *Pow Tech '83. I. Chem. E. Sym. Ser.*, **69**, 1983.
2. EEUA Handbook No. 15, *Pneumatic Handling of Powdered Materials (including fluidization)*. Engineering Equipment Users Association, Constable, London, 1963.
3. Stadler, H. 'Dense phase pneumatic conveying', *S. Afr. Mech. Eng.*, **26**, (4), 168, 1976.
4. Brown, R. W. 'Dense phase pneumatic conveying', *BHRA*, **A4**, 33, 1977.
5. Anon., 'Dense phase conveying', *Bulk*, **5**, (2), 41, 1978; 6, (2), 55, 1979.

FILLING AND PACKAGING

30.1 INTRODUCTION

The size of container varies from bottles containing 10 g in the perfumery industry to 25–50 kg sacks of fertilizer or even intermediate bulk containers (IBCs) of 2 t.

The types of containers include bottles, packets, sacks or drums and they can be made from glass, metal, plastic, paper, etc. as appropriate.

Outside the container is the label, probably a display wrapping and the over-wrapping which provides protection during transportation. The over-wrapper holds several containers and a number of over-wrappers are usually put together on a pallet for transportation.

Filling and packaging operations are carried out on arrangements varying from simple belts with manual operations to highly mechanized, automated installations. Figures 30.1 and 30.2 show the operations on typical packaging lines. Whilst the high-speed, automatic lines usually require few operatives and semi-automatic and manual lines are slower and more labour intensive, the faster line is not necessarily the most economical in all situations.

Scale and type of equipment will be dictated by:

(a) The type of product to be filled and packed is a fundamental factor, i.e. whether it is toxic or non-toxic with the possible need to protect the operator (e.g. filling and packing hormonal products, anaesthetics, etc); whether it is to be filled under sterile or clean conditions and the need to protect the product from the operator, cross-contamination from other products and the environment.
(b) The location of the plant and any special constraints (e.g. different ambient conditions and laws) of the country concerned.
(c) Size and style of package and optimum production rate.
(d) Batch size or throughput.
(e) Phase and flow properties of the materials.
(f) Packaging quantity.
(g) Finished stock investment.
(h) Capital cost.
(i) The organization of the production department, i.e. whether responsibilities are grouped on a product or operation basis.
(j) The degree of automation finally decided upon – possible use of robotics.

Fig. 30.1 Filling and packaging equipment and operation

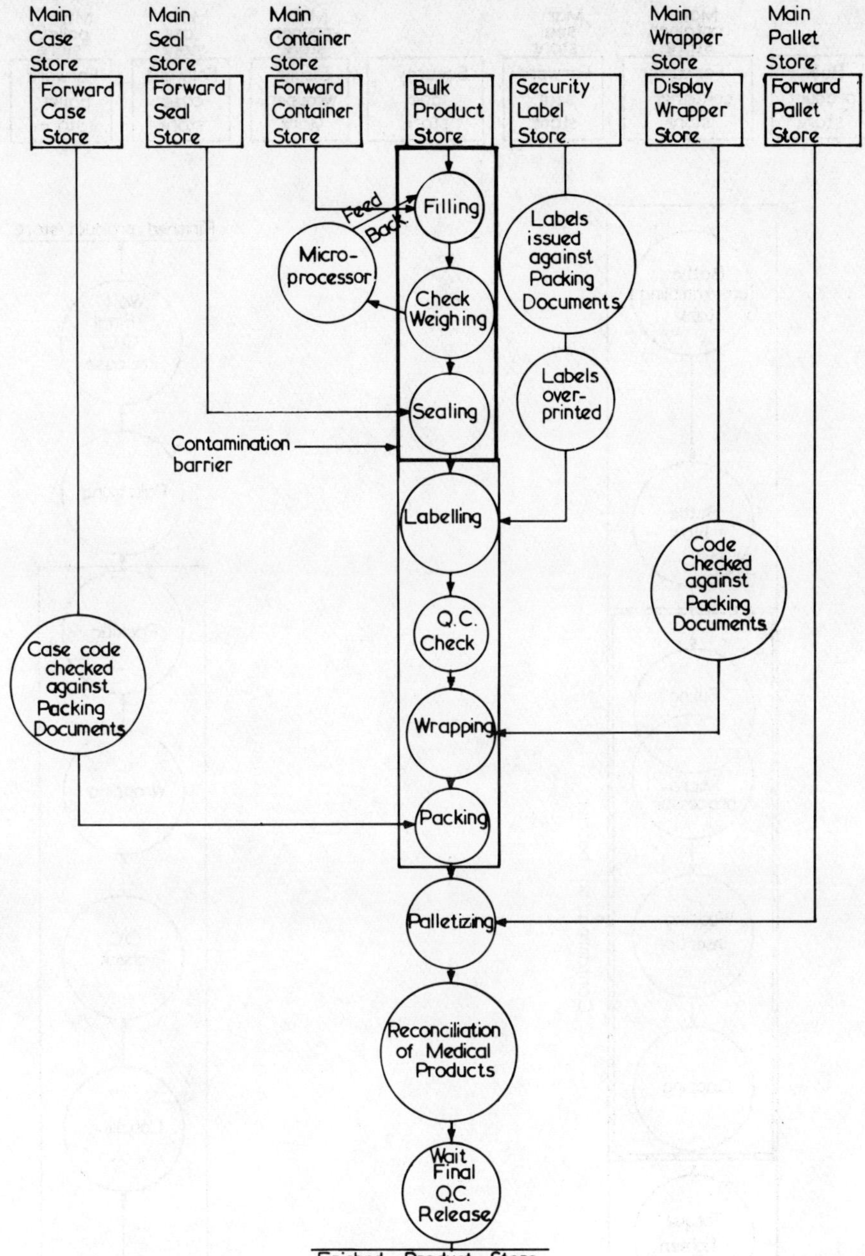

(k) The shift-working pattern, i.e. day-shift for five days, two-shift for five days or continuous shifts.

(l) Labour requirement and availability.

(m) Equipment changeover time.

(n) Reliability of demand forecasts.

Fig. 30.2   Typical 'U'-line-tablet packing belt

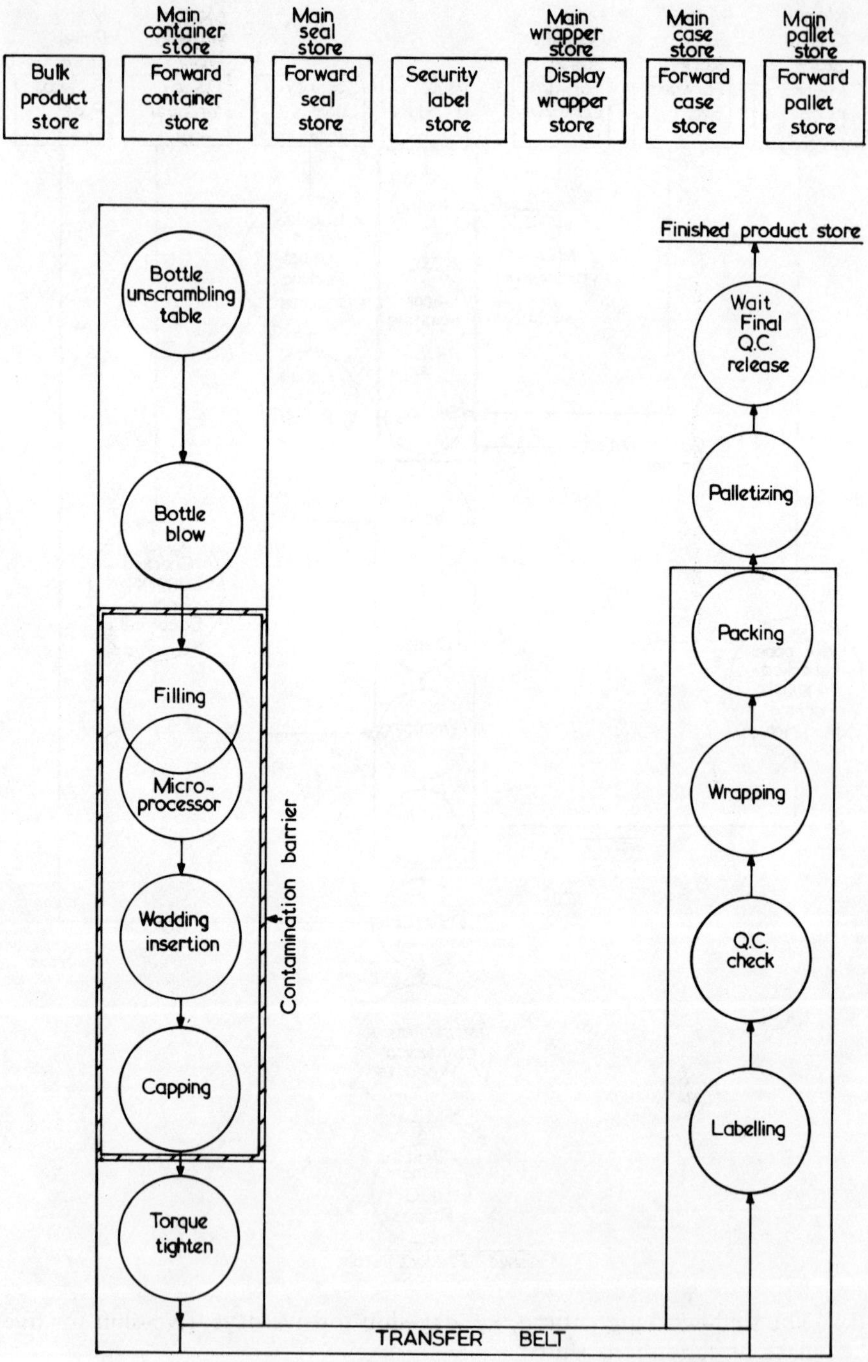

30.2 LINE LAYOUT

It is rare that a filling and packaging plant can be bought as a complete unit. Usually the layout engineer has to deal with a number of manufacturers for such operations such as weighing, dispensing, sealing, labelling, wrapping, conveying, etc. plus flow promoting devices for troublesome solids. Microprocessors and robotics are being increasingly used for control, data acquisition, documentation, etc. It is the engineer's task to see that the items are compatible with each other and with the objective of the plant. To this end the layout engineer should consult with operation and maintenance staff and should comply with the requirements of the regulatory authorities such as those for safety, fire and quality control.

The requirements of hygiene and product security are paramount in the pharmaceutical industry and short product runs with a consequent large number of changeovers are quite common. The design of equipment and layout with good access for cleaning and removal of spilled or damaged product, containers or labels is essential and should comply with the requirements of good practice as proposed by the appropriate regulatory body. Equally good design to facilitate changeover and safe adjustment of equipment between product changeovers is important.

The packaging machinery industry is one of rapidly changing technology and the style of package marketing oriented with 'fashion' plays an important role. Thus the layout of packaging areas needs to be flexible to accommodate rapid change in equipment and production patterns. The design of services to packaging lines (i.e. electricity, compressed air, vacuum, water, steam, drainage, air conditioning or ventilation and heating, etc.) needs careful consideration, i.e. underfloor ducts, overhead service rails or service bollards with plug-in connections. The needs of cleanliness, flexibility and safety should be fully taken care of (e.g. appropriate flooring, ceiling and partitioning, flame-proofing, pest control, sound-proofing, fume extraction, etc.) Some of these points are illustrated in Fig. 30.3.

Fig. 30.3 Example of a line layout (Courtesy: The Boots Company)

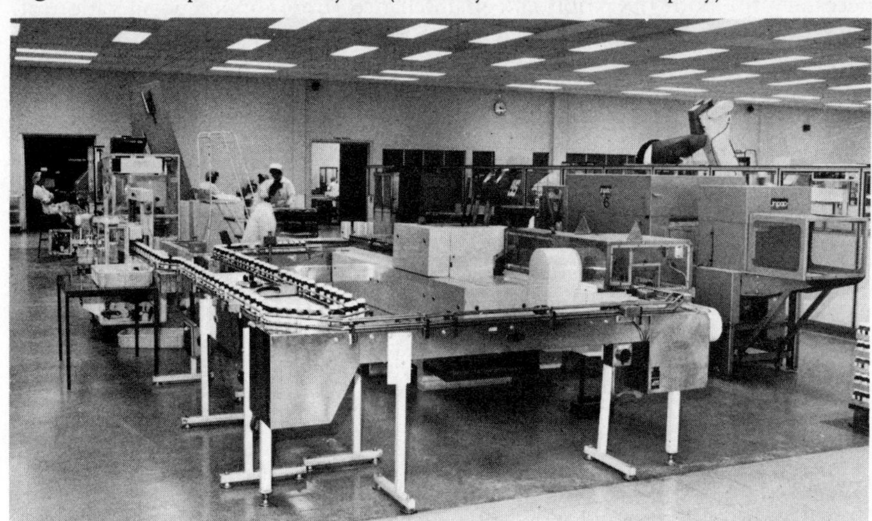

Fig. 30.4 Segregated packaging line (Courtesy: The Boots Company)

Where lines are operating adjacent to each other, they should be restricted to a product group, i.e. food, internals, medicinals, etc. and these lines should be segregated, to prevent cross-contamination as shown in Fig. 30.4.

Products such as powders – which can cause airborne contamination – should be fully screened, even within the same product group, whilst products such as liquids and tablets would, in the main, be adequately segregated by a 2 m-high screen. Ideally, screening should allow for 5 m centres approximately for straight lines, widening where 'C' or 'U'-lines are required.

Layout of packaging lines must take into account the needs of the labour necessary for both operation and maintenance. Ergonomic consideration of the working area, height of conveyors, seating (see Fig. 17.9 p. 261), etc. must be given, to minimize fatigue and, hence, risk of errors.

Location and layout of machines and controls also require study. On automated lines, use of buffering systems such as rotary or moving belt tables should be considered between machines so that the line operation can continue if one machine is out of service. It is also advisable to be able to alter machine speeds by 5 per cent along the line to avoid build-ups.

30.3 PROCESS AND PACKAGING RECONCILIATION

When the packaging plant is required to be directly in line with its process plant, bulk buffer storage is usually introduced between process and packaging to safeguard continuity of process operations if packaging is interrupted. Product may always go to the storage or only be diverted when there is interruption. The packaging rate should then be in excess of the process rate in order to recover after the interruption. The type of bulk storage may be static, e.g. hoppers and tanks or the product may be transported to the packaging line in intermediate bulk containers which can also serve as temporary storage. The latter scheme is useful if batch identity is to be preserved. For some sensitive materials, storage should be avoided if at all possible.

Where a high packaging rate can be achieved relative to process production rates, de-coupling of the two operations will reduce operation costs. This is particularly important where packaging and subsequent warehousing is labour intensive, and can be condensed to one-shift packaging of continuously made products (Fig. 26.2, p. 364). The packaging plant becomes available for daily maintenance or cleaning or for switching to different products.

These de-coupled operations require bulk storage in tanks, silos or bulk warehouses prior to packaging. Advantage should be taken of the residence time to apply quality control techniques, thus avoiding wasted packaging. Product recovered from bulk storage should be screened immediately prior to weighing or proportioning, to remove trash acquired during storage, breakdown or otherwise.

Provision must be made for receipt and storage of containers, seals, labels, etc. with means of selection and transfer to the filling zone. In order to achieve a balanced flow of these components and the process material, use of local stores should be considered. A supply route direct from main stores can result in either line stoppages while awaiting components, or excess of materials stored local to the packaging line. Forward stores should be positioned near to, but not at, the packaging line site. If forward stores are not possible, then adequate area must be allowed local to the packaging line for buffer storage of components. A service accessway down the side of each line is required for feeding-in the components to the appropriate station on the packaging line. Consideration needs to be given to whether this is for hand pallet trucks or fork-lift trucks or other automated means (e.g. conveyor or chute).

30.4 FILLING EQUIPMENT

Operators and equipment at the filling point should be protected by the inclusion, if necessary, of dust and spillage removal facilities. Filling rooms should be dry, clean and insulated against plant noise. Weighing and proportioning equipment should be kept in a warm and dry environment, free from vibration. Many types of machine are available, each designed to supply a measured quantity taking into account the flow properties of varying bulk materials and required speeds. Examples of drum- and bag-filling layout are given in Figs. 30.5 and 30.6.

Great care should be taken when selecting these often expensive machines, to ensure that they complement the specifications of the intended containers. With an acceptance quota level (AQL) of 1 per cent commonly specified by

Fig. 30.5 Paper bag filling layout (Courtesy: Shell Chemicals (UK))

container manufacturers the filling line could receive 1 in 100 defective containers.

The choice of container, whether it be glass, metal, plastic or paper can greatly affect the noise generated by the operating packing line.

Fig. 30.6 Drum filling layout (Courtesy: Shell Chemicals (UK))

30.5 LABELLING

The importance of ensuring that all products are accurately labelled cannot be over-stressed and label security should be a feature of the layout of the packing system.

Where on-line labelling is employed a high degree of operator discipline is required and if a packed product reconciliation is required to ensure an accurate balance of input and finished packs, the same reconciliation may be required for labels. Good ergonomic layout will help in this.

Labels supplied in packs, on reels or pre-printed on the container can be checked individually by running through or passing by verification machines which use light beams and light pens to identify pre-printed identifying blocks or bar codes. Care, however, must be taken in the layout to aid avoidance of mix-ups after verification.

Batch numbers plus other information such as date of expiry needs to be added, ideally at the time of labelling using on-line printing.

30.6 WRAPPING AND PALLETIZING

There are many considerations which govern the requirements for a finished pack in order to store and distribute given quantities, giving it adequate protection, whilst keeping costs to a minimum. These include:

(a) The nature of the product.
(b) The product container.
(c) The average quantity ordered by customers.
(d) Strong packs for stacking (neat stacking saves warehouse space and reduces damage).
(e) Whether or not the pack outer is intended for use as an advertising display medium.

Finished packs will vary from the simple brown paper parcel to boxes, or shrink-wrapped packs. In some instances, there is no further embellishment of the sack or drum container, particularly where the product is going to another industrial user.

The choice of pack will affect the workplace layout and the length of the line. Shrink-wrapping involves the use of bulky equipment and hence increases the space required.

The forming of packages into unit loads for economic handling storage and transport is also a function of packaging. Units are normally made up onto pallets for fork-lift truck handling, either by hand or by fully automatic palletizing machines. Units can be shrink- or stretch-wrapped with polythene film for stability. Shrink-wrapping techniques can be applied to make-up unit loads which are then completely waterproof and can be handled by fork-lift trucks without the use of pallets (Fig. 30.7).

Conveyors handling packages from the filling plant to palletizing station should have an accumulator magazine (generally made up from roller track) which allows for the intermittent operation of the palletizer whilst maintaining continuous operation of the packaging plant.

With a fork-lift truck warehouse distribution system, an accumulator conveyor should be provided for unit loads when dispensed from the palletizer to allow for intermittent truck operation. Space must be allowed for trucks to collect these loads, to return empty pallets to the palletizer and to handle pallet-covering materials.

Figure 30.8 illustrates a typical palletizer system.

Some operations permit direct dispatch from the packaging plant to road or rail vehicles as an alternative, or in addition to warehousing. In these in-

Fig. 30.7 Palletless unit loads (Courtesy: Moellers (UK))

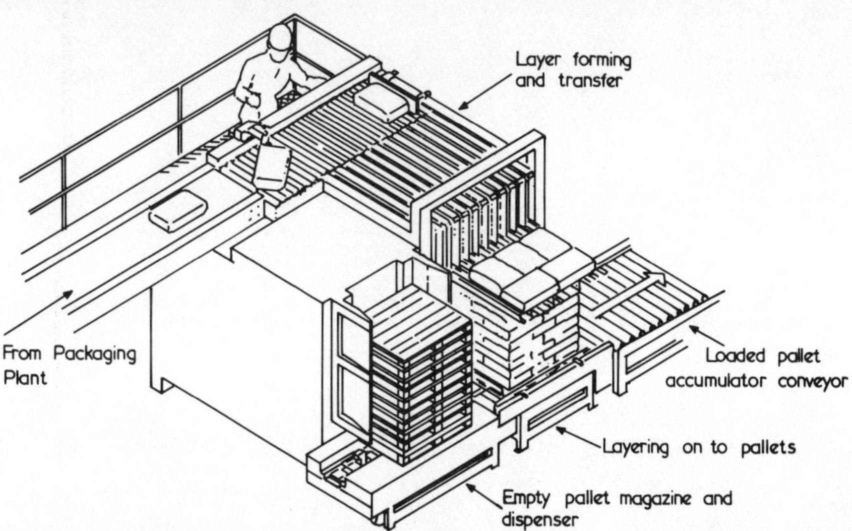

Fig. 30.8 Typical automatic palletizer

stances, high-level plants enable packages to be delivered directly to the vehicle by roller track or inclined chute. Plant capacities should be related also to traffic movement, to avoid vehicle congestion and delay.

Warehousing is discussed in Chapter 12.

30.7   PRODUCT CHECKING

Techniques for securing the quality of products are improving constantly and becoming more sophisticated. In the medical industry, for example, the health authorities are demanding higher standards every year. Checking quantity and quality of all materials used should be carried out by line supervision and on-line quality assurance inspectors throughout a packing run, and it may be necessary to allow space for a quality assurance station. Where ethical medical products are concerned, a packed product reconciliation may be necessary, to ensure an accurate balance of input and finished pack.

If quality assurance demands that pallets of finished goods are not transferred to warehouse until they are given final release, appropriate space at the end of the line or in the packing hall will be necessary.

The quarantining of packed stock awaiting clearance for final issue can be a major problem in a large production organization. The use of efficient storage arrangements such as an automated warehouse for finished packed stock can give great benefits here as the control documentation necessary to remove the stock from the warehouse can be held by the quality assurance department until final clearance is given. In this way a separate quarantine storage area is not required.

PIPING

31.1 GENERAL CONSIDERATIONS

Piping has a major influence on the design of a plant since it can constitute up to about 30 per cent of the costs. Provision should be made for a simple, neat and economical layout allowing for easy supporting and adequate flexibility. Careful routeing can obviate the use of expansion loops and bellows to take up thermal expansion and so save on fitting, fabrication and erection costs. Due regard should be given to erection, insulation, maintenance, safety, operation, etc. Grouping of pipelines where possible reduces the number of structural supports, limits space isolation and improves access. Cost comparisons ensure economic choice of valves, fittings, etc. and a careful study of the flow diagrams often reveals unnecessary valves, over-sized and over-specified pipelines.

31.2 SELECTION OF PIPING

The design, materials and fabrication of piping systems must comply with codes of practice applicable to the particular plant and country. There are several well-established national codes and ones widely used are the American code for pressure piping, ANSI B 31.3[1] and the British equivalent, BS 3351.[2] Piping connected to power boilers or unfired steam generating equipment should be designed in accordance with a code such as section 1 of the ASME Code[3] and for other piping within boiler houses, section 1 of ANSI B 31.1 is applicable.[4] Reference should be made to these codes for interpretation of design criteria; permissible variations in pressure/temperature conditions; material selections and stress data; pipe wall calculations; piping flexibility; fabrication and non-destructive testing.

   The selection of piping materials considers both performance and economics in relation to the nature of the fluid conveyed (e.g. whether it is corrosive or not) and its design pressure and temperature. The atmospheric surroundings, the soil and, occasionally, the quantity of fluid, can be equally important. Pipe routeing should take account of the specified materials, accommodating thermal movement at reasonable cost.

The nominal wall thickness of a pipe is determined generally from allowable hoop stress levels. Reference is made to the applicable code to find the:

(a) Minimum design thickness for the operating pressure and temperature.
(b) Corrosion and erosion allowance.
(c) Manufacturing tolerance.

The corrosion allowance (b) may also be fixed by the special conditions appertaining in a particular project. The additive thickness thus obtained is compared with the nearest thicker pipe wall commercially available, except where the need for special alloys justifies non-standard pipework. Otherwise, the use of the 'non-preferred' nominal pipe sizes (in mm) DN 10, 32, 65, 90, 112, 125, 175 should be avoided except where required to connect to equipment or to achieve a critical fluid velocity. Normally no piperun should be smaller than DN 15, except for instrument piping. Pulled bends are commonly used for steel in pipe sizes up to DN 100 at a minimum radius of 5 times the nominal pipe diameter. However, modern methods enable tighter bends to be pulled on piping up to DN 600. Consideration should be given to the thinning of the pipe wall during the bending process in choosing the required wall thickness. To save space, for pipe sizes above DN 150 welding elbows are generally used with a radius equal to 1.5 times the nominal pipe diameter. Mitre elbows are sometimes employed, particularly in the larger sizes. These are fabricated from pipe with one or more mitre cuts. A single cut elbow should only be used for low-pressure vent piping or piping working near atmospheric pressure.

Branch connections can be made by forged fittings available from manufacturers, by suitable cutting and welding the two pipes or, in some cases, the use of special fabrication techniques, to forge a branch in the parent pipe. The type of branch does not usually have any direct effect on layout although access is needed for installation (and for any fabrication in position).

In general, the use of flanges is limited to connections at flanged equipment and valves. Cast iron and similar special piping sections are joined by flanges and for sizes of DN 50 and above require careful space and support considerations. Flanges may also be essential to cater for:

(a) Frequent dismantling for cleaning.
(b) Field assembly of shop fabricated piping requiring shop heat treatment.
(c) Piping which cannot be field welded.
(d) Dismantling of equipment such as reactor heads, compressors, etc.
(e) Frequent rodding out or inspection.

Flanges should not be put over roads, walkways and cable trays because of the risk of leak. The choice of flange is governed by allowable pressure/temperature ratings and in respect of the American ANSI B 16.5,[5] and the British counterpart BS 1560,[6] are classified as 125, 150, 300, 600, 900, 1,500 and 2,500 lbs. These classifications are pressure/temperature related for ease of selection to the specific pressure/temperature application. Metric rated flanges can be specified to BS 4504[7] but this standard has not received universal acceptance by all manufacturers of flanged components. New standards being developed are intended for a new international design code incorporating both systems.

31.3   PIPE STRESSING

31.3.1   GENERAL

A piping layout should provide adequately for expansion and contraction due to temperature changes, without placing excessive stresses on piping materials, or excess forces and moments on equipment or anchors. Each time a plant is put into or taken out of service a stress cycle occurs. If repeated often enough a failure could occur due to fatigue stress. The layout of piping which will be subjected to significant temperature changes is frequently modified after flexibility analysis has revealed excessive stresses which could cause loss of containment and unacceptable risk of hazard.

Some equipment types readily satisfy process requirements within acceptable pipe stress limitations. Other types of equipment require careful consideration. These include:

(a)   Rotating machinery such as pumps, centrifugal compressors, steam and gas turbines, fans and blowers.
(b)   Reciprocating compressors and pumps.
(c)   Heaters, dryers and furnaces, particularly startup heaters.
(d)   Aircooled exchangers (fin fans).
(e)   Cold box refrigeration systems where aluminium is employed.
(f)   Brick or refractory lined thin-walled vessels or pipe.
(g)   Cast-iron equipment of any kind, including valves.
(h)   Glass-lined vessels and pipe.
(i)   Shell and tube exchangers with a bellows in the shell.
(j)   Glass reinforced plastic vessels.
(k)   Graphite heat exchangers and vessels.
(l)   Vessels on load cells.
(m)   Storage tanks over 30 m diameter.
(n)   Steam jacketed piping.
(o)   Safety and relief valve discharge pipes.

In particular, there are usually stressing problems with the following equipment.

Pumps

The casing of a centrifugal pump is usually much stronger than the pipe connecting to it. However, quite moderate pipe loadings (in terms of allowable pipe stress) can produce casing strain causing sufficient misalignment between pump and driver shafts to bring about premature bearing failure. A significant feature of process pump piping is that it is generally one or two sizes larger than the pump nozzle size so that the pipe leg becomes a comparatively rigid lever acting on the pump.

Since pumps are generally mass produced in a competitive market, a high power/weight ratio ensues and this results in low acceptable forces and movement levels on nozzles. Centrifugal pumps can accept reasonable piping stress loads but positive displacement pumps cannot because the smaller internal clearance makes casing distortion unacceptable. The American Petroleum Institute standard API 610 contains guidelines which are usually acceptable for nozzle loadings.[8]

Turbines

On industrial steam turbines for a given nozzle size allowable forces and moments are considerably lower than those for pumps. The limiting formulae applied to turbines can be found in the National Electrical Manufacturers Association (NEMA) standards.[9] These apply to American equipment and their acceptance of use needs to be clarified at the start of a project.

Compressors

Large centrifugal compressors can present difficult stress analysis problems. Good alignment of piping at nozzles is essential in view of their size coupled with adequate pipe support in the vicinity of the compressor. It is not sufficient to design to code allowable pipe stress levels because the associated loading is significantly higher than the load limitations for the compressor nozzle. In general, large installations require a full stress analysis.

Fired heaters

The high metal temperatures experienced with fired heaters severely limit the allowable stresses, forces and moments. Restrictions are applied to piping nozzles, translations and rotations to preserve the design clearance between the refractory lining and the tubes, and also to suit the design of the gas seals.

Heat exchangers

The design of shell and tube heat exchangers is governed by the ASME Code[10] and designed to Tubular Exchanger Manufacturers Association (TEMA) standards.[11] The manufacturers limit the stress values in the exchanger shell local to the nozzles, which governs the allowable forces and moments from piping.

Piping

Thermal movement of pipe or equipment is generally accommodated and/or controlled by use of one or more of the following methods.

(a) Careful positioning of equipment taking advantage of the inherent flexibility of pipework.
(b) Decreased pipe wall thickness to reduce thermal forces and moments.
(c) A designed expansion loop or offset legs.
(d) Expansion bellow units.
(e) An expansion joint.
(f) Accommodating the stresses by control of expansion direction via supports, guides, anchors, piles, etc.

   Expansion loops can require considerable space (see Figs. 31.1, 31.2, 4.2 (p. 30) and 16.5 (p. 228)).

   When sliding expansion joints are used special consideration must be given to axial thrust effects. They will only work satisfactorily within strict tolerances for axial alignment, lateral deflection, angular rotation and system pressure/temperature conditions. Systems in which sliding expansion joints

Fig. 31.1 Steam superheaters showing expansion loops (Courtesy: Babcock Woodall-Duckham)

are installed are controlled by rigid anchors, guides, supports or stops positioned such that the joint operates within its designed limits.

Pipe supports

The function of pipe supports is to prevent the following:

(a) Exceeding allowable pipe stresses.

Fig. 31.2 Expansion loops for a rack (Courtesy: The Boots Company)

(b) Leakage at joints.
(c) Excessive thrusts and moments on connected equipment.
(d) Resonance, when the piping is subject to vibrations.
(e) Piping distortion or sag which could impair its operation.

Pipes of size DN 50 and below are usually supported at site by the erecting contractor without any special design. Those of size DN 80 and above conform where applicable to the requirements of the national codes of practice. If DN 65 is employed either scheme can be used depending on piping complexity. Pipe supports normally consist of shoes, clamps, 'U'-bolts, and hangers. Special pipe supports will require some of these to be modified or supplied with additional components for anchor and guide requirements. Spring supports are used to allow piping movement, usually thermal expansion. They are designed so that the load carried either varies with the deflection of the spring or remains constant with the deflection. The latter system is more expensive. If friction loads are to be minimized. Teflon (polytetrafluoroethene, or PTFE) slide plates may be used (Fig. 31.3 and see Sherwood[12], Kellogg[13] and Institute of Petroleum[14]). Except for steam tracing, pipes should not normally be supported from other pipes.

31.3.2   FLEXIBILITY ANALYSIS OF PIPEWORK

The flexibility of a pipework system can be defined as its ability to absorb safely the strains imposed upon it from all sources of loading. A flexible system is thus one which operates at stress levels less than the maximum safe stress for the pipe material over its design temperature range for the specified life of the system.

Strain is the result of stressing the pipe material which in general arises from the following.

(a) Fluid pressure in the bore or external to the pipe.
(b) Change in pipe temperature (emanating from either internal fluid temperature or conditions external to the pipe) when terminals are anchored.
(c) Movement of the terminal vessel in a direction other than that of the free expansion of the pipeline.
(d) Dead weight of piping, fluid (operating or test), lining, ice, lagging, covering, valves, fittings, etc.
(e) Pulsating flow in the fluid, or vibration from a mechanical source, e.g. a reciprocating machine in the system.
(f) Thermal stress due to uneven temperature along the line or in partial circumference.

A pipeline can also be loaded continuously or in transient, from a number of less common sources, e.g.:

(a) Fluid change of state.
(b) Support method.
(c) Wind loading.
(d) Structure deflection.
(e) Support failure.

The assessment of a pipeline's flexibility takes account of all sources of loading in the system as accurately as demanded or defined by design criteria for that particular system. In practice, the amount of work required to check analytically every conceivable form of loading would be excessive and therefore it is necessary to concentrate on the main forms of loading whilst bearing in mind, when examining calculation results, that they merely give an *indi-*

Fig. 31.3 Typical pipe support methods

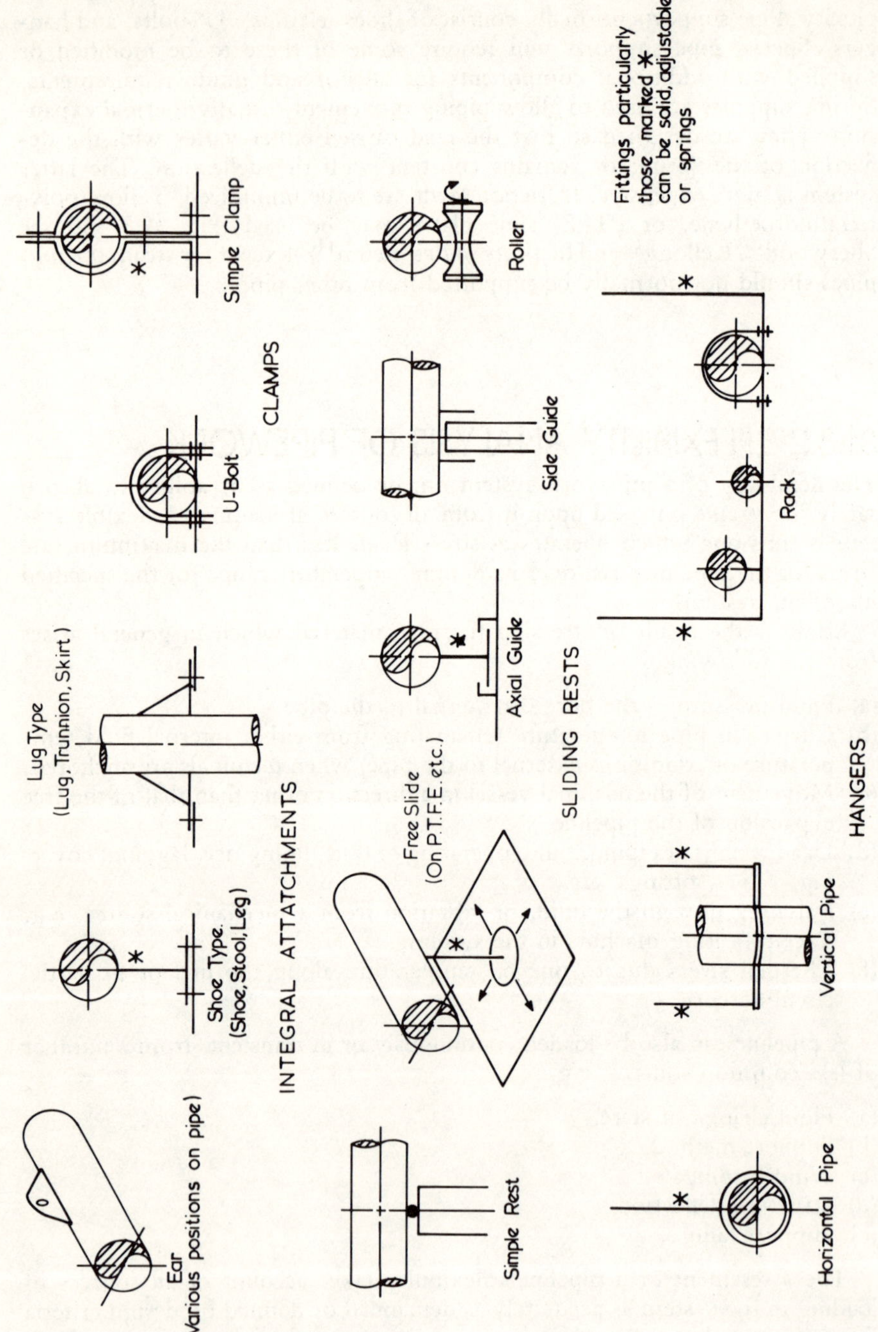

cation of conditions pertaining to a practical application. Effects which are not calculated must be recognized and minimized.

The ultimate aim of flexibility calculations is to produce for the designer sufficient knowledge for him to pronounce the pipework:

(a) *Safe*, for the design life of the system.
(b) *Practical*, i.e. allowing ease of manufacture and fabrication, installation, inspection and testing, operation and maintenance, and
(c) *Economical*, i.e. with regard to the use of material and labour at any stage from conception to operation and maintenance.

Pipe flexibility analysis must comply with the codes and regulations called for in the particular project. The American ANSI B 31.3 is a widely used guide for stress-analysing the piping systems[1] and it contains tables of allowable stresses in pipe walls for a wide range of pipe materials. In practice, though, it is found that more calculation work is undertaken to establish pipe terminal loading on strain-sensitive equipment (usually rotating machinery such as pumps, turbines and compressors) than to establish safe stress levels in the pipe material itself.

The separate criteria for flexibility calculations are:

(a) Establishing equipment loading within limits agreed with equipment suppliers.
(b) Proving stress levels in the piping are within stress level limits of the contractual code.
(c) Providing the civil engineer with the loads necessary for design of structures, structural members and foundations.
(d) Establishing loading on pipe supports, guides, restraints, anchors, etc.
(e) Establishing the movements due to thermal expansion/contraction which have to be accommodated by supports and guides.
(f) Establishing expansion loop positions and dimensions.
(g) Proving the necessity for expansion bellows and determining the flexibility characteristics of bellows units (or other expansion/absorption devices).

Flexibility analysis is usually applied by selecting one or more of the following methods.

Visual Analysis

This involves:

(a) Estimating the main thermal movements in the piping system.
(b) Assessing the stress effects of bending moments, always ensuring that assumption errs on the side of safety.
(c) Comparing the stress level estimated, with the design code allowable (DCA) stress.

If the estimated stress is greater, then usually a more exact calculation is made.

Approximate calculation

These methods include the use of nomographs which take account of configuration geometry and enable an overestimation of stress levels in a short space of time. Desk-top computer programs are also used for approximate calculations.

Both visual and approximate methods are applicable only when the design criteria are at pipe-stress level, i.e. these methods are not sufficiently accurate for estimating pipe loading on strain-sensitive equipment; however, they may yield sufficient information to allow design of structures and foundations. In such instances initial estimates provided should err on the high side to prevent redesign at a later stage. The cost of a few extra piles is minimal by comparison, particularly when it is likely that the piling contractor will leave the site early in construction. The structural department also require to know at an early stage the locations of the anchor bays for the piperack and the estimated axial anchor loads and side thrusts which are additional to the wind load.

Comprehensive calculation

This involves computer calculation using the most up-to-date pipe stressing programs available.[15,16]

The procedure applies to sections of pipe contained between constraints which either allow no pipe movement (e.g. anchor) or restricted movement (tank nozzle). Thus there could be two or more sections defined for stress analysis in one physical length of pipe, because of intermediate anchors.

The data input consists of dimensions for piping components, piping material, pressures and temperatures throughout the section, restrictions of movement by the restraints, terminal equipment movements, etc.

The results include:

(a) Stress levels at terminals, and at the piping components such as tees, bends, etc.
(b) Forces and moments at each terminal, and at other preselected points in the piping.
(c) Movements (deflections) of the piping throughout the system.

A set of results is obtained for each selected load (usually temperature/pressure) condition defined to simulate the operation, foreseen fault, relief valve blowoff, etc. Using these results, the pipe layout and support system can be amended and the new system re-analysed until satisfactory. The results can be summarized on pipe stress sketches and diagrams (Fig. 31.4).

31.4   PIPING LAYOUT

Normally all process and utility piping should be located above ground except water mains which may be buried for protection against freezing or to free ground for greater general access. Wherever possible, piping should be arranged in horizontal banks for ease of support. Banks running North and South should be run at different elevations to those running East and West

Fig. 31.4 Pipe stressing drawings (a) bending stress diagram (b) pipe stress sketch

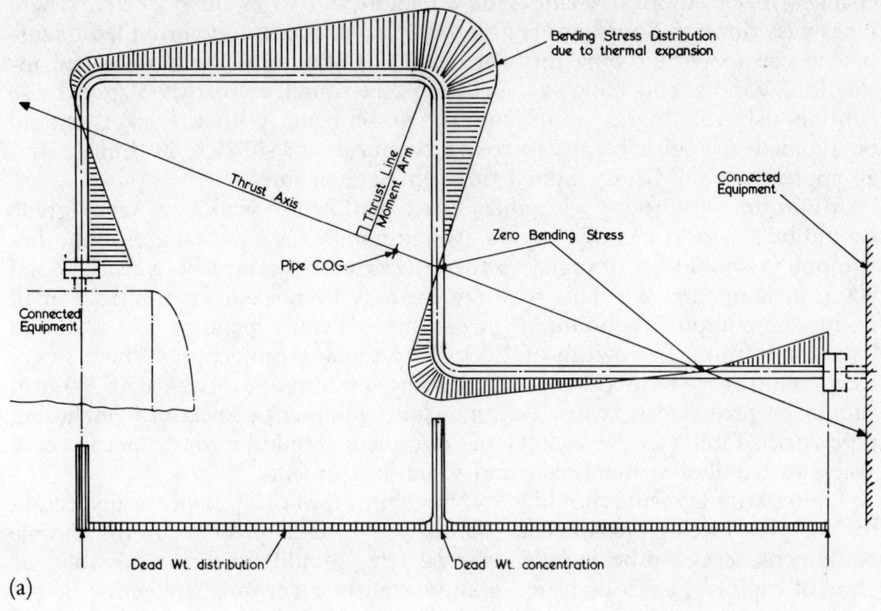

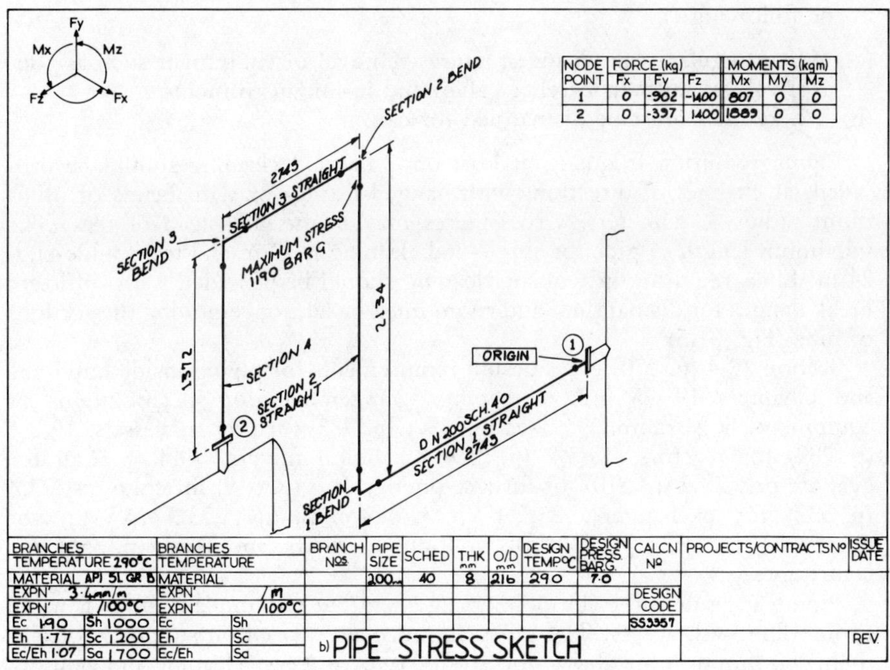

NODE POINT	FORCE (kg)			MOMENTS (kgm)		
	Fx	Fy	Fz	Mx	My	Mz
1	0	-902	-1400	807	0	0
2	0	-397	1400	1889	0	0

BRANCHES TEMPERATURE 290°C		BRANCHES TEMPERATURE		BRANCH Nºs.	PIPE SIZE	SCHED	THK mm	O/D mm	DESIGN TEMP°C	DESIGN PRESS. BAR G.	CALCN Nº	PROJECTS/CONTRACTS Nº	ISSUE DATE
MATERIAL API 5L GR B		MATERIAL			200mm	40	8	216	290	7.0			
EXPN' 3.4mm/m		EXPN'	/m								DESIGN CODE		
EXPN' /100°C		EXPN' /100°C											
Ec 190	Sh 1000	Ec	Sh								SS3357		
Eh 1.77	Sc 1200	Eh	Sc										REV.
Ec/Eh 1.07	Sa 1700	Ec/Eh	Sa										

b) PIPE STRESS SKETCH

to avoid clashes. Exceptions should be made so as to avoid unnecessary changes in elevation at changes of direction and to avoid pockets. Where pipes pass through floors, roofs or walls, pipe sleeves can be provided of sufficient size to permit pipe movement and/or accommodate flanges and insulation. Vapour-collecting systems should be routed so that the vapours rise continuously to a higher point from the vessel being vented. Pockets should be avoided in lines carrying corrosive chemicals, or slurries, in drains, etc., an appropriate fall being applied throughout each run.

Minimum headroom clearance over platforms, walkways and grade should be 2.1 m. Generally, piping around pumps, heat exchangers and other equipment should be arranged so that a clear accessway of between 750 and 900 mm is maintained. This requirement may be relaxed for multiple small pumps or equipment mounted on common bases. Operating aisles should have a minimum free width of 1.5 m between any projection. Where structural columns are located in operating aisles, a minimum clearance of 750 mm should be provided between columns and equipment projections (including pipework). Piping in the vicinity of equipment should be run to permit easy access to handholes, manholes, and visual instruments.

Piping arrangements should allow for removal of equipment for inspection testing or servicing. Maintenance areas provided in plant areas for mobile equipment access, tube bundle removal, etc. should (as far as possible) be clear of piping. Horizontal piping supported in a vertical bank must be capable of having each pipe removed independently. Supports should be provided where practicable so as to reduce the necessity for temporary supports in the following cases:

(a) Where regular maintenance requires removal of equipment such as control valves, safety and relief valves and in-line instruments.
(b) Where lines must be dismantled for cleaning.

Piping requiring frequent (at least once a week) cleaning should be provided, at changes of direction, with flanged fittings or with bends of minimum radius 5 diameters. Cross-pieces may be fitted instead of tees. The maximum length of pipe for single-end cleaning is 12 m and for double-end, 24 m. Lines requiring infrequent cleaning should be provided with sufficient break flanges for dismantling and room must be left for removing the sections of pipe (Fig. 31.5).

Section 18.4 (p. 270) gives design requirements for piping inside buildings and Chapters 19–25 indicate piping arrangements for specific items of equipment. For example, see section 19.1 (p. 283) for process vessels, 19.2.1 (p. 289) for reactors, 19.3.2 (p. 299) for liquid mixers, 19.4 (p. 299) for evaporators, 20.5 (p. 310) for furnace piping, 21.3 (p. 319) for columns, 22.4 (p. 332) for exchangers, 23.2.1 (p. 336) for pumps, 23.3 (p. 342) for compressors, 24.3 (p. 350) for continuous filters and 25.3 (p. 356) for centrifuges.

Piping at grade is the cheapest arrangement but within plant limits is liable to interfere with access. The pipes are normally placed on supports to raise them 300 mm or more above, the ground to permit easy cleaning and painting (Fig. 31.6). Requirements for trapping and draining will frequently determine the support height. Where the crossing of walkways is unavoidable the lines should be provided with stiles.

Open pipe trenches may be used in-between process units when there is

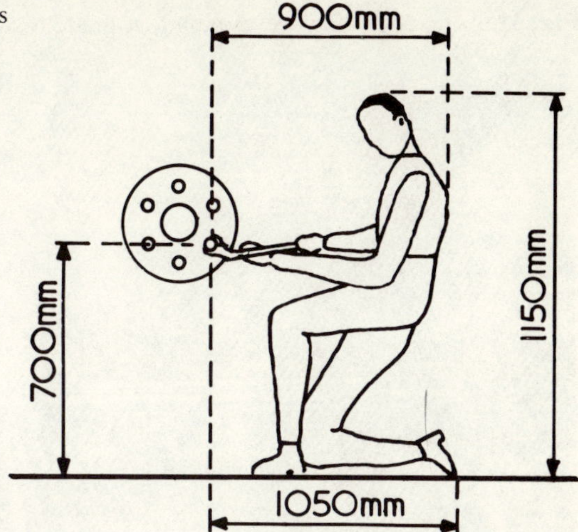

Fig. 31.5   Access to pipe flanges

no risk of flammable vapours collecting, or of the material freezing, e.g. for steam mains. They should not be used where there is a danger of flooding. It is often convenient to run open trenches alongside roadways at such an elevation that the pipes can run under the road in sleeves or culverts with no change in elevation (Fig. 31.11, p. 446). The minimum width of pipe trenches should be 600 mm. A minimum clearance of 100 mm should be provided between pipe projections and walls and 50 mm to the point of trench bottoms.

Buried piping between process units is acceptable providing the contents cannot solidify or block. They are buried to a minimum depth of 750 mm and where they go under roads and other concreted areas they should be laid in ducts or solidly encased in concrete. Buried piping should never go under buildings. Where valves, meters and similar fittings are used they should be housed in a suitable brick or concrete chamber with proper access to the surface (Fig. 31.7). Room must be allowed for anchors at pipe ends and changes of direction.

Buried gas piping should not be laid adjacent to potable water piping or plastic or asbestos materials. Where buried lines are laid near to, or across, buried electric power cables they should always be laid beneath the cables. Pipes carrying hot liquids should be laid as far away from the cables as possible. If underground piping and cables are used, it is essential that the pipe and cable is put into position at the same time as foundation work is being undertaken. If the soil is acidic, the pipes may have to be protected against corrosion.

31.5   PIPERACKS

Piperacks provide the main system by which the interconnecting long process pipelines and utility headers are carried through a plant area to serve their

Fig. 31.6 (a) & (b) Layout of piping at grade (Courtesy: Humphreys & Glasgow)

(a)

(b)

Fig. 31.7 (a) & (b) Valve accessibility for buried piping (Courtesy: British Gas)

(a)

(b)

designated items of equipment.[17] Blowdown, flare and relief headers are located on the racks, as are often instrument and electrical cable trays. The choice between bridges or racks is one of economics of construction although bridges may be employed where ground space is at a premium and greater clearances are required.

The piperack layout should commence as soon as the engineering flow-sheets; utility flowsheets; and plot plan are available. Using these documents, a schematic or routeing diagram is prepared by initially routeing piping directly between equipment and only using racks where necessary. It is then possible to identify the densest section, thus establishing where racks are desirable and thence the rack width and/or number of tiers of piping (see below). Some reconciliation between rack and equipment layout can then be made (Fig. 31.8) to reduce the amount of racking. The rack may be straight through, or 'L'- 'T'- or 'U'-shaped according to needs (Fig. 31.9). (See also Kern.[17])

Fig. 31.8   Interaction between process and piperacks (Kern[17])

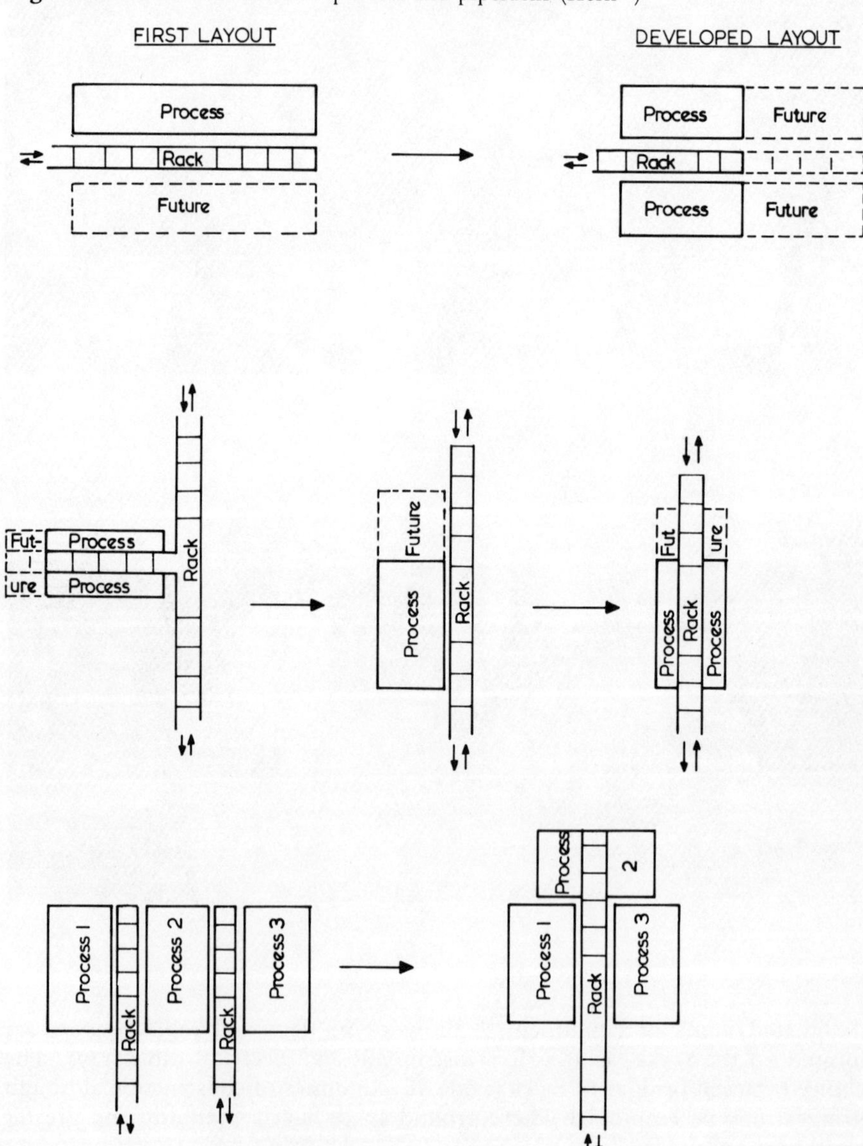

Fig. 31.9 Typical arrangements of piperacks (Kern[17])

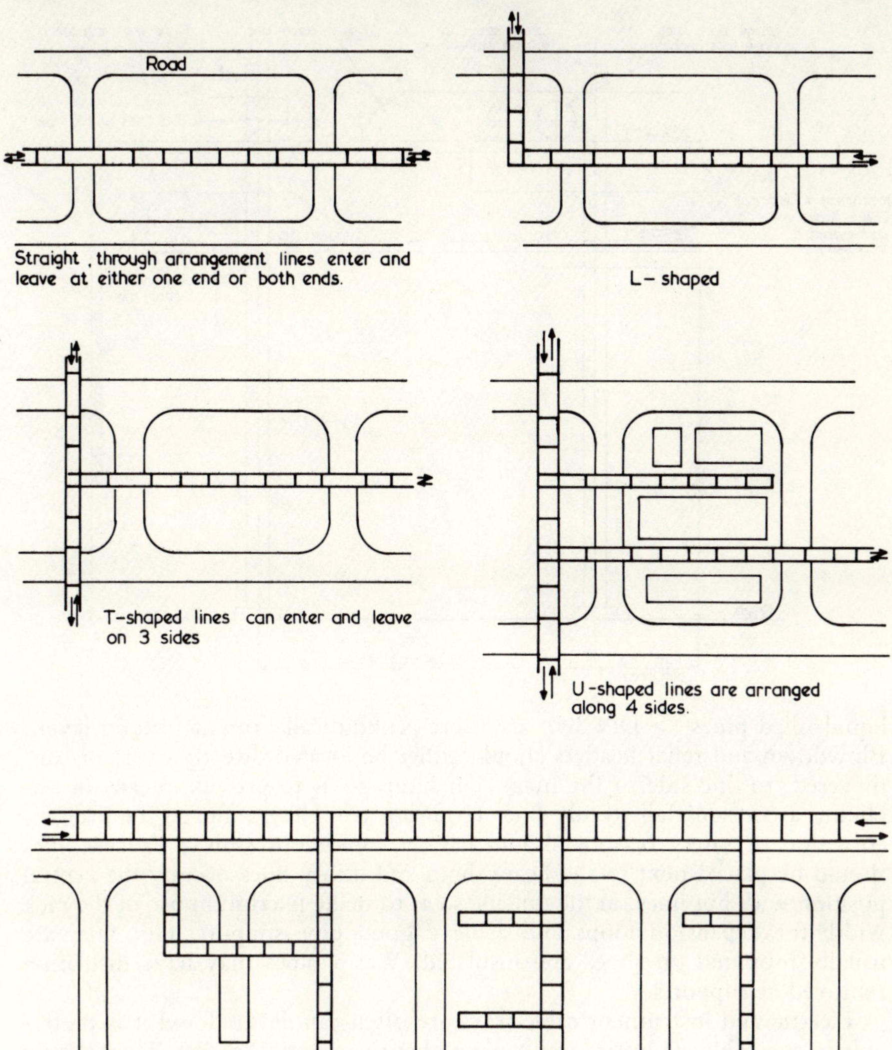

Piperacks running in the same direction should preferably be at the same elevation. The spacing of stanchions should be regular, although the matching of adjacent equipment structure columns should be examined for combined foundations. Support points of large-diameter lines should be placed directly over or as near to the supporting stanchions as possible (Fig. 31.10) in order to minimize bending moments in the piperack members. (Heavy

Fig. 31.10 Typical cross-section of layout at piperacks

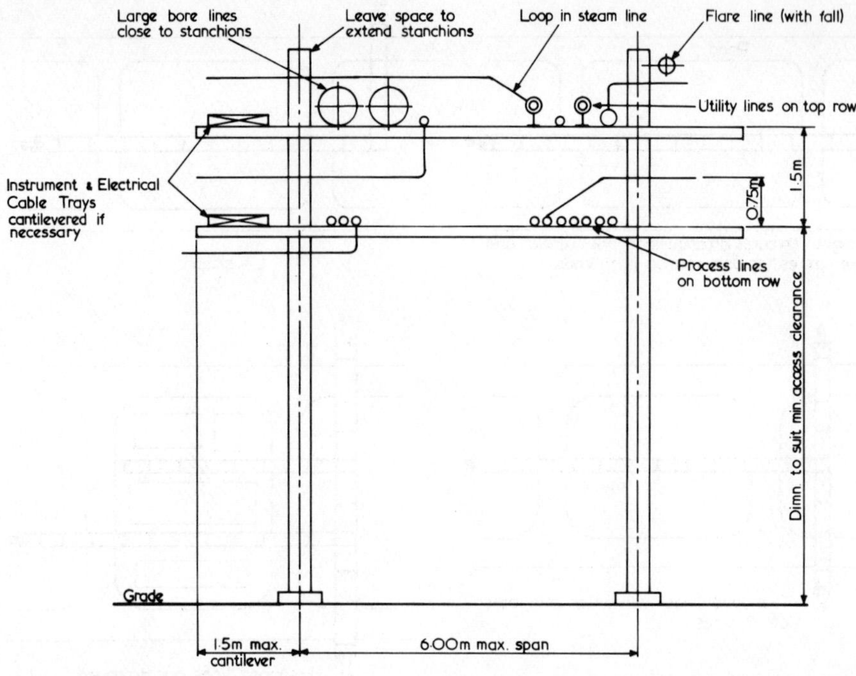

liquid-filled pipes (> DN 300) are more economically run at ground level.) Blowdown and relief headers should either be located directly over, or cantilevered, to one side of the main stanchions so as to provide means of obtaining a constant fall. (Only lines requiring complete drainage for process, corrosion or safety reasons should have a continuous slope.) Process lines should be placed next to the heavy lines and utility lines occupy the central position with hot lines on the outside so as to make maximum use of the rack width for expansion loops and achieve good pipe support. Hot lines are usually mounted on shoes and insulated. Warm pipes may have insulation removed at supports.

Electric and instrument cable trays are often cantilevered and it is preferable to run cables in horizontal banks rather than vertical banks. They should not be put next to pipes carrying hot fluids or solvents. Fire protection may be required where critical cables are carried by the piperack. Methods include sprays, shields (if the direction of radiation can be predicted), fire resistant trunking and fire resistant cables.

Using pipeline spacing charts[18] and taking note of insulation thicknesses the actual rack width required can be calculated. Typical horizontal clearances, which are governed by the outside diameter of the pipe or insulation, are as follows:

(a) Pipe-to-pipe separation: 50 mm.
(b) For cold insulated lines the clearance should be 75 mm.
(c) When piping is flanged the clearance will be between the outside surface of the uninsulated flange (whether insulated or not) of the larger pipe and the outside surface of the pipe or insulation of the smaller pipe.

444

(d) Pipe-to-steel clearance: 50 mm. This should be from flange or insulation, whichever is the larger.

When change in direction of a horizontal pipe is necessary, it is best also to make any elevation changes at the same point at this leaves more space for future expansion. Allowance should be made for future space requirement for pipes and cables; this is commonly 20–30 per cent extra width and strength. Pipes should not run over stanchions if there is the possibility of adding another layer of piping in the future.

The calculated width must then be compared with the plot space available. On congested sites and in high-density piping areas racks are likely to contain two layers of pipework (Fig. 31.10). Triple-layer pipeways should be avoided except for cable trays, expansion loops or very short runs. For major plants to obtain reasonably compatible beam sections, the width of piperacks should generally be 6 m but in special cases this can be increased, provided the cost of deeper traverse beams is acceptable. Equipment placed under the rack can also influence the width, for instance where a double row of pumps is planned. If the choice is for a two-tier rack, then it is the general rule to place utility lines on the top level and all process lines together with lines containing corrosive liquids on the lower level.

Pipelines should not change position in the rack unless required to do so to:

(a) Avoid fouling.
(b) Economize on the supporting steelwork.
(c) Provide stressing flexibility.

A right-angled change of direction of the whole piperack offers an opportunity to change the order of lines.

Takeoff elevations should be at a constant level to common bottom of pipe level since it is preferable to support offtake piping from below rather than with hangers. Room must be left between the rack and adjacent buildings or structures for takeoffs coming down to pumps, etc. Branch connections are normally at 90° to the main pipe. For single pipes crossing a bank of pipes, clearance should be 75 mm; and for banks of pipes crossing a bank of pipes, clearance should be 450 mm.

There must be sufficient access all around a pipe for initial and subsequent operations such as radiography, hydraulic testing, painting, insulating or coating with bitumen, etc. if required, and for subsequent maintenance.

To determine the elevation of the rack levels, consideration must be given to required headroom:

(a) Over railways and access roads.
(b) For access to equipment under the rack, e.g. by mobile lifting equipment (as in Figs. 16.1 (p. 222) and 31.11).
(c) Under lines connecting the rack to adjacent equipment.

Rack width and distances are kept to a minimum between stanchions (see Fig. 31.10) to keep beams to a reasonable size and this should be taken into account when considering headroom requirements. A useful initial guide for use within a plot area is a minimum height above concrete of 4.5 m. This figure will generally allow for adequate clearance under a rack. If a further level is required this is likely to be 1.5 m above the 4.5 m level. Typical clear-

Fig. 31.11 Pipebridges pipe trenches and access (Courtesy: British Petroleum)

ances over roads could be 6 m and over railway tracks 7 m. Generally, pipe-racks are given greater headroom than simple piperuns. Pipebridges may have longer distances than racks betwen supports (up to 30 m) and can have any reasonable headroom (Figs 16.1 (p. 222) and 31.11).

Where valves are called for at battery limits on a piperack it is usual to supply a platform with ladder access. The valves should be staggered in each alternate line with the platform passing between the vertical spindles for hand-wheel or level access.

The drainage near a piperack must be such that spills from the rack run away from the rack to the appropriate sewer.

31.6   UTILITY SYSTEMS

Utility piping systems are specified on a separate utility line diagram for each utility. These diagrams indicate where utility pipelines connect to process lines or equipment and show the interconnecting headers. Utilities consist mainly of steam, air, cooling water and nitrogen systems but also include fuel oil, fuel gas, refrigeration and firefighting water. Piping carrying utilities used directly in a process (e.g. town or demineralized water), should be treated as part of the process piping.

In this section, utilities are considered with the plots as the site layout of utilities was discussed in Chapter 14.

31.6.1 STEAM PIPING AND TRACING

The layout of steam piping should be examined for adequate flexibility of thermal movement. Steam main headers are kept to the outside edges of the top level of the piperack to allow space for expansion loops where necessary. Each main should have shutoff valves at or near the battery limits of the unit so as to isolate it from the rest of the plant. Distribution of steam from a main header should preferably be made by running subheaders, each serving a number of steam users and provided with an isolation valve at the main header. Supply and return branch lines should be connected to the top of the steam header with the branch isolation valve located in the horizontal. Side connections should be avoided unless clearance above the headers prevents the use of top connections. All steam piping should be run to avoid condensate pockets and to save steam traps. Steam and condensate lines are not normally laid to a pitch or fall. All steam lines should be provided with adequate drip legs, drains or steam traps and none of these should make a direct connection with the sewers. The discharge condensate from steam traps should normally be connected to the top of a condensate header. Where steam traps are to discharge to grade, the drain lines should be routed to a local catch point.

Steam tracing may be provided for pipelines and equipment to:

(a) Prevent the freezing of fluids.
(b) Maintain viscous fluids in a fluid condition.
(c) Preheat process lines to prevent solidifying of liquid in cold lines on startup.

Each tracer pipe should have a separate takeoff from the steam tracing header or subheader, with its individual isolation valve and steam trap. Tracer lines are installed touching the main pipe or equipment with shared insulation. Tracer pipes are attached with tie wires or straps and start at the high point of the line and end at the low point regardless of the direction of flow in the traced line. Spiral winding of valve bodies, level controllers, etc. are arranged so that the tracer is self-draining. Where flanges and valves occur in the line being traced, break points are provided in the tracer to permit easy removal. Single straight tracers should be run on the bottom of the pipe but off the centreline to avoid pipe supports. Steam jacketed pipes are expensive and are only used when it is vital to maintain process fluids in a liquid state. Examples are lines containing fluids such as sulphur, bitumen, heavy fuel oil, etc. which are likely to solidify at atmospheric temperature and can be extremely difficult to remelt after solidifying.

31.6.2 AIR PIPING

Connections for plant and instrument air piping should be made at the top of the header except in the case of a branch being less than the header size when the connection may be made at the side. Plant and instrument air headers should be normally located on the top level of the piperack. Isolation valves for plant air should be located at the equipment, whereas for instrument air they should be at the header. Instrument air must always be oil-free and dry. Dead legs in air piping should be avoided to prevent the possibility of collected water freezing.

31.6.3 COOLING WATER PIPING

Cooling water is either distributed to the various users as an underground system or above ground on elevated piperacks. The type of system employed is decided at the start of a project or follows existing arrangements. Underground piping is most likely to exist already outside the plot limits, run below the normal frost line or 750 mm if greater. A system will include both the flow and return lines each supplied with isolation valves at the plot limits and, if underground, these valves should be located in concrete boxes. Dead legs in water piping should be avoided to prevent the possibility of freezing in cold weather. Where this possibility exists, drain points must be provided to permit the sealed section to be drained or steam tracing applied. For safety reasons, the supply of water to exchangers, condensers, coolers, etc. should be arranged so that the equipment is kept filled in the event of water failure. If this is not possible then a check valve is fitted in the supply line to prevent the water draining back to the header. Water lines for users of large quantities of water should each be provided with a bypass between inlet and outlet lines to enable continuous flow if the equipment is shut down during freezing weather. All branch connections from water headers should preferably be taken from the bottom of the header.

31.6.4 FUEL PIPING

When fuel gas is supplied to the plot limit it should be dry and at constant pressure. Fuel gas and fuel oil branch connections should be made at the top of the header. Isolation valves should be provided at plant battery limits and placed on all header branches to permit shutoff in the event of fire. Fuel gas piping should be arranged so as to eliminate any pockets or seals in which condensate would collect. Fuel oil piping should normally be installed as a circulating system with strainers located in the suction line of each set of fuel oil pumps. These and their piping should be located away from fan inlets to eliminate the possibility of leaking oil being sucked into the fan inlet.

31.6.5 REFRIGERANT PIPING

Refrigerant piping should not be located in corridors, stairways or lift shafts and should always be accessible. When routed inside buildings, headroom should be at least 2.25 m above floors except when the piping is run against the ceiling. The pressure drop in refrigerant liquid lines should be sufficiently small to maintain the system pressure at a high enough level to prevent vaporization of the liquid.

31.6.6   FIREFIGHTING WATER PIPING

Provision of fire fighting water systems is discussed in section 17.7 (p. 255).

31.6.7   VENT CONNECTIONS

Vent and blowdown piping runs should be as short as possible avoiding pockets and unnecessary bends. Branches should preferably drop into the top of the header, declined in the direction of the flow. Discussion of flare stacks is contained in section 13.3 (p. 198) and relief devices in section 31.7.7.

31.6.8   WASHING DOWN FACILITIES

For general plant maintenance and service, utility hose stations consisting of water, steam, air and nitrogen points should be provided at convenient points throughout a process plant. They should be located so that all parts of the plant can be reached with a 15 m-length hose. For points above grade in structures and buildings steam and air services are generally provided, also with a 15 m-length hose.

For cleaning some types of process fouling in heat exchangers, a high-pressure water washdown facility is needed. It is often useful to provide a permanent installation, i.e. a walled (1–2 m-high) well-drained enclosure with good road and crane access.

Emergency showers and eye baths should be placed inside heated buildings wherever possible. If an outdoor location is necessary, heaters to warm the water should be provided and advantage should be taken of positions where freezing will be avoided. Steam tracing should not be used on the piping since the water may become dangerously hot.

31.7   INSTRUMENTS

Instruments are connected to vessels, equipment or piping and include pressure, temperature and level instruments, flowmeters, analytical instruments, control valves and relief valves.

31.7.1   PRESSURE INSTRUMENTS

All local reading gauges should be installed so that they can be easily read from grade, platforms or ladders with faces inclined downward rather than upward. Ideally, the gauges should be at eye height (1.7 m) above the floor (Fig. 31.12). Gauges affected by operation of particular valves should be read-

Fig. 31.12   Height of instruments

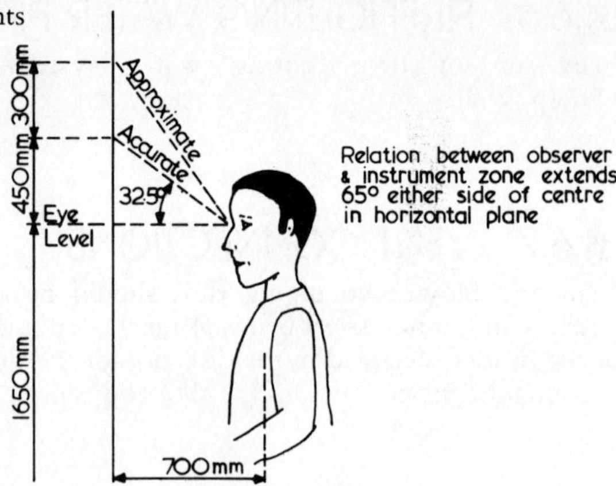

able from those valves. All gauges, especially transmitting types, should be easily accessible from platforms, etc. for testing and maintenance purposes. Care should be taken to ensure that gauges are mounted away from sources of vibration. The connections for vapours and gases in horizontal lines should be on the top half of the lines, while those for liquids in horizontal lines should be located in the lower half of the lines to ensure adequate venting of entrained vapours. On no account should connections be made from the bottom dead centre of a line, to avoid the collection of silt, tar, etc. in the connections.

31.7.2   TEMPERATURE INSTRUMENTS

All thermowells should be installed at points where they can be conveniently maintained and checked. Local instruments should be fitted where they can be easily read from valves. Clearance should be provided around the thermowell to enable easy removal of the thermocouple or other elements.

Care should be taken not to install the wells in dead zones or in pockets of still process fluid. Temperature instruments in vessels and piping are normally required to measure the average temperature of the flowing fluid. They are therefore usually placed in the liquid space in vessels, close to outlet nozzles, or at the bottom of the downcomer in a distillation column. When required, temperature connections are positioned close to inlet and outlet at pumps, exchangers and control valves. These temperature instrument locations are usually downstream of pressure connections, and upstream of additive-injection points. Thermowells should be placed well ahead or just after the straight length of orifice piping, in order to avoid flow disturbances. Where two streams of different temperatures meet, a minimum of 10 pipe diameters should be allowed for mixing the flow before sensing the temperature. On furnaces and heaters thermocouples should not be installed at points where false readings can result through reflection or radiation from brickwork or burner flames.

In line sizes below DN 80 a local enlargement of the pipe is required in order to accommodate the thermowell, although some companies' standards

450

call for enlargements in pipe sizes DN 100 and below. Where sensing elements are located in elbows or angled branches the tip of the well should face the flow direction to ensure quick response.

31.7.3   LEVEL INSTRUMENTS

Where possible, level gauges should be located so that they can be read from the level control instrument or transmitters. All level instruments, i.e. gauges, controllers, alarms and switches should be located for ease of access from grade, platform or permanent ladder to the instruments themselves and to their associated shutoff, vent and drain valves. Level gauges and controllers should not penetrate tower or structure platforms.

31.7.4   FLOW INSTRUMENTS

Flow detectors include orifice plates, venturi tubes, pitot tubes and rotameters. To ensure accurate flow measurement, minimum lengths of straight pipe before (20 diameters) and after (10 diameters) orifice plates and flow detectors are required to ensure steady flow conditions. In liquid service, an orifice can only operate properly when it is completely flooded and free from flashing and gas entrainment; this applies to most flow-sensing elements. The flooded condition normally prevails in pump-discharge and pressurized lines, and so orifices can still work effectively in vertical, horizontal or slanted piping. For vertical pipes the flow usually must be upwards to ensure flooding.

To measure flow in gravity-flow lines:

(a) A static head must be provided in front of the orifice.
(b) The orifice plate must be positioned in a horizontal section of the pipe unless feeding to a seal pot such as a barometric leg.
(c) A slight upward slope should be provided after the plate to ensure the pipe is fully flooded.

A commonly used ratio of orifice diameter/pipe bore is 0.7 as this usually gives a reasonable pressure differential. Care must be taken to ensure ample clearance around orifice plates and tappings for cleaning and rodding out. A minimum clearance of 600 m is required between walls and orifice flanges. They should be accessible from platforms, walkways, etc. without special provisions. No orifice should be used in lines less than DN 50 but, if required the pipeline size should be increased to DN 50 for the straight measuring length. The orifice plate is generally located between a pair of orifice flanges, certainly in the smaller sizes, say up to DN 300. In the larger sizes the plate is located between a pair of standard flanges as set down in the piping specification and tapping points located at distances equal to the nominal diameter upstream of the plate and half this nominal diameter downstream. For liquid service the tappings should be located on or below the horizontal centreline. This ensures that the lines are self-venting and any collected air or vapours can return into the line. On air or gas service tappings should be located in the top section of the pipe to allow drainage of condensate back into the line.

While the installation of venturi tubes requires no special consideration, pitot tubes require removal clearance and accessibility. Rotameters must always be installed in the truly vertical position with the flow upwards. A suggested 600 mm dimension above and below is recommended for cleaning purposes. As rotameters are often of glass construction mounted at eye level, they must be put behind transparent screens in case of breakage.

Access is needed to all flow detectors but, in addition, clearance for cover removal should be allowed for the inspection and adjustment of differential-pressure cells, when they are associated with the flow detectors. The most economical support for these cells is on existing structures or walls, but local supports can be provided with tubular stanchions, flanged to the floor. These should not be placed in accessways. Electric-starter supports can be combined with those for differential-pressure cells. Though the preferred arrangement is close to the orifice, differential-pressure cells often have to be remote. Where necessary, drip legs, sediment chambers and air chambers are provided between the orifice and cell. These separators should be close to the cell and connections to them should be close to the cell and connections to them should be self-draining.

31.7.5 ANALYTICAL INSTRUMENTS

Complex analytical instruments such as pH, infra-red, gas phase chromatography (GPC), refractive index, etc. often used in conjunction with a microprocessor, should be accessible for *in situ* maintenance from a floor. For complete removal of instruments weighing more than 9 kg a hitching point or davit should be considered unless near the floor when access should be left for a truck to be wheeled underneath. Some instruments may have to be enclosed (sometimes under slight positive pressure) in their own local environment and room must be left for such enclosures. This may be either for protection of the instruments themselves or the instrumentation may have ignition sources.

Instrument mechanisms are delicate and instruments should not be located where they can be damaged by the removal of other items during maintenance. They should also not be installed near vibrating machinery and may require electrical screening.

31.7.6 CONTROL VALVES

Control valve bodies are similar to hand control valves of the same generic type, usually globe, ball, plug or butterfly design. A control valve is fitted with an actuator or power unit in place of the handwheel or lever. Actuators are powered by pneumatics or electrics, facilitating remote control.

Control valves should be installed with their spindles vertical and so that they are readily accessible from platforms, walkways or grade. For sizes DN 80 and above, they should be mounted with at least 400 mm clearance above the floor or access platform. Sufficient space must be allowed for the removal of the top works and bottom covers of control valves without removing the valves from the line. In outdoor plants, it is usual to locate all control valves about 600 mm above grade, and thus provide convenient access for operation and maintenance. Block and bypass valves follow the same criterion. Generally, control valves are heavy items which should be located at points where they can be easily supported and maintained without the need for scaffolding or lifting gear. Pipe supporting should enable the control valve to be removed whilst leaving the pipe fully supported.

Control valves which are operated in conjunction with remote instruments should be located so that the gauge or transmitter is visible from the control valve station. Extra consideration should be given to valves in hot or cold service with extended bonnets or cooling fins. Where control valves have handwheels, greater space is required around the valve to ensure that the handwheel is accessible.

In nearly all cases control valves are provided with block valves and a valved bypass. The manifold should be arranged to permit easy removal of the complete control valve without excessive springing of the piping.

31.7.7 RELIEF DEVICES

Conventional relief valves must be installed with the stem in the vertical position. Care should be taken to ensure that the valves are not subject to significant thermal expansion stresses or support significant weight of piping. Otherwise, valve tightness could be affected and pocketing of the discharge in depressed portions of the line may result.

All relief valves, safety valves or bursting discs, discharging to atmosphere should be piped to safety as discussed in section 13.3 (p. 198) and illustrated in Fig. 31.13. Relief valves discharging into a closed system should be installed so as to prevent liquid being trapped on the outlet side of the valve. If locating a relief valve below the header is unavoidable then a manually operated drain valve should be installed at the valve outlet and piped away to the appropriate effluent system. Relief valve headers should always slope towards the blowdown drum. Reaction forces from relieved fluids escaping to atmosphere may require special support of discharge piping.

Relief valve inlet piping must be kept to a minimum and should never be smaller than the valve inlet size. No valve or restriction of any kind, excepting a bursting disc, is permissible between single relief valves and the equipment they protect. When twin relief valves are used, a positive isolation arrangement must be provided before both relief valves. This is usually in the form of two interlocked block valves so that one relief is always available for operation. It must be capable of the total duty. Interlock arrangements cannot be applied for relief duty spread over more than two valves.

If vent lines can be piped to an accessible but safe point, then visual observation can make testing and operation easier. Access for inspection to relief valves may be by fixed ladder for DN 80 or smaller bores but preferably by platform or floor for larger bores. In buildings, mobile platforms can be used. For maintenance purposes a hitching point or davit should be considered for larger valves, e.g. weighing over 9 kg.

Fig. 31.13 Piping vents to a stack (Courtesy: The Boots Company)

31.8   VALVES AND BLEED POINTS

31.8.1   VENTS, DRAINS AND SAMPLE POINTS

Relevant pipelines should be provided with a vent at each high point and a drain connection at each low point. Process vents and drains should be indicated on the flowsheet but additional points, not so shown, may be needed for hydraulic test purposes. Drain and vent valves should be located as close as possible to the pipeline and should stand clear of any insulation. They should not be piped away unless there is a process requirement. Drains that require piping away should be terminated such that the end of the pipe is visible from the drain point. If vents are required to be piped away care must be taken to avoid pockets. Advantage should be taken of venting or draining a pipeline through equipment wherever feasible and where contamination is not possible.

Similarly, sample point connections should be shown on the flowsheet. Where conditions permit, samples should be taken from existing drain/vent connections. The sample point is located in the portion of pipe which is subject to continuous flow and not in bypass, dead branches or pockets. Sample points on gas lines should be taken above the horizontal centreline of the pipe and for liquid lines from the side or at 45° below the horizontal. In no case may a sample be taken from the underside of the line. The sample line should be kept as short as possible with no pockets or dead legs. Sample outlets in piping on equipment in hot services should be provided with means for cool-

454

ing the sample. Connections should be ideally located at an elevation 1 m above any floor or platform level and not at eye level or above with good surrounding access for handling the sample bottle over the tundish. Wherever possible sampling points should be grouped together with good venting and draining facilities. For some dangerous cases, an enclosed sample cabinet may be needed and appropriate space left for the cabinet in the layout.

31.8.2 VALVE LOCATION

Valves which are frequently operated or which require servicing should be readily accessible from grade, platforms, stairways. It is bad practice to locate valves which have to be operated regularly, in such a position that they have to be reached from fixed or portable ladders. This should only be allowed if there is no reasonable alternative, the valve is small, slight force is needed and the valve operation is only required occasionally (not more than once per month). Valves which are operated very rarely (say once per year or for a future extension) may be positioned inaccessible to a process operator provided it is unnecessary for them to be operated in an emergency. The maximum distance above operating floor or platform level to the centreline of valve handwheels (except when extension operating gear is used) should be 2.2 m. Those which are infrequently operated or cannot reasonably be located or are not readily accessible can be chain operated or provided with extension stems. It is important that the chains hang within 1 m of the operating level. They should be arranged so that they can be attached to supports or walls and do not obstruct accessways. The use of chains and extended spindles, however, should be kept to an absolute minimum. Valve handwheels should be located so they do not interfere with access or maintenance equipment. Handwheels should be orientated so that they are in an operable position with valve stems positioned as follows, in order of preference:

(a) Vertically upward,
(b) Horizontal,
(c) Upward 45°,
(d) Downward 45°.

Stems sloping downward should be avoided unless required for process reasons and there is no chance of deposited solids getting into the gland. Even where this is not a possibility, a downward orientation should not be used without approval from the valve supplier.

A valve with a horizontal spindle is easiest to use when its handwheel spindle is between 750 mm (small valves) and 1.5 m (large valves) above the operating level (Fig. 31.14). For valves mounted with the spindle vertical, the optimum height of the handwheel is approximately 1.1 m above the operating level. If the valve is used infrequently, heights up to 1.5 m are acceptable. Isolating valves on lines entering or leaving the plant at battery limits should be grouped together so that they can be conveniently operated from a single platform.

Fig. 31.14 Valve height

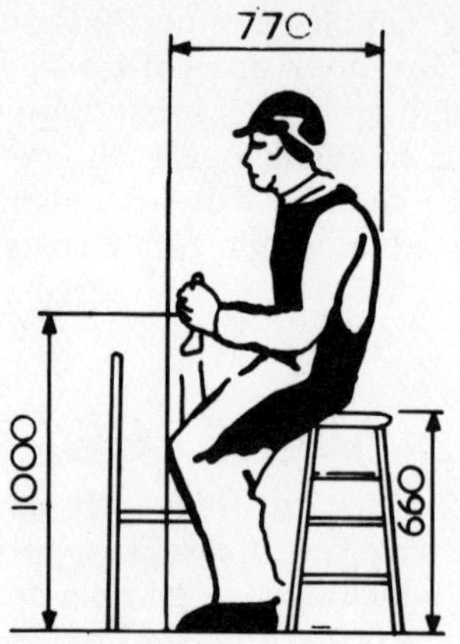

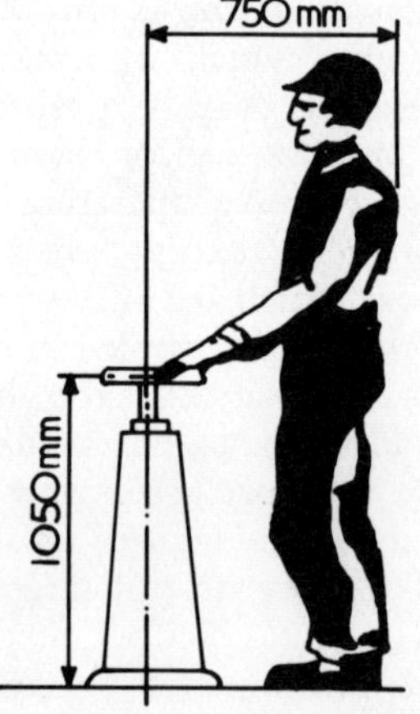

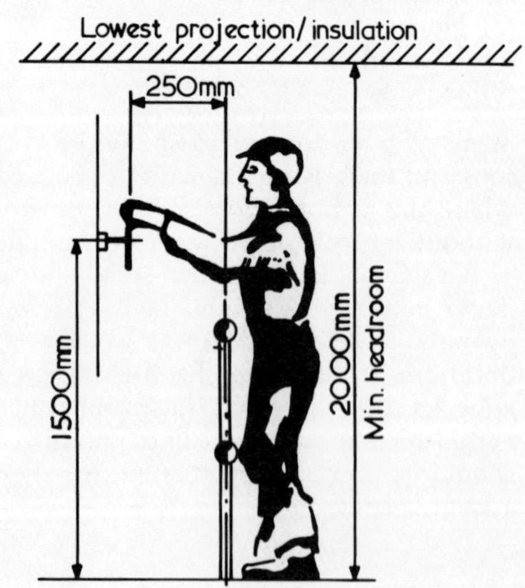

31.8.3   SHUTOFF VALVES

Gate valves are the commonest type of shutoff valve used for isolation. They may have non-rising stems, inside-screw rising stems or outside-screw rising stems. Rising stems require more space but the position of the stem indicates the gate position. The outside-screw is the simplest to maintain but is more expensive, heavier, and requires more space than the inside-screw design. Gate valves handling fluids with suspended solids should be installed so as to prevent a build-up of solids in the bottom of the gate. Valves should never be installed with their stems below the horizontal in case leakage occurs due to spindles being scored by deposited solids.

Diaphragm valves are cheap and easy to maintain and are suitable for many applications and can be used for corrosive, volatile, toxic or suspended solid fluid service. A diaphragm seals off the bonnet preventing the fluid from contacting the inner bonnet or stem. The choice of diaphragm material is limited to rubber compound or plastic materials dependent on service but in any case restricted to conditions below 200° and 4 bar gauge. The diaphragm can be replaced without removing the valve from the line. The valve can be placed in any position or orientation.

Ball valves are often for positive shutoff and are generally wrench-operated with a 90° open to shut movement. They consist of a straight through-ported metal ball, making them particularly suitable for low-pressure loss requirements. However, the ball sits on soft seals such as polytetrafluoro-ethene (P T F E) which introduces a temperature limitation. Top-entry ball valves are the easiest to maintain since the ball and seats are inserted from above without disturbing the pipework. The other type of ball valve has a split body which has to be taken away from the piping to get at the internals.

Plug valves likewise have a more positive shutoff than gate valves. They each have a tapered plug which has a hole of the same shape and size as the interior of the valve. There are three body shapes. The short pattern has the same face to face dimension as that of the gate valve and is preferred for most services. The regular and venturi patterns produce less pressure drop and are specified where this is important. Plug valves, like ball valves, each have a 90° open to shut movement and are manufactured as either 'lubricated' or 'non-lubricated' types. The lubricated type is easier to operate and is less prone to seizure and can be used in any service where the lubricant does not contaminate the piped fluid. Convenient access to the valve spindle is needed for lubricant renewal. The non-lubricated type can be used at higher temperatures.

Butterfly valves are generally used for services such as cooling water or in larger gas lines, where tight shutoff is not a requirement. Closure of the valve is by a disc trunnion mounted through the body and lever operated through a 90° turn. Their slimness and light weight provide advantages over other shutoff valves where space is at a premium. Tight shutoff can be achieved with soft seated valves.

Fluid pressure tends to close the valve and locking devices may be installed on the handle. Large valves may require operating mechanisms, usually worm gearing.

The open/shut movement requires a disc rotation of 60° or 90° and is often executed with remotely controlled pneumatic cylinders.

31.8.4 THROTTLING VALVES

Globe valves can be used for all services where throttling or control is required. It has been established that cost and throttling efficiency become unfavourable above DN 150. They are designed either as inside or outside rising stem types and the strictures about 'below horizontal installation' and 'solids' apply as with gate valves. In addition, the larger simpler kinds should not be installed near elbows as the eddying can cause stem misalignment.

Ball and butterfly valves may also be used for less precise throttling duties.

31.8.5 CHECK VALVES

Check valves are introduced into pipelines to prevent backflow. Three types of valve are manufactured, namely swing, piston, and ball.

The swing check valve is used where minimum pressure drop is required and they are particularly suitable for liquids. The flow keeps the swing gate open but gravity or flow reversal closes it. This type is unstable in lines subject to pulsating flow. However, improvements in operation can be achieved with outside levers and weights. Swing checks are generally located in horizontal pipelines but have been located in vertical legs.

The flow pattern for a piston check valve is the same as for a globe valve. They are suitable in vapour, steam and water service and for pulsating flow conditions but must not be used for fluids with suspended solids. Although piston checks are available for both horizontal and vertical lines, they cannot be interchanged as the piston itself must remain vertical for efficient operation. Piston checks are not normally specified above size DN 150.

A ball check valve is a lift type which stops flow reversal more rapidly than the others. Limited to up to size DN 150, they are not suitable for pulsating flow but recommended for viscous fluids which deposit solid residues.

Non-slam tilting disc check valves are used when there may be a possibility of a pressure surge. These each consist of the check valve plus an external dashpot.

31.9 TESTING AND INSPECTION

The majority of process plant piping is installed for moderate service conditions and is usually hydraulically tested to moderate pressures, with few requirements for *in situ* inspection. The preceding paragraphs contain simplified guidance on the necessity for providing adequate connections to the piping for connecting to the hydraulic test supply, pressurizing equipment, air vents, drains and anti-freezing precautions.

Layout consideration for the testing/inspection operations will normally relate to the provision of isolating blinds and valves to allow either group system testing or to allow piping systems to be tested independently of the connected vessels when high vessel test pressures or the introduction of water into these vessels is to be avoided. Access for the temporary isolating valves and blinds, together with access to the vent and drain points must be allowed for in the layout. Space may be needed for local hydraulic pressurizing equipment and the provision of water supplies and safe drainage of test water. The weight of water contents in large piping systems must not be forgotten during the layout stage – the additional test conditions weight may well require re-routing the pipes adjacent to substantial structures for support during the test conditions.

Whilst water is the preferred test medium on economy and safety considerations, the requirements for testing with other fluids may arise, particularly where process materials would react with residual water or moisture in the pipework. Testing with compressible gases may require special consideration at the layout stage to minimize the consequences of loss of containment during testing. Vacuum testing procedures may have layout implications to provide temporary vacuum equipment or access to pipe joints for local leak-in detection equipment.

Inspection of piping *in situ* would normally be carried out by non-destructive testing methods to check the integrity of welds and individual piping components. In-service inspection on high-integrity systems must also be allowed for and may require additional layout consideration because of the requirements for checking wall thickness of pipework systems or other physical features of the piping during its service life. Equipment for wall thickness measurement or crack detection is normally portable, usually employing ultrasonic or dypenitrant techniques and access for the operator only is normally required. Where in-service X-rays are necessary, the space around pipework and welds required for passage of the X-ray equipment must be allowed in the layout and it may be necessary to provide access routes for the X-ray equipment from ground level to the usage points. An on-site photographic laboratory may be required adjacent to the plant and space allowed for this unit, together with the necessary services. Location of test equipment must take account of the electrical classification of the plants to prevent hazards from non-flameproof electrical and electronic equipment arising during testing.

REFERENCES

1. ANSI B 31.3, *Chemical Plant and Petroleum Refinery Piping*. American National Standards Institute, New York, 1980.
2. BS 3351, 'Piping systems for petroleum refineries and petrochemical plants', British Standards Institution, London, 1971.
3. ASME *Boiler and Pressure Vessel Code*. American Society of Mechanical Engineers, 1980.
4. ANSI B 31.1 *Power Piping System*. American National Standards Institute, New York, 1980.
5. ANSI B 16.5 *Steel Pipe Flanges and Flanged Fittings*. American National Standards Institute, New York, 1977.

6. BS 1560, Part 2, 'Metric Dimensions', 'Steel pipe flanges and flanged fittings (nominal sizes $\frac{1}{2}''$ to 24") for the petroleum industry', British Standards Institution, 1970.

7. BS 4504, 'Flanges and bolting for pipes, valves and fittings, metric sizes', Part 1, 1969, 'Ferrous'; Part 2, 1974, 'Copper alloy and composite flanges', British Standards Institution, London.

8. API 610 *Centrifugal Pumps for General Refinery Service*. American Petroleum Institute, 1980.

9. NEMA *Turbines*. National Electrical Manufacturers Association. Washington.

10. ASME *Unfired Pressure Vessel Code*. American Society of Mechanical Engineers.

11. TEMA *Standards of TEMA*. Tubular Exchanger Manufacturers Association, New York.

12. Sherwood, D. R. and Whistance, D. J. *Piping Guide*. Syntex Books, San Francisco, 1973.

13. The M. W. Kellogg Co., *Design of Piping Systems*. Wiley, New York, 1956.

14. Institute of Petroleum, *Pipeline Safety Code, Model Code of Safe Practice, Part 6*, (4th edn.), John Wiley, London, 1982.

15. ASME 'Pressure vessels and piping computer program evaluation and qualification', Energy Technology Conference, Houston (1977). American Society of Mechanical Engineers, New York.

16. Pearl, R. 'Process piping problems solved in 15 minutes', *Process Engng*, 57, Feb. 1979.

17. Kern, R. 'Piperack design for process plants', *Chem. Engng.*, 30, Jan. 105, 1978.

18. Holmes, E. H. *Handbook of Industrial Pipework Engineering*. Halsted Press/Wiley, 1973.

19. Kern, R. 'Instrument arrangements for ease of maintenance and convenient operation', *Chem. Engng*, 10 April, 127, 1978.

APPENDICES

PROGRAMS AND FORMALIZED METHODS

A.1 PROGRAMS FOR COSTING AND OPTIMIZING LAYOUTS

A.1.1 COMPUTERIZED PIPE-ESTIMATING SYSTEM

Fine[1] has suggested a computer program for estimating the piping cost of a chemical plant based on a simple piping model, which assumes that each plant item can be approximated by a line whose end co-ordinates are fixed. One end of this line is called the reference end, the other end is the far end. Nozzles are assumed to lie on this line at a specified distance from the reference end in the direction of the far end. Some items, tees, reducers, etc. are assumed to be point items: for these items reference ends and far ends coincide. The length of a pipe connecting a nozzle at co-ordinates (X_1, Y_1, Z_1) to a nozzle at (X_2, Y_2, Z_2) is calculated by the equation:

$$L = |X_1 - X_2| + |Y_1 - Y_2| + |Z_1 - Z_2|$$

providing the pipe does not run outside the box established by the nozzles as diagonally opposite vertices in the principal axes framework. If the pipe runs onto a pipe track outside the bounding box, then the pipe length is increased by twice the distance from the track to the box.

The model has been extended to enable pipe cost estimates for as many alternative layout arrangements as required to be produced. In order to guard against placing two items in the same position, each plant item is given a characteristic radius and characteristic clearance value. The required check is that the separation of centrelines of any two plant items must exceed the sum of the characteristic radii and the larger clearance value.

The importance of this computerized pipe-estimating system lies in the ability to produce reasonable estimates much earlier than otherwise and the manhours taken to produce the estimates are also reduced.

A.1.2 OPTIMIZED LAYOUT OF PLANT MODULES

Gunn[2] has described a program for the optimal layout of a chemical plant by decomposing the plant into plant modules. The objective function considered is the cost of interconnecting piping between modules and the cost of the building or site. The co-ordinates of the nozzle connections within the modules are input to the program together with the datum co-ordinates of the modules, cost for unit volume of the building or cost for unit area of the site, pipe descriptions and costs per unit length of pipe. A method of direct search is used to find the optimal positions of the plant modules that minimize the objective function. Minimum and maximum constraints are set on the co-ordinates of the modules so that they do not interpenetrate. Optimal positions of the plant modules and the costs of interconnecting piping and building are available at the end of the program and may be used for further design work.

The calculation of the length of the interconnecting piping is based on the piping model given in the equation:

$$L = |X_1 - X_2| + |Y_1 - Y_2| + |Z_1 - Z_2| \pm R_{12}$$

where (X_1, Y_1, Z_1) are the space co-ordinates of a nozzle in module 1 from which the pipe runs, (X_2, Y_2, Z_2) are the space co-ordinates of a nozzle in module 2 to which the pipe runs, while R_{12} is the pipe redundancy caused by reversals in the direction of the pipe, e.g. branched pipe may be considered to give rise to a negative redundancy.

Recently Al-Asadi[3] has added a computer subroutine for an automatic setting of constraints on module positions. Also, he has introduced the piping module, or pipe track, so that the technique has been extended to include the optimal routeing of the main pipes in a chemical plant. Moreover, a computer subroutine, based on a fairly simple procedure, has been developed for finding all the practical feasible and shortest routes of other pipes, not included in the piping modules. The pipes are defined by the length of the straight pipe parts and the number of bends, which together form the routed pipe. Each pipe part is defined by its start and end co-ordinates. The subroutine also generates clash messages if the pipes run through plant items or any other volumes in the plant prohibited for pipe routeing.

Al-Asadi's program produces mainly numerical output and the results are presented on a physical model (see Figs 18.1 (p. 266) and 18.4 (p. 269) or as floor plans (Fig. 18.2 (p. 267)).

A.1.3 THE COMPUTER-AIDED DEVELOPMENT OF PLANT LAYOUTS

Shuqair[4] was concerned with developing a software package to tackle the problem of chemical plant layout design. Because automating an overall solution for the problem of processing plant layout design is, as yet, impracticable, the design process was, like that of Gunn, partitioned into a series of smaller problems, which could be tackled sequentially and independently, using the appropriate quantitative or qualitative approach. Several computer programs and algorithms were developed for the partitioned problem of plant layout at various stages.

The procedure through which these programs and algorithms approach in tackling the problem can be summarized as follows:

(a) The whole plant is divided into activity groups and plant blocks.
(b) Every plant block is considered in turn and the following issues are determined:
   (i) the priority sequence of locating pieces of equipment inside a plant block;
   (ii) the location of plant block item;
   (iii) pipe routeing;
   (iv) cost of the proposed plant block layout.

This system is revised as required as illustrated in Fig. A.1.

When the layout in all the plant blocks constituting the plant under consideration is decided upon, a computer program is used for the optimal arrangement of these plant blocks around the central piperack as specified by the designer.

A.1.4 OPTIMAL PROCESS LAYOUT BY A BRANCH AND BOUND TECHNIQUE

Mustacchi[5] considered the optimization of plant layout by applying a branch and bound search method. The algorithm developed tackles plants with items

Fig. A.1 Computer-aided block development of layout by Shuqair[4] (a) plant block no. 3 plot plan (b) isometric view of plant block no. 3 (c) plant block no. 3 layout system (d) complete plant final layout system

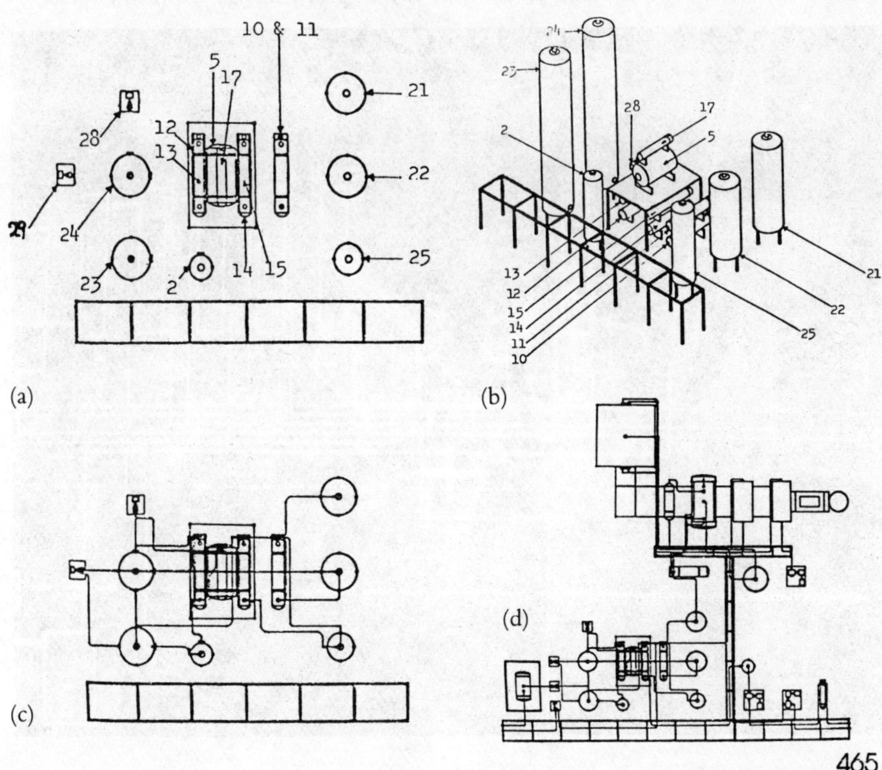

on one side of the piperack and minimizes the cost of pipes travelling along the piperack.

The limitations of the technique developed are that the technique considers the minimization of piping cost only: no building or operating costs are considered. Also, the technique handles plants with items on one side of the piperack only.

A.2   PROGRAMS FOR COMPUTER VISUALIZATION OF LAYOUTS

A.2.1   THE COMPUTER-AIDED PRODUCTION OF UNIT PLOT PLANS

Bush and Wells[6] used a computer graphics screen and a light pen for the generation of plot plans and pipe works. The designer directs and assesses visually the progression of the layout according to the cost which appears on the screen for each alternative layout.

The procedure is to assume the shape of the central piperack which is drawn to scale on the screen. The designer may then input the information

Fig. A.2 The computer-aided production of unit plot plans by Bush and Wells[6]

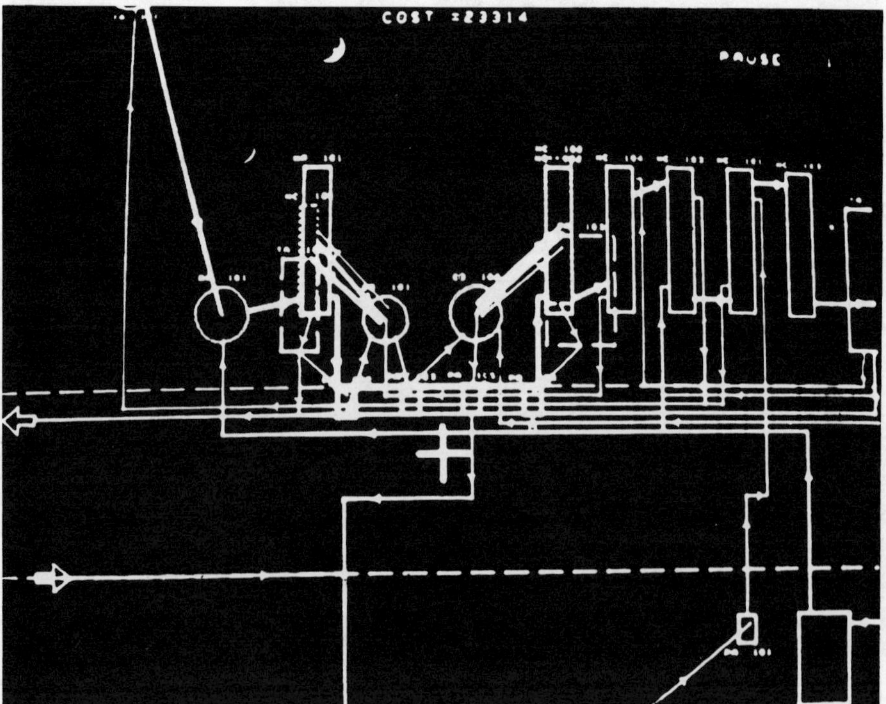

about any item to be drawn. This information consists of a code specification denoting the type and function of the item and also a number denoting the particular item intended. This should correspond with that used on the process flow diagram. Each type of plant item is represented by a circle or rectangle (Fig. A.2). The computer calculates the scale size of the item representation and displays this together with the access requirement which is represented also as a circle or rectangle surrounding the item. Using the light pen the designer moves the tracking cross to a desired position while the machine regenerates the item on the cross. The designer appears to 'pull' the item around the screen using the light pen. In this way each plant item is displayed and positioned.

Three types of pipework representations were used to simulate various types of connection by pipes as follows:

(a) Vessel-to-vessel piping along the piperack.
(b) Pipe-to-vessel piping.
(c) Direct piping.

As the designer positions each plant item on the screen, the computer displays any pipe connected to the item according to the type of piping required, if there is any. Upon relocation of a plant item, the pipework is automatically regenerated at the new item position. As the pipeline is displayed and positioned by the designer the estimated overall cost is shown at the top of the screen. Thus the designer may immediately assess the cost effectiveness of his decision. In order to assist the designer in selecting the item to move, the pipework appears at different brightness levels according to installed cost.

A.2.2   THE DETERMINATION OF PLANT LAYOUT BY INTERACTIVE COMPUTER METHODS

Three separate problem areas were studied by Newell[7]: vessel layout, pipe routeing and the visualization of a digital model of a chemical plant.

Vessel layout was tackled by modelling each vessel as a collection of simple geometric shapes which have relative positional constraints imposed in order to represent differing vessel sizes and shapes (Fig. A.3). Intelligent search procedures were used to optimize the layout, and because of the combinatorial explosion inherent in such problems, facilities were provided for the user to specify constraints which reduce the solution time.

Pipe routeing was tackled by using a shortest route algorithm that generated minimal routes on an automatically generated network. This network was so constructed that all obstacles in space were avoided by the routed pipes. Algorithms for routeing branched pipes with three or more terminal nozzles were presented. The routeing algorithms were designed such that constraints on a routeing scheme could be incorporated by the user defining favourable or unfavourable volumes for routeing pipes. An algorithm for jointly minimizing the cost of pipe and bends was also presented.

The method concerned with visualization proposed a novel hidden-line removal algorithm which was capable of handling a reasonably complex scene. Figure A.3 illustrates some of Newell's methods.

Fig. A.3 Computer-aided layout development and visualization by Newell[7] (a) pipes routed to avoid walkway between vessels (b) example of pipes encouraged to run along the same pipetrack (c) three views of a reformer

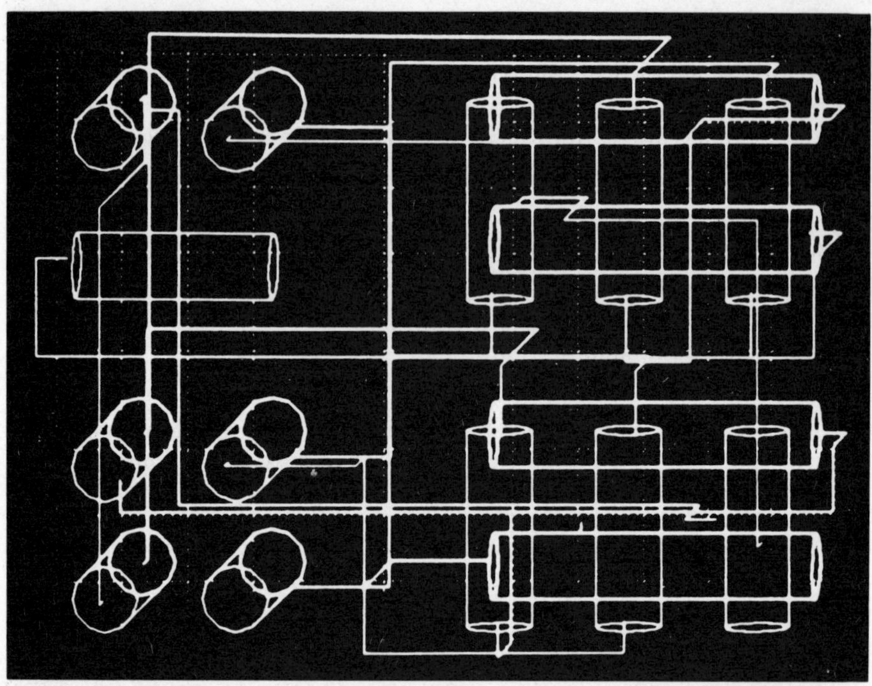

(a)

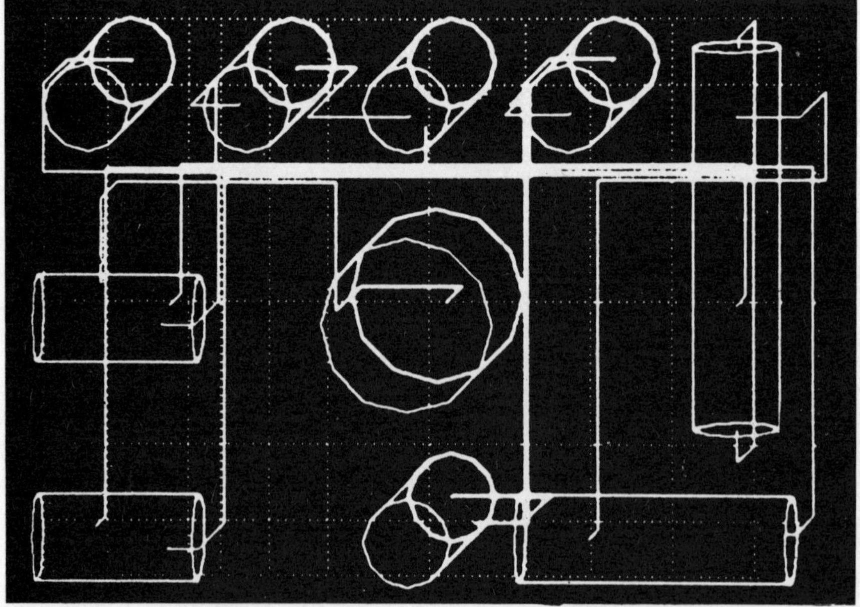

(b)

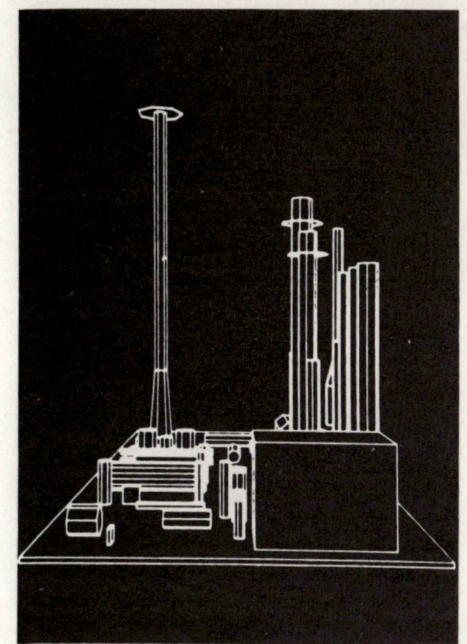

c(i)

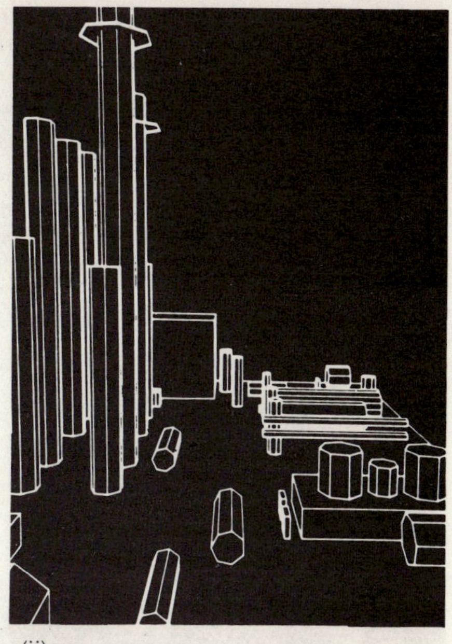

c(ii)

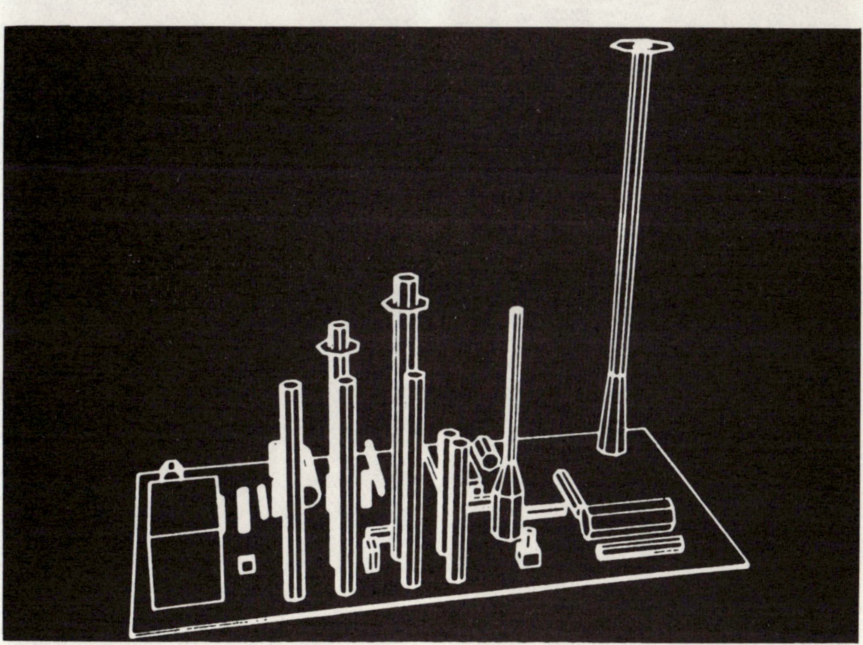

c(iii)

In further development of this work Leesley and Newell[8] used a database interactive graphics system with a display unit to create the image of the plant in the database, visualize it three-dimensionally and manipulate it on the screen using certain viewing facilities till a satisfactory layout was achieved (Fig. A.4).

Fig. A.4 Computer-aided determination of plant layout by Leesley and Newell[8] (a) system comprising the three subsystems one of which is a vacuum distillation unit (b) vacuum distillation unit after the two other subsystems have been removed (c) another view of unit (d) sketching facility used to fit pipes to pumps of vacuum distillation unit

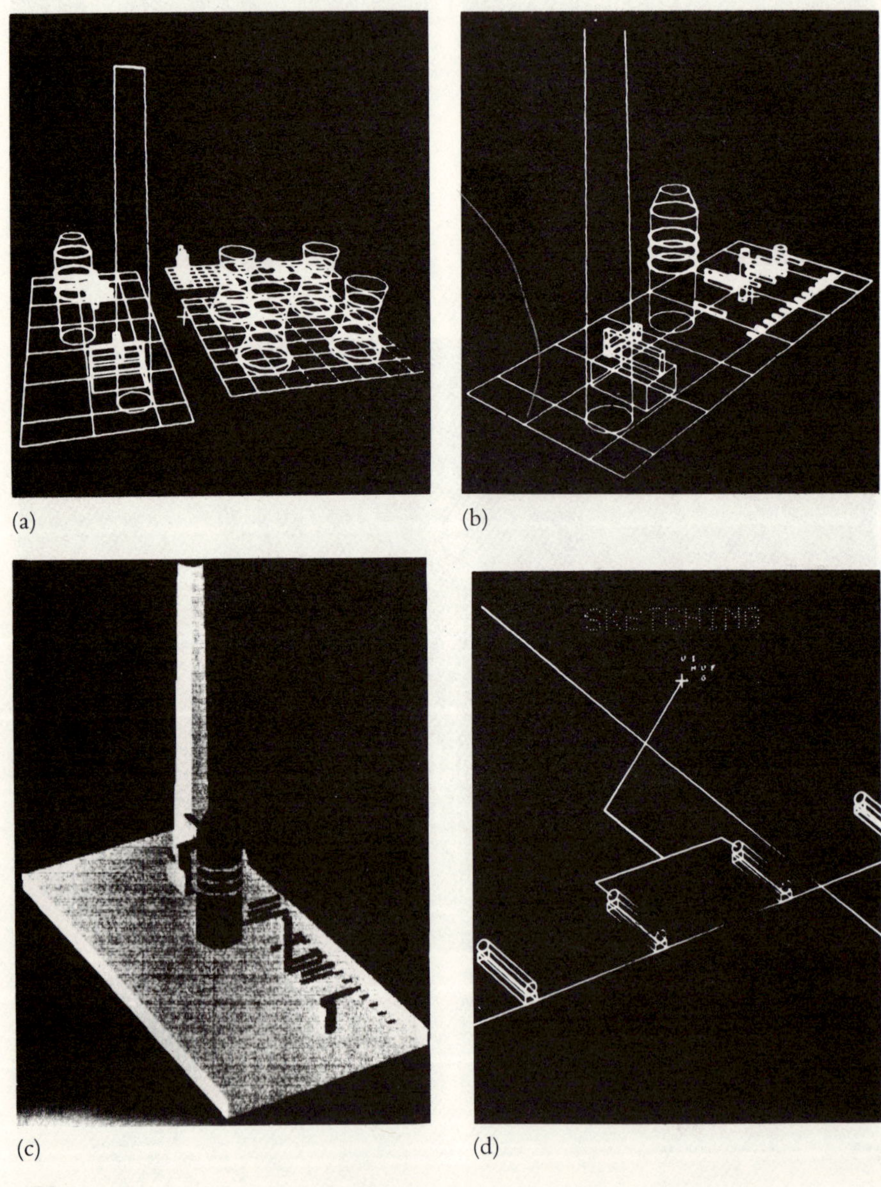

(a)

(b)

(c)

(d)

Information about the plant, its vessels, and their nozzles was given to the program. To make this task simple, a hierarchical input structure was developed in the following way:

(a) The total plant complex was referred to as a set of subsystems.
(b) Within each subsystem there are vessels each one of which is made up of shapes.
(c) Each vessel has a number of nozzles.

A number of facilities exist for manipulation of the picture in order to provide a comprehensive viewing procedure. These facilities are:

(a) Control of picture parts.
(b) Magnification.
(c) Choice of viewpoint.
(d) Hidden-line removal.

The above facilities can be used by the engineer to examine the results of a first guess of the spatial positions of the vessels. At this stage, vessels might be moved around within each subsystem in turn till all subsystems give a satisfactory layout. Having done so, the pipes must be added to connect vessels according to other input instructions which are prepared separately.

A facility known as 'sketching' is used to sketch pipes to the picture. Although the designer sketches on the two-dimensional screen surface, the program would be able to convey his intentions correctly to the three-dimensional database of the plant.

The work of Leesley and Newell was developed along with that of Madden and Taylor[9] to produce a package entitled Plant Design and Management System (PDMS)[10] now in wide commercial use. The central feature of PDMS is a large database in which are stored the identity, size, shape, location, etc. of all plant items and components in the plant. The database is built up by designers as decisions and information on equipment positions, pipe routes, structures, etc. are made available. The design team work interactively on separate parts of the database, using traditional design information received from vendors and other design disciplines. Modifications, or additions, to the developing design are simplified by database information access strategies and the interactive working procedures.

In effect, the database is a full-size three-dimensional model of the plant which is always up-to-date with the current state of the design. Special-purpose application programs use the information stored in the database to carry out specific tasks which can be used at any time to aid designers in different aspects of the plant design. These application programs can be grouped into four types of task: building up a design model, design aids, design validation and design visualization.

Building up a design model

Interactive methods are provided for creating, positioning, moving or modifying models of items such as equipment, structure, piping, etc. There are facilities for the creation and storage of a master catalogue of standard piping components.

Selective retrieval of catalogued piping components according to the rules set out in project piping specificatons can be made to ensure correct component selection during piping layout.

Design aids

There are automatic pipe-routeing facilities to assist comparison of initial plant layout proposals. Using these, the designer can build up simple equipment models, position the equipment in a proposed layout, add favourable areas for piping such as pipe tracks or unfavourable areas for piping such as access spaces and identify equipment nozzles to be joined by pipes of given sizes and specifications. The facility will then route the pipes orthogonally within the available space according to a 'least penalty' algorithm and add the pipes to the database. Drawings and pipe quantities can then be recorded; the model or the constraints on piping can then be changed by the designer and the automatic pipe routine repeated with the drawings and pipe quantities updated. The proposals thus generated can be compared against other project constraints using the piping quantities as a quantitative aid to converging upon an acceptable overall layout.

Automatic facilities for the layout and spacing of pipes on piperacks according to designers' instructions and project design standards.

Design validation

Checks can be made for physical interference between any components in the design during the design stage.

The pipework can be checked on a component-to-component basis for connectivity and continuity to detect any logical, connective or physical connection errors.

Design visualization

Pictorial views of the whole plant, or any part of the plant, can be drawn from any viewpoint to convey overall pictures (see Fig. 7.16 (p. 98) and Fig. 7.7 (p. 78) or highlight individual aspects of equipment layout (Fig. A.5).

Conventional engineering plans (Fig. A.6), elevations (Fig. A.7), sections, etc. can be drawn complete with automatic generation and updating of dimensions and notes.

All drawings can be produced with fully automatic removal of hidden lines to enhance the readability of the drawings or pictures.

Printed reports can be made of any aspect of the design stores in the database. Familiar examples include equipment lists, nozzle schedules, pipe lists, piping materials takeoff (MTO) etc. These reports are available in any format specified by the designer to suit his, or management's, needs.

At present (mid-1984) PDMS represents the most successful and sophisticated use of computer aided design (CAD) techniques in plant layout. Further developments of computer-aided three-dimensional modelling and visualization can be expected to explore ways in which layout-related design constraints can be expressed within application programs to provide designer aids for more generation, analysis and evaluation.

Fig. A.5 Computer-aided development of equipment layout (Courtesy: Babcock Woodall-Duckham (see Fig. 7.16(b) (p. 98) for final result)

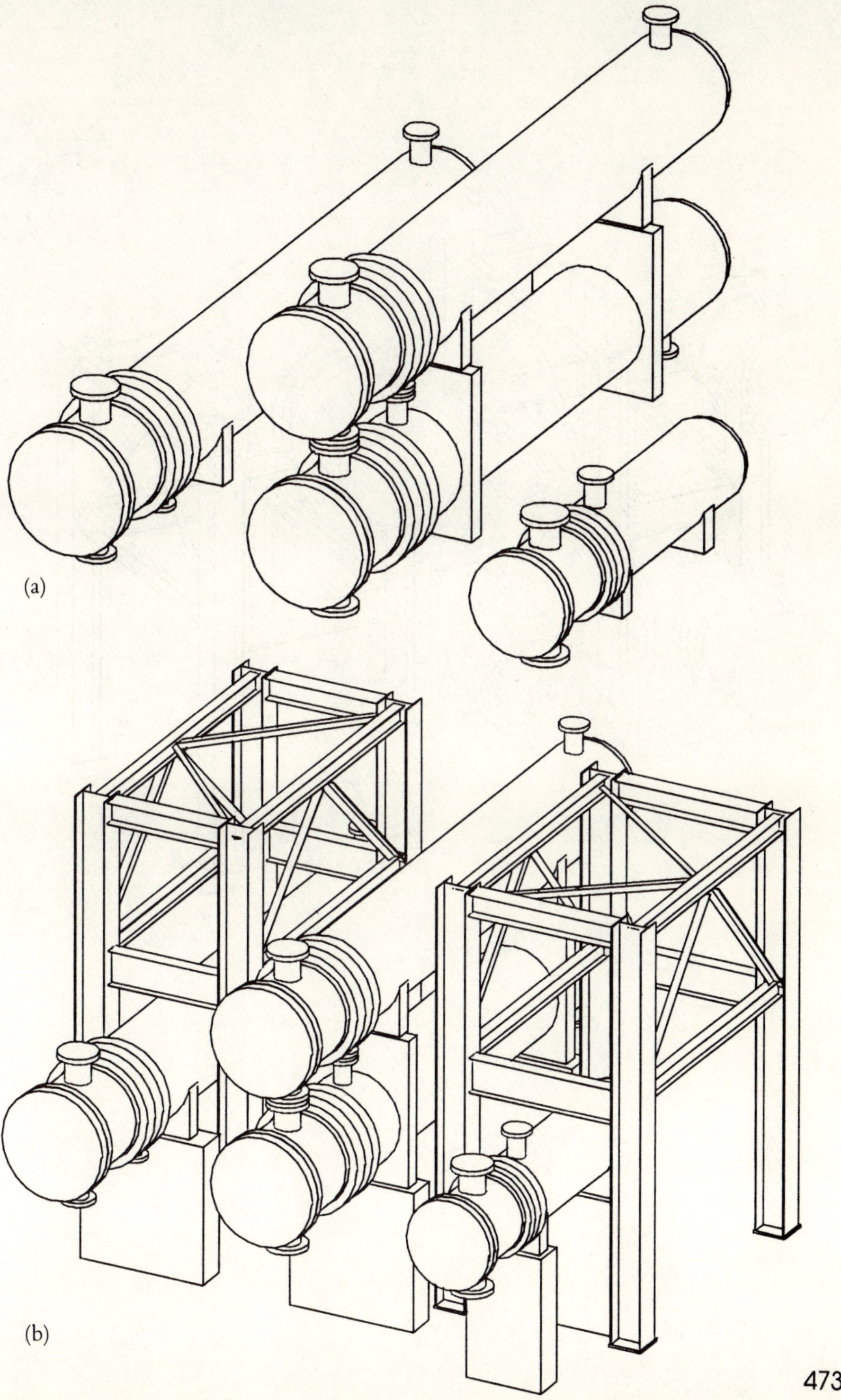

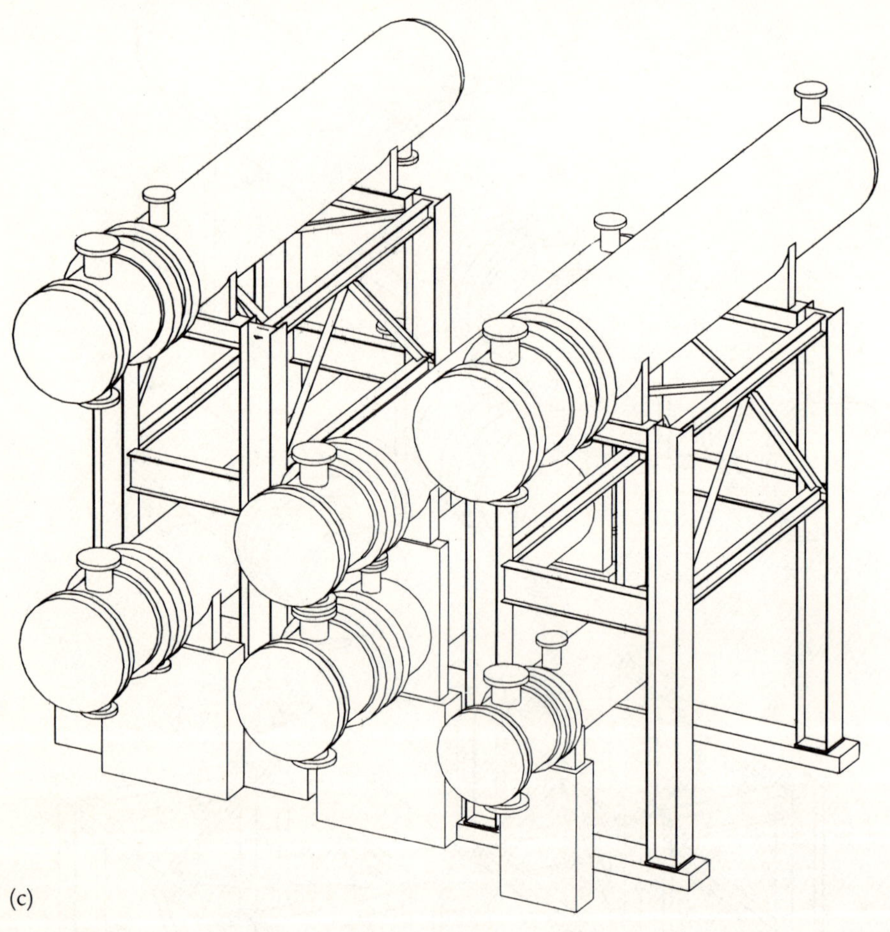

(c)

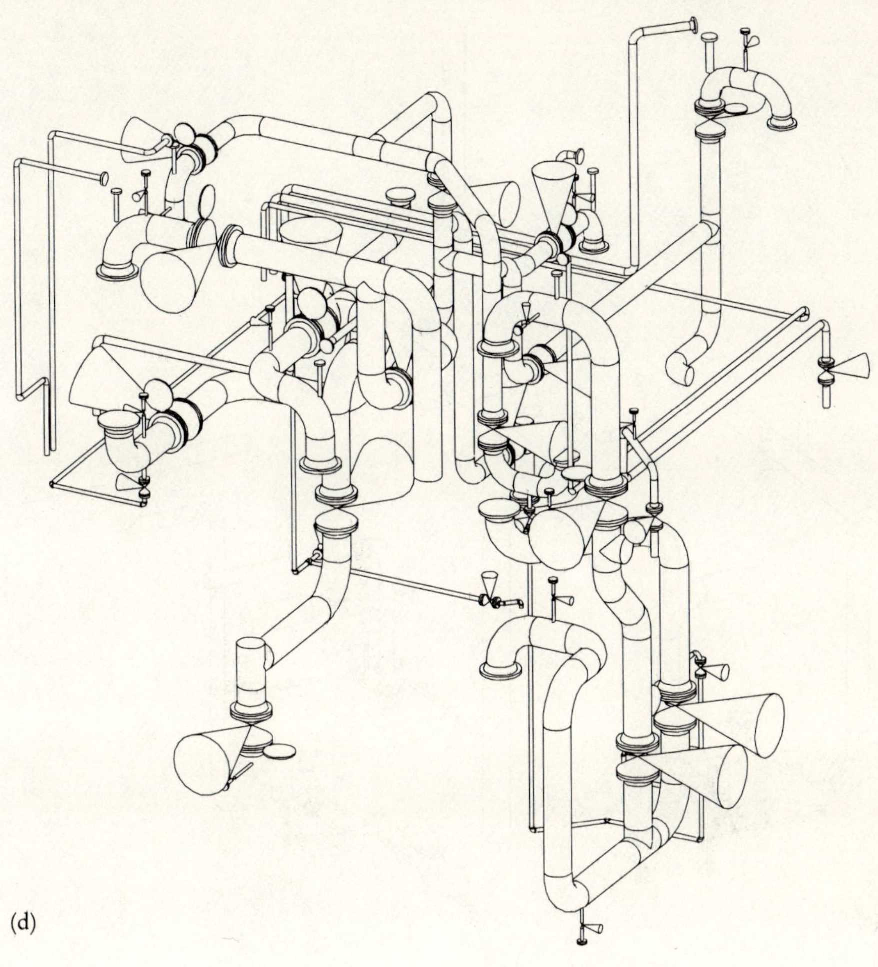

(d)

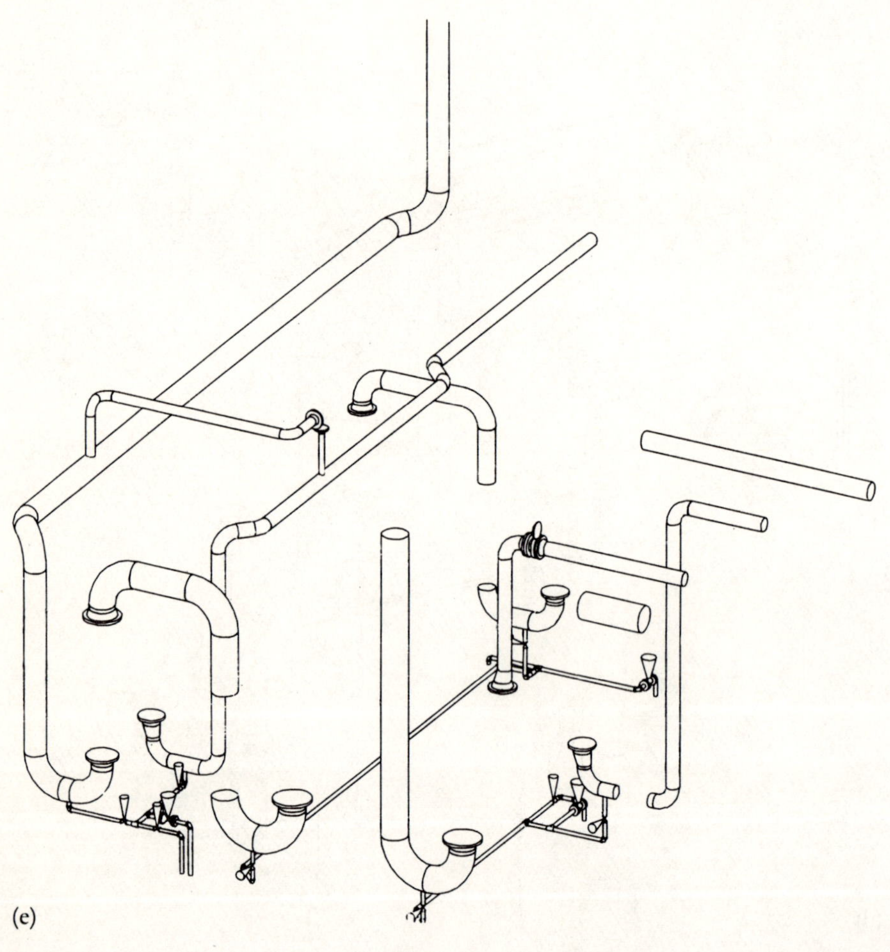

(e)

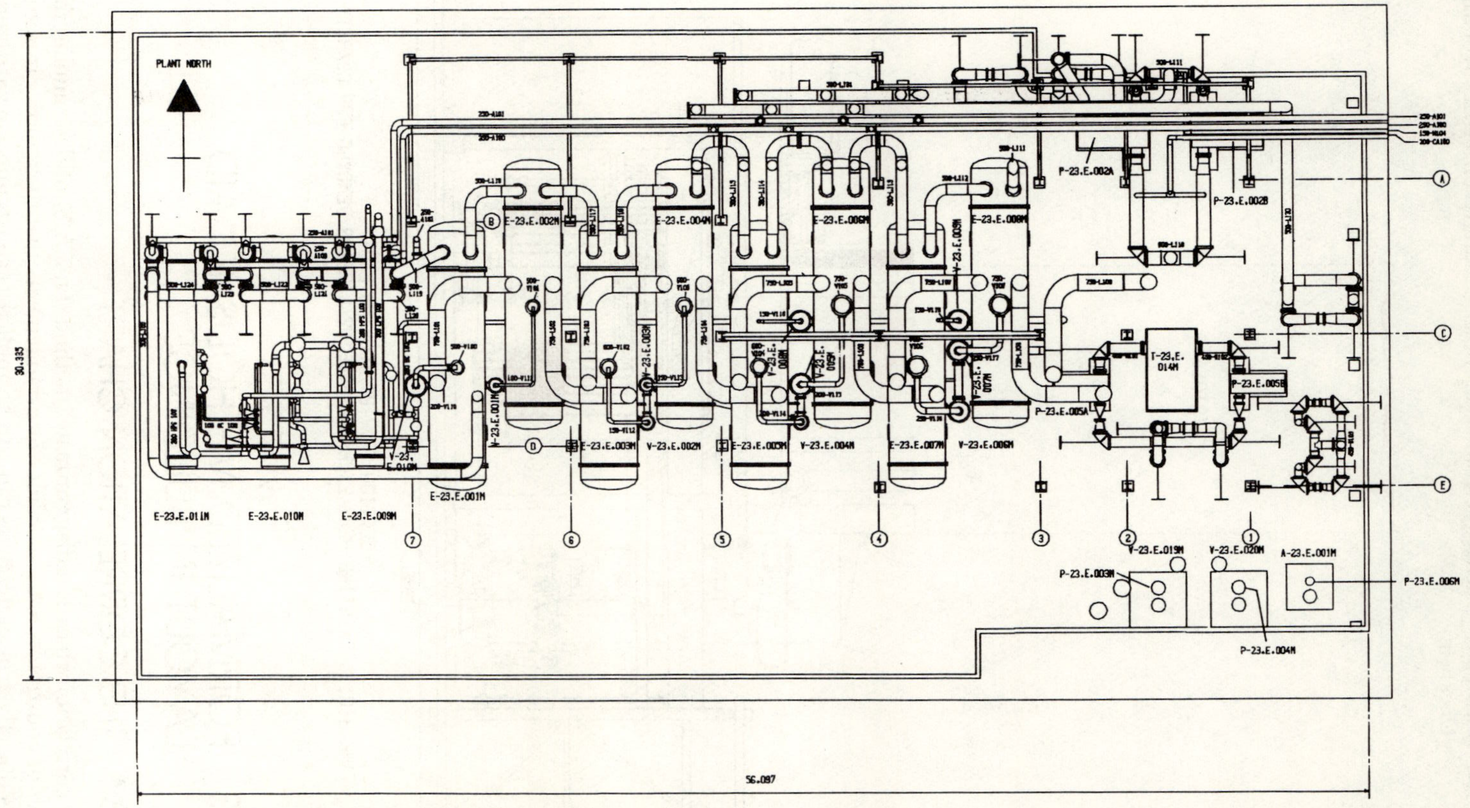

Fig. A.6 Computer-generated plot plan (Courtesy: Babcock Woodall-Duckham)

Fig. A.7 Computer-generated elevation drawing (Courtesy: Babcock Woodall-Duckham)

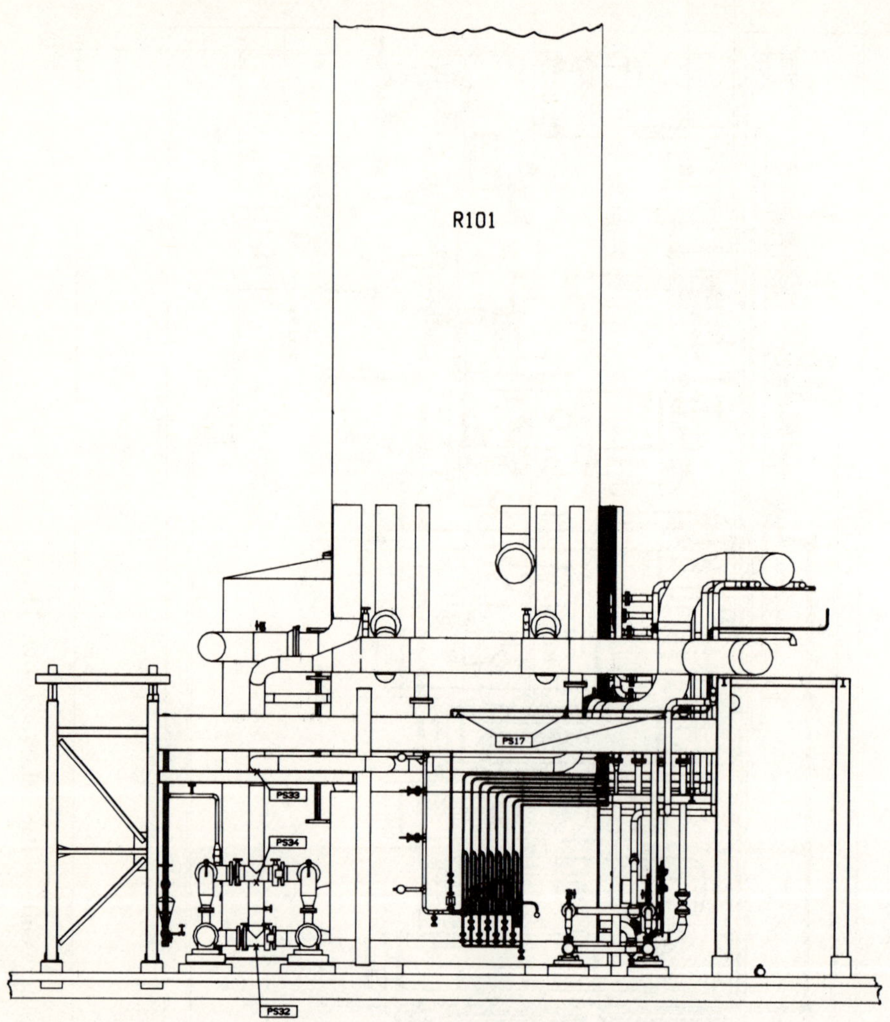

A.3   FORMALIZED METHODS TO AID LAYOUT IN BUILDINGS

A.3.1   THE CORRELATION CHART

The correlation chart is a diagrammatic method (Fig. A.8) of determining the effect of constraints and recording the arrangements that they allow. In some cases objectives or preferences can also be applied as constraints, so narrowing

478

Fig. A.8 Correlation chart method (a) progress chart (b) grid showing absolute exclusions (c) grid showing permissible layouts

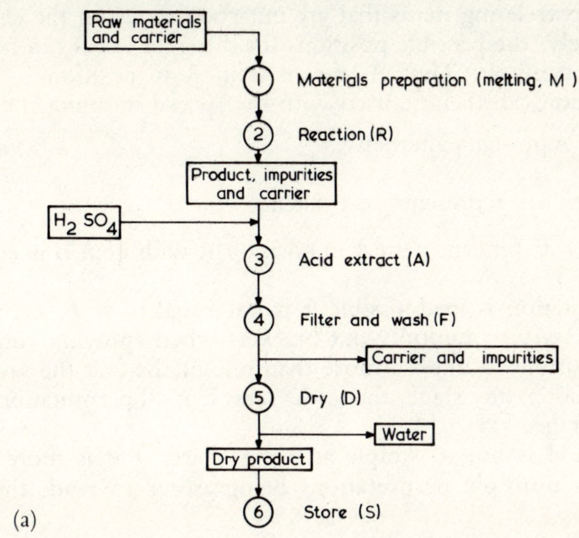

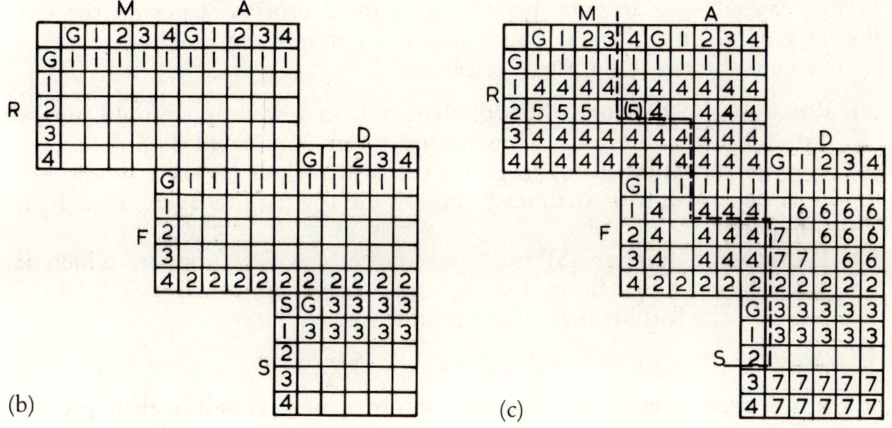

the field still further. The method only gives guidance on positional relationships, not spacings.

A grid is drawn with the rows representing possible positions – such as floors in a building, or numbered positions in an area – of one plant item, and the columns representing possible positions of another plant item. Constraints are recorded and labelled, say, X, Y, Z . . . . . If any constraint prohibits an item going into a particular position then the appropriate square is struck out by writing in it the reference label of that constraint. Vacant squares thus show permitted combinations.

The sets of lines of the grid can be extended and crossed by rows or columns representing other items, and prohibitions or preferences again applied. Then the only feasible permutations are those that can be traced through the rectangular network using only vacant squares. The advantages of this method are its visual presentation and the ease with which it can be learned and used; the disadvantages are the amount of drawing-up required, and the difficulty in correlating items that are not contiguous on the chart.

Alternatively, the possible positions for different items can be represented by algebraic symbols. Thus A_1 means item A in position 1. Ordinary algebraic notation can then be used with the special meaning, i.e.:

addition – represents alternatives
and
multiplication – represents co-existence

Thus $A_1 (B_2 + B_3)$ means item A in position 1, with item B in either position 2 or position 3.

If a permutation is inadmissible it is put equal to zero, i.e. it is deleted. It is not necessary to multiply-out brackets when applying constraints, and it may be possible to impose more than one of these at the same time. On multiplying-out at any stage, the terms represent all permutations still allowable at that stage.

This method is not so simple as using charts, but is more powerful in dealing with multiple permutations being easier to read, though not to visualize.

As an example of this method consider the following process, the outline process diagram for which is shown in Fig. A.8a. The problem is one of vertical layout – to arrange the vessels of these process stages on the five floors (ground, first, second, third, fourth) of an existing building.

The constraints can be listed as follows:

(a) Reaction and pressure filters discharge from below and should not be put on the ground floor. Thus R_G and F_G are eliminated (Fig. A.8b).
(b) The pressure filter is not to go on the fourth floor, as there would not be enough room to withdraw the spindle for maintenance. Thus F_4 is eliminated.
(c) The storage hopper (S) must be above a weigh hopper, which is situated on the first floor. Thus S_G and S_1 are eliminated.

So all the feasible layouts are represented by:

$$(M_{G,1\dots 4}) (R_{1\dots 4}) (A_{G,1\dots 4}) (F_{1,2,3}) (D_{G,1\dots 4}) (S_{2,3,4})$$

(d) It is desired to leave reaction and acid treatment vessels in their present positions on the second and first floors respectively. Thus R_2 and A_1 are to be chosen if possible.
(e) For reasons of heat transfer and gravity flow, melting should be adjacent to reaction and above it. This means that M_3 or M_4 should be chosen: of these, M_3 is nearer to R_2, and so is to be preferred.
(f) For ease of handling the cake from the pressure filter, it should be dropped directly into the dryer. This implies

$$F_1 D_G + F_2(D_G + D_1) + F_3(D_G + D_1 + D_2).$$

(g) Subject to all other conditions, the cost of lifting materials is to be minimized. Lifting may be necessary for moving slurry from A to F, and/or moving powder from D to S. Thus, S_3 and S_4 are ruled out as not minimal. Similarly it rules out F_2D_G, F_3D_G, F_3D_1, all of which impose more lifting than the other three relations between F and D.

480

The schemes to be evaluated in terms of the costs of moving slurry and powder are then reduced to:

$M_3R_2A_1F_1D_GS_2$: slurry moved horizontally on first floor; powder to be raised from ground to second floor (Fig. A.8).

$M_3R_2A_1F_3D_2S_2$: slurry to be raised from first to third floor; powder moved horizontally on second floor.

$M_3R_2A_1F_2D_1S_2$: slurry and powder each raised one floor from first to second.

Should it transpire that there is congestion on any of the floors, then additional lifting costs may be accepted. Terms like M_4 or S_3 could be brought back into consideration.

Note: European floor notation is used above. For US G → 1, 1 → 2, 2 → 3, 3 → 4.

A.3.2   TRAVEL CHARTS

These were originally used for finding the optimum linear arrangement of machines in jobbing shops, and the name related to the record of the amount of travel of different jobs between the machines.[11]

An effective process-unit layout can be determined using this technique by establishing the magnitude of materials handling relationships between all combinations of items of the plant. The theoretical optimum layout would then be one in which each item would be adjacent to every other item with which it has relationships. In most cases, this theoretical optimum is very difficult to achieve, but one should attempt to approach it.

The chart is set up in squares. Both horizontally and vertically, the number of squares needed is equal to the number of items represented in the layout. These items are listed horizontally across the top of the chart as contributing items and in the same order, vertically, down the chart as using items. A diagonal line is drawn across the chart. The total cost per unit distance of all linkages between any two items is then inserted in the square which is the meeting point of a row and a column each representing an item. In the case of neighbouring items this cost will lie in a square next to the main diagonal; for items that are not neighbours, the cost will lie in a square away from the main diagonal. The total cost of linking all items is found by multiplying the cost in each square by the distance of that square from the main diagonal and summing. Visual inspection of the columns then shows which interchanges (permutations of the items) would reduce this total cost by bringing certain costs nearer the main diagnonal. The process is repeated until an optimum arrangement has been reached.

This method can be modified to compensate in part for different plant sizes. As a method of approximating, certain simplifications can be made; e.g. it can be used with just significant costs; again, groups of items that are to be sited near one another can each be treated as a single item with 'group' linkage to other items.

If space utilization, travel time or distance is a critical item of consideration, this technique can be used to advantage. Moreover, the technique can be used in determining the most economic position of a new unit in relation to the whole plant, or the location of a piece of equipment within an existing plant. However, the main disadvantage of the technique is that it only leads to an optimum linear solution which is seldom wanted; on the other hand, it measures the relative importance of having different pairs of items close to one another, which is a useful first step to the two- or three-dimensional problem.

Some computational routines have been devised for two- and three-dimensional proximities (see section A.4). The simplest manual method is probably one of trial and error in which items are situated near one another as suggested by significant costs, and the arrangement modified by reference to the travel chart.

A.3.3 SEQUENCING TECHNIQUES

Although mainly used to find feasible sequences of operations in a process, sequencing techniques may sometimes be useful in layout problems. They are based on a grid similar to the travel chart, but usually simpler, e.g. costs may refer only to vertical transports, or costs may be omitted and symbols used to indicate inadmissibility or preference. Instead of permutating the items, however, chain diagrams are prepared (similar to outline process charts), starting from some key item of plant or some key position (e.g. ground level); these are prepared on a basis of 'preferred links' till all items have been included in one chain or another. The next step is to consider simplification by combining items at the same level or in neighbouring positions and schematic diagrams may be prepared, showing other links such as access, etc. Finally, the most promising chains may be rough-costed, or otherwise compared, and the best selected for study in greater detail. Sequencing techniques are essentially a procedure for analysing the effect of individual layout constraints or preferences as a means of comparing layout ideas and converging towards an optimum.

A.3.4 ERGONOMIC METHODS

Ergonomic computer methods, now being developed for use in production engineering situations, do not appear applicable to process engineering problems except, perhaps, for packaging and warehousing.

A.4 PROGRAMS FOR FACTORY LAYOUT CONCEPTION

A.4.1 CRAFT 'COMPUTERIZED RELATIVE ALLOCATION OF FACILITIES TECHNIQUE'

This is a program by Armour, Buffa and Vollman[12] which uses a simulation algorithm to arrange departments in a given plant area. Input into the algorithm includes the area of the individual departments or plant units, handling volume in unit loads moved between departments and handling cost per unit load per unit distance between pairs of departments. The program determines distances between departments in any arrangement and by using the handling cost matrix, seeks minimum cost of exchanges of department locations. The exchange is repeated until the minimum cost is found.

The measure of effectiveness suggested in CRAFT makes sense when the layout planning is limited to production departments; however, it does not take into consideration the problems involved in planning service facilities, such as washrooms, maintenance shops and the like, in the layout. On the other hand, the unit cost is established in advance and is independent of the plant arrangement, which is often not true in practice. Furthermore, CRAFT is capable only of handling a problem where the building is specified. In the design of a new building, most experts agree that a building plan should be designed around the layout, not the layout around the building. CRAFT is not designed to tackle this problem.

A.4.2 ALDEP 'AUTOMATED LAYOUT DESIGN PROGRAM'

Seehof and Evans[13] developed a program which applied a heuristic programming method to create block layouts. Input to the program includes the area requirements of each department or unit, and preference table, which is a matrix of weighting factors that indicates the desirability of units to be near each other. These weighting factors can be given values by the user to represent his view of their relative importance.

The design program will lay out up to three floors. It performs a two-step process for each floor of a building. Firstly, any non-assigned department's area is allocated to specific floors and, second, those departments assigned to a floor are given a specific location on that floor. A department is randomly selected and processed, then the preference table for that department is searched to find any department with a performance of highest priority which is then selected. When all departments have been processed the resulting layout is scored and printed out. By selecting block layouts with highest scores the best layouts are obtained for further analysis.

This program can be used with non-production departments as well as production departments. It has the same limitation as CRAFT regarding the requirement for a building outline in a program input. Therefore, it is not geared to resolving the new plant problem.

A.4.3 CORELAP 'COMPUTERISED RELATIONSHIP LAYOUT PLANNING'

This program by Lee and Moore[14] requires input in the form of a relationship (REL) chart as described by Muther.[15] The data from the REL chart must be in matrix form for handling in the computer memory. Other components of the input information are the area requirements for each of the departments included in the REL chart, the size of the unit square to be manipulated, and the maximum ratio of building length to width. This ratio is necessary to allow the layout to take an irregular shape. The main algorithm consists of a heuristic program which adds departments in a logical fashion to generate a block plan layout. The output consists of a matrix printout representing the block plan.

REFERENCES

1. Fine, B. 'Piping design in chemical plant', *I.Chem.E./A.I.Ch.E. Sym. Ser.* **4**, 107, 1965.
2. Gunn, D. J. 'The optimized layout of a chemical plant by digital computer', *Computer Aided Design,* **2**, (3), 111, 1970.
3. Al-Asadi, H. *Computer Aided Layout of Chemical Plant.* Ph.D. thesis, University of Wales, Swansea, 1980.
4. Shuqair, M. M. *Studies on Plant Layout.* Ph.D. thesis, Sheffield, 1978.
5. Mustacchi, C. 'Optimal process layout by a branch and bound technique', *Ing. Chem. Ital.,* **10**, (12), 203, 1974.
6. Bush, M. J. and Wells, G. L. 'The computer aided production of unit plot plans for chemical plant', *I.Chem.E. Sym. Ser.* **35**, 2:15, 1972.
7. Newell, R. G. *Algorithms for the Design of Chemical Plant Layout and Pipe Routing,* Ph.D. thesis, London, 1974.
8. Leesley, M. E. and Newell, R. G. 'The determination of plant layout by interactive computer methods', *I.Chem.E. Sym. Ser.* **35**, 2:20, 1972.
9. Madden, J. and Taylor, V. T. 'Design data requirements in the process industries', *Computer Aided Design,* **11**, (3), 142, 1979.
10. Computer Aided Design Centre Publications, Madingly Road, Cambridge (Now (1984) CAD Centre Ltd).
11. Wild, B. *Mass-Production Management.* Wiley, London, 1972.
12. Armour, G. C., Buffa, E. S. and Vollman, T. E. 'Allocating facilities with CRAFT', *Harvard Business Review,* **42**, (2), 130, 1964; *Management Science,* 294, Jan. 1963.
13. Seehof, J. M. and Evans, W. O. 'Automated layout design program', *J. Industrial Engrg,* **18**, (12), 690, 1967.
14. Lee, R. C. and Moore, J. M. 'CORELAP – Computerized relationship layout planning', *J. Industrial Engrg,* **18**, (3), 195, 1967.
15. Muther, R. *Systematic Layout Planning,* Ind. Educ. Inst., Boston, Mass. 1961.

HAZARD ASSESSMENT CALCULATIONS

CONTENTS

B.1   INTRODUCTION

B.1.1   SCOPE

The scheme and equations presented in this Appendix are given to aid the assessment procedure outlined in section 8.5 (p. 117) and are intended only for in-house assessment of layouts. Simple methods not needing a computer are given so that order of magnitude calculations can be made quickly. The results of the calculations will, hopefully, give indications to the engineer of how he can improve the safety of a layout.

The calculations, however, are not meant to be used for justification of layouts. For this a more detailed assessment is required. The subject of hazard assessment is rapidly increasing in knowledge so it would be pointless to try and include comprehensive methods in this book as they would become quickly out of data (e.g. see 4th International Loss Prevention Symposium[1] and Safety and Reliability Directorate[2]). Thus the engineer needs to consult the current literature before carrying out a detailed hazard assessment. Such an assessment can require considerable effort and take some time to prepare. Nevertheless, it is important that this detailed assessment should be undertaken; indeed, the safety authorities will increasingly require it.

Further discussion on calculation methods may be found in Lees,[3] TNO,[4] Wells[5] and Lihou.[6]

B.1.2   GENERAL PHILOSOPHY

When undertaking hazard assessment there are a number of uncertainties, such as:

(a)   Position, size and direction of leak.
(b)   Direction and speed of wind.
(c)   Weather conditions.
(d)   Position and number of potential victims.

In addition, when the assessment is carried out in the early stages of design the topography and degree of development of both the site and the environs are uncertain. Thus to be consistent, the accuracy of the calculation methods should be of the order of the level of uncertainty.

486

It is assumed that the engineer will only select items for assessment where loss of containment is likely. The Appendix therefore aims to give the simplest equations as close as possible to the worst feasible situation which is often the escape travelling directly from source to target. In this way it is hoped that the accuracy is consistent with the uncertainty. This may not always be the case so, if in any particular assessment the engineer feels unhappy about the use of the methods in this Appendix, then a more detailed assessment should be undertaken.

The cause of leaks is not considered, only the consequences of the leak, i.e. the simple question is asked: 'What happens if there is a loss at this particular item from whatever cause?' The Appendix also assumes that only instantaneous releases give clouds large enough for explosions or fireballs to occur though this is not strictly true. A criterion to distinguish between instantaneous and continuous releases is given.

Other restrictions, limitations, etc. are given in the respective sections.

It should be stressed that values calculated in this Appendix or read from Appendix C are not to be incorporated into a design without first using engineering judgement.

B.2   INSTANTANEOUS RELEASE OF GAS OR VAPOUR

B.2.1   SIZE OF CLOUD

Table B.1 shows how to calculate the size of the cloud, assuming simple thermodynamic flash and Example B.1 illustrates its use.

As the amount of entrainment depends on the particular manner of failure, it is very uncertain. So the suggestion of Kletz[7] is used, i.e. that the amount of entrainment is equal to the amount of vapour.

Table B.1   Size of instantaneous cloud

Weight in vessel	$= W_o$
Fraction flashed	$m = s\left(\dfrac{T_o - T_b}{L_b}\right),\ m \leqslant 1$
	where s = liquid specific heat L_b = latent heat at atmospheric boiling point T_b T_o = temperature of vessel contents
Fraction entrained	$e = m$ for $m \leqslant 0.5$, $e = 1 - m$ for $m > 0.5$
Weight of cloud	$W = (e + m)\, W_o$ (flashing liquid) $W = W_o$ (gas)
Volume of cloud at atmospheric pressure P_a and air temperature T_a	$V = \dfrac{WRT_a}{MP_a} = \dfrac{WM_a}{\rho_a M}$ where R = gas constant M = molecular weight ρ_a = air density

Example B.1

Twenty tonnes of ethylamine at 100 °C are held in a vessel 3.5 m diameter by 3.5 m high.

Find the size of the instantaneous cloud given the following data on ethylamine $C_2H_5NH_2$:

Molecular weight	45
Vapour pressure/atm.	1 2 5 10 20
Temperature/°C	17 36 65 92 124
Liquid density	683 kg m^{-3}
Liquid specific heat	2.92 kJ kg^{-1} K^{-1}
Gas specific heat	1.77 kJ kg^{-1} K^{-1}
Latent heat at b.p.	623 kJ kg^{-1}
Heat of combustion (gas)	35.3 MJ kg^{-1} (298 K)
Gas diffusion coefficient in air	1.1×10^{-5} m^2 s^{-1}
Lower flammable limit	0.0355 v/v
Long-term exposure limit	10 p.p.m. v/v (TWA − 8h)
Immediately dangerous to life or health level	4,000 p.p.m. v/v (for 30 min.)

ANSWER

Fraction as vapour

$$m = \frac{2.92(373 - 290)}{623} = 0.389$$

Allow the same for fraction entrained $e = 0.389$

Total flashed

$$= 0.778 \times 20 = 15.6\,\text{t}$$

$$= \frac{15{,}600 \times 29}{1.22 \times 45} = 8{,}240\ \text{m}^3$$

Liquid left $= 4.4\,\text{t (at b.p.} = 17\,°\text{C)}$

B.2.2   DISPERSION OF CLOUD TO LOWER FLAMMABLE LIMIT

The region at risk from a fireball or vapour cloud explosion is determined by the distance the cloud expands before dispersing below the LFL. Ideally, there should be no permanent ignition sources within this region, though this may not be realistic for large releases and/or for releases that are unlikely to occur.

There is some conservatism built into the use of LFL because those quoted in the literature are usually the lowest values found. If the question at issue is explosion overpressure (see next section) then the mixture will have to be some way above the LFL before significant overpressure arises.

In Table B.2, two areas have been considered: one within the site with plenty of site equipment, buildings, etc. causing turbulence, and the second outside the site consisting of arable land with scattered housing, etc. causing little turbulence.

To account for the changeover at the site boundary, a virtual source is introduced. The distance of this source inside the boundary is calculated as that which would still produce, under off-site dispersion conditions, the same site boundary concentration found with on-site conditions (see Fig. B.1). Weather conditions are assumed to be the most frequently occurring, i.e. neutral stability D (see section 8.2.3, p. 105).

Dispersion is assumed to be governed by neutral buoyancy.[8] For a release at ground level the concentration C at a point $X + x, y, Z$, relative to the source of release, is given by:

$$C = \frac{2V}{(2\pi)^{3/2}} \cdot \frac{1}{\sigma_x \sigma_y \sigma_z} \exp. \left[-\frac{X^2}{2\sigma_x{}^2} - \frac{y^2}{2\sigma_y{}^2} - \frac{Z^2}{2\sigma_z{}^2} \right]$$

where x is the horizontal distance travelled by the centre of the cloud from the source

y is the horizontal distance (perpendicular from the centreline of the cloud path)

Z is the vertical distance above the cloud centre

The dispersion coefficients σ_x, σ_y, σ_z are functions of x given in Table B.2 and in section 9 of this Appendix. The correlations quoted are those based on the work of TNO[4] which itself derives from the work of Pasquill[8] and others.[9] They have been changed to give instantaneous concentrations (see section B.3.4). One advantage of TNO's equations is that they allow an explicit relation to find distances. However, the disadvantage is that they are not intended to be used below 100 m. However since there seems to be no other suitable scheme (except Katan's[10] which is limited) they have to be used below 100 m but only for rough guidance.

In the above equation the release has been assumed at ground level and also in Table B.2 concentrations are mostly assumed to be at ground level ($Z = 0$) and concentrations are usually considered at the cloud centre ($X = 0$, $y = 0$).

A satisfactory mechanism for the initial mixing in release does not appear to be available, so the initial cloud centre concentration C_a is conservatively taken as 100 per cent v/v instead of the infinite concentration the dispersion model implies. To allow for this, a virtual source is employed (Fig. B.1). The distance of this source from the release point is that distance the cloud would have to drift in order for the concentration to reduce from infinity to 1 v/v. A further virtual source is assumed to allow for the change of surface roughness at the site boundary (Fig. B.1).

The distances calculated to the LFL (and to the toxic limits − section B.2.5) depend much on the amount of initial mixing assumed. So the reader is advised to use any better model that might be published subsequent to this book. Similarly dense phase dispersion has been ignored because there appear to be no simple equations available. Again, the reader

Fig. B.1 Sources and virtual sources (a) initial mixing of cloud (b) site boundary for cloud (c) site boundary for plume (d) jet mixing for plume

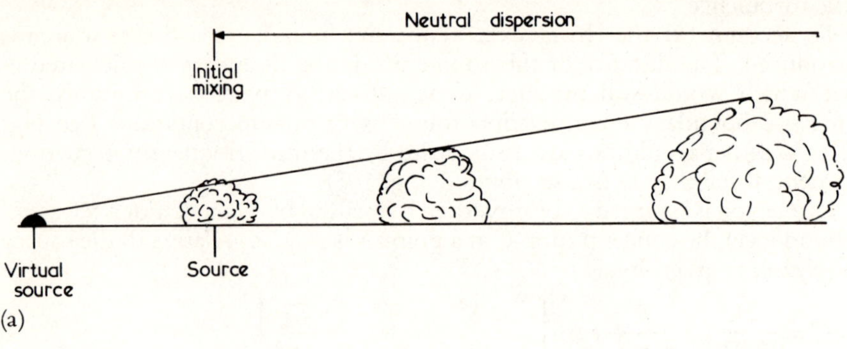

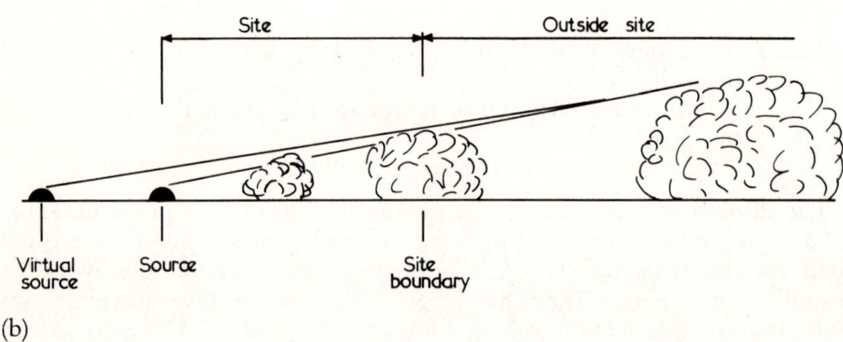

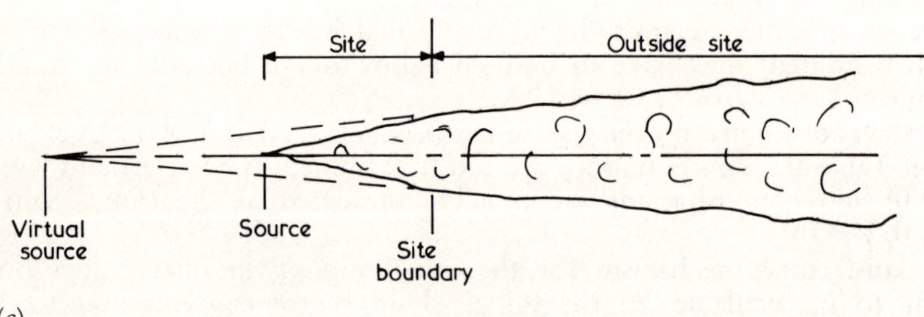

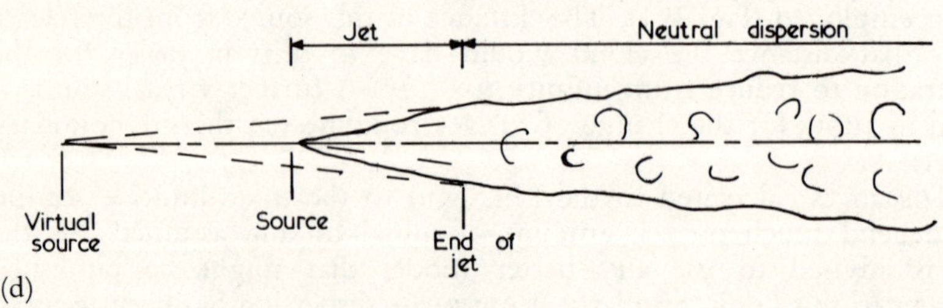

should employ suitable equations that may be published in the future or he may choose to use existing dense phase models in a computer calculation[2]. Further discussion on vapour dispersion calculations had been given by Slater,[11] Cox,[12] Blackmore[13] and Britter and Griffiths.[14] Note that, although the quantitative aspects of dense clouds may be ignored, the qualitative behaviour — such as seeking out slopes, hollows and channels — should always be considered.

Example B.2 illustrates the method given in Table B.2.

Table B.2   Distance to LFL

Distance, a, of virtual source	$$\sigma_x\sigma_y\sigma_z = \frac{2V}{(2\pi)^{3/2}C_a}$$ $a\ (\equiv x)$ found from σs below C_a = assumed initial centre concentration (say 1 v/v, which is conservative)
Distance, x, within site to LFL, C	$$\sigma_x\sigma_y\sigma_z = \frac{2V}{(2\pi)^{3/2}C}$$ $x_v\ (\equiv x)$ found from σs below $x = x_v - a$
Concentration at site boundary	$$C_b = \frac{2V}{(2\pi)^{3/2}\sigma_x\sigma_y\sigma_z}$$ σs found from below with $x \equiv x_b + a$
Distance, b, of virtual source	$$\sigma_x\sigma_y\sigma_z = \frac{2V}{(2\pi)^{3/2}C_b}$$ $b\ (\equiv x)$ found from σs below
Distance x outside site to LFL, C	$$\sigma_x\sigma_y\sigma_z = \frac{2V}{(2\pi)^{3/2}C}$$ $x_v\ (\equiv x)$ found from σs below $x = x_v - b + x_b$
Dispersion coefficients	$\sigma_x = 0.13x$ $\sigma_y = c_y x^{n_y}$ $\sigma_z = c_z x^{n_z}$

Values of c_y, c_z, n_y, n_z, depend on atmospheric conditions and ground surface roughness. A complete list is given in section B.9 but a suitable set to use is (in m):

	On site	*Off site*
σ_x	$0.13x$	$0.13x$
σ_y	$0.064x^{0.905}$	$0.064x^{0.905}$
σ_z	$0.395x^{0.701}$	$0.200x^{0.760}$
$(\sigma_y\sigma_z)^{1/2}$	$0.159x^{0.803}$	$0.113x^{0.833}$
$(\sigma_x\sigma_y\sigma_z)^{1/3}$	$0.149x^{0.869}$	$0.118x^{0.888}$
Lapse conditions	neutral (category D)	neutral (category D)
Roughness, z_o	urban (1 m)	cultivated (0.1 m)

Example B.2

Find the distance to the LFL for the cloud released in Example B.1, assuming the site boundary is 100 m from the tank.

ANSWER

Distance of virtual source

$$\sigma_x\sigma_y\sigma_z = \frac{2 \times 8,240}{(2\pi)^{3/2} \times 1} = 1,046 \text{ m}^3$$

From Table B.2, on site
$a = 129$ m

Distance to LFL

$$\sigma_x\sigma_y\sigma_z = \frac{2 \times 8,240}{(2\pi)^{3/2}0.0355} = 29,475 \text{ m}^3$$

From Table B.2, on site
$x_v = 470$
or $\quad x = 470 - 129$
$\quad\quad\quad = 341$ m from actual source
This is over the site boundary

Concentration at site boundary

$$C_b = \frac{2 \times 8,240}{(2\pi)^{3/2}\sigma_x\sigma_y\sigma_z}$$

$$= \frac{2 \times 8,240}{(2\pi)^{3/2}4,695} = 0.223 \text{ v/v}$$

using Table B.2, on site with
$x_b + a = 229$ m

Distance of virtual source

$\sigma_x\sigma_y\sigma_z = 4,695$ m³
From Table B.2, off site, $b = 265$ m

Distance to LFL

As above $\sigma_x\sigma_y\sigma_z = 29,475$ m³
From Table B.2, off site $x_v = 528$ m
$x = 528 - 265 + 100 = 363$ m
If off site conditions throughout
$a = 151$ m
$x = 528 - 151 = 377$ m

B.2.3   EXPLOSION OVERPRESSURE

Table B.3 relates the size of the vapour cloud with distance and overpressure based on a procedure by Kletz.[7] This uses the TNT deflagration equivalent model which only approximates what occurs. (It is thought that the model overestimates pressures near the epicentre and underestimates pressures further afield). However, it gives an idea of what type of building, etc. should be placed at certain distances from the loss of containment. This TNT model does not apply to clouds below 5 t hydrocarbon equivalent, but they can be ignored as experience has shown that small clouds are not likely to generate much blast pressure on ignition. Similarly, clouds over 50 t explode inefficiently and may be considered as 50 t. Example B.3 illustrates the method. For additional guidance, further notes on blast damage effects are given in section B.7 and in Giesbrecht *et al.*[15] and Baker *et al.*[16]

492

Table B.3   Explosion overpressure (surface burst)

Effective weight		
$W_e = 0,$	$W_c < 5$ t	
$W_e = WH_C/H_{HC},$	5 t $< W_c < 50$ t	
$W_e = 50$ t,	$W_c > 50$ t	

H_c = heat of combustion
H_{HC} = heat of combustion of hydrocarbon = 46,000 kJ kg^{-1}
x = distance

Incident overpressure (bar)	$\dfrac{x}{W_c^{1/3}}$ $(m/t^{-1/3})$	Items should not be subject to overpressure stated
0.02	220	High concentrations of people, e.g. schools, hospitals
0.04	125	Domestic housing[17]
0.05	105	Public roads
0.07	80	Ordinary plant buildings
0.10	65	Buildings with shatter-resistant windows. Fixed roof tanks containing highly flammable or toxic materials
0.20	40	Floating roof tanks, other fixed roof tanks. Cooling towers. Utility areas. Site roads
0.30	33	Plants with large atmospheric pressure vessels or units having large superficial area
0.40	27	Other hazardous plants
0.70	20	Non-hazardous (if unoccupied) plants. Control rooms designed for blast resistance.[18]

Example B.3

Determine the distances various items should be safely placed assuming explosion of the cloud in Example B.1.

ANSWER

Hydrocarbon equivalent
$$W_c = 15.6 \times \frac{35.3}{46.0} = 12 \text{ t}$$

Using Table B.3:

Item	Incident overpressure (bar)	Distance from source (m)
Schools	0.02	500
Housing	0.04	290
Public roads	0.05	240
Offices	0.07	180
Shatter-resistant windows	0.10	150
Site roads, utilities	0.20	90
Hazardous plants	0.30–0.40	75–60
Protected control room	0.70	50

B.2.4   FIREBALL SIZE

If a flammable cloud ignites but fails to explode, it forms a fireball. This is of short duration (10–15 s) and only harms people and ignites emergency vents or pools of flammable liquids.

Table B.4 shows how to find the properties of a fireball. The two regions at risk from a fireball are: the one inside the fireball which will cause ignition of vents, etc. and in which personnel will be killed, and the one inside a certain flux in which people may be injured. This flux depends on exposure time roughly to the inverse power of 2/3 (see API[19] and Tan[20]). In finding the safe distance the worst cases have been taken. As unconfined vapour cloud explosions (UVCEs) do not have much overpressure for releases below 5 t, and the heat effect can be more important, the pre-mixed stoichiometric case has been given. Above 5 t (no 50 t limitation) the diffusion case has been presented for the fireball as the pre-mixed case is assumed to give a UVCE.

As most of the work[19–25] on fireballs has been concerned with hydrocarbons, the hydrocarbons equivalent has been used in Table B.4.

Example B.4 gives a typical calculation.

Table B.4   Fireball properties

Flame radius	$r_F = 3.2 W_c^{1/3}$
Time of burning	$t_B = 1.1 W_c^{1/6}$, $W_c < 5{,}000$ kg (pre-mixed stoichiometric) $t_B = 2.6 W_c^{1/6}$, $W_c > 5{,}000$ kg (diffusion controlled)
Distance to flux I	$x = r_F(I_o/I)^{1/2}$
Distance to where	$I < 47/t_B^{2/3}$ (safe dose) $x = 10 W_c^{7/18}$, $W_c < 5{,}000$ kg when $I_o = 450$ kW m^{-2} $x = 12 W_c^{7/18}$, $W_c > 5{,}000$ kg when $I_o = 350$ kW m^{-2}

Example B.4

Determine the fireball time and radius and the distance of safe dose flux for the vapour cloud in Example B.1.

ANSWER

Radius of fireball for 12 t hydrocarbon equivalent	$= 73$ m
Time of burning	$= 2.6 \times 12{,}000^{1/6}$ $= 12$ s
Safe dose flux	$= 47/12^{2/3} = 9$ kW m^{-2}
Distance to safe dose flux	$= 463$ m

B.2.5   DISPERSION OF TOXIC CLOUD

Table B.5 shows distances to the safe concentration, both within and without buildings. In order to take advantage of the protection of a building, it should be vented or evacuated after the cloud has passed. Thus Table B.5 also gives times of passage, etc. It is also important to know the time of exposure as this is usually specified with the safe concentrations.

Table B.5 makes use of the dispersion coefficients in Table B.2 which depend on whether dispersion takes place on or off site. In the latter case the site boundary correction (Fig. B.1) calculated in Table B.2 should be used. The method is used in Example B.5.

In Table B.5 the effect of a building has been found by solving:

$$\tau \frac{dC_l}{dt} = C_o \exp.\left[-\frac{x^2}{2\sigma_x^2}\right] - C_l \rightarrow C_o \exp.\left[-\frac{x^2}{2\sigma_x^2}\right]$$

i.e. only small toxic ingresses have been considered. The solution has been found by assuming no variation in the dispersion coefficients σ_x.

Thus

$$C_l = \frac{C_o}{\tau w} \int_{-\infty}^{\infty} \exp.\left[-\frac{x^2}{2\sigma_x^2}\right] dx = \frac{V}{\tau w \sigma_y \sigma_z \pi}$$

The building ventilation constant τ is equal to the building volume/(mass transfer coefficient × ventilation area). Some possible values are given in section B.9.

The windspeed of 2 m s^{-1} is suggested as a value to be used. This will be conservative as higher speeds (5 m s^{-1}) are more typical of stability category D (see section 8.2.3, p. 105 and Table 8.1, p. 107).

Table B.5   Dispersion of toxic cloud

Distance to safe concentration
C_s

$$\sigma_x \sigma_y \sigma_z = \frac{2V}{(2\pi)^{3/2}C_s}$$

where σs are functions of distance x,
(see Table B.2) and $C_s = $ IDLH

Effective dosage time

$$t_d = \sqrt{(2\pi)}\frac{\sigma_x}{w}$$

where $w = $ windspeed $= 2$ m s^{-1}

Distance to safe concentration
in building

$$\sigma_y \sigma_z = \frac{V}{\pi w \tau C_s}$$

where $\tau = $ building ventilation constant $\gg t_d$

Time to reach peak
concentration

$$t_p = \frac{x}{w}$$

Peak cloud concentration

$$C_o = \frac{2V}{(2\pi)^{3/2}\sigma_x \sigma_y \sigma_z}$$

Time for cloud to pass with
cone $> C_s$

$$\Delta = \frac{2\sigma_x \sqrt{2}}{w}\left(\ln \frac{C_o}{C_s}\right)^{1/2}$$

Outside site $\qquad x = x_v - b + x_b$

Inside site $\qquad x = x_v - a$

Correction for finite source

Length correction $\qquad$ Use $\dfrac{x_s}{\sqrt{(2\pi)}}$ instead of σ_x if former is larger

where x_s = source length

Width correction $\qquad$ Use $\dfrac{y_s}{\sqrt{(2\pi)}}$ instead of σ_y if former is larger

where y_s = source width

Example B.5

Find for the cloud in Example B.1, the distance to the safe concentration and the protection provided with a building of 1 h building ventilation constant. Also calculate the time to clear the boundary at 100 m.

ANSWER

Distance to safe concentration

$$\sigma_x \sigma_y \sigma_z = \frac{2 \times 8{,}240}{(2\pi)^{3/2} \times 4 \times 10^{-3}} = 261{,}600$$

From Table B.2, off site
$x_v = 1{,}199$ m
Allowing for initial mixing $x = 1{,}048$ m
or also allowing for dispersion within site,
$x = 1{,}034$ m
(see Example B.2)
From Table B.2, for cloud completely dispersed on site, $x_v = 1{,}070$ m
or allowing for initial mixing (see Example B.2)
$x = 941$ m

Distance to safe concentration in building

$$\sigma_y \sigma_z = \frac{8{,}240}{\pi \times 2 \times 3{,}600 \times 4 \times 10^{-3}} = 92 \text{ m}^2$$

From Table B.2, off site, $x_v = 205$ m
Allowing for initial mixing $x = 205 - 151$
$= 54$ m
or also allowing for dispersion within site,
$x = 205 - 265 + 100 = 40$ m!
For cloud completely dispersed on site, x_v
$= 164$ m
or $x = 164 - 129 = 35$ m (see Example B.2).

Dosage time

From Table B.2, off site
$\sigma_x = 0.13 \times 205 = 27$ m

$$t_d = \sqrt{(2\pi)}\,\frac{27}{2} = 33 \text{ s which} \ll \tau$$

$\therefore$ effective protection provided by buildings

Time to reach peak concentration at boundary

$$t_p = \frac{100}{2} = 50 \text{ s}$$

| Peak concentration at boundary | From Example, B.2, $C_b = 0.22$ v/v |
| Time for cloud to pass boundary with concentration $> 4{,}000$ p.p.m. v/v | σ_x at $x = 129 + 100$ m from Table B.2 is 30 m
$\Delta = \dfrac{2 \times 30}{2} \sqrt{2} \left(\ln \dfrac{0.22}{4 \times 10^{-3}} \right)^{1/2} = 85$ s |

B.3 STEADY LEAK OF GAS OR VAPOUR

B.3.1 LEAKAGE RATES

Table B.6 lists the discharge rates and other conditions for gases and two-phase flow. When the vessel pressure is above about 2 atm., the gases reach sonic velocities in the orifice choke. After the choke it is assumed that as the pressure reduces to atmosphere there is a series of shock waves that maintain the gas velocity at about sonic values.[26] It is also assumed that no air is entrained while this happens though this is not necessarily true.[27]

The work of Fauske[28] and Cude[29] indicates that with boiling liquids escaping directly from the side of the vessel, all the flashing takes place after the orifice. However, if the break is above the liquid level or in a pipe leading from the vessel then flashing does occur in the choke. The equations presented in Table B.6 have been derived assuming thermodynamic equilibrium is achieved in sufficiently long pipes ($L_p/D_L > 12$) and that no vaporization takes place in short pipes ($L_p/D_L < 2$). The accommodation for non-equilibrium in intermediate ranges is empirical. The effects of both the vena contracta and friction have been lumped into the discharge coefficient c_D. It is likely that better models will result from work in progress[1] and models for mixtures are needed.

Example B.6 illustrates the method for a flashing liquid leaking from a pipe running from the vessel.

Table B.6 Discharge rates

Gas (subsonic)

Condition	$\dfrac{P_a}{P_o} > \left(\dfrac{2}{1+\gamma} \right)^{\gamma/(\gamma-1)}$
Discharge temperature	$T = T_o \left(\dfrac{P_a}{P_o} \right)^{(\gamma-1)/\gamma}$
Discharge density	$\rho = \dfrac{P_a M}{RT}$
Discharge velocity	$u = c_D \left[\dfrac{2\gamma R'}{(\gamma-1)M} (T_o - T) \right]^{1/2}$

Mass flow

$$G = \frac{\pi D_L^2}{4} \rho u$$

Time when fraction ψ remaining

$$t = \frac{W}{G} \int_\psi^1 \left(\frac{1 - T/T_o}{\psi^{\gamma-1} - T/T_o} \right)^{1/2} d\psi$$

$$\approx \frac{2W}{G} \left(\frac{1 - T/T_o}{\gamma - 1} \right) \left[1 - \left(\frac{\psi^{\gamma-1} - T/T_o}{1 - T/T_o} \right)^{1/2} \right]$$

Conditions after time t in vessel

$$P = P_o \psi^\gamma$$
$$T = T_o \psi^{\gamma-1}$$

Gas (sonic)

Condition

$$\frac{P_a}{P_o} < \left(\frac{2}{1 + \gamma} \right)^{\gamma/(\gamma-1)}$$

Discharge temperature

$$T = \frac{2T_o}{1 + \gamma}$$

Discharge density

$$\rho = \frac{P_a M}{RT}$$

Discharge velocity

$$u = c_D \left[\frac{2\gamma R' T_o}{(1 + \gamma)M} \right]^{1/2}$$

Mass flow

$$G = c_D \frac{\pi D_L^2}{4} P'_o \left[\frac{\gamma M}{R' T_o} \left(\frac{2}{\gamma + 1} \right)^{(\gamma+1)/(\gamma-1)} \right]^{1/2}$$

Time when fraction ψ remaining

$$t = \frac{W}{G} \left(\frac{2}{\gamma - 1} \right) (\psi^{(1-\gamma)/2} - 1)$$

Condition after time t in vessel

$$P = P_o \psi^\gamma$$
$$T = T_o \psi^{\gamma-1}$$

Flashing liquid (subsonic)

Condition

$$\ln \frac{P_o}{P_a} < \frac{0.5}{1 - \dfrac{R T_b}{LM}}$$

Discharge condition

$$T = T_b, \quad P = P_a$$

Fraction vapour

$$m = \frac{s(T_o - T_b)}{L}$$

Velocity

$$\frac{u^2}{2c_D^2} = \frac{T_o - T_b}{2T_b} \cdot L'm + \frac{P'_o - P'_a}{\rho_L}$$

Vapour density

$$\rho_v = \frac{P_a M}{R T_b}$$

Mean density

$$\frac{1}{\rho} = \frac{m}{\rho_v} + \frac{1 - m}{\rho_L}$$

Mass flow
$$G = G_1 = \frac{\pi D_L{}^2}{4}\,\rho u,\ \text{for } L_p > 12 D_L$$

i.e. thermodynamic equilibrium has been achieved
For lower L_p see under sonic flow
L_p, D_L are length and diameter of discharge pipe

Discharge time
$$t_D = \frac{W}{G},\ \text{ignoring pressure variation, etc.}$$

Condition in vessel when fraction ψ remaining
$$\left(\frac{V_T}{\psi W} - \frac{1}{\rho_L}\right)\frac{PM}{RT} = \left(\frac{V_T}{W} - \frac{1}{\rho_L}\right)\frac{P_o M}{RT_o} + s\frac{T_o - T}{L}$$

where P is vapour pressure at T and V_T = volume of vessel

Flashing liquid (sonic)

Condition
$$\ln\left(\frac{P_o}{P_a}\right) \geqslant \frac{0.5}{1 - \dfrac{RT_b}{LM}}$$

Discharge conditions
$$T = T_b,\ P = P_a$$

Fraction vapour
$$m = s\frac{T_o - T_b}{L}$$

Velocity
$$\frac{u^2}{2c_D{}^2} = \frac{0.5 L' m}{\dfrac{LM}{RT_b} - 2 + \dfrac{T_o/2}{T_o - T_b}} + \frac{P'_o - P'_a}{\rho_L}$$

Vapour density
$$\rho_v = \frac{PM}{RT_b}$$

Mean density
$$\frac{1}{\rho} = \frac{m}{\rho_v} + \frac{1 - m}{\rho_L}$$

Choke conditions
$$\ln\left(\frac{P_o}{P_N}\right) = \frac{0.5}{1 - \dfrac{RT_N}{LM}} \cong \frac{0.5}{1 - \dfrac{RT_o}{LM}}$$

P_N is vapour pressure at T_N

Fraction vapour
$$m_N = s\frac{T_o - T_N}{L}$$

Velocity
$$\frac{u_N{}^2}{2c_D{}^2} = \frac{T_o - T_N}{2T_N} \cdot L' m_N + \frac{P'_o - P'_N}{\rho_L}$$

Vapour density
$$\rho_{vN} = \frac{P_N M}{RT_N}$$

Mean density
$$\frac{1}{\rho_N} = \frac{m_N}{\rho_{vN}} + \frac{1 - m_N}{\rho_L}$$

Mass flow
$$G = G_1 = \frac{\pi D_L^2}{4} \rho_N u_N, \quad L_p > 12 D_L$$

i.e. thermodynamic equilibrium achieved

$$G = G_2 = \frac{\pi D_L^2}{4} c_D [2\rho_L(P'_o - P'_a)]^{1/2}, \quad L_p < 2D_L$$

i.e. no flashing has occurred

$$G = G_2 + (G_1 - G_2)(L_p/D_L - 2)/10, \quad 2D_L < L_p < 12D_L$$

i.e. flashing started but thermodynamic equilibrium not achieved
L_p and D_L are length and diameter of discharge pipe

Minimum discharge time
$$t_D = \frac{W}{G}, \text{ ignoring pressure changes, etc. in vessel}$$

Conditions in vessel   See flashing liquid (subsonic) (p. 498)

Gas and flashing liquid

Apparent diameter
$$D_a = \sqrt{\left(\frac{4G}{\pi \rho u}\right)}$$

Discharge coefficient c_D (0.8 in default)

Volume flow as vapour at atmospheric conditions
$$Q = \frac{GRT_a}{P_a M} = \frac{GM_a}{\rho_a M}$$

Weight released $= W$

Vapour released expressed as vapour at atm. cond.
$$V = \frac{WM_a}{\rho_a M}$$

Example B.6

Find the leakage rate through a 25 mm–diameter pipe broken 1 m away from the tank in Example B.1.

ANSWER

Condition
$$\ln\left(\frac{P_o}{P_a}\right) = \ln 12 = 2.49$$

$$\frac{0.5}{1 - \dfrac{RT_b}{LM}} = \frac{0.5}{1 - \dfrac{8.314 \times 290}{623 \times 45}} = 0.55$$

Thus flow is sonic

Discharge conditions $T_b = 290$ K, $P_a = 1$ atm.

Factor vapour
$$m = \frac{2.92(373 - 290)}{623}$$

$$= 0.389$$

Velocity

$$\frac{u^2}{2 \times 0.8^2} = \frac{0.389 \times 623{,}000/2}{\dfrac{623 \times 45}{8.314 \times 290} - 2 + \dfrac{373/2}{373 - 290}}$$

$$+ \frac{(12 - 1)}{683} \times 10^5$$

$$u = 123 \text{ ms}^{-1}$$

Vapour density

$$\rho_v = \frac{100 \times 45}{8.314 \times 290} = 1.87 \text{ kg m}^{-3}$$

Mean density

$$\frac{1}{\rho} = \frac{0.389}{1.87} + \frac{1 - 0.389}{683}$$

$$\rho = 4.79 \text{ kg m}^{-3}$$

Discharge pipe

$$\frac{L_p}{D_L} = \frac{1}{0.025} = 40$$

Thus vaporization will occur in choke

Choke conditions

$$\ln\left(\frac{12}{P_N}\right) = \frac{0.5}{1 - \dfrac{8.314 T_N}{L \times 45}} \cong \frac{0.5}{1 - \dfrac{8.314 \times 373}{L \times 45}}$$

Use Watson's correlation for latent heat

$$L = L_b\left(\frac{1 - T_N/T_c}{1 - T_b/T_c}\right)^{0.38}$$

$$= 623\left(\frac{1 - T_N/456}{1 - 290/456}\right)^{0.38}$$

$$\text{(critical temperature} = 456 \text{ K)}$$

With $T = 373$ K, $L = 479$ kJ kg^{-1}, $P_N = 6.69$ atm. and $T_N = 349$ K

Further calculation gives
$P_N = 6.79$ atm., $T_N = 349.5$ K, $L_N = 526$ kJ kg^{-1}

Fraction vapour

$$m_N = \frac{2.92 \times (373 - 349.5)}{526}$$

$$= 0.130$$

Velocity

$$\frac{u_N^2}{2 \times 0.8^2} = \frac{(373 - 349.5) \times 0.130 \times 526{,}000}{2 \times 349.5} + \frac{(12 - 6.79)}{683} 10^5$$

$$u_N = 62.6 \text{ m s}^{-1}$$

Vapour density

$$\rho_{vN} = \frac{6.79 \times 100 \times 45}{8.314 \times 349.5} = 10.5 \text{ kg m}^{-3}$$

Mean density

$$\frac{1}{\rho_N} = \frac{0.130}{10.5} + \frac{1 - 0.130}{683}$$

$$\rho_N = 73.2 \text{ kg m}^{-3}$$

Mass flow	$G = \dfrac{\pi}{4} \times 0.025^2 \times 73.2 \times 62.6$
	$= 2.25 \text{ kg s}^{-1}$
Discharge time	$t_D = \dfrac{20{,}000}{2.25 \times 3{,}600} = 2.47 \text{ hr}$
Apparent diameter	$D_a = \left(\dfrac{4 \times 2.25}{\pi \times 4.79 \times 123}\right)^{1/2} = 70 \text{ mm}$
Volume flow as vapour at atmospheric condition	$Q = \dfrac{2.25 \times 29}{1.22 \times 45} = 1.19 \text{ m}^3 \text{ s}^{-1}$
Volume released as vapour at atmospheric condition	$V = \dfrac{20{,}000 \times 29}{1.22 \times 45} = 10{,}565 \text{ m}^3$

B.3.2 DISPERSION OF A JET TO LOWER FLAMMABLE LIMIT

Normally if the flow is sonic then the lower flammable limit is reached before the initial momentum of the jet is lost. Table B.7 presents the relevant equations developed by Cude.[30] The momentum is considered spent when the jet velocity drops to the wind velocity. The equations are used to locate safely furnaces and other ignition sources (see Example B.7).

Cude's model was developed for one-phase flow but it is assumed to hold where two phases are escaping. It is also assumed that all entrained liquid has isothermally evaporated (and not precipitated) by the time the lower flammable limit is reached. These assumptions will be suspect particularly if issuing velocity is low.

Where there is a substantial mean concentration of the material in the surrounding air, the version of the equation given in Table B.17 should be used. One example of this is the release of liquid oxygen.

Table B.7   Dispersion of jets

Air fuel weight ratio	$\theta = \left(\dfrac{1}{C} - 1\right)\dfrac{M_a}{M}$
	where mean concentration C is half the centreline concentration[30]
Velocity	$v = \dfrac{u}{1 + \theta},\ v > w$
Jet diameter	$D^2 = \dfrac{4Q}{\pi C v}$
Length of jet	$r = (D - D_a)\dfrac{\cot \lambda}{2}$
	with $\cot \lambda = 6.25$

Example B.7

Find the distance to the LFL of the jet discussed in Example B.6 and of a similar jet from a 100 mm-diameter hole.

ANSWER

For 25 mm-diameter hole

Mean concentration

$$C = \frac{0.0355}{2}$$

Air/organic

$$\theta = \left(\frac{2}{0.0355} - 1\right) \times \frac{29}{45} = 35.66 \text{ kg/kg}$$

Velocity

$$v = \frac{123}{1 + 35.66} = 3.35 \text{ m s}^{-1}$$

This is greater than the windspeed 2 m s^{-1}

Jet diameter at LFL

$$D^2 = \frac{4 \times 1.19 \times 2}{\pi \times 0.0355 \times 3.35} = 25.48$$

$$D = 5.05 \text{ m}$$

Distance to LFL

$$r = (5.05 - 0.07) \times \frac{6.25}{2}$$

$$= 15.5 \text{ m}$$

For 100 mm-diameter hole

Jet diameter and length increased four times (20 and 62 m).

B.3.3   SIZE OF JET FLAME

Table B.8 gives equations[31] to find the dimensions and intensity of an ignited jet. Example B.8 illustrates their use.

It is assumed that the flame can be in any direction. So the received intensity is calculated for the 'end-on' case with a horizontal flame. Criteria[7,31-33] for heat fluxes are also given, in Table B.8, and the methods for finding surface temperatures are given in section B.5. More accurate methods than those given in Table B.8 of calculating flame temperatures which allow for dissociation are available in standard texts, e.g. by Perry and Chilton.[34] Similarly, better estimates of the fraction of heat radiated F are given in Brzustowski and Sommer[17] and API.[19]

A similar case is that of an emergency vent or flare. The situation is assumed where the wind makes the flame blow directly at the receiver (Fig. B.2a). However, at large distances the radiation received from the side of the flame may be larger (Fig. B.2b) and so this is included in Table B.8.

It is assumed in Table B.8 that the ratio of radiation received to radiation from the source is given sufficiently accurately by the inverse of $1 + \pi \times$ (distance)^2/facing source area.

Table B.8   Jet flame characteristics

Air/fuel weight ratio	$\theta = \dfrac{N_a M_a}{M}$
	N_a = stoichiometric mol. air per mol. from nozzle
Velocity at flame tip	$v = u/(1 + \theta)$
Flame diameter	$D^2 = \dfrac{4 Q N_p T_F}{\pi v T_a}$
	N_p = stoichiometric mol. products per mol. from nozzle
Flame temperature	$T_F \approx \dfrac{H_c + sT + \theta s_a T_a}{\bar{s}_a(1 + \theta)} > 2{,}300 \text{ K}$
	$\bar{s}_a$ = mean sp. ht $\approx 1.3 \text{ kJ kg}^{-1} \text{ K}^{-1}$
Flame length	$l = (D - D_a)\dfrac{\cot \lambda}{2} \approx D\dfrac{\cot \lambda}{2}$
	with $\cot \lambda = 10.6$
Flame area	$A = \dfrac{(D^2 - D_a{}^2)\pi}{4 \sin \lambda} + \dfrac{\pi D^2}{4} \approx \dfrac{\pi D^2}{4}\left(1 + \dfrac{1}{\sin \lambda}\right)$
	with $\sin \lambda = 0.094$
Radiant flux	$I_o = \dfrac{G H_c F}{A}, \; F = 0.2$
Received flux from jet	$I = \dfrac{I_o}{1 + 4(r - l)^2/D^2}$ (end-on)
	r = distance from nozzle
Received flux from flare or vent	The greater of the above and
	$I = \dfrac{I_o}{1 + 2\pi\left(r^2 - \dfrac{l^2}{4}\right)\Big/ (Dl)}$ (side-on)

Tolerable intensities (kW m^{-2})

Drenched tank	38
Special buildings (no windows, fireproof doors)	25
Normal buildings	14
Vegetation	10–12
Escape routes	6 (up to 30 s)
Personnel in emergencies	3 (up to 30 min.)
Plastic cables	2
Stationary personnel	1.5

Example B.8

Find the flame size and the horizontal distance to the various critical fluxes for the jets considered in Example B.7. Assume air temperature is 283 K.

ANSWER

Stoichiometry

$$C_2H_5NH_2 + 3.75O_2 = 2CO_2 + 3.5H_2O + 0.5N_2$$

$$N_a = 3.75 \times 4.76 = 17.85$$
$$N_p = 3.75 \times 3.76 + 2 + 3.5 + 0.5 = 20.10$$

For 25 mm-diameter hole

Air/organic

$$\theta = \frac{17.85 \times 29}{45} = 11.50 \text{ kg/kg}$$

Velocity at flame tip

$$v = \frac{123}{1 + 11.5} = 9.84 \text{ ms}^{-1}$$

Flame temperature

$$T_F = \frac{35,300 + 2.92 \times 373 + 11.50 \times 1 \times 283}{1.3 \times 12.50}$$

$$= 2,438 \text{ K}$$

So take $T_F = 2,300$ K

Flame diameter

$$D^2 = \frac{4 \times 1.19 \times 20.10 \times 2,300}{\pi \times 9.84 \times 283}$$

$$D = 5.01 \text{ m}$$

Flame length

$$l = (5.01 - 0.07) \times 10.6/2$$
$$= 26.18 \text{ m}$$

Flame area

$$A = \frac{\pi \times 5.01^2}{4}\left(\frac{1}{0.094} + 1\right) = 229 \text{ m}^2$$

Radiant flux

$$I_o = \frac{2.25 \times 35,300 \times 0.2}{229} = 69 \text{ kW m}^{-2}$$

Received flux

$$I = \frac{69}{1 + 4(r - 26.18)^2/5.01^2}$$

$$r = 26.2 + 2.5\sqrt{\left(\frac{69}{I} - 1\right)}$$

Item	Flux/kW m⁻²	Distance from leak (m)
Drenched storage tanks	38	29
Special buildings	25	30
Normal buildings	14	32
Vegetation	12	32
Escape routes	6	35
Personnel in emergencies	3	39
Plastic cables	2	41
Stationary personnel	1.5	43

For 100 mm-diameter hole

Flame diameter and length and distances increased by a factor of 4.

Fig. B.2 Flame/receiver orientation (a) end-on (b) side-on

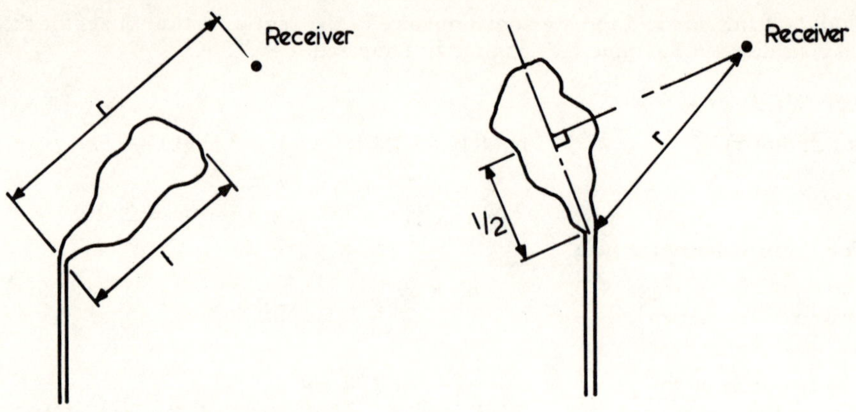

B.3.4   DISPERSION OF TOXIC PLUME

Table B.9 presents the equations[8] to determine the distance to safe concentration, both inside and outside buildings. However, buildings afford less protection than for an instantaneous cloud because of the longer duration of the effect. The use of the equations are illustrated in Example B.9.

The toxic limits are usually in the neutral phase and for a continuous release the concentration at a point x, y, Z relative to the source of release, is given by:

$$C = \frac{Q}{\pi w \sigma_y \sigma_z} \exp. - \left[\frac{y^2}{2\sigma_y{}^2} + \frac{Z^2}{2\sigma_z{}^2} \right]$$

where x is the horizontal distance along the centreline of the plume, y is the horizontal distance perpendicular from the centreline of the plume and Z is the height above the centreline of the plume.

The dispersion coefficients are functions of x given in Table B.2 for the instantaneous case.

The strictures about ignoring dense clouds given in section B.2.2 also apply here. Correction can be made for the jet stage by first assuming the jet is horizontal and then finding a virtual source for the neutral stage based on equating centreline concentrations. Similarly, a virtual source is derived for the neutral stage outside the site boundary (Fig. B.1).

Away from the jet momentum region the plume fluctuates about the wind direction (Fig. B.3) and so the mean concentrations at any point in space are effectively less than the concentration within the plume. Allowance for this has been made in calculating building effects by using the method given by TNO.[4]

$$\frac{1}{t} \int_0^t C dt \equiv C \left(\frac{18.75}{t} \right)^{0.2} , \quad t > 18.75 \text{ s}$$

$$\equiv C \qquad\qquad , \quad t < 18.75 \text{ s}$$

On this basis the 10 min. mean value is half that of the instantaneous

Fig. B.3 Plume fluctuations

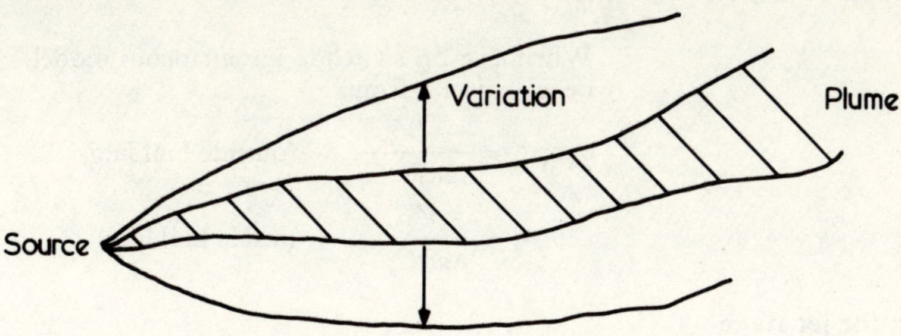

value. The correlations for the σs quoted in Table B.2 and section B.9.1 (p. 544) are instantaneous values calculated from the time-averaged TNO values.

This allowance can also be made outside buildings, as toxic materials can take some minutes to build up in the body. Ten minutes is a time often assumed. However, for very toxic materials and for flammability considerations, the instantaneous plume concentration should be used for outside buildings.

It should be noted that other work (e.g. NRPB[9]) suggests that the 0.2 power only holds for the time range of between 30 s and 6 min. and that the power is 0.5 above 7 min. Also the 10 min. mean is reported to be one-fifth the instantaneous value for weather conditions E and F.

As only instantaneous and steady dispersion and not dispersion from releases of finite time are considered, a criterion is used to separate finite cases into either instantaneous or steady models. The rule is that for the continuous model to hold, the time of passage of a cloud over a point should be much smaller than the discharge time. Since the width of a cloud increases with distance from the source this means the time for a cloud to reach a point should be much greater than the discharge time. Thus, near the source the continuous model may be valid but the instantaneous model applies well away from the release point.

The dispersion equations assume point sources. Approximate corrections for finite width and length of sources are given in Table B.9. Near the source the width is considered infinite and the length correction assumes no dispersion within the release area. Clancey[35] discusses these aspects further.

Table B.9   Dispersion of toxic plume

Distance to safe concentration	$\sigma_y \sigma_z = \dfrac{Qv}{\pi w C_s}$
	$v = 0.5$ for 10 min. mean concentration
Dosage time = discharge time	$t_d = \dfrac{V}{Q}$
Distance to safe concentration in building	$\sigma_y \sigma_z = \dfrac{V}{\pi w \tau C_s}, \ \tau \gg t_d < 18.75 \text{ s}$
	$= \dfrac{V}{\pi w \tau C_s}\left(\dfrac{18.75}{t_d}\right)^{0.2}, \ \tau \gg t_d > 18.75 \text{ s}$

Time for plume to reach x

$$t_p = \frac{x}{w}$$

When $t_p > 3t_d$ switch to instantaneous model i.e. calculate x from

$$\sigma_x\sigma_y\sigma_z = \frac{2V}{(2\pi)^{3/2}C_s} \qquad \text{(outside building)}$$

$$\sigma_y\sigma_z = \frac{V}{\pi w\tau C_s} \qquad \text{(inside building)}$$

Correction for jet stage

Concentration at end of jet

$$C_j = \frac{2}{1 + \left(\dfrac{u}{w} - 1\right)\dfrac{M}{M_a}} \approx \frac{2M_a w}{Mu}$$

Length of jet

$$x_j = 6.25\left(\frac{2Q}{\pi C_j w}\right)^{1/2}$$

Virtual source

$$\sigma_y\sigma_z = \frac{Q}{\pi w C_j}$$

Correction

$j\,(\equiv x)$ found from Table B.2 replace x by $x - x_j + j$

Correction for finite sources

Length correction

measure x from leeward edge of source

Width correction

use $\dfrac{y_s}{\sqrt{(2\pi)}}$ instead of σ_y if former larger

where $y_s =$ width of source

Example B.9

Find the distance to plume disappearance, and to the safe concentration both within and without building for the releases described in Example B.7. The ventilation constant for the building is to be taken as 1 h.

ANSWER

For 25 mm-diameter hole

Concentration at end of jet

$$C_j = \frac{2 \times 29}{45} \times \frac{2}{123} = 0.021 \text{ v/v}$$

Length of jet

$$x_j = 6.25\sqrt{\left/\left(\frac{2 \times 1.21}{\pi \times 0.021 \times 2}\right)\right.} = 27 \text{ m}$$

Virtual source

$$\sigma_y\sigma_z = \frac{1.19}{2\pi \times 0.021} = 9.0 \text{ m}^2$$

From Table B.2, on site, $j = 39$ m and off site $j = 51$ m

508

Distance to safe concentration $\quad \sigma_y\sigma_z = \dfrac{1.19 \times 0.5}{\pi \times 2 \times 4 \times 10^{-3}} = 24 \text{ m}^2$

From Table B.2, on site,
Distance ignoring jet stage $= 72$ m
Correction for jet stage $x = 72 - 39 + 27 = 60$ m
that is well within the site
From Table B.2, off site conditions,
Distance ignoring jet stage $= 92$ m
Or corrected $x = 92 - 51 + 27 = 68$ m

Discharge time $\qquad t_d = 2.47\ h$
This is greater than ventilation constant so the building provides no effective protection.

Time to reach safe
  concentration from $\qquad t_p \cong \dfrac{72}{2} = 36 \text{ s} \ll 3t_d$
  virtual source

so continuous model applies

For 100 mm-diameter hole

Concentration at end of jet $\qquad C_j = 0.0021 \text{ v/v}$

Length of jet $\qquad x_j = 4 \times 26.6 = 106$ m

Virtual source $\qquad \sigma_y\sigma_z = 16 \times 9 = 144 \text{ m}^2$
From Table B.2, on site $j = 219$ m, off site
$j = 271$ m

Concentration at site boundary $\quad$ From Table B.2 on site, ignoring jet

$\sqrt{(\sigma_y\sigma_z)} = 0.159 \times 100^{0.803}$
$\sigma_y\sigma_z = 41 \text{ m}^2$

$C_b = \dfrac{16 \times 1.19}{\pi \times 41 \times 2} = 0.0748 \text{ v/v}$

Virtual source $\qquad \sigma_y\sigma_z = \dfrac{16 \times 1.19}{2 \times 0.0748 \times \pi} = 41 \text{ m}^2$

From Table B.2, off site $x_o = 128$ m

Distance to safe concentration $\quad \sigma_y\sigma_z = 16 \times 24 = 384 \text{ m}^2$
From Table B.2 on site,
x (ignoring jet stage) $= 403$ m
Correction for jet stage
$x = 403 - 219 + 106 = 290$ m
Similarly, off site, $x = 489$ m without jet and
324 m with jet stage.
The former, allowing for site boundary, $x = 461$ m

Discharge time $\qquad t_d = \dfrac{10{,}565}{16 \times 1.21} = 546 \text{ s} = 0.151 \text{ hr}$

Distance to safe concentration
  in building $\qquad \sigma_y\sigma_z = \dfrac{10{,}565}{\pi \times 2 \times 3{,}600 \times 4 \times 10^{-3}} \left(\dfrac{18.75}{546}\right)^{0.2}$

$\qquad\qquad = 59.5 \text{ m}^2$

From Table B.2, on site, $x = 126$ m, off site $x = 159$ m
These are less than the jet virtual sources so within range of jet. Application of Table B.7 gives distance as 87 m where building concentration reaches 4,000 p.p.m. with central concentration outside building of $0.004 \times 3600/546 = 0.026$ v/v, jet velocity 2.4 ms^{-1} and jet diameter 28 m.

Time to reach point on site for safe concentration from virtual source

$$t_{\mathrm{p}} = \frac{403}{2} = 202 \text{ s}$$

$$3t_{\mathrm{d}} = 1,637 \text{ s}$$

Thus $t_{\mathrm{p}} \ll 3t_{\mathrm{d}}$ and continuous model is applicable at 403 m.

B.4.   LOSS OF LIQUID

B.4.1   LEAKAGE RATE, POOL SIZE AND EVAPORATION RATE

With a catastrophic failure of a tank roof, the resultant open tank behaves as a pool. Any failure in the sides, base or pipes of a tank will result in a liquid pool. The diameter of the pool is usually determined by the local topography. However, for flat ground, Table B.10 gives the variation of radius with time, though unconfined spreading should, and can be, avoided. The rate of flow through a non-catastrophic leak is also given.

The evaporation from a pool forms a plume. If the pool temperature is above the flash point the dispersion to the LFL determines the position of ignition sources. The dispersion to the toxic limit determines the position of offices, etc.

The equations in Table B.10 determine the amount of evaporation.[36-38] For non-cryogens and cryogens on water a steady evaporation rate is achieved. This rate cannot be greater than the leakage rate. The smaller of the two rates is used in the steady neutral cloud in Table B.9 to find the distance to either the LFL or the toxic limit. However, cryogenic spills on water should be avoided as large clouds can be formed.

With cryogens on land the rate of evaporation rapidly falls off as the ground underneath cools. The amount evaporated in the first minute is calculated. This cannot be greater than the amount leaked in the first minute. The lower amount is used in the Table B.5 instantaneous neutral cloud to find the positions of the LFL and toxic limit. After the first minute the pool then evaporates as with a non-cryogenic liquid.

As pools and tanks are finite sources, it is necessary to correct for their size in the dispersion calculations (see Tables B.9 and B.5). The consequences of evaporation are illustrated in Example B.10.

Table B.10   Liquid leakage rates and pool sizes

Leakage rate of liquid	$$G = c_{\mathrm{D}} \frac{\pi D}{4} L^2\, \rho_{\mathrm{L}} \left(\frac{P'_{\mathrm{o}} - P'_{\mathrm{a}}}{\rho_{\mathrm{L}}} + gH \right)^{1/2} \cdot \sqrt{2}$$ $$Q = GM_{\mathrm{a}}/M\rho_{\mathrm{a}}$$
Pool size, unconfined instantaneous spill	$$D_{\mathrm{p}}^2 = D_{\mathrm{o}}^2 + t\left(\frac{128 g\delta W_{\mathrm{o}}}{\pi \rho_{\mathrm{L}}} \right)^{1/2}$$ $$< \frac{4 W_{\mathrm{o}}}{\pi \rho_{\mathrm{L}} h_{\mathrm{m}}}$$
Pool size, unconfined continuous spill	$$D_{\mathrm{p}}^2 = \left(\frac{512 g\delta G t^3}{9\pi \rho_{\mathrm{L}}} \right)^{1/2}$$ $$= \frac{4 G t}{\pi \rho_{\mathrm{L}} h_{\mathrm{m}}}$$ smaller value of D, larger of t
Relative density	$\delta = 1$ (ground) $= 1 - \dfrac{\rho_{\mathrm{L}}}{\rho_{\mathrm{w}}}$ (on water)
Minimum pool depth	$h_{\mathrm{m}} = 25$ mm, rough sandy soil 20 mm, farm land 10 mm, smooth sand, gravel 5 mm, concrete, stone 1.8 mm, calm water
Non-cryogen evaporation rate	$$Q = \lambda_2 w D_{\mathrm{p}}^2\, \frac{\pi}{4} \frac{P_{\mathrm{v}}}{P_{\mathrm{a}}} \left(\frac{2}{w^2 D_{\mathrm{p}}} \right)^{n_e} \quad \text{(for use in Table B.9)}$$ where, for stability criterion D, $\lambda_2 = 1.7 \times 10^{-3}$ ms units and $n_c = 0.130$ Section B.9 gives λ_2 for other criteria *Note*: Q must be less than leakage rate
Evaporation of cryogen on water	$$G = \frac{\pi D_{\mathrm{p}}^2}{4} h \frac{T_{\mathrm{w}} - T_{\mathrm{b}}}{L}$$ $$h = 0.6 \text{ kW m}^{-2}\text{K}^{-1}$$ $$Q = \frac{GM_{\mathrm{a}}}{\rho_{\mathrm{a}} M} \quad \text{(for use in Table B.9)}$$ *Note*: G must be less than leakage rate
Evaporation of cryogen on land	$$W = \frac{2\beta k}{L} (T_{\mathrm{g}} - T_{\mathrm{b}}) \left(\frac{t}{\pi \alpha} \right) \cdot \frac{\pi D_{\mathrm{p}}^2}{4}$$ $$V = \frac{W M_{\mathrm{a}}}{\rho_{\mathrm{a}} M} \quad \text{(for use in Table B.2)}$$ $$t = \text{value when cryogen rate} = \frac{V}{2t} = \text{non-cryogen rate}$$ $\beta = 1$ for non-porous ground $\beta = 3$ for porous ground $$\alpha = \frac{k}{\rho s} = \text{thermal diffusibility}$$

Substrate	$k/(\alpha)^{1/2}$
Concrete	$1.43 \text{ kWs}^{1/2} \text{ m}^{-2} \text{ K}^{-1}$
Soil (average)	1.42
Soil (dry sandy)	0.58
Soil (80% moisture and sand)	1.02

Note: W must be less than amount spilt in time, t

Example B.10

Find the distance to plume disappearance, to the safe concentration outside buildings and to the LFL for evaporation from the tank in Example B.1 after the roof is lost. The cases are: (a) from the tank itself, and (b) from an unconfined pool on concrete formed from a 10 cm–diameter leak in tank base.

ANSWER

(a) Evaporation from tank

Evaporation rate

$$Q = 1.7 \times 10^{-3} \times 2 \times 3.5^2 \frac{\pi}{4} \times \frac{1}{1} \times \left(\frac{2}{2^2 \times 3.5}\right)^{0.130}$$

$$= 0.0254 \text{ m}^3 \text{ s}^{-1}$$

Distance to safe toxic concentration

$$\sigma_y \sigma_z = \frac{0.0254 \times 0.5}{\pi \times 2 \times 4 \times 10^{-3}} = 0.50 \text{ m}^2$$

From Table B.2, on site $x = 6$ m i.e. near tank

so replace σ_y by $\dfrac{3.5}{\sqrt{(2\pi)}}$

Thus $\quad \sigma_z = 0.50 \dfrac{\sqrt{(2\pi)}}{3.5} = 0.36$ m

From Table B.2, on site $x = 0.9$ m
Hence, effective $x = 0.9 + 1.75$
$$= 2.65 \text{ m}$$
i.e. very near tank

(b) Evaporation from pool

Height left in tank

$$H = \frac{4{,}400 \times 4}{683 \times \pi \times 3.5^2} = 0.67 \text{ m}$$

Leakage rate

$$G = 0.6 \times \frac{\pi 0.1^2}{4} \times 683 \sqrt{(9.807 \times 0.67 \times 2)}$$

$$= 11.7 \text{ kg s}^{-1}$$

i.e. emptying time $> \dfrac{4{,}400}{11.7} = 377$ s

Volumetric leakage rate

$$Q = \frac{11.7 \times 29}{1.22 \times 45} = 6.16 \text{ m}^3 \text{ s}^{-1}$$

Evaporation rate

$$Q = 1.7 \times 10^{-3} \times 2\pi \frac{D_p^2}{4} \times \frac{1}{1} \times \left(\frac{2}{2^2 D_p}\right)^{0.130}$$

$$= 6.16 \text{ m}^3 \text{ s}^{-1}$$

diameter of pool $= 66$ m

Distance to safe toxic concentration

$$\sigma_y \sigma_z = \frac{6.16 \times 0.5}{\pi \times 2 \times 4 \times 10^{-3}} = 123 \text{ m}^2$$

From Table B.2, $x = 246$ m if outside site

Effective width

$$\sigma_y = 9.3 \text{ m}, \quad \frac{D_p}{\sqrt{(2\pi)}} = 26 \text{ m}$$

So correction needed for width of source

$$\sigma_z = \frac{123}{26} = 4.7 \text{ m}$$

From Table B.2, on site $x = 34$ m, off site 63 m
Hence, on site, from pool centre $x = 33 + 34$
$$= 67 \text{ m}$$
and of site effective $x = 63 + 33$
$$= 96 \text{ m}$$

Distance to LFL

$$\sigma_y \sigma_z = \frac{6.16}{\pi \times 2 \times 0.0355} = 27.62 \text{ m}^2$$

From Table B.2, $x = 100$ m if outside site

$$\sigma_y = 4 \text{ m}, \quad \frac{D_p}{\sqrt{(2\pi)}} = 26 \text{ m}$$

so correction needed for width of source

$$\sigma_z = \frac{27.62}{26} = 1.0 \text{ m}$$

$x = 4.0$ m for on site
from pool centre $x = 4 + 33 = 37$ m
Similarly $x = 41$ m for off site

Time of spreading ignoring evaporation

$$66^2 = \left(\frac{512 \times 9.807 \times 11.7 \times t^3}{9\pi 683}\right)^{1/2}$$

$$t = 184 \text{ s}$$

$$66^2 = \frac{4 \times 11.7 \times t}{\pi \times 683 \times 0.005}$$

$t = 1{,}001$ s $=$ time applicable
Note, however, that minimum discharge time is
377 s so pool may not reach 66 m. Thus, effects
may be less severe than calculated.

B.4.2   POOL AND TANK FIRES

Table B.11 shows how to calculate the size and intensity of a pool or tank fire.[39-41] The effect of the heat on a plant and buildings is discussed in section B.5.

The wind can bend a flame.[42] Accordingly, the flame orientations chosen are the ones giving the greatest received flux. Close to, this is when the flame points directly at the receiver. For large separations, more flux is received from the side of the fire. This is shown in Example B.11 and Fig. B.2.

Table B.11   Pool and tank fires

Mass burning velocity	$B = 1.3 \times 10^{-6} \rho_L \dfrac{H_c}{L}$
Burning rate	$B \dfrac{\pi D_p{}^2}{4} \leqslant G$
Flame height	$\dfrac{l}{D_p} = 42 \left[\dfrac{B}{\rho_a (g D_p)^{1/2}} \right]^{0 \cdot 61}$
Flame area	$A = \pi l D_p$
Radiant flux	$I_o = \dfrac{B H_c F}{4 l / D_p}, \; F = 0.2$
Received flux	$I = \phi I_o$ ϕ is the larger of ϕ_F and ϕ_S
Configuration factor (end-on)	$\phi_F = \dfrac{1}{1 + \dfrac{4}{D_p{}^2} [(x^2 + z^2)^{1/2} - l]^2}$
Configuration factor (side on)	$\phi_S = \dfrac{1}{1 + \dfrac{\pi}{D_p l} \left[\sqrt{\left(x^2 + z^2 - \dfrac{l^2}{4} \right)} - \dfrac{D_p}{2} \right]^2}$
Flame limit	$x^2 + z^2 \leqslant l^2$

Example B.11

Find the flame size and distance to various critical flux for fires concerned with the tank in Example B.1 after the roof is lost. The fires are: (a) from the tank itself, and (b) from the pool on concrete ground, discussed in Example B.10.

ANSWER

(a) Tank fire

Mass burning velocity	$B = 1.3 \times 10^{-6} \times \dfrac{683 \times 35{,}300}{623}$
	$= 0.05 \text{ kg s}^{-1} \text{ m}^{-2}$
Mass burning rate	$= 0.05 \times \pi \dfrac{3.5^2}{4}$
	$= 0.484 \text{ kg s}^{-1}$
Time to burn	For 4.4 t left,
	$\text{time} = \dfrac{4400}{0.484} = 9{,}100 \text{ s} = 2.5 \text{ hr}$

Flame height
$$l = 3.5 \times 42\left[\frac{0.05}{1.22(9.807 \times 3.5)^{1/2}}\right]^{0.61}$$
$$= 7.15 \text{ m}$$

Flame area
$$A = 7.15 \times 3.5 \times \pi = 78.6 \text{ m}^2$$

Radiant flux
$$I_o = \frac{0.05 \times 35{,}300 \times 0.2}{4 \times 7.15/3.5} = 43 \text{ kW m}^{-2}$$

Configuration factor (end-on)
$$\phi_F = \frac{1}{1 + 4[\sqrt{(x^2 + z^2)} - 7.15]^2/3.5^2}$$
$$\text{or } x^2 = \left[7.15 + 1.75\left(\frac{1}{\phi_F} - 1\right)^{1/2}\right]^2 - z^2$$

Configuration factor (side-on)
$$\phi_S = \frac{1}{1 + \dfrac{\pi}{3.5 \times 7.15}\left[\sqrt{\left(x^2 + z^2 - \dfrac{7.15^2}{4}\right)} - \dfrac{3.5}{2}\right]^2}$$
$$x^2 = \left[1.75 + 2.82\left(\frac{1}{\phi_S} - 1\right)^{1/2}\right]^2 - z^2 + 12.78$$

Flame limit
$$x^2 = 7.15^2 - z^2 = 51.12 - z^2$$

Item	Flux $(kW\ m^{-2})$	Config. factor I/I_o	Horizontal distance from tank centre (m)		
			Level with flame centre	Level with tank rim	On ground
Flame limit		1	6.2	7.2	6.2
Drenched tanks	38	0.88	6.9	7.8	7.0
Special buildings	25	0.58	7.9	8.6	7.9
Normal buildings	14	0.32	9.0	9.7	9.0
Vegetation	12	0.28	9.3	10.0	9.4
Escape routes	6	0.14	10.9	11.5	11.0
Personnel in emergencies	3	0.069	13.1	13.6	13.1
Plastic cables	2	0.046	14.7 (end-on)	15.1	14.7
Stationary personnel	1.5	0.035	16.6 (side-on)	17.0	16.6

(b) Pool fire

Leakage rate

From Example B.10
$$G = 11.7 \text{ kg s}^{-1}$$

Mass burning velocity

(see part (a)) $B = 0.05 \text{ kg s}^{-1} \text{ m}^{-2}$
$$\text{Area of pool} = \frac{11.7}{0.05} = 233 \text{ m}^2$$
$$\text{Diameter} = 17.23 \text{ m}$$

Mass burning rate

$$G = 11.7 \text{ kg s}^{-1}$$

Flame height
$$l = 17.23 \times 42\left[\frac{0.05}{1.22(9.807 \times 17.23)^{1/2}}\right]^{0.61}$$
$$= 21.57 \text{ m}$$

Flame area

$$A = 21.57 \times 17.23 \times \pi = 1{,}168 \text{ m}^2$$

Radiant flux

$$I_o = \frac{0.05 \times 35{,}300 \times 0.2}{4 \times 21.57/17.23} = 70.5 \text{ kW m}^{-2}$$

Configuration factor
  (end on)

$$\phi_F = \frac{1}{1 + 4[\sqrt{(x^2 + z^2)} - 21.57]^2/17.23^2}$$

or

$$x^2 = \left[21.57 + 8.62\left(\frac{1}{\phi_F} - 1\right)^{1/2}\right]^2 - z^2$$

Configuration factor
  (side on)

$$\phi_S = \frac{1}{1 + \dfrac{\pi}{17.23 \times 21.57}\left[\sqrt{\left(x^2 + z^2 - \dfrac{21.57^2}{4}\right)} - \dfrac{17.23}{2}\right]^2}$$

or

$$x^2 = \left[8.62 + 10.88\left(\frac{1}{\phi_S} - 1\right)^{1/2}\right]^2 - z^2 + 116.3$$

Flame limit

$$x^2 + z^2 = 21.57$$

Item	*Flux (kW m^{-2})*	*Config. factor (I/I_o)*	*Distance from pool centre (m)*	
			Level with flame centre	*On ground*
Flame limit		1	19	22
Drenched tanks	38	0.54	28	30
Special buildings	25	0.35	31	33
Normal buildings	14	0.20	37	39
Vegetation	12	0.17	39	41
Escape routes	6	0.085	49	50
Personnel in emergencies	3	0.043	62 (end–on)	62
Plastic cables	2	0.028	72 (side–on)	73
Stationary personnel	1.5	0.021	82	83

Time of spreading (ignoring
  burning)

$$17.2^2 = \left(\frac{512 \times 9.807 \times 11.7 \times t^3}{9\pi 683}\right)^{1/2}$$

$$t = 31 \text{ s}$$

$$17.2^2 = \frac{4 \times 11.7 \times t}{\pi \times 683 \times 0.005}$$

$$t = 69 \text{ s} = \text{time applicable}$$

Note: The flow rate will reduce as the tank empties so the pool size and fire effects
will decrease with time.

B.5 FIRE DAMAGE AND PROTECTION

The effect of fire on a receiver is largely determined by the intensity
received and the temperature reached by the receiver.

    Table B.12 shows how to calculate this temperature depending on such
factors as immersion in flame, insulation, above or below the liquid level,

etc. Example B.12 illustrates some of these points. Important temperatures to consider are the ignition temperature[43] and yield temperature of receivers.[44] It should be noted that an increase in temperature can have two adverse effects on equipment: firstly by raising the temperature of the contents causing a pressure rise, and secondly by raising the temperature of the material of construction therefore reducing the strength.

Table B.12 Thermal effect on receivers

Surface temperature due to radiation

$$I = \sigma(T_s^4 - T_a^4) + \frac{h_c}{\varepsilon}(T_s - T_a)$$

$$h_c = 0.01 \text{ kW m}^{-2}\text{K}^{-1} \text{ for no wind}$$

Surface	Temperature (°C)	Emissivity (ε)
Aluminium, rough	26	0.055
Iron, oxidized	100	0.735
Cast iron	925–1,115	0.87–0.95
Zinc, oxidized at 400 °C	400	0.11
Galvanized sheet iron, grey oxidized	24	0.276
Red brick	20	0.93
Flat black lacquer	40–95	0.96–0.98
Oil paints, coloured	100	0.92–0.96
Aluminium paint after heating to 325 °C	150–315	0.35
White paint	40–95	0.12–0.26
Water surface	0–100	0.95–0.963

Surface temperature in flame $T_s = T_F$

Surface temperature in burning liquid $T_s = T_b$

Water drench rate

$$S = \frac{I}{\eta L_w + s_w(T_s - T_a)}, \quad T_s = T_{bw}$$

$$\eta = \text{fraction evaporated}$$

$$S = \frac{I}{s_w(T_s - T_a)}, \quad T_s < T_{bw}$$

Temperature under insulation

$$\frac{T_m}{T_s} = 1 - \left[\exp. -\frac{tk_l}{x_l}\bigg/\left(\frac{W_R s_R}{A_R} + \frac{x_l \rho_l s_l}{2}\right)\right] \times \left[1 - \frac{T_o}{T_s}\right]$$

$T_m \not> $ boiling point of contents if below liquid level

Tolerable ignition temperatures

Receiver	Tolerable temperature (°C)	Emissivity	Tolerable intensity (kW m⁻²)
Water drench	90	1	38
Equipment	550	1	30
Special buildings	500	1	25
Normal buildings	390	1	14

Receiver	Tolerable temperature (°C)	Emissivity	Tolerable intensity (kW m^{-2})
Vegetation	330	1	10
Fluid chemicals[43]	> 230 (most 350–500)	—	—
Carbon disulphide (CS$_2$)	120	—	—
Plastics	120	1	2
Escape routes (30 s)	65	0.1	6
Emergency work (30 min.)	40	0.1	3
Safe	25	0.1	1.5

Example B.12

(a) Consider a large steel vessel, wall thickness 16 mm protected by 40 mm of vermiculite cement. Find the metal temperature after 1 and 2 h in a flux of 100 kW m^{-2} for surface emissivities of 1 ($\sim$black), 0.3 ($\sim$aluminium paint) and 0.1 ($\sim$white). Properties of steel are density 7,800 kg m^{-3}, specific heat 0.5 kJ kg^{-1} K^{-1} and properties of vermiculite are density 430 kg m^{-3}, specific heat 0.84 kJ kg^{-1} K^{-1} and thermal conductivity 9.5 $\times$ 10^{-5} kW m^{-1} K^{-1}.

ANSWER

Surface temperature

$$100 = 5.6697 \times 10^{-11}(T_s^4 - 293^4) + \frac{0.01}{\varepsilon}(T_s - 293)$$

$$T_s = 856, 797, 640\,°C \text{ for } \varepsilon = 1, 0.3 \text{ and } 0.1$$

Insulation time constant

$$t_c = \left(\frac{W_R s_R}{A_R} + \frac{x_I \rho_I s_I}{2}\right)\frac{x_I}{k_I}$$

$$= \left(\frac{7{,}800 \times 0.016 A_R \times 0.5}{A_R} + \frac{0.04 \times 430 \times 0.84}{2}\right)$$

$$\times \frac{0.04}{9.5 \times 10^{-5}} = 2.93 \times 10^4 \text{ s}$$

Metal temperature

$$T_m = T_s - (T_s - T_0)\,e^{-t/t_c}$$

$$= 0.12 T_s + 18 \text{ (after 1 hr) (in °C)}$$
$$= 0.22 T_s + 16 \text{ (after 2 hr) (in °C)}$$

For $\varepsilon = 1$, $T_m = 117$ and $202\,°C$ after 1 and 2 hr
$\varepsilon = 0.3$, $T_m = 110$ and $189\,°C$
$\varepsilon = 0.1$, $T_m = 92$ and $155\,°C$

(b) A 3.5 m–diameter $\times$ 3.5 m–high vessel at 20 °C is protected by a water drench. Find the water needed if the received flux is 38 kW m^{-2} and the water just reaches 100 °C.

ANSWER

Tank area

$$= \text{area of side} + \text{top}$$
$$= \pi 3.5^2 + \pi 3.5^2/4$$
$$= 48.1 \text{ m}^2$$

Heat received (if $\varepsilon \approx 1$)

$$= 48.1 \times 38 = 1{,}828 \text{ kW}$$

Water rate

$$= \frac{1{,}828}{4.19(100 - 20)} = 5.45 \text{ kg s}^{-1}$$

B.6   IMPLICATIONS FOR LAYOUT

B.6.1   RISK CRITERIA

Table B.16 summarizes society's apparent criteria of risk. The aspect having most effect on layout is the one in which society tolerates a limited risk of up to 10 fatalities but is reluctant to accept the risk, however small, of 1,000 fatalities or more.[45] This means, in practice, that if a dense population extends up to the site fence, then there can be no significant risk beyond the boundary (see Example B.13).

Where there is a sparse population density the finite risk can be entertained. Society expects the risk to a single exposed person to be the same as the risk to an individual in general life. Society goes on to expect that the risk of killing N people should be no greater than $1/N$th that of killing one person (Provinciale Waterstaat[45] gives $1/N^2$). Society will allow a higher risk (about 10 times greater, see Ch. 8) to employees than the public, as it presumes that they know the risks and have been trained to cope with them.

Table B.13 uses the general risk to an individual member of the public (10^{-4} per year) as the maximum acceptable risk and suggests risks $1/100$th of this as the target. Since this Appendix assumes that all persons inside the critical intensity will be killed, the risk is fulfilled in one incident. This means that frequency of plant failure is the same as the appropriate risk.

It follows that the more people at risk, the more reliable the plant must be and loss of containment must occur less frequently. Consequently, the risk to the individual from the plant goes down as the number at risk increases. The expression in Table B.13 for individual risk is derived using as a basis the conclusions of the WASH 1400 study[46] that the frequency of n or more persons being killed in natural and man-made disaster varies as the inverse of n.

Table B.13   Risk criteria

Acceptable risk to individual y^{-1}	$C_A < 10^{-6}$ (public, near site) $C_A < 10^{-5}$ (employees)
Unacceptable risk to individual y^{-1}	$C_u > 10^{-4}$ (public, near site) $C_u > 10^{-3}$ (employees)
Acceptable rate of failure of plants y^{-1}	$F_A < \dfrac{C_A}{N},\ N < 10$ $= 0,\ N \geqslant 10$ $N =$ no. of people at risk
Unacceptable rate of failure of plants y^{-1}	$F_u > \dfrac{C_u}{N},\ N < 1000$ $> 0,\ N \geqslant 1000$
Risk to individual from plant failures y^{-1}	$C \cong \dfrac{F}{N}\left[\ln\left(\dfrac{N+1}{2}\right) + \dfrac{5N+3}{4(N+1)}\right]$

Number at risk when an incident gives alternative effects (e.g. fire or toxic)	$N = \sum C_c N_c$ where C_c = chance of effect, N_c = number at risk from effect
Overall risk from a number of independent sources	$C = \sum F_i N_i$ where F_i = failure rate of ith source N_i = number at risk from ith source
Radius around site containing N people	$x = \dfrac{1}{\sqrt{\pi}} \left(\dfrac{N}{\rho_D} + A_s \right)^{1/2}$

A_s = site area

ρ_D = population density

e.g. 10^4 km^{-2} very dense, 17 m between houses

$\qquad 10^3$ km^{-2} normal urban, 55 m between houses

$\qquad\qquad$ 1 km^{-2} farms, 1.7 km between houses

B.6.2 APPLICATION OF CALCULATION RESULTS

In interpreting the results of any calculations, the assumptions and limitations of the equations must be borne in mind. A conclusion may be that more detailed analysis is needed.

The considerations should be separated into off site, on site but off plot, on plot and, finally, overall.

This is illustrated in Example B.13.

Example B.13

Interpret the results of Examples B.1–B.11.

ANSWER

A. Off-site effects – summary of distances (in m)

1(a)   Instantaneous release

	All built-up area	*100 m built-up, then country*	*All country*
LFL	341	363	377
Fireball, safe dose	463	463	463
Blast, schools	500	500	500
housing	290	290	290
roads	240	240	240
Safe toxic, open	941	1,034	1,048
Safe toxic, building	35	—	54

1(b)   Open tank after instantaneous release

LFL	At tank	—	At tank
Fire, 1.5 kW m^{-2}	17	17	17
Safe toxic	Near tank	—	Near tank

520

1(c)   Unconfined pool from 10 cm leak after instantaneous release

Fire 1.5 kW m^{-2}	83	—	83
Pool radius (fire)	9	—	9
LFL	37	—	41
Safe toxic	67	—	96
Pool radius (evap.)	33	—	33 (concrete)

2.   2.5 cm steady release under pressure

LFL (no jet)	16 (28)	—	16 (38)
Fire 1.5 kW m^{-2}	43	—	43
Safe toxic (no jet)	60 (72)	—	68 (92)

3.   10 cm steady release under pressure

LFL (no jet)	62 (159)	62 (172)	62 (200)
Fire 1.5 kW m^{-2}	172	172	172
Safe toxic in open (no jet)	290 (403)	324 (461)	324 (489)
Safe toxic in building (no jet)	87 (126)	87 (131)	87 (159)

B.   Off-site effects – conclusions

1.   Instantaneous release

   (a) Reduce inventory or pressure.

   (b) Failing that, have a larger site so that the ethylamine tank is 500 m from the site boundary and warnings to shut windows, etc. can be given for beyond 500 m.

   (c) Failing having a large site, but keeping the toxic warning system, it is acceptable to have ten people in the 100–500 m radius region providing the chance of tank failure is less than 10^{-7} per annum. The population density would be 9 (km)$^{-2}$ (scattered).

   (d) The plant could not be tolerated if there were more than 1,000 in the 100–500 m region (density 909 (km)$^{-2}$, normal urban) or the risk of failure were greater than $10^{-4}/N$ per annum where N = population at risk. For a dense population 10^4 (km)$^{-2}$ the boundary fence would have to be at least 467 m from plant. This would put 1,000 people at risk in region 467–500 m, assuming a failure rate of less than 10^{-7} per annum. However, 467–500 m is within the error of the method, which means that a very densely populated area could not be tolerated inside the 500 m radius.

   (e) The effects of evaporation and fires after the roof failure do not generally affect the public if the plant is 100 m inside the site.

2.   2.5 cm-diameter steady release

This does not affect the public if the plant is 100 m behind the fence.

3.   10 cm-diameter steady release

This failure would affect the public from, mainly, toxic effects in the open if the plant were less than 500 m inside the boundary fence and the jet hit something close to the release point.

C. On-site effects – summary of distances (in m)

1(a) Instantaneous release

LFL	341
Fireball, safe dose	463
Fireball radius	73
Blast-resistant control rooms	50
Hazardous plants	60–75
Shatter-resistant windows	150
Offices	180
Safe toxic in buildings	35
Safe toxic in open	941

1(b and c)   After instantaneous release

	Open tank (m)	Unconfined pool ex. 10 cm leak (m)
LFL	Close	37
Pool radius (fire)	—	9
Fire, drenched tanks	8	30
Special buildings	9	33
Normal buildings	10	39
Vegetation	10	41
Escape routes	12	50
Personnel in emergencies	14	62
Plastic cables	15	73
Stationary personnel (1.5 kW m^{-2})	17	83
Safe toxic limit	3	67
Pool radius (evaporation)	—	33

2 and 3   Steady releases under pressure

	2.5 cm	10 cm
LFL	16 (28, no jet)	63 (159, no jet)
Fire, drenched tanks	29	116
Special buildings	30	120
Normal buildings	32	128
Vegetation	32	128
Escape routes	35	140
Personnel in emergencies	39	156
Plastic cables	41	164
Stationary personnel	43	172
Safe toxic limit in open	60 (72, no jet)	290 (403, no jet)
Safe toxic limit in building	60 (72, no jet)	87 (126, no jet)

D. On-site but off-plot effects – conclusions

1.   Instantaneous release

(a) Offices and plants with ignition sources should be 400 m away from the ethylamine vessel and the next plot 70 m away providing it has no ignition source. This indicates that a 100 m-radius site is too small, unless the inventory is reduced.

(b) Beyond the 70 m radius, employees in unprotected buildings and the open are at risk from blast and toxicity up to about 150 m, and employees in the open are at risk from toxicity up to 1,000 m and from a fireball flux up to 500 m. As an example, assume that there is a 25 per cent chance of blast and 25 per cent chance of a fireball and a toxic warning system for above 500 m. An acceptable situation would be to have two persons in the open for 70–150 m, five in the open 150–500 m, and seven in unprotected buildings in 70–150 m. This would give an average of 7.5 fatalities per tank failure, so the acceptable frequency of tank failure is less than $10^{-5}/7.5$ per year, assuming no other failure.

2.   *2.5 cm steady release under pressure*
This poses no danger beyond 70 m.

3.   *10 cm steady release under pressure*
Over 70 m the dangers are to people in the open from thermal and toxic effects up to 150 m with possible toxic effects up to 400 m. An acceptable solution would be to have 5 people in the 70–150 m range and 4 people in the 150–400 m range. If it is assumed that there is no jet stage in dispersion and that there is a 50 per cent chance of ignition, the mean number of people killed would be seven so the acceptable frequency of failure would be about $10^{-5}/7$, assuming no other failure.

E.   On-plot effects – conclusions

1.   *Instantaneous release*
The position of the control room should be at least 50 m away to withstand the blast. However, if there is no ignition on failure, and the 10 cm line breaks and the resultant pool is unconfined, it could engulf the control room. The pool must therefore be confined.

2.   *2.5 cm steady release under pressure*
This is a size of leak that can determine Zone 2 electrical classification (LFL at 16 m). The thermal criteria give distances of 40 m. The safe toxic distance of 60–70 m suggests that the control room would have to have its own air supply.

3.   *10 cm steady release under pressure*
This leak swamps the plot in all respects including, if it ignites, the possibility of the flame playing on the control room. Thus there should not be ideally more than nine personnel on the plot with a risk of failure less than $10^{-5}/9$, assuming no other failure.

F.   Overall conclusions and decisions

Assume that the site is 500 m radius with a toxic warning system so that the public is safe. Also assume that the plot radius is 70 m and the unconfined pool has been eliminated. However, the ethylamine tank details remain unchanged. Let there be the following maximum distribution of employees in the open.

30 → 70 (m)	2
70 → 150	4
150 → 200	10
200 → 250	10
250 → 400	40
400 → 500	80

Further, let there be 4 personnel in the plot control room at 50 m and 7 in unprotected buildings between 70–150 m. Allow the risk from the instantaneous, 2.5 cm and 10 cm releases to be equal, i.e. each acceptable risk is $10^{-5}/3$ and unacceptable risk $10^{-3}/3$ year^{-1} to individuals.

1. *Instantaneous release*
    Assuming 25 per cent chance of explosion and 25 per cent of fireball
    Personnel killed in fireball (500 m) = 146
    Personnel killed in explosion (150 m) = 13
    ★ Personnel killed in open by toxicity (500 m) = 146
    Mean personnel killed = $146 \times 0.25 + 13 \times 0.25 + 146 \times 0.5$
    $$= 112.75$$

    Acceptable failure rate = 0 as $N > 10$

    $$\text{Unacceptable failure rate} = \frac{10^{-3}}{3 \times 112.75} = 3 \times 10^{-6} \text{ year}^{-1}$$

2. *2.5 cm leak*
    Assuming 50 per cent chance of ignition
    Personnel killed in fire (30 m) = 0
    ★ Personnel killed by toxicity (70 m) = 2
    Mean personnel killed = $2 \times 0.5 = 1$

    $$\text{Thus acceptable failure rate} = \frac{10^{-5}}{3} = 3 \times 10^{-6} \text{ year}^{-1}$$

    $$\text{and unacceptable failure rate} = \frac{10^{-3}}{3} = 3 \times 10^{-4} \text{ year}^{-1}$$

3. *10 cm leak*
    Assuming 50 per cent of ignition.
    Personnel killed in fire (130 m in buildings, 150 m in open) = $4 + 6$
    $= 10$
    ★ Personnel killed by toxicity (400 m in open, 130 m in buildings)
    $\simeq 66 + 6 = 72$
    Mean personnel killed = $10 \times 0.5 + 72 \times 0.5 = 41$
    Thus acceptable failure rate = 0 as $N > 10$

    $$\text{and unacceptable failure rate} = \frac{10^{-3}}{3 \times 41} = 10^{-5} \text{ year}^{-1}$$

★ Note as clouds drift directionally, the above toxicity fatalities may be overestimates.

G.   Next moves

As about 160 people are at risk, more detailed analysis is required to see whether this preliminary analysis is otpimistic or pessimistic. The engineers will be asked if the above failure rates are feasible. The chemical engineers will be asked again to reduce the inventory, pressure and temperature.

B.7.   FURTHER NOTES ON BLAST EFFECTS

B.7.1.   HUMAN AND BUILDING TOLERANCES

Human tolerance of blast pressure is surprisingly large, especially if the rate of pressure increase is moderate. In the open, a person is more likely to be injured by being thrown to the ground and dragged along by the succeeding wind than injured by blast pressure alone.[47,48] Indications of injury from overpressure are:

	Overpressure (bar)
50% mortality	> 1.4
Lung damage	> 0.35
Eardrum damage	> 0.17
Penetrating small glass fragments	> 0.04

For conventional buildings, i.e. domestic housing, and buildings not specially designed for blast resistance the following structural damage may be expected at the respective overpressures (in bars).

	Walls	Roof tiles	Glass
Completely destroyed	0.8	> 0.1	0.06
Over 90% substantially broken and displaced	0.4–0.6	0.07	0.04
Up to 50% broken but no displacement	0.3–0.4	0.04–0.07	0.016
Cracked and distorted	0.15–0.3	0.03–0.04	> 0.01
Undamaged	≤ 0.15	≤ 0.03	≤ 0.01

Of particular interest in respect of glass damage is the pressure level of 0.04 bar.[49] Although one would expect over 90 per cent glass breakage, tests have shown that the fragments do not acquire enough energy from the shockwaves to have a significant probability of penetrating bare or lightly clothed skin. Above an incident pressure level of about 0.06 bar, flying glass becomes the most probable cause of injury until the building itself starts to collapse.

B.7.2   PLANT COMPONENTS

Table B.14 gives the effect of blast overpressure on vulnerable plant components. It has been based[50] on the response of targets to the blast effects from nuclear devices and the extremities of each horizontal line represents a 1 per cent and 99 per cent probability of failure respectively.

B.7.3   CONTROL ROOMS

In this section, only guidance on preliminary siting can be given, based on the work of Kletz,[7] CIA[18] and ACMH.[51] Detailed calculations by the

Table B.14 Typical damage caused by overpressure effects on explosion (Walker[50] Courtesy: Stanford Research Institute)

Overpressure (bar)

Equipment	0.03	0.07	0.10	0.14	0.17	0.21	0.24	0.28	0.31	0.34	0.38	0.41	0.45	0.48	0.52	0.55	0.59	0.62	0.66	0.69	0.83	0.97	1.10	1.20	1.40	>1.40
Control house steel roof	a	c	d				n																			
Control house concrete roof	a	ep	d				n													o						
Cooling tower	b		f				o																			
Tank cone roof		d				k							u										—			
Instrument cubicle				a		lm						t														
Fired heater					g	i				t																
Reactor chemical					a			i					p					t								
Filter					h				f										v		t					
Regenerator							j			ip					t											
Tank floating roof							k						u													d
Reactor cracking							j							i							t					
Pipe supports							p				so															
Utilities: gas meter									q																	
Utilities electric transformer									h						l					t						

Equipment	Damage markers (increasing overpressure →)
Electric motor	h ... l ... v
Blower	q ... t
Fractionation column	r ... t
Pressure vessel horizontal	pi ... t
Utilities: gas regulator	i ... mq
Extraction column	i ... v t
Steam turbine	i ... m s ... v
Heat exchanger	i ... t
Tank sphere	j ... i t
Pressure vessel vertical	i t
Pump	i ... v

a Windows and gauges break
b Louvres fall at 0.02–0.03 bar
c Switchgear damaged by roof collapse
d Roof collapses
e Instruments damaged
f Inner parts damaged
g Brick cracks
h Debris-missile damage occurs

i Unit moves and pipes break
j Bracing fails
k Unit uplifts (half-filled)
l Power lines severed
m Controls damaged
n Block walls fail
o Frame collapses
p Frame deforms

q Case damage
r Frame cracks
s Piping breaks
t Unit overturns or is destroyed
u Unit uplifts (filled)
v Unit moves on foundation

chemical and structural engineer are needed in each case to take account of the blast characteristics of particular chemicals, the structural design of the room and the current knowledge and state of the art of these types of calculation.

The room should be outside the flammable confines of the cloud, which means at least 20 m from the epicentre of an explosion. Exact position depends on plant inventories.

(a) For inventories equivalent to 15 t or more of hydrocarbon:
    (i) between 20–30 m from the epicentre the control room should withstand 1 bar for 30 ms;
    (ii) for distances between 30–100 m the structure should withstand 0.7 bar for 20 ms;
    (iii) for distances between 100–150 m the criteria are 0.2 bar for 100 ms;
    (iv) for distances between 150–250 m the control room should withstand a maximum pressure of 0.1 bar;
    (v) above 250 m no special protection is necessary.
(b) With inventories of between 2–15 t:
    (i) the control room should withstand a peak pressure of 0.1 bar if it is situated within a distance of the epicentre given by 20 < distance in m < 15 × inventory in t + 25;
    (ii) above the upper limit in (i) no special protection is needed.
(c) Below 2 t no special protection is needed assuming the room is at least 20 m from the plant.

The epicentre of an explosion is assumed to be at the plant centre for small plots. With large plots, the various possible epicentres are explored and the nearest used to consider the control site.

The precautions to be taken for (a)(i) are to build virtually a blast-proof control room. For (a)(ii) walls and roofs are specially designed. Windows are permitted but should be few in number, not more than 1 m² each in area, rebated, fitted with catch bars and located so that, if they blow in, they are unlikely to injure people. The glass should be shatter-resistant or coated with plastic film. Wired glass should not be used.[3] For (a)(iii), (a)(iv) and (b)(i) walls and roof sill need some special design and windows should be shatter-resistant or plastic-coated, though there is no restriction in size. With (a)(v), (b)(ii) and (c), ordinary brick-built and normal window construction can be used.

If it is accepted that for operational reasons control rooms should not be more than 35 m from a plant and nearer, if possible, the basis for siting becomes:

(a) Above 15 t, 35 m, extra special structure, special small windows.
(b) 2–15 t, 20 m, some special structural features, protected windows.
(c) Below 2 t, 20 m, no protection.

Again it should be iterated, these are only rough guidance principles for preliminary layout.

B.8   AREA CLASSIFICATION ZONE SIZES

Preliminary zone sizes can be fixed by reference to Appendix C. Subsequently (see section 8.7.5 p. 130) they need to be checked, so this section presents calculation schemes for estimating the size of zones and ventilation requirements. The calculations, *per se,* do not consider the class of zone (0, 1, 2) which has to be decided on the likelihood of vapour occurring as defined in BS 5345[52] and given in section 6.4.1 (p. 57). The results of the calculations should be interpreted in conjunction with that standard.

This section also considers toxicity and fire as well as flammability with which BS 5345 deals solely. However, the principles of zoning are similarly applied to these two extra properties. One advantage of considering the extra properties is that it prompts discussion on reaction to emergencies such as escape and firefighting, and another is that it can lead to balancing the cost of fire with the cost of flameproofing, etc.

For toxicity it is suggested that IDLH values are associated with Zones 1 and 2 and LTEL values with Zone 0 in accordance with the following definitions.

Toxic Zone 0
A zone in which a gas–air mixture having a toxic concentration above the LTEL value is continuously present or present for long periods.

Toxic Zone 1
A zone in which a gas–air mixture having a toxic concentration above the IDLH value is likely to occur for short periods in normal operation.

Toxic Zone 2
A zone in which a toxic gas–air mixture having a toxic concentration above the IDLH value is not likely to occur in normal operation, but if it occurs will exist only for a short time.

With thermal flux the definitions suggested are as follows.

Thermal Zone 0
A zone in which a particular class of target would be harmed from a fire resulting from the ignition of a gas–air mixture that is continuously present or present for long periods.

Thermal Zone 1
A zone in which a particular class of target would be harmed from a fire resulting from the ignition of a gas–air mixture that is likely to occur for short periods in normal operation.

Thermal Zone 2
A zone in which a particular class of target would be harmed from a fire resulting from the ignition of a gas–air mixture which is not likely to occur

often in normal operation but if such a mixture occurs, it will exist only for a short time.

Examples of types of target are the receivers listed in Table B.12 so the extent of thermal zones will vary, being the largest for personnel.

On small sites, as with flammability, it may not be advantageous to differentiate between Zones 0, 1 and 2, but just have, for example, 'toxic zones' and 'non-toxic' zones.

B.8.1   SMALL CONTINUOUS RELEASE OF GAS OR VAPOUR IN THE OPEN

The procedures follow mainly those for large releases (section B.3). The equations for calculating the rate of release and jet and flame behaviour are the same. However, for the dispersion stage, distances will usually be less than the 100 m limit specified for the TNO[4] scheme. Values of the various dispersion coefficients have therefore been taken from Katan's equation[10] together with Gifford's[53] assumption. It should be noted that Katan's work was concerned with refuelling aircraft with aviation spirit and the use of his results in wider contexts must lead to large uncertainties. However, it appears to be the only short-range data available.

Katan's equation is:

$$C = \frac{18.4Qv_1}{wx^{1\cdot81}}$$

where $v_1 = 1$ for above-ground release and $v_1 = 2$ for surface release

This corresponds to $\sigma_y\sigma_z = 0.000865x^{1\cdot81}$

With Gifford's relationship $\sigma_y = 1.8\sigma_z$

it follows that

$$\sigma_y = 0.125x^{0\cdot905}$$
$$\sigma_z = 0.069x^{0\cdot905}$$

In addition, the equation for a neutral plume

$$C = \frac{Qv_1}{\pi w\sigma_y\sigma_z}\exp.\left[-\frac{Z^2}{2\sigma_z^2}\right]$$

has been used to find the maximum height to which a particular concentration rises, i.e. when $\sigma_z = z_p/\sqrt{2}$

$$\text{so that } C = \frac{2Qv_1}{1.8\pi z_p^2 w2.718}$$

Table B.15 and Fig. B.4 show how zone sizes are calculated and Example B.15 illustrates the procedures.

530

Table B.15 Minor continuous release of gas or vapour in the open

Rate of release (Q) expressed as volume flow of vapour	see Table B.6
Distance to LFL (r)	see Table B.7
Zone size, uncertain direction	distance r all round
Zone size, vertical direction (e.g. vent)	cone r high, diameter D at top $\Big\}\; u \gg w$
Concentration at end of jet (C_j) Length of jet (x_j) $\Big\}$ see Table B.9	

Virtual source

$$j = \left(\frac{18.4Qvv_1}{wC_j}\right)^{0.55}$$

$v_1 = 2$ for ground release, else 1
$w = 2\ \mathrm{m\ s}^{-1}$

Apparent distance to concentration C

$$x_v = \left(\frac{18.4Qvv_1}{wC}\right)^{0.55}$$

$v = 0.5$ for 10 min. mean (toxic), else 1 (flammability)

Maximum height that C occurs above plume centre

$$z_p = \left(\frac{Qvv_1}{7.7wC}\right)^{0.5}$$

Zone size around source

Horizontal $x = x_v + x_j - j$
Vertical $z = z_p + x_j$
(or horizontal value if greater)

Distance to a given thermal flux (r)	See Table B.8
Zone size	distance r all round

Example B.15

Undertake area classification calculations for the pressure tank in Example B.1 assuming a 2.5 mm hole below the liquid level but above ground.

ANSWER

Most of the conditions are the same as in Example B.6.

Rate of release expressed as volume flow of vapour	1/100th of that in Example B.6, namely 0.012 m³ s⁻¹
Apparent diameter	1/10th of that in Example B.6, namely 7 mm
Distance to LFL	1/10th of that in Example B.7, namely $\approx 1.6\ \mathrm{m}$
Zone size	1.6 m all round leak
Concentration at end of jet	As in Example B.9, namely 0.021 v/v
Length of jet	1/10th of that in Example B.9, namely 2.7 m

Virtual source

$$j = \left(\frac{18.4 \times 0.012 \times 1}{2 \times 0.021}\right)^{0.55} = 2.5 \text{ m}$$

Distance to IDLH

$$x_v = \left(\frac{18.4 \times 0.012 \times 0.5 \times 1}{2 \times 4 \times 10^{-3}}\right)^{0.55} = 4.2 \text{ m}$$

Height to IDLH

$$z = \left(\frac{0.012 \times 0.5 \times 1}{7.7 \times 2 \times 4 \times 10^{-3}}\right)^{0.5} = 0.3 \text{ m}$$

Zone size   horizontal from leak $x = 4.2 - 2.5 + 2.7 = 4.4$ m
vertical up and down from leak
$z_p = 2.7 + 0.3 = 3$ m

Distance to LTEL

$$x_v = \left(\frac{18.4 \times 0.012 \times 0.5 \times 1}{2 \times 10^{-5}}\right)^{0.55} = 114 \text{ m}$$

Height to LTEL

$$z_p = \left(\frac{0.012 \times 0.5 \times 1}{7.7 \times 2 \times 10^{-5}}\right)^{0.5} = 6.2 \text{ m}$$

Zone size   horizontal from leak $x = 114 - 2.5 + 2.7 = 114$ m
vertical up and down
from leak $z = 6.2 + 2.7 \approx 9$ m

Flame length   1/10th of that in Example B.8, namely 2.6 m

Distances for thermal flux   1/10th of those in Example B.8, namely 3 $\rightarrow$ 4.3 m

Zone size   3 $\rightarrow$ 4.3 m all round depending on receivers

Conclusions

	Zone radius (m)	Zone height (m)
LFL	1.6	1.6
IDLH	4.2	3
LTEL	114	9
Fire (equipment)	3.2	3.2
Fire (personnel)	4.3	4.3

B.8.2   SMALL RELEASE OF LIQUID IN THE OPEN

The equations are given in Table B.16 and used in Example B.16. They follow closely the major release equations in Tables B.10 and B.11 of section B.4 except dispersion values from Katan's[10] work are used and an allowance is made for the finite concentration above the pool as the dispersion model assumes an infinite concentration. This allowance is based on the virtual source concept:

$$p = \left(\frac{36.8Q}{wP_v/P_a}\right)^{0.55} \approx 0.18D_p \text{ (on eliminating } Q, \text{ see Table B.10)}$$

where p is the distance of the virtual source from the leeward edge of the pool.

Fig. B.4 Zone dimensions for gas discharge (a) jet dispersion from uncertain direction (b) jet dispersion from vertical relief valve, etc. (c) jet and neutral dispersion

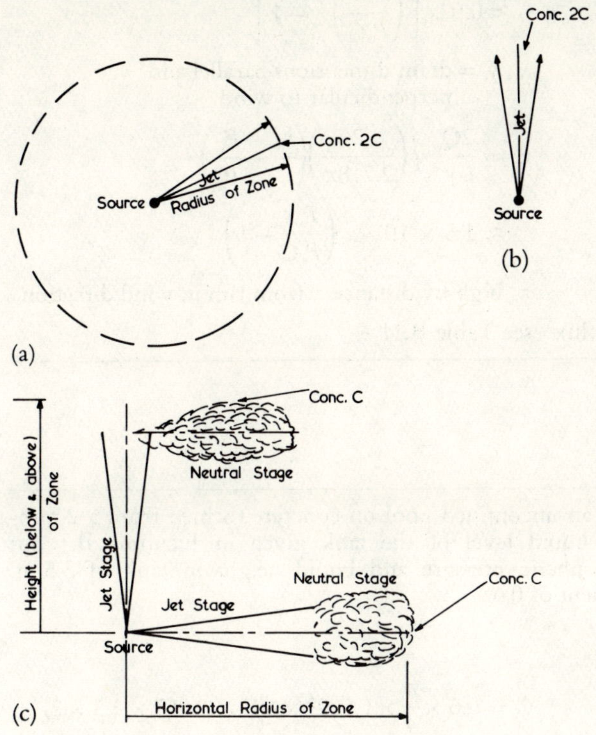

The equations for pools are based on treating the source as infinite but the following ones for drains allow for finite size (section B.4.1).

Table B.16   Minor release of liquid in the open

Release rate Pool diameter (D_p) Evaporation rate (Q)	see Table B.10
Distance from edge of a circular pool to concentration C	$x = \left(\dfrac{36.8Q}{w}\right)^{0.55}\left[\left(\dfrac{v}{C}\right)^{0.55} - \left(\dfrac{P_a}{P_v}\right)^{0.55}\right]$ $\approx 0.18D_p\left[\left(\dfrac{P_v v}{P_a C}\right)^{0.55} - 1\right]$ $v = 1$ for LFL $v = 0.5$ for 10 min. mean (toxic)
Max. height where C occurs	$z_p = \left(\dfrac{Q}{3.85w}\right)^{0.5}\left[\left(\dfrac{v}{C}\right)^{0.5} - \left(\dfrac{P_a}{P_v}\right)^{0.5}\right]$ $\approx 0.018D_p\left[\left(\dfrac{P_v v}{P_a C}\right)^{0.5} - 1\right]$
Zone size	$2x + D_p$ diameter by z_p high

Distance from rim of a long drain to concentration C	$x = \left(\dfrac{18Q}{w\gamma_s}\right)^{1\cdot1}\left[\left(\dfrac{v}{C}\right)^{1\cdot1} - \left(\dfrac{P_a}{P_v}\right)^{1\cdot1}\right]$

$$\approx 0.02x_s\left[\left(\dfrac{P_v v}{P_a C}\right)^{1\cdot1} - 1\right]$$

x_s, y_s = drain dimensions parallel and perpendicular to wind

Max. height that C occurs above drain	$z_p = \dfrac{2Q}{w\gamma_s}\sqrt{\left(\dfrac{2}{2.718\pi}\right)\left(\dfrac{v}{C} - \dfrac{P_v}{P_a}\right)}$

$$\approx 1.5 \times 10^{-3}x_s\left(\dfrac{P_v v}{P_a C} - 1\right)$$

Zone size	z_p high by distance x from rim in wind direction
Distance to given thermal flux	see Table B.11

Example B.16

Find the zone sizes due to an unconfined pool on concrete formed from a 2 mm-diameter leak below the liquid level of the tank given in Example B.1 but assuming storage at atmospheric pressure and liquid height in tank of 3.5 m. Assume a discharge coefficient of 0.6.

ANSWER

Release rate	$G = 0.6 \times \dfrac{\pi}{4} \times 0.002^2 \times 683 \sqrt{(9.807 \times 3.5 \times 2)}$

$$= 0.011 \text{ kg s}^{-1}$$
$$Q = 0.0056 \text{ m}^3 \text{ s}^{-1}$$

Pool diameter	$0.0056 = 1.7 \times 10^{-3} \times \dfrac{2\pi D_p^2}{4} \times \dfrac{1}{1} \times \left(\dfrac{2}{2^2 D_p}\right)^{0\cdot130}$

$$D_p = 1.6 \text{ m}$$

Time of spreading (ignoring evaporation)	$1.6^2 = \left(\dfrac{512 \times 9.807 \times 1 \times 0.011 \times t^3}{9\pi683}\right)^{0\cdot5}$

$$t = 13 \text{ s}$$

$$1.6^2 = \dfrac{4}{\pi} \times \dfrac{0.011t}{683 \times 0.005}$$

$$t = 624 \text{ s} = \text{time applicable}$$

Distance from pool edge to LFL	$x = 0.18 \times 1.6\left[\left(\dfrac{1}{1} \times \dfrac{1}{0.0355}\right)^{0\cdot55} - 1\right]$

$$= 1.5 \text{ m}$$

Height above pool to LFL	$z_p = 0.018 \times 1.6\left[\left(\dfrac{1}{1} \times \dfrac{1}{0.0355}\right)^{0\cdot5} - 1\right]$

$$= 0.1 \text{ m}$$

Zone size	diameter $= 1.6 + 2 \times 1.5 = 4.6$ m height $= 0.1$ m

Distance from pool edge to IDLH

$$x = 0.18 \times 1.6\left[\left(\frac{1}{1} \times \frac{0.5}{0.004}\right)^{0.55} - 1\right]$$

$$= 3.8 \text{ m}$$

Height above pool to IDLH

$$z_\text{p} = 0.018 \times 1.6\left[\left(\frac{1}{1} \times \frac{0.5}{0.004}\right)^{0.5} - 1\right]$$

$$= 0.3 \text{ m}$$

Zone size	diameter $= 1.6 + 2 \times 3.8 = 9.2$ m height $= 0.3$ m

Distance from pool edge to LTEL

$$x = 0.18 \times 1.6\left[\left(\frac{1 \times 0.5}{1 \times 10^{-5}}\right)^{0.55} - 1\right]$$

$$= 110 \text{ m}$$

Height above pool to LTEL

$$z_\text{p} = 0.018 \times 1.6\left[\left(\frac{1 \times 0.5}{1 \times 10^{-5}}\right)^{0.5} - 1\right]$$

$$= 6 \text{ m}$$

Zone size	diameter $= 1.6 + 2 \times 110 = 222$ m height $= 6$ m
Mass burning velocity	See Example B.11, namely $B = 0.05$ kg s^{-1} m^{-2}

Pool diameter

$$D_\text{p} = \left(\frac{4G}{\pi B}\right)^{0.5} = \left(\frac{4 \times 0.011}{\pi \times 0.05}\right)^{0.5} = 0.53 \text{ m}$$

Flame height

$$l = 0.53 \times 42\left[\frac{0.05}{1.22 \times (9.807 \times 0.53)^{1/2}}\right]^{0.61}$$

$$= 1.9 \text{ m}$$

Radiant flux

$$I_\text{o} = \frac{0.05 \times 35{,}300 \times 0.2}{4 \times (1.9/0.53)} = 24.4 \text{ kW m}^{-2}$$

Configuration factor ϕ_F (end-on)

$$= \frac{1}{1 + 4(r - 1.9)^2/0.53^2}$$

$$\text{or } r = 1.9 + \left(\frac{1}{\phi_\text{F}} - 1\right)^{1/2} \times 0.265$$

Item	Flux (kw m^{-2})	Configuration factor	Distance from pool centre (m)	
Flame limit			1.9	
Drenched tank	38	1.6	0	
Equipment	30	1.2	0	
Normal buildings	14	0.57	2.1	i.e. walls
Vegetation	12	0.49	2.2	i.e. wood fittings
Escape routes	6	0.25	2.4	
Personnel in emergency	3	0.12	2.6	
Plastic cables	2	0.082	2.8	
Personnel Stationary	1.5	0.062	2.9	

Zone size	2–3 m radius from pool centre

Time of spreading (ignoring burning)

$$0.53^2 = \left(\frac{512 \times 9.807 \times 1 \times 0.011 \times t^3}{9\pi 683} \right)^{1/2}$$

$$t = 3 \text{ s}$$

$$0.53^2 = \frac{4}{\pi} \times \frac{0.011 \times t}{683 \times 0.005}$$

$$t = 68 \text{ s} = \text{time applicable}$$

Conclusions

	Zone radius (m)	Zone height (m)
LFL	2.3	0.1
IDLH	4.6	0.3
LTEL	110	6
Fire (equipment)	2.1	2.1
Fire (personnel)	3	3

B.8.3 SMALL CONTINUOUS RELEASE OF GAS OR VAPOUR IN A BUILDING

Before computing, the layout should be arranged so that all parts of the building are well ventilated.

Usually it is the policy to make the whole room in which a release can occur part of the zone. The remaining parts are the areas just outside the room by the doors, louvres and other openings (see Fig. C.4, p. 572). However, this may be uneconomical especially for large indoor areas so the equations of Table B.17 are presented to be able to estimate the zone sizes.

A useful quantity to calculate first is the mean concentration C in the room based on the ventilation constant (see section B.9.3). This may indicate that forced ventilation is needed to reduce the concentration to a tolerable level. However, where fans are installed the consequences of their failure must also be considered. Another quantity that gives an order of magnitude appreciation is the mean space temperature reached in a fire.

It is difficult to know the flow patterns of air in a room so the following scheme can only be regarded as approximate.

Dispersion is initially taken as by jet action until the jet velocity equals the room air velocity or, as more likely, the jet hits equipment or a wall. The room air velocity is calculated as the room dimension in the direction of the draught divided by the ventilation constant, i.e. the air is assumed to blow uniformly parallel. An allowance for a background concentration is incorporated into the jet equations for inside a building.

After the end of the jet stage, either molecular diffusion in a uniform velocity field is assumed for laminar flow, or, mixing is assumed complete if the flow is turbulent. The transition is taken at the Reynolds number of 2,300 based on equivalent diameter perpendicular to the flow. A feature of laminar diffusion is that the vapour persists along the centreline of the plume. To allow for any effect of equipment on mixing in the turbulent region, it is suggested it is assumed that the vapour is only completely mixed in the free volume surrounding the source.

Marshall[26] has given equations for gravity dispersion in still air. However, it is likely that the ventilation will destroy the gravity effect.

Table B.17 has assumed that the air flow is horizontal with a low ceiling. The equations can be used with height and length, interchanged if the building is tall and air flow is mainly upwards (see BS 5345[52]).

The calculated thermal fluxes can only be regarded as approximate. A fire in a building can either be starved of oxygen or alternatively fed by the chimney effect. Also the effect of toxic fumes from the fire may be more damaging to personnel than thermal flux.

Table B.17   Minor continuous release of gas or vapour in a building

Rate of release (Q)	see Table B.6
Mean room and vent exit concentration	$\bar{C} = \dfrac{Q\tau}{V_B}$ τ = ventilation constant V_B = volume of room ignoring dead spaces
Mean air velocity	$w = \dfrac{\text{room length}}{\tau} = \dfrac{L_B}{\tau}$
For jets, distance (r) to concentration C	$r = 6.25 \dfrac{D - D_a}{2}$ $D = D_a \dfrac{2(1 - \bar{C})M_a}{(C - \bar{C})M} \left[\dfrac{\rho}{\rho_a} \left(1 - \dfrac{C}{2} + \dfrac{MC}{2M_a} \right) \right]^{1/2}$ $v = \dfrac{u(C - \bar{C})/2}{(1 - \bar{C})\left[\dfrac{C}{2} + \dfrac{M_a}{M}\left(1 - \dfrac{C}{2} \right) \right]}$ providing $v > w$
Zone size (jets)	Distance r all round
For non-jets	$C = \dfrac{Qv_1}{4\pi \mathscr{D}x} \exp.\left[-\dfrac{(\gamma^2 + Z^2)w}{4\mathscr{D}x} \right]$ when $w\bar{D} < 0.032\ \mathrm{m^2\,s^{-1}}$ $C = V_B\bar{C}/V_F$, when $w\bar{D} > 0.032\ \mathrm{m^2\,s^{-1}}$ where V_F is the free volume around the source, $\bar{D} = 4 \times$ flow area/perimeter and $\mathscr{D}$ is the diffusion coefficient
Zone size (non-jets)	$\sqrt{(\gamma^2 + Z^2)}$ at point x, when $w\bar{D} < 0.032\,\mathrm{m^2\,s^{-1}}$ Free volume V_F, when $w\bar{D} > 0.032\ \mathrm{m^2\,s^{-1}}$
Dispersion outside building	Use Table B.15 with $C = \bar{C}$ to define a virtual source

Mean room and exit vent temperature (T_B) due to jet fire	$T_B = T_a + \bar{C}\dfrac{MH_c}{\bar{s}_a M_a}$
Distance to given thermal flux (r)	See Table B.8 assuming $T_B \approx T_a$
Zone size	distance r all round

Example B.17

Repeat Example B.15 for inside a room, size 20 m × 10 m × 4 m with a natural ventilation constant 1 h

ANSWER

Rate of release	$Q = 0.012$ m³ s⁻¹ (see Example B.15)

Rate of release $\quad Q = 0.012$ m^3 s^{-1} (see Example B.15)

Apparent diameter $\quad D_a = 7$ mm

Vapour density of discharge $\quad 1.87$ kg m^{-3} (see Example B.6)

Mean room concentration
$$\bar{C} = \frac{0.012 \times 3{,}600}{800} = 0.054$$

This is too high, decrease τ to 270 s by installing fans

$$\bar{C} = \frac{0.012 \times 270}{800} = 0.004 \text{ v/v (IDLH value)}$$

Mean air velocity in room
$$w = \frac{20}{270} = 0.074 \text{ m/s}$$

$$\bar{D} = 4 \times 10 \times 4/(10 + 10 + 4 + 4) = 5.7 \text{ m}$$
$w\bar{D} = 0.074 \times 5.7 = 0.4$ m^2 s^{-1} which is greater than 0.032 m^2 s^{-1} so turbulent flow

Horizontal distance to LFL
$$D = 0.007 \frac{(1 - 0.004) \times 2}{(0.0355 - 0.004)} \times \frac{29}{45}$$
$$\times \left[\frac{4.79}{1.22} \times \left(1 - \frac{0.0355}{2}\left(1 - \frac{45}{29} \right) \right) \right]^{1/2}$$

$$= 0.568 \text{ m}$$
$$r = (0.568 - 0.007) \times 6.25/2$$
$$= 1.8 \text{ m}$$

$$v = \frac{123(0.0355 - 0.004)/2}{(1 - 0.004)\left[\dfrac{0.0355}{2} + \dfrac{29}{45}\left(1 - \dfrac{0.0355}{2} \right) \right]}$$

$$= 3 \text{ m s}^{-1}$$

$> w = 0.074$ m s^{-1}, the mean air velocity
If jet is stopped by impingement,
free volume required by LFL $= 800$ $\times 0.004/0.0355$
$= 90$ m$^3 = (2x)^2 h_B = 16x^2$
assuming equally spread sideways and lengthways
$x = 2.4$ m, $h_B = 4$ m

Zone size	Take as 2.4 m horizontal radius × 4 m high
Distance outside building to LTEL	Virtual source is the apparent distance to $\bar{C}_0 = 0.004$ which from Example B.15 is 4.2 m (distance to IDLH) From Example B.15 apparent distance to LTEL is 114 m So distance outside building = 110 m

Mean room temperature in fire while ventilation operating

$$T_{\mathrm{B}} = 290.4 + 0.004 \times \frac{35{,}300 \times 45}{1.3 \times 29}$$

$$= 459 \text{ K } (186\,^\circ\text{C})$$

From Table B.12, this would be intolerable to operators and would damage plastics, but probably not otherwise be dangerous

Distance for thermal flux $3 \to 4.3$ m (see Example B.15)

Conclusions

	Zone radius (m)	Zone height (m)
LFL	2.4	4
IDLH	Whole room	
LTEL	Whole room	
Fire (equipment)	3.2	3.2
Fire (plastics/ personnel)	Whole room	

B.8.4   SMALL RELEASE OF LIQUID IN A BUILDING

Before attempting any calculations, the layout should be arranged so that all liquid leaks do not collect on the floor but are conducted to safety in covered drains. In this way the calculation scheme given in Table B.18 and illustrated in Example B.18 should indicate that zones will be small.

As with the previous section, dispersion is assumed to be by molecular diffusion with uniform velocity in the laminar region and complete in the turbulent region. The equations for evaporation are based on those in Perry[54] but are reconciled to the simple diffusion model by putting

$$\frac{0.664}{(\text{Schmidt no.})^{1/6}} \cong 1/\sqrt{\pi}$$

This assumption implies with laminar flow that there is a stationary layer over a pool or drain which effectively halves the concentration at the surface. Downwind, the laminar dispersion model assumes a point source for a circular pool or a linear source for a drain.

The diffusion equations indicate that the vapour plume from the pool or drain hugs the floor. If this is the case then pump plinths keep equipment out of the vapour as well as above liquid spills.

As with gas leaks, it is suggested that for turbulent flow, the mixing is assumed to be confined to the volume free of equipment around the source.

Table B.18   Minor release of liquid in a building

Release rate $\left.\right\}$ Pool diameter	See Table B.10

Evaporation rate

$$Q = \frac{P_v}{P_a} \sqrt{\left(\frac{w\mathscr{D}x_s}{\pi}\right)} \gamma_s, \quad wx_s < 0.21 \text{ m}^2 \text{ s}^{-1}$$

$$Q = 0.9\left(\frac{P_v}{P_a}\right)\mathscr{D}^{1/2}(wx_s)^{0.8}\gamma_s, \quad wx_s > 0.21 \text{ m}^2 \text{ s}^{-1}$$

$$x \equiv D_p \text{ and } \gamma_s \equiv D_p\pi/4 \text{ for pool}$$

Mean room and vent exit concentration

$$\bar{C} = \frac{Q\tau}{V_B}, \text{ should be } \ll \frac{P_v}{P_a}$$

For horizontal airflow

Concentration at edge of spill

$$C = \frac{1}{2}\frac{P_v}{P_a} \text{ erfc}\left[\sqrt{\left(\frac{w}{\mathscr{D}x_s}\right)}\frac{z}{2}\right],$$

when $w\bar{D} < 0.032$ m²/s

$C = Q/(w\gamma_s h_B)$, when $w\bar{D} > 0.032$ m²/s where h_B = free height above spill and $\bar{D} = 4 \times$ flow area/perimeter

Concentration away from edge of drain

$$C = \frac{1}{\gamma_s}\frac{Q}{(\pi w\mathscr{D}x)^{1/2}} \exp.\left[\frac{-wz^2}{4\mathscr{D}x}\right]$$

when $w\bar{D} < 0.032$ m² s⁻¹

providing for $x < \dfrac{wz^2}{2\mathscr{D}}$, C is greater than concentration at drain edge at same z.

$$C = \frac{V_B\bar{C}}{V_F}, \text{ when } w\bar{D} > 0.032 \text{ m}^2 \text{ s}^{-1}$$

where V_F = free volume around drain

Maximum height at which C occurs

$$z = \frac{Q}{Y_sCw} \sqrt{\left(\frac{2}{\pi \exp. (1)}\right)}$$

positioned at $x = \dfrac{wz^2}{2\mathscr{D}}$ when $w\bar{D} < 0.032$ m²/s⁻¹

Concentration away from edge of a circular pool

$$C = \frac{Q}{2\pi\mathscr{D}x}\exp. -\left[\frac{(\gamma^2 + z^2)w}{4\mathscr{D}x}\right],$$

when $w\bar{D} < 0.032$ m² s⁻¹

providing $D_p < 2\left(\dfrac{\mathscr{D}x\pi}{w}\right)^{1/2}$,

---

and if $x < \dfrac{(y^2 + z^2)w}{4\mathscr{D}}$, C is greater than concentration at pool edge at same z.

$$C = \frac{V_B \bar{C}}{V_F}, \text{ when } w\bar{D} > 0.032 \text{ m}^2 \text{ s}^{-1}$$

where V_F is free volume around pool

Maximum radius at which
C occurs

$$y^2 + z^2 = \frac{Q}{Cw} \frac{2}{\pi \exp. (1)}$$

positioned at

$$x = \frac{(y^2 + z^2)w}{4\mathscr{D}}, \text{ when } w\bar{D} < 0.032 \text{ m}^2 \text{ s}^{-1}$$

For vertical airflow

Concentration directly above
pool or drain

$$C = \frac{1}{2} \frac{P_v}{P_a}, \text{ when } w\bar{D} < 0.032 \text{ m}^2 \text{ s}^{-1}$$

$$C = \frac{Q}{wx_s y_s}, \text{ when } w\bar{D} > 0.032 \text{ m}^2 \text{ s}^{-1}$$

$\bar{D} = 4 \times$ floor area/wall perimeter

Concentration away from
drain or pool

replace Q by $Q/2$ and interchange x and z in above equations

For all directions of flow

Dispersion outside building   see Table B.17

Rate of burning (G)   see Table B.11

Mean room and vent exit
temperature

$$T_B = T_a + \frac{GH_c \tau}{\bar{s}_a \rho_a V_B}$$

Distance to given flux
$(T_B \approx T_a)$ see Table B.11

Example B.18

(a) Repeat Example B.16 for inside the room discussed in Example B.17 with the increased ventilation. (The calculation ought to be done with the natural ventilation constant as well.) The diffusion coefficient of ethylamine in air can be taken as 1.1×10^{-5} m^2 s^{-1}.

ANSWER

Release rate $Q = 0.0056$ m^3 s^{-1} (see Example B.16)

Pool diameter

$$0.0056 = \frac{1}{1}\left(\frac{0.074 \times 1.1 \times 10^{-5}}{\pi}\right)^{0.5} \frac{\pi}{4} D_\mathrm{p}^{1.5}$$

$D_\mathrm{p} = 5.8$ m
$wD_\mathrm{p} = 0.074 \times 5.8 = 0.43 > 0.21$ m^2 s^{-1}
This indicates a turbulent boundary layer, so recalculate

$$0.0056 = 0.9 \times \frac{1}{1}\sqrt{(1.1 \times 10^{-5})}\,0.074^{0.8}\frac{\pi}{4}D_\mathrm{p}^{1.8}$$

$D_\mathrm{p} = 5.2$ m

Time of spreading

$G = 0.011$ kg s^{-1} (see Example B.16)

$$5.2^2 = \left(\frac{512 \times 9.807 \times 1 \times 0.011t^3}{9\pi683}\right)^{1/2}$$

$t = 63$ s

$$5.2^2 = \frac{4}{\pi} \times \frac{0.011t}{683 \times 0.005}$$

$t = 6{,}600$ s $=$ time applicable

Mean concentration

$$\bar{C} = \frac{0.0056 \times 270}{800} = 0.0019 < \text{IDLH}$$

Concentration variations

$$\bar{D} = \frac{4 \times 4 \times 10}{8 + 20} = 5.7 \text{ m}$$

$w\bar{D} = 5.7 \times 0.074 = 0.42$ m^2 s^{-1}
so turbulent flow

Free height at pool
    edge

$= 4$ m
so concentration at pool edge

$$C = \frac{0.0056}{0.074 \times \pi/4 \times 5.2 \times 4}$$

$$= 0.0046 \text{ v/v}$$

Free volume required by
    LFL

$$V_\mathrm{F} = 800 \times \frac{0.0019}{0.0355}$$

$$= 42.8 \text{ m}^3$$
$$= 10 \times 4 \times x$$

assuming spread over cross section, perpendicular to flow
$x = 1.07$ m
Make zone 3.6 m radius from centre of pool by 4 m high

Free volume required
    by IDLH

$$= 800 \times \frac{0.0019}{0.004}$$

$$= 380 \text{ m}^3$$
$$= 10x \times 4$$
$$x = 9.5 \text{ m}$$

Mean temperature	$T_{\mathrm{B}} = 290 + \dfrac{0.011 \times 35{,}300 \times 270}{1.22 \times 1.3 \times 800}$	
	$= 373\ \mathrm{K}\ (100\,°\mathrm{C})$	
Pool diameter of fire	0.53 m (see Example B.16) taking 68 s to spread	
Safe distances for equipment in fire	$\sim$2.1 m (see Example B.16)	
Flame limit	1.9 m (see Example B.16)	

Conclusions

	Zone radius (m)	Zone height (m)
LFL	3.6	4
IDLH	Half the room	4
LTEL	Whole room	
Fire (equipment)	2.1	2.1
Fire (personnel)	Whole room	

(b) Instead of forming a pool, the ethylamine could run along a 250 mm × 10 m open mesh drain across the middle of the room.

ANSWER

Mean air velocity	$0.074\ \mathrm{m\ s^{-1}}$ (see Example B.17)
Evaporation rate	$wx_{\mathrm{s}} = 0.074 \times 0.25 = 0.019 < 0.21\ \mathrm{m^2\ s^{-1}}$ so laminar boundary layer

$$Q = \frac{1}{1}\sqrt{\left(\frac{0.074 \times 1.1 \times 10^{-5} \times 0.25}{\pi}\right)} \times 10$$

$$= 0.0025\ \mathrm{m^3\ s^{-1}}$$

Mean room concentration	$\bar{C} = \dfrac{0.0025 \times 270}{800}$
	$= 0.00086\ \mathrm{v/v} < \mathrm{IDLH}$
Free height at drain edge	$= 4\ \mathrm{m}$ so concentration at drain edge

$$C = \frac{0.0025}{0.074 \times 10 \times 4} = 0.00086\ \mathrm{v/v}$$

Free volume required by LFL

$$= \frac{800 \times 0.00086}{0.0355} = 19.4\ \mathrm{m^3}$$

$= 10x^2$ if spread equally lengthways and upwards
$x = 1.4$ m
So zone is 10 m wide × 1.4 m high × 0.7 m either side

Free volume required by IDLH

$$= \frac{800 \times 0.00086}{0.004} = 172\ \mathrm{m^3}$$

$= 40x$ if spread over cross-section
$x = 4.3$ m
i.e. 2.2 m either side of drain

Burning velocity	$B = 0.05$ kg s^{-1} m^{-2} (see Example B.11)

$$G = 0.05 \times 10 \times 0.25$$
$$= 0.125 \text{ kg s}^{-1}$$

The release rate is 0.011 kg s^{-1} so only 0.88 m of drain will be on fire

$$D_p = 4 \times \text{area/perimeter}$$

$$= \frac{2 \times 0.88 \times 0.25}{0.88 + 0.25} = 0.39 \text{ m}$$

Mean temperature   373 K as for pool

Flame height

$$l = 0.39 \times 42 \left[\frac{0.05}{122 \times (9.807 \times 0.39)^{1/2}} \right]^{0.61}$$

$$= 1.55 \text{ m}$$

Radiant flux

$$I_o = \frac{0.05 \times 35,300 \times 0.2}{4 \times 1.55/0.39}$$

$$= 22.2 \text{ kW m}^2$$

Configuration factor (end-on)

$$\phi_F = \frac{1}{1 + \dfrac{4}{0.39^2}(r - 1.55)^2}$$

$$r = 1.55 + 0.195\left(\frac{1}{\phi_F} - 1\right)^{1/2}$$

Item	Flux (kW m^{-2})	Configuration factor	Distance from drain centre (m)
Drenched tanks	38	—	
Equipment	30	—	
Normal buildings	14	0.63	1.7 (walls)
Vegetation	12	0.54	1.7 (wood fittings)

Conclusions

	Zone radius (m)	Zone height (m)
LFL	0.7	1.4
IDLH	2.2	4
LTEL	Whole room	
Fire (buildings, not plastics)	1.7	1.7
Fire (personnel)	Whole room	

B.9 DATA

B.9.1 DISPERSION COEFFICIENTS

These are instantaneous coefficients calculated from TNO[4] (see section B.3.4). Other values may be found in NRPB.[9]

Horizontal, in wind direction
$$\sigma_x = 0.13x \text{ m}$$

Horizontal, perpendicular to wind direction
$$\sigma_y = c_y x^{n_y} \text{ m}$$

Stability (see Table 8.1, p. 107)

		c_y	n_y
Very unstable	(A)	0.2635	0.865
Unstable	(B)	0.1855	0.866
Slightly unstable	(C)	0.1045	0.897
Neutral	(D)	0.0640	0.905
Stable	(E)	0.0490	0.902
Very stable	(F)	0.0325	0.902

Note: The c_y given is the instantaneous value which is assumed to be half the 10 min. mean value.

Vertical
$$\sigma_z = c_z x^{n_z}$$

c_z and n_z depend on surface roughness z_0 which relates to the type of terrain thus:

Surface type	Example	Roughness (z_0) (m)
Flat lands	Fens with few trees	0.03
Farmland	Airfield, arable land	0.10
Horticultural	Glasshouses, strong crops, scattered houses	0.30
Residential	Dense but low buildings, forests	1.00
Urban	Dense with high buildings	3.00

$$c_z = c_z(z_0 = 1) . z_0^{0.3010}$$

$$n_z = n_z (z_0 = 1) - 0.059 \log_{10} z_0$$

where values of c_z and n_z at $z_0 = 1$ depend on the stability parameter.

Stability parameter	$(z_0 = 1 \text{ m})$	
	c_z	n_z
(A)	0.550	0.842
(B)	0.455	0.792
(C)	0.441	0.740
(D)	0.395	0.701
(E)	0.296	0.671
(F)	0.236	0.611

B.9.2 EVAPORATION PARAMETERS

Stability	Sutton parameter n_s	λ_2	$n_c = \dfrac{n_s}{2 + n_s}$
(A)	0.17	1.0×10^{-3}★	0.078
(B)	0.20	1.2×10^{-3}	0.091
(C)	0.25	1.5×10^{-3}	0.111
(D)	0.30	1.7×10^{-3}	0.130
(E)	0.35	1.8×10^{-3}★	0.149
(F)	0.44	1.8×10^{-3}★	0.180

★ (A), (E), (F) values extrapolated from (B), (C), (D) values using n_s.

B.9.3 VENTILATION CONSTANTS

Approximate Values for Natural Ventilation into Buildings (Consult also CIBS Code.[55])

Construction	Time for air change/hr
Multi-storey, brick or concrete construction	
Lower and intermediate floors	1.0
Top floor with flat roof	1.0
Top floor with sheeted roof (lined)	0.8
Top floor with sheeted roof (unlined)	0.7
Single-storey unpartitioned spaces	
Brick or concrete construction	
Up to 300 m²	0.7
300–3,000 m²	1.3
3,000–10,000 m²	2.0
Over 10,000 m²	4.0
Curtain wall or sheet construction (lined)	
Up to 300 m²	0.6
300–3,000 m²	1.0
3,000–10,000 m²	1.3
Over 10,000 m²	2.0
Sheet construction (unlined)	
Up to 300 m²	0.4
300–3,000 m²	0.7
3,000–10,000 m²	1.0
Over 10,000 m²	1.3

Notes: These times assume no large open doorways, louvred ventilators, roof ventilation nor fans.

Comfort parameters
Typical comfortable airflow for factories is 1.8 m³ h⁻¹ of air/m² of floor space.[34]
Comfortable maximum air velocities are 0.1 m s⁻¹ at 16 °C to 0.3 m s⁻¹ at 24 °C.[52]

B.9.4 TOXIC AND FLAMMABILITY LIMITS (p.p.m. by volume)

	LTEL	STEL	IDLH	LFL	UFL
Acetaldehyde	100	150	10,000	39,700	570,000
Acetic acid	10	15	1,000	54,000	160,000
Acetone	1,000	1,250	20,000	25,500	128,000
Acetonitrile	40	60	4,000	44,000	160,000
Acrylonitrile	20	30	4,000	30,500	170,000
Ammonia	25	35	500	155,000	270,000
Benzene	10		2,000	14,000	71,000

	LTEL	STEL	IDLH	LFL	UFL
Butadiene	1,000	1,250	20,000	20,000	115,000
n–Butanol	50		8,000	14,500	112,500
Carbon disulphide	10	30	500	12,500	500,000
Carbon monoxide	50	400	1,500	125,000	742,000
Carbon tetrachloride	10	20	300	N/A	N/A
Chlorine	1	3	25	N/A	N/A
Cyclohexane	300	375	10,000	12,600	77,500
Dimethyl formamide	10	20	3,500	22,000	15,200
Ethyl ether	400	500	19,000	18,500	365,000
Ethylamine	10		4,000	35,500	139,500
Ethylene oxide	5		800	30,000	800,000
Formaldehyde	2		100	7,000	73,000
Hydrogen cyanide	10	2	50	56,000	400,000
Hydrogen sulphide	10	10	300	43,000	455,000
i-Propanol	400	500	20,000	20,200	118,000
Methanol	200	250	25,000	67,200	365,000
Nitric oxide	25	35	100	N/A	N/A
Nitrogen dioxide	5	5	50	N/A	N/A
Octane	300	375	3,750	9,500	32,000
Pentane	600	750	5,000	14,000	78,000
Phenol	5	10	100		
Phosgene	0.1		2		
Styrene	100	250	5,000	11,000	61,000
Sulphur dioxide	2	5	100	N/A	N/A
Toluene	100	150	2,000	12,700	67,500
Sources	HSE[56]	Mackin-son[57]	CRC[58] Perry[59] Sax[60] Steere[61]		

Note: N/A indicates not applicable; blanks mean that data is unavailable.

REFERENCES

1. 4th International Symposium on Loss Prevention, Harrogate, *I.Chem.E. Sym. Ser.* **80**, 1983.
2. Safety and Reliability Directorate *Unclassified Reports.* United Kingdom Atomic Energy Authority, Wigshaw Lane, Culcheth, Warrington WA3 4NE.
3. Lees, F. P. *Loss Prevention in the Process Industries.* Butterworths, 1980.
4. TNO, *Methods for the calculation of the physical effects of the escape of dangerous material (liquids and gases)* (in two parts), Report of the Committee for the Prevention of Disasters, Directorate-General of Labour, Ministry of Social Affairs, Netherlands, 1979.
5. Wells, G. L. *Safety in Process Plant Design.* George Godwin/I.Chem.E., 1980.
6. Lihou, D. A. (ed.), *Hazard Identification and Control in the Process Industries.* Oyez Publishing 1981.
7. Kletz, T. A. 'Plant layout and location: some methods for taking hazardous occurrences into account', A.I.Ch.E., *Loss Prevention.* **13**, 147, 1980.

8. Pasquill, F. *Atmospheric Diffusion* (2nd edn). Ellis Horwood, 1974.

9. NRPB R–91 *A Model for Short and Medium Range Dispersion of Radio Nucleoles Released to the Atmosphere*. National Radiological Protection Board, Harwell, 1977.

10. Katan, L. L. *Fire Research Technical Paper No. 1*. HMSO, London, 1951.

11. Slater, D. H. 'Vapour clouds', *Chem. & Ind.* (8), 295, 1978.

12. Cox, R. A. 'Methods for predicting the atmospheric dispersion of massive releases of flammable vapour', *Progress in Energy and Combustion Science*, **6**, 141, 1980.

13. Blackmore, D. R., Herman, M. N. and Woodward, J. L. 'Heavy gas dispersion models', *J. Hazardous Materials*, **6**, 107, 1982.

14. Britter, R. E. and Griffiths, R. F. (eds) *Dense Phase Dispersion*. Elsevier Scientific Publishing Co., Amsterdam, 1982.

15. Giesbrecht, H. *et al.* 'Analysis of the potential explosive effects of amounts of combustible gases released into the atmosphere', *Chem. Ing. Tech.* **52**(2), 114, 1980; **53**(1), 1, 1981.

16. Baker, W. E. *et al. Explosion Hazards and Evaluation*. Elsevier Scientific Publishing Co., Amsterdam, 1983.

17. Brzustowski, T. A. and Sommer, E. C. *Predicting Radiant Heating from Flares*. Thirty-eighth Midyear Meeting, American Petroleum Institute, Philadelphia, API reprint 64–73, 17 May 1973.

18. CIA *An Approach to the Categorization of Process Plant Hazards and Control Building Design*. Safety Committee of the Chemical Industry Safety and Health Council of the Chemical Industries Association, 1979.

19. API RP 521, *Guide for Pressure Relief and Depressurizing Systems*. American Petroleum Institute, 1980.

20. Tan, S. H. 'Flare system design simplified', *Hydrocarbon Processing* **46**, 172, 1967.

21. Hardee, H. C., Lee, D. O. and Benedick, W. B. 'Thermal hazards from LNG fireballs', *Combustion Science and Technology*, **17**, 189, 1978.

22. Fay, J. A. and Lewis, D. H. 'Unsteady burning of unconfined fuel vapour clouds'. *Proceedings 16th Symposium on Combustion* 1397, 1977.

23. Fay, J. A., Desgroseilliers, G. J. and Lewis, D. H. *Radiation from Burning Hydrocarbon Clouds*. A report prepared for the United States Department of the Environment, DoE/EV–0036, May 1979.

24. High, R. W. 'The Saturn fireball', *University of New York Academy of Science*, **152**, 441.

25. Hasegawa, K. and Sato, K. *Study on the Fireball following Steam Explosion of n-Pentane*. 2nd International Symposium on Loss Prevention and Safety Promotion in the Process Industries, Heidelberg, Dechema, (Frankfurt), 297, 1978.

26. Marshall, J. G. 'The size of flammable clouds arising from continuous releases in the atmosphere', *I.Chem.E. Sym. Ser.* **49**, 99, 1977; **58**, 11, 1980.

27. Seddon, O. and Haverty, O. Technical Note No. AERO 2400, Farnborough, 1955.

28. Fauske, H. K. 'The discharge of saturated water through tubes', *Chem. Eng. Prog. Symp. Ser.* **61**, (59), 210, 1965.

29. Cude, A. L. *The Generation, Spread and Decay of Flammable Vapour*

Clouds. I.Chem.E. Course on Process Safety, Teesside Polytechnic, Middlesbrough, 1975.

30. Cude, A. L. 'Dispersion of gases vented to atmosphere by relief valves', *Chem. Engr, Lond.* **290**, 629, 1974.

31. Craven, A. D. 'Thermal radiation hazards from the ignition of emergency vents', *I.Chem.E. Symp. Ser.* **33a**, 7, 1972.

32. Craven, A. D. 'Fire and explosion hazards associated with the ignition of small scale unconfined spillages', *I.Chem.E. Sym. Ser.* **17**, 39, 1976.

33. BS 476, *Fire tests on building materials and structures,* Parts 2–8, British Standards Institution, London, 1970–81.

34. Perry, R. H. and Chilton, C. H. *Chemical Engineers Handbook* (5th edn). McGraw-Hill, 1973, **9**, p. 36.

35. Clancey, V. J. 'The evaporation and dispersion of flammable liquid spillages', *I.Chem.E. Sym. Ser.* **39a**, 80, 1974.

36. Shaw, P. and Briscoe, F. *Vaporization of Spills of Hazardous Liquids on Land and Water.* United Kingdom Atomic Energy Authority Reprint SRD R100 1978.

37. Pasquill, F. 'Evaporation from a plane, free-liquid surface into a turbulent air stream', *Proc. Roy. Soc.* (London), **A182**, 75, 1943.

38. Sutton, O. G. *Micrometeorology.* McGraw-Hill, 1953.

39. Stark, G. W. V. 'Liquid spillage fires' *I.Chem.E. Sym. Ser.* **33a**, 71, 1972.

40. Burgess, D. and Zabetakis, M. G. *Fire and Explosion Hazards Associated with Liquefied Natural Gas.* United States Bureau of Mines Report R 6099, 1962.

41. Thomas, P. H. *The Size of Flames from Natural Fires.* Ninth Symposium (International) on Combustion, Academic Press, New York, 1963.

42. Welker, J. R. and Sliepcevich, C. M. *Susceptibility of Potential Target Components to Defeat by Thermal Action.* University of Oklahoma Research Institute Report No. OURI–1578–FR (1970).

43. Perry, J. H. (ed.) *Chemical Engineers Handbook* (3rd edn). McGraw-Hill 1950, p. 1584.

44. European Convention for Constructional Steelwork, *European Recommendations for the Fire Safety of Steel Structures.* Elsevier Scientific Publishing Co., Amsterdam, 1982.

45. Provinciale Waterstaat, *Pollution Control and Use of Norms in Groningen.* Provinciale Waterstaat, Groningen, April 1979.

46. WASH 1400, *Reactor Safety Study. An Assessment of Accident Risks in United States Commercial Nuclear Power Plants.* United States Atomic Energy Commission, Washington, D.C., 1974.

47. Bowen, I. G., Fletcher, E. R. and Richmond, D. R. *Estimates of Man's Tolerance to the Direct Effect of Air Blast.* HQ DASA–2113, Lovelace Foundation, Washington D.C., 1968.

48. Fletcher, E. R., Richmond, D. R. and White, C. S. *Biological Hazards.* Sixteenth Explosives Safety Seminar. *1,* Department Pof Defense Explosives Safety Board, 1974.

49. Fletcher, E. R. and Richmond, E. R. *Characteristics and Biological Effects of Fragments from Glass and Acrylic Windows Broken by Air Blast.* Research Report 22 (ed. J. W. Kerr), United States Defense Civil Preparedness Agency, 1974.

50. Walker, F. E. *Estimating Production and Repair Effort in Blast-Damaged Petroleum Refineries.* Stanford Research Institute Report 6300–620, July 1969 SRI International: Menlo Park, California and Croydon, United Kingdom.

51. HSC *Advisory Committee on Major Hazards, Second Report.* Health and Safety Commission, HMSO, London 1979.

52. BS 5345, 'Code of practice for the selection, installation, and maintenance of electrical apparatus for use in potentially explosive atmospheres (other than mining applications or explosives processing and manufacture)', Part 2 (1983) *Classification of Hazardous Areas,* British Standards Institution.

53. Gifford, F. A. 'Peak to average concentration ratios according to a fluctuating plume dispersion model', *Int. J. Air Poll.* **3**, 253, 1960.

54. Perry, J. M. (ed.) *Chemical Engineers Handbook* (3rd edn). McGraw-Hill, 1950, p. 545.

55. CIBS *Building Energy Code, Part I.* Chartered Institution of Building Services, London 1977.

56. HSE *Occupational Exposure Limits 1984.* Guidance Note Environmental Hygiene EH 40, Health and Safety Executive, HMSO, London, 1984.

57. Mackinson, F. W., Stricoff, R. S. and Partridge, L. J. Jr. *Pocket Guide to Chemical Hazards.* United States Departments of Health and Human Services/Labor, 1980.

58. CRC *Handbook of Chemistry and Physics* (59th edn). CRC Press, 1978.

59. Perry, J. H. (ed.) *Chemical Engineers Handbook* (3rd edn). McGraw-Hill, 1950, p. 1585.

60. Sax, N. I. *Dangerous Properties of Industrial Materials* (6th edn). Van Nostrand Reinhold, 1983.

61. Steere, N. V. (ed.) *Handbook of Laboratory Safety* (2nd edn). CRC Press, 1971.

NOTATION

a	distance of virtual source, due to initial mixing, from actual source	m
A	flame area	m^2
A_R	area of receiver	m^2
A_S	site area	m^2
b	distance of virtual source, due to boundary, from actual source	m
B	burning rate	$kg\ s^{-1}\ m^{-2}$
c_D	discharge coefficient	
C	concentration	v/v
C_a	centre concentration of cloud, after initial mixing	v/v
C_b	concentration at site boundary	v/v
C_I	concentration in building	v/v
C_j	concentration at end of jet	v/v
C_L	lower flammable limit	v/v
C_o	concentration at cloud centre or axis	v/v

C_s	safe toxic concentration	v/v
c_y, c_z	dispersion constants	(m units)
C	mean concentration	v/v
C	risk to individual	year^{-1}
C_A	acceptable risk to individual	year^{-1}
C_c	chance of effect happening	year^{-1}
C_u	unacceptable risk to individual	year^{-1}
D	jet or flame diameter	m
D_a	apparent leak diameter	m
D_L	leak diameter	m
D_p	pool diameter	m
D_o	initial pool diameter	m
D	equivalent diameter	m
e	fraction entrained	
F	fraction of heat radiated	~ 0.2
F	plant failure rate	year^{-1}
F_A	acceptable rate of plant failure	year^{-1}
F_i	failure rate of ith source	year^{-1}
F_u	unacceptable rate of plant failure	year^{-1}
g	acceleration due to gravity	9.807 m s^{-2}
G	leak or evaporation rate	kg s^{-1}
h	heat transfer coefficient, spill on water	~ 0.6 kW m^{-2} K^{-1}
h_B	height of building or room or free height above spill	m
h_c	convective heat transfer coefficient	~ 0.01 kW m^{-2} K^{-1}
h_m	minimum height of pool	m
H	head of liquid	m
H_c	heat of combustion	kJ kg^{-1}
H_{HC}	heat of combustion of hydrocarbon	46,000 kJ kg^{-1}
i	ith source	—
I	intensity of radiation at receiver	kW m^{-2}
I_0	intensity of radiation at source	kW m^{-2}
j	distance of end of jet from virtual source	m
k	thermal conductivity of ground	kW m^{-1} K^{-1}
k_I	thermal conductivity of insulation	kW m^{-1} K^{-1}
l	height or length of flame	m
L	latent heat of evaporation	kJ kg^{-1}
L'	latent heat of evaporation	J kg^{-1}
L_b	latent heat of evaporation at atm. boiling point	kJ kg^{-1}
L_B	length of building or room	m
L_p	length of discharge pipe	m
L_w	latent heat of water (100 °C)	2,257 kJ/kg
m	fraction as vapour	
m_N	vapour fraction at orifice	
M	molecular weight	kg kmol^{-1}
M_a	molecular weight of air	29 kg kmol^{-1}
n	number of fatalities	
n_c	evaporation index	
n_s	Sutton's parameter	

$\left.\begin{array}{c}n_y \\ n_z\end{array}\right\}$	dispersion indices	
N	population at risk	
N_a	stoichiometric number of mol. of air per mol. of fuel	
N_c	population at risk from an effect	
N_i	population at risk from ith source	
N_p	stoichiometric number of moles of combustion products per mole of fuel	
p	distance of virtual source, due to finite concentration over pool	m
P_a	atmospheric pressure	$100\ \text{kN m}^{-2}$
P_a'	atmospheric pressure	$10^5\ \text{N m}^{-2}$
P_i	intermediate pressure in vessel	kN m^{-2}
P_N	pressure at choke	kN m^{-2}
P_v	vapour pressure	kN m^{-2}
P_o	starting pressure in vessel	kN m^{-2}
P_o'	starting pressure in vessel	N m^{-2}
Q	volume flow at atm. temperature and pressure	$\text{m}^3\ \text{s}^{-1}$
r	distance from release point	m
r_F	fireball radius	m
R	gas constant	$8.314\ \text{kJ kmol}^{-1}\ \text{K}^{-1}$
R'	gas constant	$8314\ \text{J kmol}^{-1}\ \text{K}^{-1}$
s	liquid specific heat	$\text{kJ kg}^{-1}\ \text{K}^{-1}$
s_a	air specific heat	$\sim 1\ \text{kJ kg}^{-1}\ \text{K}^{-1}$
s_a	mean air specific heat	$1.3\ \text{kJ kg}^{-1}\ \text{K}^{-1}$
s_l	insulation specific heat	$\text{kJ kg}^{-1}\ \text{K}^{-1}$
s_R	receiver specific heat	$\text{kJ kg}^{-1}\ \text{K}^{-1}$
s_w	water specific heat	$4.19\ \text{kJ kg}^{-1}\ \text{K}^{-1}$
S	drench rate	$\text{kg m}^{-2}\ \text{s}^{-1}$
t	time	s
t_B	fireball burning time	s
t_d	discharge time	s
t_p	time for cloud to reach a point	s
T	temperature after choke	K
T_a	air temperature	K
T_b	atmospheric boiling point	K
T_{bw}	boiling point of water	373 K
T_B	mean temperature in room	K
T_c	critical temperature	K
T_F	flame temperature	K
T_g	ground temperature	K
T_i	intermediate vessel temperature	K
T_m	temperature under insulation	K
T_N	temperature at choke	K
T_s	surface temperature	K
T_o	starting vessel temperature	K
u	velocity after choke	m s^{-1}
u_N	velocity at choke	m s^{-1}

v	jet velocity	m s^{-1}
V	volume of cloud at atm. pressure and temperature	m^3
V_B	volume of building or room	m^3
V_F	free volume around spill	m^3
V_T	volume of tank	m^3
w	wind velocity	m s^{-1}
W	amount leaked or evaporated	kg
W_c	hydrocarbon equivalent	kg
W_R	weight of receiver	kg
W_o	weight in vessel	kg
x	horizontal distance from source	m
x_b	distance of site boundary from source	m
x_I	thickness of insulation	m
x_j	length of jet	m
x_s	source length	m
x_v	distance from virtual source	m
X	distance from cloud centre in line of travel	m
y	horizontal distance perpendicular from the centreline of the plume or cloud path	m
y_s	source width	m
z	height above base of fire or leak source	m
z_o	roughness factor	m
z_p	max. height above plume centre at which a concentration C occurs	m
Z	height above cloud centre or centre of plume	m
α	thermal diffusivity of ground $=\dfrac{k}{\rho s}$	m^2 s^{-1}
β	ground roughness factor	
γ	ratio of gas specific heats	
δ	relative density	
$\mathscr{D}$	diffusion coefficient	m^2 s^{-1}
Δ	time for cloud to pass a point	s
ε	emissivity	
η	fraction drench water evaporated	
θ	air/fuel ratio	kg/kg
λ	jet angle/2	degrees
λ_2	evaporation constant	m s units
μ_a	viscosity of air	$\sim 1.7 \times 10^{-5}$ kg s^{-1} m^{-1}
v	ratio 10 min. mean to instantaneous concentration	0.5
v_1	allowance for release at ground level	
ρ	density after choke	kg m^{-3}
ρ_a	air density	1.22 kg m^{-3}
ρ_D	population density	m^{-2}

ρ_I	insulation density	kg m^{-3}
ρ_L	liquid density	kg m^{-3}
ρ_N	density at choke	kg m^{-3}
ρ_v	saturated vapour density at T_b	kg m^{-3}
σ	Stefan–Boltzmann constant	5.67×10^{-11} kW m^{-2} K^{-4}
σ_x	horizontal dispersion coefficient (parallel to wind)	m
σ_y	horizontal dispersion coefficient (perpendicular to wind)	m
σ_z	vertical dispersion coefficient	m
τ	building ventilation constant	s
ϕ	configuration factor	
ϕ_F	configuration factor (end-on)	
ϕ_S	configuration factor (side-on)	
ψ	fraction of original contents left	

TYPICAL LAYOUT DATA FOR PRELIMINARY LAYOUTS

Detailed design requirements are to be found in the appropriate national codes, standards and specifications for the individual classes of equipment and operation. The following tables provide some guidelines of typical restraints[1-8] which might be applied during the early development of the site and plant layout with due consideration given to safety.

Some of these typical quantities are taken from old codes of practice, etc. which have now been superseded or withdrawn because the uncritical use of standard distances is no longer accepted as good layout practice. However, these standard distances in the old publications can still be considered as good typical values for initial layout.

It must be emphasized, though, that no values in this Appendix should appear in the final layout without detailed checking that they apply to the circumstances of the plant or site being designed.

C.1 SITE AREAS AND SIZES (PRELIMINARY)

Administration	10 m²	per administration employee
Workshop	20 m²	per workshop employee
Laboratory	20 m²	per laboratory employee
Canteen	1 m²	per dining space
	3.5 m²	per place including kitchen and store
Medical centre	0.1–0.15 m² minimum 10 m²	per employee depending on complexity of service
Fire-Station (housing 1 fire, 1 crash, 1 foam, 1 generator and 1 security vehicle)	500 m²	per site
Garage (including maintenance)	100 m²	per vehicle
Main perimeter roads	10 m	wide
Primary access roads	6 m	wide
Secondary access roads	3.5 m	wide
Pump access roads	3.0 m	wide
Pathways	1.2 m	wide up to 10 people/min.
	2.0 m	wide over 10 people/min. (e.g. near offices, canteens, bus stops)

Stairways	1.0 m	wide including stringers
Landings (in direction of stairway)	1.0 m	wide including stringers
Platforms	1.0 m	wide including stringers
Road turning circles – 90° turn and 'T'-junctions		radius equal to width of road
Minimum railway curve	56 m	inside curve radius
Cooling towers per tower	0.04 m²/kW	mechanical draught
	to 0.08 m²/kW	natural draught
Boiler (excluding house)	0.002 m³/kW	(height $= 4 \times$ side)

C.2 PRELIMINARY GENERAL SPACINGS FOR PLOTS AND SITES

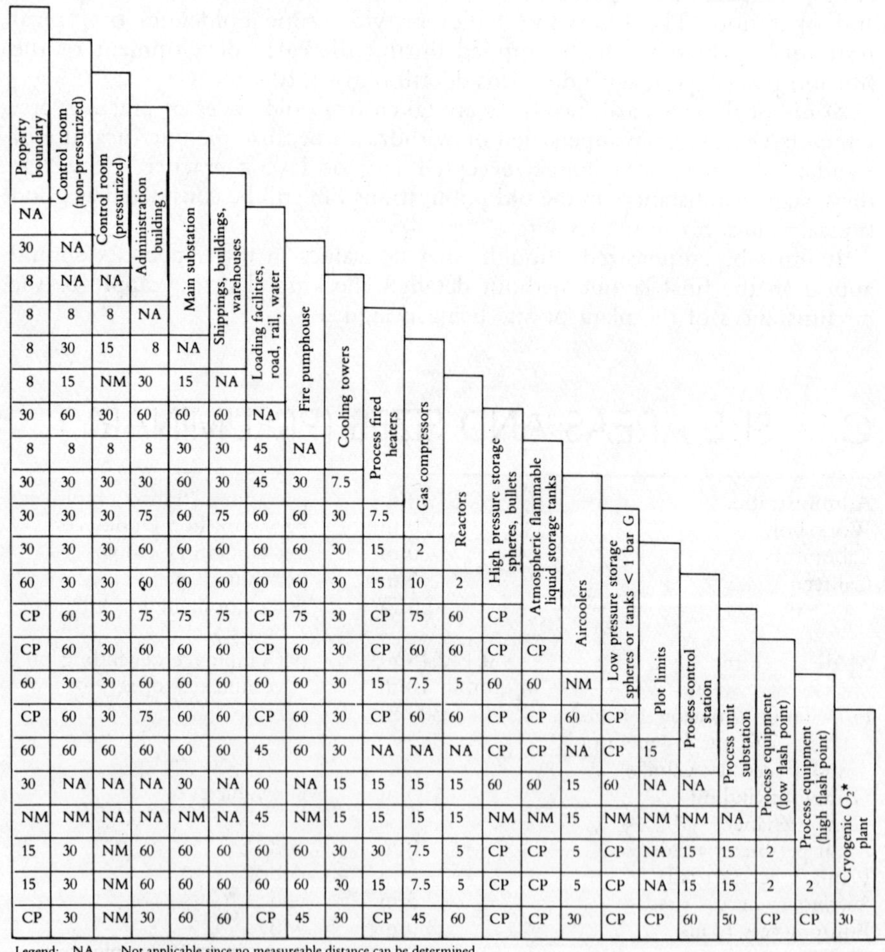

Column key (diagonal headers, left to right):

1. Property boundary
2. Control room (non-pressurized)
3. Control room (pressurized)
4. Administration building
5. Main substation
6. Shippings, buildings, warehouses
7. Loading facilities, road, rail, water
8. Fire pumphouse
9. Cooling towers
10. Process fired heaters
11. Gas compressors
12. Reactors
13. High pressure storage spheres, bullets
14. Atmospheric flammable liquid storage tanks
15. Aircoolers
16. Low pressure storage spheres or tanks < 1 bar G
17. Plot limits
18. Process control station
19. Process unit substation
20. Process equipment (low flash point)
21. Process equipment (high flash point)
22. Cryogenic O₂* plant

	1	2	3	4	5	6	7	8	9	10	11	12	13	14	15	16	17	18	19	20	21	22
1 Property boundary	NA																					
2 Control room (non-pressurized)	30	NA																				
3 Control room (pressurized)	8	NA	NA																			
4 Administration building	8	8	8	NA																		
5 Main substation	8	30	15	8	NA																	
6 Shippings, buildings, warehouses	8	15	NM	30	15	NA																
7 Loading facilities, road, rail, water	30	60	30	60	60	60	NA															
8 Fire pumphouse	8	8	8	8	30	30	45	NA														
9 Cooling towers	30	30	30	30	60	30	45	30	7.5													
10 Process fired heaters	30	30	30	75	60	75	60	60	30	7.5												
11 Gas compressors	30	30	30	60	60	60	60	60	30	15	2											
12 Reactors	60	30	30	60	60	60	60	60	30	15	10	2										
13 High pressure storage spheres, bullets	CP	60	30	75	75	75	CP	75	30	CP	75	60	CP									
14 Atmospheric flammable liquid storage tanks	CP	60	30	60	60	60	CP	60	30	CP	60	60	CP	CP								
15 Aircoolers	60	30	30	60	60	60	60	60	30	15	7.5	5	60	60	NM							
16 Low pressure storage spheres or tanks < 1 bar G	CP	60	30	75	60	60	CP	60	30	CP	60	60	CP	CP	60	CP						
17 Plot limits	60	60	60	60	60	60	45	60	30	NA	NA	NA	CP	CP	NA	CP	15					
18 Process control station	30	NA	NA	NA	30	NA	60	NA	15	15	15	15	60	60	15	60	NA	NA				
19 Process unit substation	NM	NM	NA	NA	NM	NA	45	NM	15	15	15	15	NM	NM	15	NM	NM	NM	NA			
20 Process equipment (low flash point)	15	30	NM	60	60	60	60	60	30	30	7.5	5	CP	CP	5	CP	NA	15	15	2		
21 Process equipment (high flash point)	15	30	NM	60	60	60	60	60	30	15	7.5	5	CP	CP	5	CP	NA	15	15	2	2	
22 Cryogenic O₂* plant	CP	30	NM	30	60	60	CP	45	30	CP	45	60	CP	CP	30	CP	CP	60	50	CP	CP	30

Legend:
NA — Not applicable since no measureable distance can be determined
NM — No minimum spacing established – use engineering judgement
CP — Reference must be made to relevant Codes of Practice but see section C. 6
* — Also see section C. 6 for minimum clearances

Notes:
(a) Flare spacing should be based on heat intensity with a minimum space of 60 m from equipment containing hydrocarbons
(b) The minimum spacings can be down to one-quarter these typical spacings when properly assessed

C.3 PRELIMINARY ACCESS REQUIREMENTS AT EQUIPMENT

Access	Item of equipment
Permanent ladder	1 Gate and globe valves – DN 80 (in mm) and smaller at vessels when located 3.5 m above grade 2 Checkvalves – all sizes at vessels when located 3.5 m above grade 3 Gauge glass 2 m above access surface, or inaccessible by portable ladder or platforms 4 Pressure instrument on vessels 2 m above access surface, or inaccessible by portable ladder or platform 5 Temperature instrument on vessels 2 m above access surface, or inaccessible by portable ladder or platform 6 Handholes located 3.5 m above grade
Platforms	*Items located over platform* 7 Manholes 8 Heat exchange units 9 Process blinds 10 Relief valves on vertical vessels DN 100 and larger 11 Control valves – all sizes 12 Cleanout points *Items located adjacent to platform* 13 Gate and globe valves – DN 100 and larger at vessels 14 Motor operated valves 15 Relief valves – DN 80 and smaller 16 Relief valves on horizontal vessels DN 100 and larger 17 Level controls and gauge glass on vessels 18 Sampling valves on vessels

C.4 PRELIMINARY MINIMUM CLEARANCES AT EQUIPMENT

Item	*Description*		*Clearance (m)*
Roads	1 Headroom for primary access roads for major maintenance vehicles		6.0
	2 Width of primary access roads		6.0
	3 Headroom for secondary roads and pump access roads		3.0–4.5
	4 Width of secondary roads and pump access roads		3.0–4.5
Railways	5 Headroom over through railways from top of rail		6.7
	6 Headroom over dead-ends and sidings from top of rail		5.1
	7 Clearance from track centreline to obstructions		2.4
Access, walkways and maintenance clearances	8 Headroom over platforms, walkways, accessways, maintenance areas		2.5
	9 Width of stairways, back to back of stringers		0.75
	10 Width of landings in direction of stairway		0.9
	11 Width of walkways at grade or elevated		0.75
	12 Vertical rise of stairways – one flight		4.5
	13 Vertical rise of ladders – single run		7.5
	14 Clearance under furnace burner nozzles for maintenance purposes		2.1
Platforms	15 Towers,	Distance of platform below bottom of manhole flange (side platform)	0.3
	16 vertical	Width of manhole platforms from manhole cover to outside edge of platform	0.75
	17 and	Platform extension beyond centreline of manhole flange (side platform)	0.75
	18 horizontal	Distance of platform below underside of flange (top platform)	0.2
	19 vessels	Width of platform from three sides of the manhole (top platform)	0.75
	20 Horizontal	Clearance in front of channel or bonnet flange	1.2
	21 exchangers	Clearance from edge of flanges	0.3
	22 Vertical	Distance of platform below top flange of channel or bonnet	1.5 max.
	23 exchanger	Width of platform from three sides of flange	0.6

Platforms	24	Furnaces	Width of platform at sides of horizontal and vertical tube furnace	0.75
	25		Width of platform at ends of horizontal tube furnaces	1.0
Pipeways	26		Pipeways not crossing roads	3.0

C.5 HANDLING FACILITIES FOR EQUIPMENT

Item		*Equipment and equipment part handled*		*Handling facility*
Vertical vessels	1	Manhole covers (up to DN 600) and vessel trays		Davits
	2	Bottom manholes		Hinged
	3	Internals of fixed bed reactors, catalyst, tower packings, etc.		None
Horizontal exchangers (at grade or in structure)	4	Removable tube bundles, and other removable parts except exchanger shells, shell covers, and floating head covers		Pulling beams or posts, for moving the bundle within the shell. Trolley beams for groups requiring up to four such beams. Trolley beams shall be provided with either: (a) two trolleys, one capable of handling the entire load and the other half-capacity, or (b) two half-capacity trolleys
	5	Exchanger shells		None
	6	Fixed tube sheet exchangers		None
		Shell covers and floating head covers		Shell davits or overhead hitching points
Vertical exchangers	7	Stationary	Tube bundles, channels, and channel covers	Hitching points
	8	tube sheet	Shell covers and floating head covers	Jib crane, davit, or hitching point
	9	at lower end	Entire small-size units	Hitching point or trolley beam

Vertical exchangers	10	Stationary tube sheet at upper end	Units designed for removing tube bundle from shell	Tube bundles, channels and channel covers	Trolley beam
	11			Shell covers and floating head covers	Hitching points
	12			Entire small-size units	Hitching point
	13		Units designed for removing the shell from the bundle: the entire unit or any of its component parts		Hitching point
	14	Fixed tube sheet exchangers			Shell davits or hitching points
Pumps compressors and drivers (housed or otherwise inaccessible)	15	100 kg–2t incl.	Parts of horizontal centrifugal pumps and steam drivers		Overhead hitching point or trolley beam
	16		Cylinder heads and pistons only of reciprocating pumps and horizontal reciprocating compressors		
	17	Over 2 t	Parts of centrifugal pumps, compressors and steam drivers including top halves of compressors, and turbine covers		Trolley beam or overhead travelling crane
	18		Cylinder heads and pistons only of reciprocating compressors		
	19		Power cylinders only of inclined type reciprocating compressors		
	20	Parts of vertical-type pumps and drivers			Overhead hitching point
	21	Electric motors and rotors			None

| Piping (housed or otherwise inaccessible) | 22 | Relief valves, DN 100 × 150 and larger | Hitching points or davits |
| | 23 | Blanks, blind flanges, fittings, and valves other than listed above and weighing more than 150 kg | Hitching points or davits when subject to frequent removal for operation or maintenance |

C.6   PRELIMINARY SPACINGS FOR TANK FARM LAYOUT

Notes to this section are on p. 568.

C.6.1   PRELIMINARY MINIMUM DISTANCES (NOTE 1) FOR LIQUEFIED OXYGEN[5,6]

	Distance (m)
To site boundary	30
To site roads	15
To process units and buildings containing combustible materials and ignition sources	30
To outside fixed combustible materials	5
To buildings containing flammable fluids	45
To road and rail loading areas	15
To overhead power lines and pipebridges	30
To other above-ground cables and important pipelines or pipelines containing flammables	15
To underground cables, trenches	10
To low-pressure gas storage	30
To compressed gas storage: flammable	30
: non-flammable	15
To liquefied pressure and refrigerated storage : flammable	45
: non-flammable	15
To liquid storage tanks : flammable (Note 2)	45
: non-flammable (Note 2)	30

C.6.2 PRELIMINARY MINIMUM DISTANCES (NOTE 1) FOR LIQUEFIED, FLAMMABLE GASES

Item	Material stored		
	Hydrocarbons	Non-hydrocarbons insoluble in water	Non-hydrocarbons soluble in water
Pressure storage (Notes 3,4)			
To boundary, process units, buildings containing a source of ignition, or any other fixed sources of ignition, e.g. process heaters	For example: Ethylene 60 m C_3 45 m C_4 30 m	For example: Methyl chloride 23 m Vinyl chloride 23 m Methyl-vinyl ether 23 m Ethyl chloride 15 m	For example: Methylamines 15 m
To building containing flammable materials, e.g. filling shed	15 m	15 m	15 m
To road or rail tank wagon filling points	15 m	15 m	15 m
To overhead power lines and pipebridges	15 m	15 m	15 m
To other above-ground power cables and important pipelines or pipelines likely to increase the hazard	(Note 5) 7.5 m	(Note 5) 7.5 m	See Note 6
Between pressure storage vessels	One-quarter of sum of diameters of adjacent tanks but not less than 1.8 m for ≤ 50 m^3 or less than 15 m for 750 m^3		
To low pressure refrigerated tanks	15 m from the bund wall of the low pressure tank, but not less than 30 m from the low pressure tank shell		
To flammable liquid (note 2) storage tanks	15 m from the bund wall of the flammable liquid tank		
To liquid oxygen storage	As defined above under 'Liquefied Oxygen'		
Zone 1 extent	1 m sphere around relief valve discharge		

Zone 2 horizontal extent from edge of tank	For example: Ethylene 30 m C₃'s 30 m C₄'s 20 m	For example: Methyl chloride 15 m Vinyl chloride 15 m Methyl-vinyl ether 15 m Ethyl chloride 10 m	For example: Methylamines 10 m
Zone 2 height of zone	260 × relief diameter above relief valve discharge (see note 9)		
Low pressure refrigerated storage (Notes 7, 8)			
To boundary, process units, buildings containing a source of ignition, or any other fixed sources of ignition	For example: Ethylene 90 m C₃'s 45 m C₄'s 15 m		For example: Ethylene oxide 15 m
To building containing flammable materials, e.g. filling shed	15 m		15 m
To road or rail tanker filling point	15 m		15 m
To overhead power lines and pipebridges	15 m		15 m
Between low pressure refrigerated tanks	One-half of sum of diameters of adjacent tanks		
To flammable liquid (Note 2) storage tanks	Not less than 30 m between low pressure refrigerated LFG and flammable liquid tank shells, but LFG and flammable liquids must be in separate bunds		
To pressure storage vessels	As defined above under 'Pressure Storage'		
To liquid oxygen storage	As defined above under 'Liquefied Oxygen'		
Zones 1 and 2	As defined above under 'Pressure Storage'		

C.6.3 LIQUIDS STORED AT AMBIENT TEMPERATURE AND PRESSURE

		Preliminary minimum clearance			
Dim. (Fig. C.1)	Dia. of tank	Water and non-flammable liquids	Class 'A' and 'B' products	Class 'A' and 'B' products (flash point < 32 °C)	Class 'C' products
			Fixed roof	Floating roof	
A	Up to 6 m 6–30 m Over 30 m	—	3 m Half tank dia. 15 m	6 m	3 m Half tank dia. Half tank dia.
B	All	1.5 m	Least of: half dia. of largest tank, dia. of smallest tank, 15 m. (Min. 6 m)	Least of: half dia. of largest tank, 6 m	Half dia. smallest tank. Min. 3 m
C	All	—	Dia. of largest tank. Min. 10 m	Dia. of largest tank. Min. 6 m	Dia. of largest tank. Min. 6 m
D	All	6 m	15 m	6 m	6 m
E	All	—	15 m	6 m	6 m
F	All	6 m	7.5 m	7.5 m	7.5 m
G	All	5 m	30 m	30 m	15 m

H	All	Depends on building lines	30 m	30 m	15 m
J	All	—	30 m	30 m	15 m
K	All	—	15 m	15 m	15 m
M	All	—	15 m	15 m	15 m
N	All		7.5 m	7.5 m	7.5 m
P	All	Outside bund	Outside bund	Outside bund	Outside bund
Q	All	—	Bund width	Bund width	Bund width
U }	Up to 3.5 m	—	1.5 m	3 m	—
V }	Over 3.5 m	—	3 m	3 m	—
W	All	—	All bund to wall height	All bund to wall height	—
X	Up to 3.5 m	—	5 m		—
	3.5–5 m	—	6 m		—
	Over 5 m	—	15 m	15 m	—
Y	Up to 3.5 m	—	2 m		—
	3.5–5 m	—	2.5 m		—
	Over 5 m	—	5 m	5 m	—
Z	All	—	5 m	5 m	—
Max. capacity/bund		—	60,000 m³	120,000 m³	—

The spacing and arrangement of tankage can vary with each application (note 1)
Class 'A' products have closed flash points below 23 °C
Class 'B' products have closed flash points between 23–66 °C
Class 'C' products have closed flash points above 66 °C

Fig. C.1 Preliminary tank farm layout (a) plan view (b) elevation

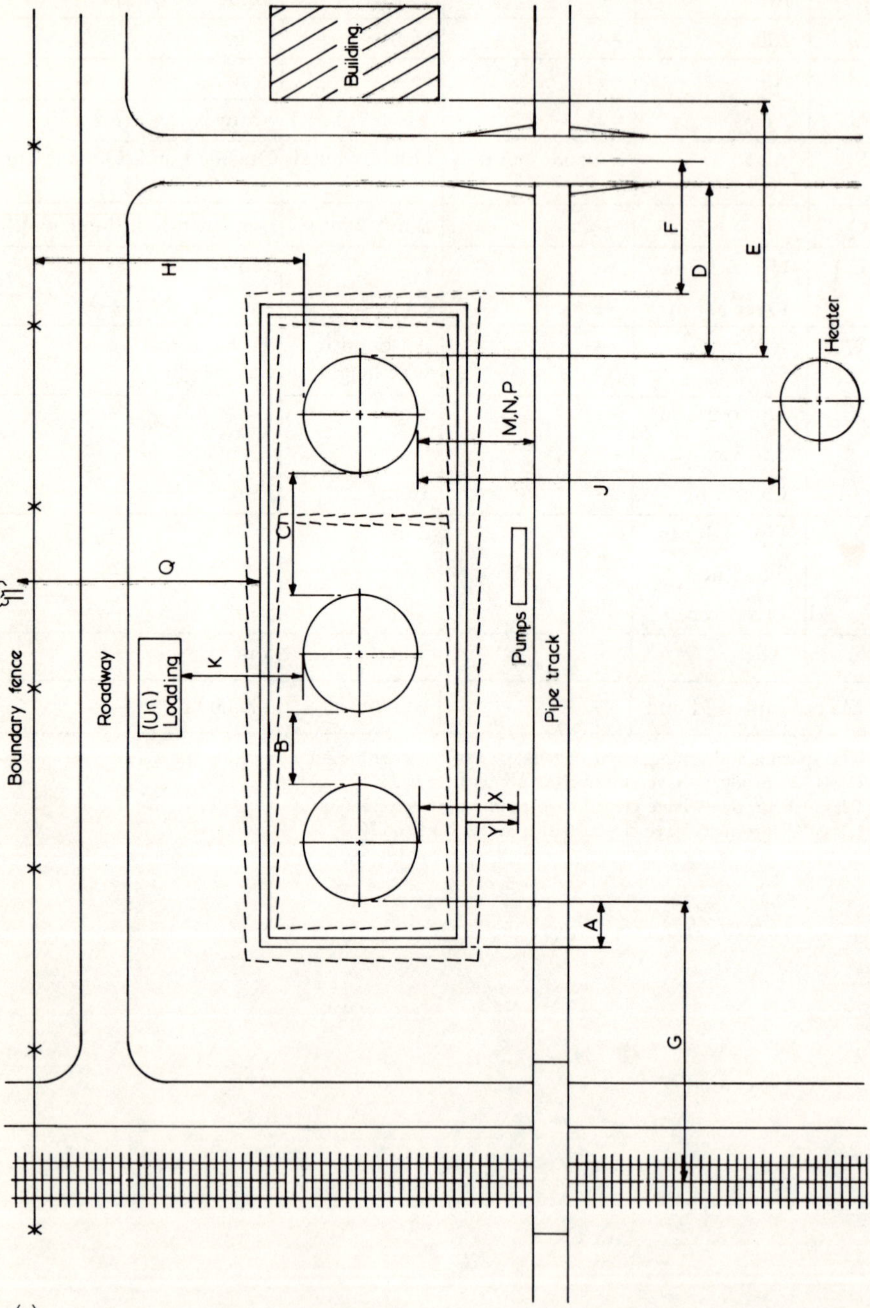

(a)

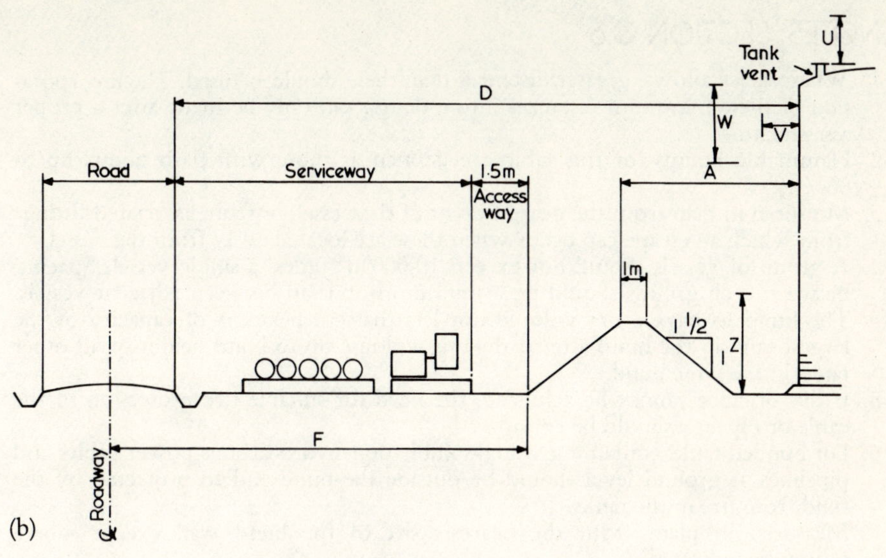

Legend	Distance
A,	from outside of tank to outside of bund at top
B,	between any 2 tanks in one tank bund
C,	between any 2 tanks in adjacent bunds
D,	from tanks to main plant roads
E,	between tanks and buildings containing flammable material
F,	from toe of bund to centre line of main plant roads
G,	from tank to centre of railway
H,	from tank to boundary fence
J,	between tank and fired heaters or ignition sources
K,	from tank to road or rail filling
M,	from tank to ground underneath power lines and pipe bridges
N,	from tank to power cables or pipelines
P,	from tank to ground above buried cables or pipes
Q,	from tank to combustible materials
U,	from tank vent to top of zone 1
V,	from outside of tank to edge of zone 1 and zone 2
W,	from tank rim to junction of zone 1 and zone 2
X,	from outside of tank to edge of zone 2
Y,	from centre line of bund wall to edge of zone 2
Z,	from ground to top of zone 2

NOTES: SECTION C.6

1. Where space allows, greater distances than these should be used. The incorporation of these minimum distances into a design can only be made after a proper
   assessment.
2. Flammable liquids for this table are defined as those with flash points up to
   66 °C.
3. Measured in plan from the nearest point of the vessel, or from associated fittings
   from which an escape can occur when these are located away from the vessel.
4. A group of vessels should not exceed 10,000 m³ unless a single vessel. Spacing
   between such groups should be a minimum of 15 m between adjacent vessels.
   The bund to have a net volume not less than 10 per cent of capacity of the
   largest tank in the bund after deducting volume up to bund height of all other
   tanks in the same bund.
5. If this distance cannot be achieved, the need for suitable fire protection of the
   cable or pipeline should be considered.
6. For bunded tanks containing water-soluble non-hydrocarbons power cables and
   pipelines at ground level should be outside the bund and so protected by the
   fund from fire in the tanks.
7. Measured in plant from the nearest part of the bund wall except where
   otherwise indicated.
8. A group of tanks should not exceed 60,000 m³. Spacing of the nearest tanks in
   any two such groups, which may have a common bund wall, should be such
   that the tank in one group should be a minimum of 15 m from the inside top of
   the bund of any adjacent group(s).
9. The zone may be bevelled across its upper corners providing all parts of the
   vessel more than 3 m from the zone edge.

C.7 PRELIMINARY ELECTRICAL AREA CLASSIFICATION DISTANCES

Note that these are for preliminary layout only in well-ventilated locations.
Definitions: Liquid = fluid below atm. b.p. (see section C.6.1 for definitions
of Class A, B and C fluids).
Gas = fluid above atm. b.p.

Fig. C.2 Preliminary extent of zone 2 around a pump seal

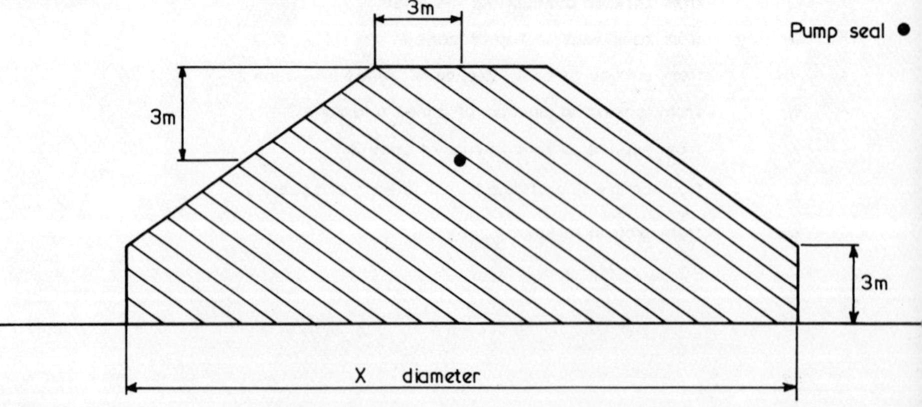

(a) CENTRIFUGAL PUMPS

Seal	Fluid conditions	Zone 1	Zone 2 X in Fig. C.2
Any (inc. reciprocating pumps)	Liquid < atm. b.p., $\geqslant$ amb. temp.	None	Diam. of pool + 6 m
Mechanical seal, external throttle bush, drain, atm. b.p. > amb. temp. No bush need be used for cases marked ★ if X doubled	Class A, > atm. b.p., temp. < 100 °C temp. < 200 °C temp. > 200 °C Class B, > atm. b.p., temp. < 200 °C temp. > 200 °C Class C, > atm. b.p., temp. < 250 °C temp. > 250 °C	0.3 m sphere around seal	20★ 40 60 20★ 50 as liquid 20
Mechanical seal, external throttle bush, vent to stack, atm. b.p. $\leqslant$ amb. temp.	Liquefied C_4's (i.e. atm. b.p. $\approx$ 0 °C) Liquefied C_3's and lighter HC (i.e. atm. b.p. $\approx$ 20 °C) (see note below) Liquefied non-hydrocarbons	0.3 m sphere around seal	40 60 20–30

Note:
Zone 1 for C_3's which may be up to 3 m depending on seal performance.

(b) EQUIPMENT OTHER THAN PUMPS

Item	Condition	Zone 1	Zone 2
Compressors in open-sided house	Gases	See note below	See Fig. C.3

Note: Zone 1 is 0.5 m around any gland, seal, drain parts, vents except 1 m is allowed around a seal oil lid and vent or a seal oil trap

Item	Condition	Zone 1	Zone 2
Equipment in normal buildings		Outdoor distances as shown in Fig. C.4	
Joints and flanges on pipes, fittings and process equipment	Liquid	None	X = diam. of pool + 6 m in Fig. C.2
	Gas lighter than air	None	3 m horizontal radius, 7.5 m above, 5 m below
	Gas heavier than air	None	7.5 m horizontal radius, 5 m above and down to floor

Note: Valve glands can be treated as pump seals

Item	Condition	Zone 1	Zone 2
Relief valves, vents, etc.	High velocity, gas lighter than air	1 m sphere	See Fig. C.5 H = 100, R = 60
	High velocity, gas heavier than air	1 m sphere	See Fig. C.5 H = 260, R = 120
	Low velocity, frequent release	1.5 m sphere	3 m sphere
	Low velocity, infrequent release	None	
Sample points <6 mm diam.	Liquids near amb. temp. into open	None	See Fig. C.6
	Other liquids into closed system	None	15 m radius, 3 m up, down to floor
	Gases into closed system	None	See 'Joints, and flanges on pipes, etc.' above
Process water drain point into open, at grade used regularly	Liquids	See note below	X = diam. of pool + 6 m in Fig. C.2
	C_3 under pressure		3 m high × 45 m radius
	C_4 under pressure	See note below	3 m high × 30 m radius
	Other gases under pressure		3 m high × 20 m radius

Note: Zone 1 is a cylinder 1 m radius and 1.5 m high for liquids and 5 m radius and 1.5 m high for gases

Instruments, etc. near or at grade	Liquids Gases	See note below See note below	X = diam. of pool + 6 m in Fig. C.2 Flanges as pipe joints Drains as sample points

Note: Zone 1 is not needed for infrequent spills but otherwise is a cylinder 3 m high by radius of 3 m if below atm. b.p. and 5 m if above atm. b.p.

Road or rail (un)loading	Liquids Gases	See Fig. C.7 for zones 1 and 2 H = 1 m See Fig. C.7 for zones 1 and 2 H = 3 m	
Ship (un)loading Unloading only		20 m around × ∞ high None	None 20 m around except seaside × 10 m high
Fixed roof tank	Liquids	See Fig. C.8 for zones 0 → 2 See also section C.6	
Floating roof tank	Liquids	See Fig. C.9 for zones 0 → 2 See also section C.6	
Pressure storage vessel Low pressure refrigerated tank	Gases	See 'Joints and flanges on pipes, etc.', 'Relief valves', 'Processwater drain point', as appropriate See also section C.6	
Open topped oil water separator	Liquids	See Fig. C.10 for zones 0 → 2	
Open topped drains and effluent pits	Liquids	See Fig. C.10 for zones 1 → 2	
Drums in open	Liquids	See Fig. C.11 (only if being filled)	3 m around drum area

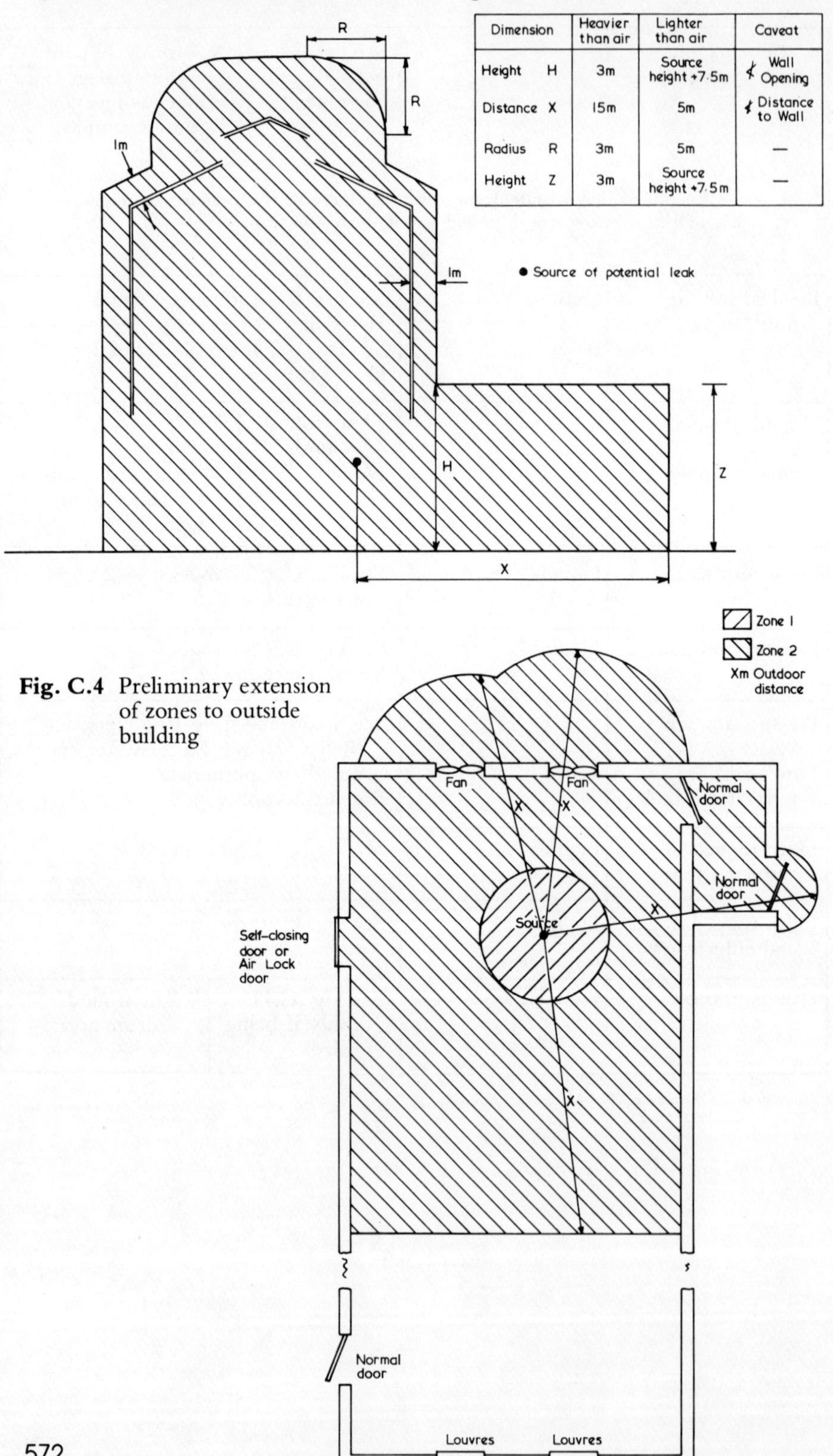

Fig. C.4 Preliminary extension of zones to outside building

Fig. C.5 Preliminary extent of zone 2 around a relief valve, etc.

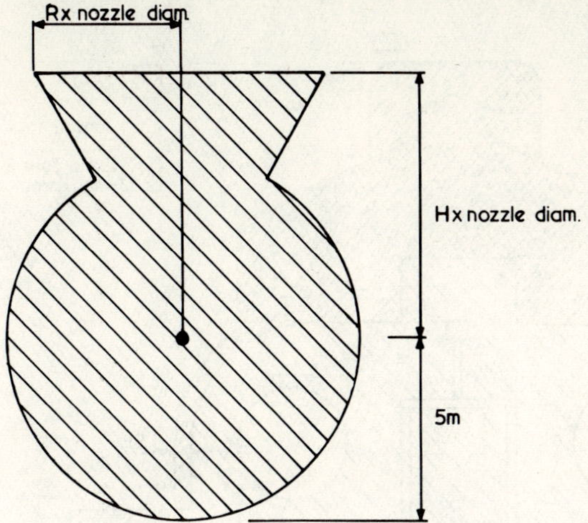

● Nozzle position

Fig. C.6 Preliminary extent of zone 2 around a liquid sample point

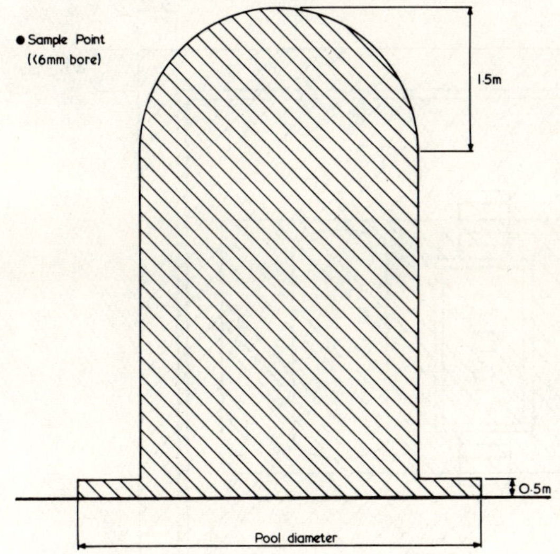

Fig. C.7 Preliminary extent of zones 1 and 2 for road or rail (un)loading areas (a) elevation (b) plan view

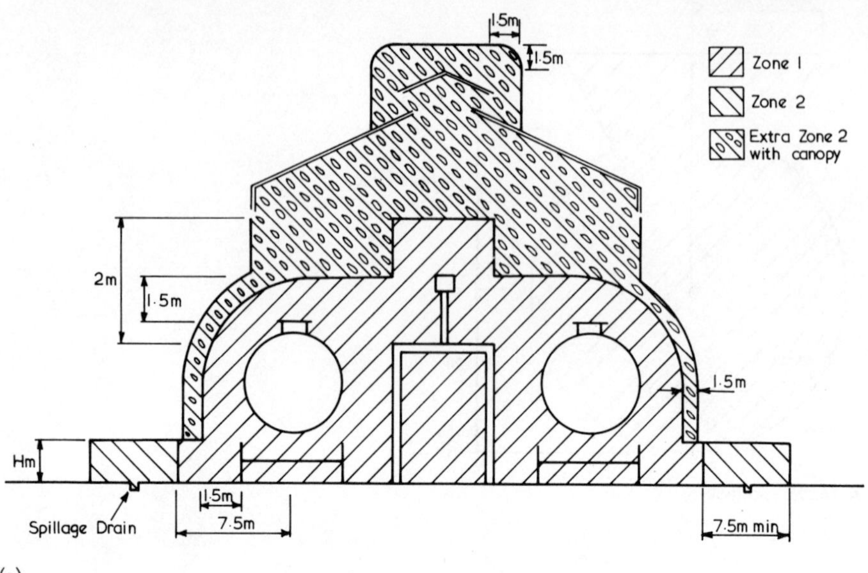

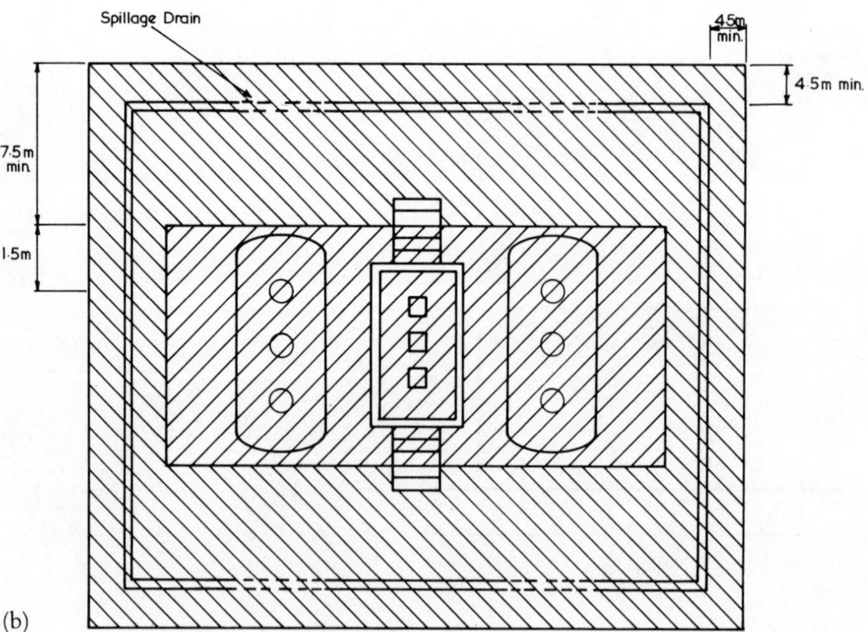

Fig. C.8 Preliminary extent of zones 0, 1 and 2 for a fixed roof tank (a) double walled tank (b) single walled tank

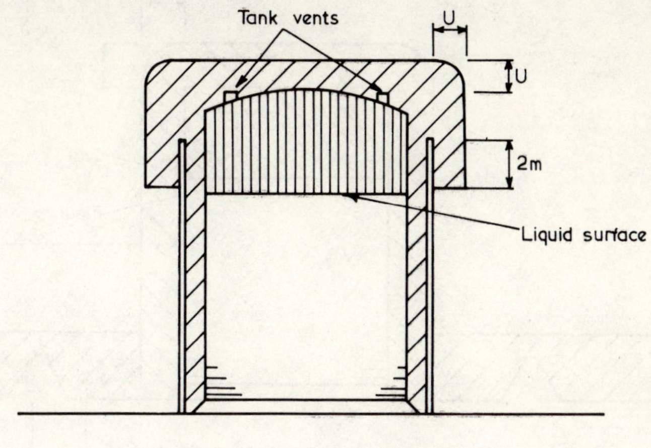

		TANK DIAM.	U	X	Y
▥	Zone 0	UP TO 3·5m	1·5m	5m	2m
▧	Zone 1	3·5 TO 5m	3m	6m	2·5m
▨	Zone 2	OVER 5m	3m	15m	5m

Fig. C.9 Preliminary extent of zones 1 and 2 for a floating roof tank

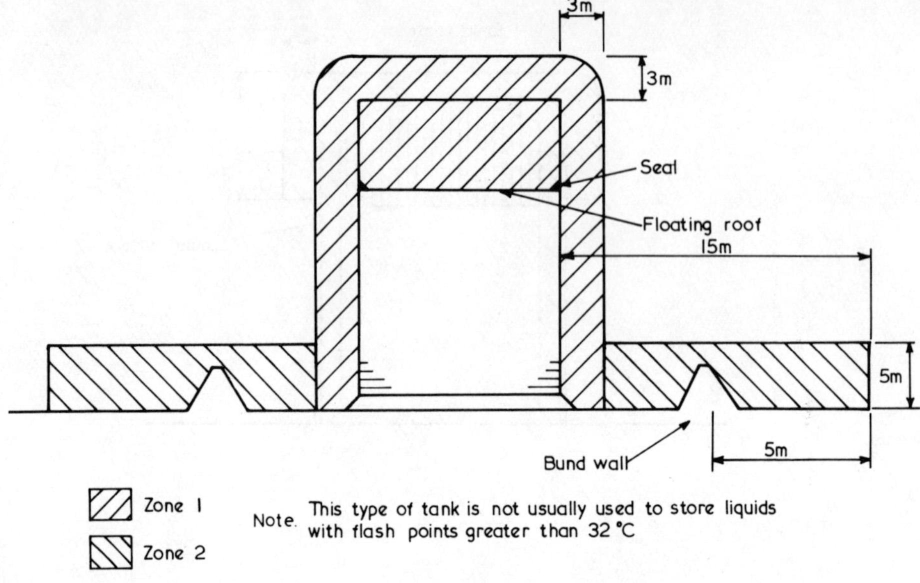

Fig. C.10 Preliminary extent of zones 0, 1 and 2 in open-topped constructions (a) open-topped oil/water separator (b) quench drain channel or effluent intercepter pit

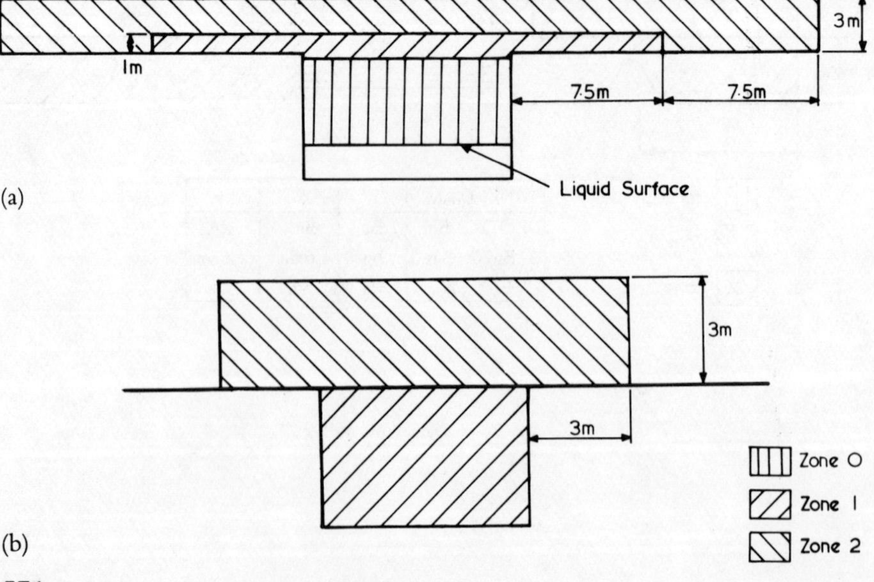

576

Fig. C.11 Preliminary extent of zone 1 for drum filling in open

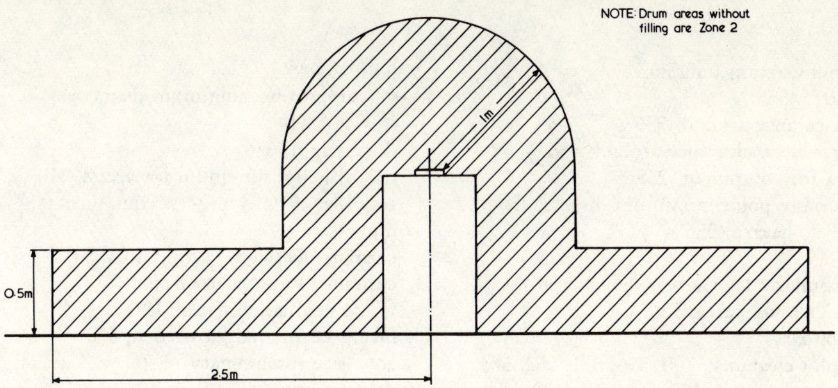

C.8   SIZE OF STORAGE PILES

1. The height (h, in m) of a right conical pile (see Fig. 11.5, p. 170) is given by:

$$h = \left(\frac{3V \tan^2 \theta}{\pi}\right)^{1/3}$$

where V = volume (m³) and θ = angle of repose (commonly 37°). If the conveyor angle is ϕ, the horizontal length of the conveyor (L_1 in m) is $L_1 = h \cot \phi$. The angle ϕ is commonly 18°. The radius (r, in m) of the bottom of the pile is $r = h \cot \theta$. It follows that the minimum length (L_2, in m) required for a conveyor and pile in one straight line on plan is:

$$L_2 = L_1 + r = h \left(\cot \phi + \cot \theta\right)$$

2. Approximate volume (V, in m³) of a straight conical pile (see Fig. 11.6, p. 171) is:

$$V = h^2 L_3 \cot \theta + \frac{\pi}{3} h^3 \cot^2 \theta$$

where L_3 (in m) is the length of the top of the pile

3. Approximate volume (V, in m³) of a curved conical pile is:

$$V = h^2 R \alpha \cot \theta + \frac{\pi}{3} h^3 \cot^2 \theta$$

where R = radius of curve (in m) and α = size of arc in radians

4. Approximate volume of closed warehouse (V, in m³) is:

$$V = h^2 L_4 \cot \theta$$

where L_4 (in m) is the length of the pile. This equation assumes fully triangular cross-section and no spaces around the piles for conveyors or mechanical unloading equipment. Thus the equation can be used as it is for underground conveying, but for unloading from one side, add 5 m to the width of the store. Also add 10–20 per cent to the length to allow for dead spaces.

C.9   INDEX TO CLEARANCES, SIZES, ETC.

REFERENCES

1. ROSPA/ICI *Electrical Installations in Flammable Atmospheres*. Report IS 91 Royal Society for Prevention of Accidents, London 1972.
2. IP *Model Code of Safe Practice – Electrical,* Part I. Institute of Petroleum/ Applied Sc. Publishers, 1965.
3. *European Model Code of Safe Practice in the Storage and Handling of Petroleum Products.* Part II: Design Layout and Construction.
4. BS 5345 'Code of Practice for the Selection, Installation and Maintenance of Electrical Apparatus for use in Potentially Explosive Atmospheres' Part 2 'Classification of Hazardous Areas', British Standards Institution, 1983.
5. IGC *Recommendations for Bulk Liquid Oxygen Storage Installations at Production Sites.* Industrial Gases Committee, Paris 1975.
6. HSE *A Code of Practice for the Bulk Storage of Liquid Oxygen at Production Sites.* HMSO, London 1974.
7. ICI *Liquefied Flammable Gases – Storage and Handling.* ICI Ltd, London, 1970.
8. Anderson, F. V. 'Plant layout', in: Kirk Othmer *Encyclopedia of Chemical Technology* (3rd edn). **18**, 1982, p. 23.

CONVERSION FACTORS FOR SOME COMMON UNITS

An asterisk (*) denotes an exact relationship.

Length	*1 in	:	25·4 mm
	*1 ft	:	0·304 8 m
	*1 yd	:	0·914 4 m
	1 mile	:	1·609 3 km
	*1 Å (angstrom)	:	10^{-10} m
Time	*1 min	:	60s
	*1 h	:	3·6 ks
	*1 day	:	86·4 ks
	1 year	:	31·5 Ms
Area	*1 in^2	:	645·16 mm^2
	1 ft^2	:	0·092 903 m^2
	1 yd^2	:	0·836 13 m^2
	1 acre	:	4046·9 m^2
	1 are	:	100 m^2
	1 mile^2	:	2·590 km^2
Volume	1 in^3	:	16·387 cm^3
	1 ft^3	:	0·028 32 m^3
	1 yd^3	:	0·764 53 m^3
	1 UK gal	:	4546·1 cm^3
	1 US gal	:	3785·4 cm^3
	1 litre	:	10^{-3} m^3
Mass	1 oz	:	28·352 g
	*1 lb	:	0·453 592 37 kg
	1 cwt	:	50·802 3 kg
	1 ton	:	1016·06 kg
	1 tonne	:	1000 kg
Force	1 pdl	:	0·138 26 N
	1 lbf	:	4·448 2 N
	1 kgf	:	9·806 7 N
	1 tonf	:	9·964 0 kN
	*1 dyn	:	10^{-5} N
Temperature difference	*1 deg F (deg R)	:	$\frac{5}{9}$ deg C (deg K)

Energy (work, heat)	1 ft lbf	:	1·355 8 J
	1 ft pdl	:	0·042 14 J
	*1 cal (internat. table)	:	4·186 8 J
	1 erg	:	10^{-7} J
	1 Btu	:	1·055 06 kJ
	1 Chu	:	1·9004 kJ
	1 hp h	:	2·684 5 MJ
	*1 kW h	:	3·6 MJ
	1 therm	:	105·51 MJ
	1 thermie	:	4·185 5 MJ
Calorific value (volumetric)	1 Btu/ft^3	:	37·259 kJ/m^3
Velocity	1 ft/s	:	0·304 8 m/s
	1 mile/h	:	0·447 04 m/s
	1 knot	:	0·5148 m/s
Volumetric flow	1 ft^3/s	:	0·028 316 m^3/s
	1 ft^3/h	:	7·865 8 cm^3/s
	1 UK gal/h	:	1·262 8 cm^3/s
	1 US gal/h	:	1·051 5 cm^3/s
Mass flow	1 lb/h	:	0·126 00 g/s
	1 ton/h	:	0·282 24 kg/s
Mass per unit area	1 lb/in^2	:	703·07 kg/m^2
	1 lb/ft^2	:	4·882 4 kg/m^2
	1 ton/sq mile	:	392·30 kg/m^2
Density	1 lb/in^3	:	27·680 g/cm^3
	1 lb/ft^3	:	16·019 kg/m^3
	1 lb/UK gal	:	99·776 kg/m^3
	1 lb/US gal	:	119·83 kg/m^3
	1 g/cm^3	:	1000 kg/m^3
Pressure	1 lbf/in^2	:	6·894 8 kN/m^2
	1 tonf/in^2	:	15·444 MN/m^2
	1 lbf/ft^2	:	47·880 N/m^2
	*1 standard atmos	:	101·325 kN/m^2
	*1 at (1 kgf/cm^2)	:	98·066 5 kN/m^2
	*1 bar	:	10^5 M/m^2
	1 ft water	:	2·989 1 kN/m^2
	1 in water	:	249·09 N/m^2
	1 in Hg	:	3·386 4 kN/m^2
	1 mm Hg (1 torr)	:	133·32 N/m^2
Power (heat flow)	1 hp (British)	:	745·70 W
	1 hp (metric)	:	735·50 W
	1 erg/s	:	10^{-7} W
	1 ft lbf/s	:	1·355 8 W
	1 Btu/h	:	0·293 07 W
	1 Chu/h	:	0·52754 W
	1 ton of refrigeration	:	3516·9 W

Moment of inertia	1 lb ft^2	:	0·042 140 kg m^2
Momentum	1 lb ft/s	:	0·138 26 kg m/s
Angular momentum	1 lb ft^2/s	:	0·042 140 kg m^2/s
Viscosity dynamic	*1 P (poise)	:	0·1 N s/m^2
	1 lb/ft h	:	0·413 38 mN s/m^2
	1 lb/ft s	:	1·488 2 N s/m^2
Viscosity, kinematic	*1 S (stokes)	:	10^{-4} m^2/s
	1 ft^2/h	:	0·258 06 cm^2/s
Surface energy (surface tension)	1 dyn/cm^2 (1 erg/cm)	:	10^{-3} J/m^2 (10^{-3} N/m)
Mass flux density	1 lb/h ft^2	:	1·356 2 g/s m^2
Heat flux density	1 Btu/h ft^2	:	3·154 6 W/m^2
	1 Chu/h ft^2	:	5·6784 W/m^2
	*1 kcal/h m^2	:	1·163 W/m^2
Heat transfer coefficient	1 Btu/h ft^2 °F	:	5·678 3 W/m^2 K
	1 Chu/h ft^2 °C	:	5·678 W/m^2 K
Specific enthalpy (latent heat, *etc.*)	*1 Btu/lb	:	2·326 kJ/kg
Heat capacity (specific heat)	*1 Btu/lb °F	:	4·186 8 kJ/kg K
Thermal conductivity	1 Btu/h ft °F	:	1·730 7 W/m K
	1 kcal/h m °C	:	1·163 W/m K

REFERENCE AND ACKNOWLEDGEMENT

Mullin, J. W. 'SI units in chemical engineering', *Chem. Eng. Lond.*, **211**, 176, 1967; *AIChE J*, **18**, 222, 1972.

ANGLO-AMERICAN GLOSSARY

CHEMICAL ENGINEERING TERMS

US	UK
agitator	mixer or stirrer
blind	slip-plate
carbon steel	mild steel
carrier, York, Lewis etc	refrigeration plant
check valve	non-return valve
clogged (of filter)	blinded
consensus standard	code of practice
conservation vent	pressure/vacuum valve
discharge valve	delivery valve
dike	bund
division (in electrical area classification)	zone
downspout	downcomer
expansion joint	bellows
faucet	tap
fibreglass reinforced plastic (FRP)	glass reinforced plastic (GRP)
figure 8 plate	spectacle plate
flame arrestor	flame trap
flashlight	torch
fractionation	distillation
gaging (of tanks)	dipping
gasoline	petrol
generator	dynamo or alternator
ground	earth
hose	flex
hydro (Canada)	electricity
ice-box	refrigerator
install	fit
insulation	lagging
inventory	stock
interlock	trip*

lift truck	fork lift truck
loading rack	gantry
manway	manhole
mill water	cooling water
nozzle	branch
pedestal, pier	plinth
pipe diameter (internal)	pipe bore
pipe rack	pipe bridge
plugged	choked
rupture disk or frangible	bursting disc
scrutinize	vet
shutdown	permanent shutdown
sieve tray	perforated plate
siphon tube	dip tube
spade	slip-plate
sparger or sparge pump	spray nozzle
spigot	tap
spool piece	bobbin piece
stack	chimney
stator	armature
tank car	rail tanker or rail tank wagon
tank truck	road tanker or road tank wagon
torch	cutting or welding torch
tower	column
tow motor	fork lift truck
tray	plate
turnaround	shutdown
utility hole	manhole
valve cheater	wheel dog
water seal	lute
wrench	spanner
written note	chit
C-wrenth	adjustable spanner
$M	thousand $
$MM	$M or million $
STP	60°F, 1 atm
32°F, atmosphere	STP
NTP	32°F, 1 atm
Rube Goldberg	Heath Robinson

*In the UK 'interlock' is used to describe a device which prevents someone opening one valve while another is open (or closed). 'Trip' describes an automatic device which closes (or opens) a valve when a temperature, pressure, flow etc reaches a pre-set value

FIRE-FIGHTING TERMS

US	UK
dry chemical	dry powder
dry powder	dry powder for metal fires
excelsior (for fire tests)	wood wool
egress	escape
sprinkler systems:	
feed main	main distribution pipe
cross main	distribution pipe
branch pipe	range pipe
rate density	application rate
fire stream	jet
standpipe	dry riser
evolutions	drills
nozzle	branchpipe
tip	nozzle
Siamese connection	collecting breeching
wye connection	dividing breeching
open butt	hose without branchpipe
fire classification:	
Class A: solids	Class A: solids
Class B: liquids and gases	Class B: liquids
Class C: electrical	Class C: gases
Class D: metals	Class D: metals

MANAGEMENT TERMS

Job	US	UK
Operator of plant	Operator	Process worker
Operator in charge of others	Lead operator	Chargehand or Asst foreman or Junior supervisor
Highest level normally reached by promotion from operator	Foreman	Foreman or Supervisor
First level of professional management (usually in charge of a single unit)	Supervisor★	Plant Manager★

Job	*US*	*UK*
Second level of professional management	Superintendent	Section Manager
Senior manager in charge of site containing many units	Plant Manager★	Works Manager
	Craftsman or mechanic	Fitter, electrician etc

★Note particularly the different meanings of the terms 'Supervisor' and 'Plant Manager' in the US and UK.

REFERENCE AND ACKNOWLEDGEMENT

Kletz, T. A. 'Anglo-American Glossary', *Chem Engr, Lond.* **409**, 33, 1984.

INDEX

Check lists, in non-alphabetical order, are in *italic type*. References to paragraphs and sections are in **bold type**.